LAND DEVELOPMENT
FOR CIVIL ENGINEERS

LAND DEVELOPMENT FOR CIVIL ENGINEERS

Thomas R. Dion
The Citadel

A WILEY-INTERSCIENCE PUBLICATION

JOHN WILEY & SONS, INC.

New York　·　Chichester　·　Brisbane　·　Toronto　·　Singapore

This text is printed on acid-free paper.

Copyright © 1993 by John Wiley & Sons, Inc.

All rights reserved. Published simultaneously in Canada.

Reproduction or translation of any part of this work beyond
that permitted by Section 107 or 108 of the 1976 United
States Copyright Act without the permission of the copyright
owner is unlawful. Requests for permission or further
information should be addressed to the Permissions Department,
John Wiley & Sons, Inc., 605 Third Avenue, New York, NY
10158-0012.

This publication is designed to provide accurate and
authoritative information in regard to the subject
matter covered. It is sold with the understanding that
the publisher is not engaged in rendering legal, accounting,
or other professional services. If legal advice or other
expert assistance is required, the services of a competent
professional person should be sought.

Library of Congress Cataloging in Publication Data

Dion, Thomas, 1946–
    Land development for civil engineers / by Thomas Dion.
       p.   cm.
    Includes bibliographical reference and index.
    ISBN 0-471-54743-3 (alk. paper)
    1. Civil engineering—Handbooks, manuals, etc. 2. Building sites—
Handbooks, manuals, etc. I. Title.
TA151.D56 1993
624—dc20   93-2341

Printed in the United States of America

10 9 8 7 6 5 4 3 2 1

*To my wife Susan, my sons Tommy and Richard,
and my family*

PREFACE

I have had the distinct privilege over the past decade and a half of teaching a course on land development to various senior civil engineering students. This course, which builds on most basic civil engineering courses in the nonstructural area, includes the integrated design of roads, storm drainage systems, potable water systems, sanitary sewer systems, site layout, and regulatory agency conformance. The value of this classroom experience is that it allows students to be exposed to the design process prior to graduation, and thus illustrates to them practical application of learned theory in an academic setting. Students interested in structural engineering also find this course helpful, because structures are located on improved sites with utilities, parking, and other types of infrastructure.

One major problem associated with teaching a course on land development has been the lack of an adequate textbook. As a result, I have attempted to set forth in a single volume information pertaining to land development that will not only introduce the prospective graduating engineer to practical integrated design applications, but also enlighten the student that there is much to be learned after graduation.

It is emphasized that this book is not an end unto itself, and certainly it is not intended to be a substitute for the individual courses from which a foundation for the use of this book must be acquired. Therefore, application of the principles presented in this book can be applied only after adequate related course work has been completed in Graphic Science, Land Surveying, Photogrammetry, Transportation Engineering, Highway Engineering, Fluid Mechanics, Geotechnical Engineering, Environmental Engineering, and Engineering Administration. Also, throughout the text numerous standards and procedures have been referenced. A prudent engineer should thoroughly review the complete source before making a determination of its applicability for a particular design application. The examples included in this book are provided only to illustrate a way to approach solving various problems. In many cases, other approaches can be used satisfactorily.

I feel that this book will be a valuable reference for practicing engineers because integrated design and analysis procedures are included. In addition, other persons involved professionally with various aspects of land development should find this book a useful reference.

June 1993   T. R. DION
Summerville, S.C.

ACKNOWLEDGMENTS

The author is indebted to many persons who have assisted in the preparation of this book. Unfortunately, space limitations preclude all names from being included here; however, thanks are given to my colleagues at The Citadel whose guidance has helped to shape the scope of this book. They include Thomas J. Anessi, Kenneth P. Brannan, Dennis J. Fallon, Loring K. Himelright, John H. Murden, Charles Lindbergh, Arnold B. Strauch, Russell H. Stout, and Michael S. Woo.

Special thanks are given to the following persons who have reviewed various portions of the book and have supplied the author with valuable comments and suggestions. They are: William B. Beauchene, East Coast Development Corporation; Ronald E. Benson, Jr., Hole, Montes & Associates; Neale E. Bird, S.C. Coastal Council; James A. Broody, First Trident Savings and Loan Company; Robert D. Carr, U.S. Department of Housing and Urban Development; J. Edwin Clark, Clemson University; George L. Craft, American Water Works Association; Stephen P. Dix, National Small Flows Clearing House; Jack M. Ellis, C.R.S. Sirrine; Craig K. Haney, Introspec; Gene E. Hardee, U.S. Department of Agriculture—SCS; Stephen F. Hutchinson, East Coast Development Corporation; Mark A. Isaak, Post, Buckley, Schuh & Jernigan; Edward J. McKay, National Geodetic Survey; D. Sherwood Miler, III, Pine Forest Properties; Sidney C. Miller, S.C. Geodetic Survey; Joseph R. Molinaro, National Association of Home Builders; J. Herbert Moore, Virginia Polytechnic Institute; James K. Nelson, Clemson University; Joseph L. Rucker, S.C. Department of Health and Environmental Control; Harry C. Saxe, Norwich University; Edward F. Straw, ISO Commercial Risk Services, Inc.; H. Fred Waller, Jr., the Asphalt Institute; Glenn A. Weesies, National Soil Erosion Laboratory; Donald E. Woodward, U.S. Department of Agriculture—SCS; and David B. Zilkoski, National Geodetic Survey.

Thanks are also given to Thomas S. Mc Canless for preparing most of the graphical displays used in this book, and to William L. Walker, Jr. for his encouragement and guidance. Finally, appreciation is also given to Daniel R. Sayre for his continued support as editor throughout the preparation of this book.

CONTENTS

PART I
Overview of Activities

1 Introduction

Land development is a complex, highly coordinated, planned undertaking that requires for success a profit motive on the part of the developer as well as an appreciation of human and municipal needs. Land development depends on the availability of land that conforms with area wide plans, codes, and regulations.

The development of property usually requires the services of a team of professionals. Development activities relating to the planning and design of a project can be provided by civil engineers, land planners, landscape architects, and/or architects working together. Other team members can include soil scientists, biologists, and archaeologists to inventory and analyze the site, while financial analysts, lenders, and marketing specialists provide economic analysis and support.

1.1 DEFINING LAND DEVELOPMENT

Land development normally embodies the conversion of unimproved property into a site that possesses features capable of supporting a desired activity. The three general categories of land development activities that describe intended usage are commercial, industrial, and residential.

1.1.1 Commercial Development

Commerce, which is defined as the "interchange of goods or commodities" (Barnhart and Stein, 1963),* is supported by commercial development. This support includes the conversion of strategically located property that minimizes consumer transportation requirements. Commercial developments can provide for office, retail, wholesale, and financial activities and also provide facilities for general services and traveler needs. The success of a commercial development is usually dependent upon the purchasing power of the supporting population within the anticipated trade area. Commercial developments can take place within the downtown central business district (CBD) of the trade area or along thoroughfares that lead to the CBD. Other types of commercial developments include outlying shopping districts and neighborhood facilities. Planned shopping centers can accommodate neighborhood, community, and regional needs.

1.1.2 Industrial Development

In contrast, industrial development, which can vary from heavy to light, focuses on providing suitable property for activities associated with processing of materials and

*Taken from the *American College Dictionary*, by permission. Random House, Inc. Copyright © 1963 by Random House, Inc. Reprinted by permission of Random House, Inc.

3

Figure 1.1   An example of a single-family development.
Courtesy of R. H. Stout.

production of consumer goods. Land that is improved for industrial use can be located within an industrial park or urban area, or can exist as a stand-alone facility. These developments are predicated on favorable social and political factors, affordable property, and the availability of material and labor. An attractive industrial site also is dependent upon climate, topography, transportation access, the availability of sewer and water, future expansion potential, and environmental compatibility. Today, most industrial developments place emphasis on aesthetics and the preservation of natural open space while minimizing noise, smoke, odor, dust, noxious gases, heat, glare, fire hazards, explosions, and vibrations (Claire 1969; So, 1979).

1.1.3   Residential Development

The conversion of property for production of safe, serviceable, and affordable residences along with associated facilities is within the scope of residential land development. Although commercial and industrial developments have individualized and specialized

Figure 1.2   An example of a multifamily development.

Figure 1.3   An example of a manufactured housing development.

needs, many aspects are common to all three types, including transportation, utility, and drainage considerations. Because the largest percentage of land area is developed for residential use, the main focus of this book is on that activity, beginning with conformance to area wide planning activities.

Typical types of residential developments include single-family, multifamily, and manufactured housing developments. Single-family developments entail a single dwelling unit (either detached or attached) on each building lot, whereas multifamily includes more than one. Figure 1.1 illustrates a typical single-family setting, where each dwelling unit has direct access to a street. Multifamily developments might encompass townhouses, apartments, or other configurations, and are typically represented as in Figure 1.2. Manufactured housing (or factory-built housing) developments, similar to that shown in Figure 1.3, include completed dwelling units, called manufactured homes, which are factory assembled and towed on their own chassis to the site, rather than constructed on site. Another type of manufactured housing includes factory-built modular units that are transported to the site where they are erected on a site-prepared foundation. This type of manufactured housing is usually classified with site-constructed dwellings.

Some jurisdictions allow for mixed-use or integrated-use developments, integrating residential areas with commercial activities that harmonize with the development plan. Examples of this integrated approach are given in Section 2.5.3.

1.2 AREA WIDE DEVELOPMENT PLANNING

To establish an overall process that considers an area's present and future growth while fostering a livable environment that safeguards property values, area wide planning must be undertaken. Regulations have evolved as tools to guide the developer in attaining satisfactory project completion for the good of the public. The resulting product serves the populace through benefits reaped by many public service activities involved with varying aspects of society. To the tax official, adequate land title records are made available for public record. For the municipal engineer, properly designed and adequately constructed road and drainage systems are provided to meet the public's needs. For health department officials, clean water supplies and adequate sewage/solid waste disposal are available to the public. For the fire chief, standard hose fittings, with proper plugs and water pressures, are available to provide effective fire protection to the citizens. The educator and recreation official can preserve and secure land to serve residents. To the residents, it means having an investment that will maintain its value and be adequate for everyday needs. To the land planner, the result is a coordinated system of public activities that otherwise would be unrelated in regard to such items as thoroughfares, major utility systems, and school/recreational areas. To the developer, regulations protect projects from unscrupulous operators who might competitively undertake substandard development activities resulting in a degradation of product marketability and use. To the local governing body, regulations are a deterrent to making premature and additional taxpayer expenditures as they require minimum land improvements that will provide a reasonable service life to the public. Finally, regulations allow for discriminatory acquisition of land made available for public use by dedication.

1.2.1   Planning Objectives

One major objective of area wide planning can be to improve the quality of life, which can result in lower rates of crime, suicide, and disorders associated with urbanization. This process takes into consideration the needs of the individual, family, neighborhood, and community. Different age groups have different interests, which can be addressed through proper planning, resulting in increased enjoyment of life.

A second major planning objective can be to provide safety to the inhabitants—safety in travel, safety from fire, safety from crime, safety from disease, and safety from peril. Safety from disease is associated with sanitation, which is a primary objective in planning a healthy environment.

A third major objective can be to provide a reasonably quiet environment. Noise associated with airports, manufacturing, railroads, and truck and bus traffic may require costly abatement programs, if proper planning is not undertaken.

Many other planning objectives can also be incorporated into area wide planning. These other objectives can include affordable housing, economic development, historic preservation, environmental protection, aesthetics, urban design issues, and so forth. The scope of the objectives is established based on local needs.

1.2.2   Purpose of Land Planning

Land planning addresses ways to provide services to the area in an efficient manner. Public funds must be expended, necessitating that a budget be formulated by the jurisdictional authority and followed, not only for day-to-day activities but also for long-range capital improvements. If economic responsibility is practiced in the planning process, improved employment, a larger tax base, and improved services will result.

1.2.3   Land Planning Types

Local planning is normally undertaken through state enabling legislation empowering governing bodies to investigate needs and promulgate plans to shape the area's growth. Planning can be categorized as metropolitan planning, community planning, and neighborhood planning. Metropolitan planning normally includes metropolitan areas with populations of 50,000 or more, and areas defined by general land use, population distributions, and economic activities. In addition to area wide transportation planning, such as that stipulated in the Urban Mass Transportation Act of 1964, as amended, drainage, water, sewer, and air purity planning may be undertaken. Other planning activities might include area wide water quality management planning (208 Planning) as set forth in the Clean Water Act (P.L. 92-500), as amended, to control pollution from point and nonpoint source discharges that might degrade the environment.

Community planning focuses on subareas within the metropolitan planning area, or possibly on small cities and municipalities. Community planning is similar to metropolitan planning, but is carried out on a more localized basis. It also addresses local land use regulation. Neighborhood planning includes activities of segmental areas within communities, addressing the coordination of community needs and capital expenditures while protecting the integrity of planning units, as illustrated in Figure 1.4.

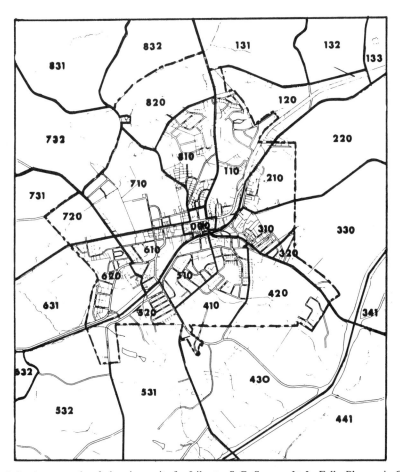

Figure 1.4   An example of planning units for Liberty, S.C. Source: E. L. Falk, Planner in Charge. *Liberty, S.C. Land Use Plan.* Clemson University, 1972. Reprinted by permission of Clemson University.

1.2.4   Forming a Plan

The formulation of a general plan includes the establishment of citizen and governmental area goals over a period of 20 to 25 years. When formulating an area plan, one must consider the community and surrounding area, especially in cases where historic and natural resources are present and require protection. For example, resort communities have different needs than do metropolitan areas. Consideration must be given to the amount of economic activity in the area because gainful employment is necessary for the quality of life of the residents. Population trends must be established to assess future impacts on housing, transportation, utilities, and other urban systems. Adequate community facilities and services in police, fire, and health protection must be anticipated. Educational and socioeconomic needs must be addressed. With urban planning, projected area needs can be translated into a capital improvements program that will allow staged construction of municipal improvements to accommodate anticipated growth, usually in five-year increments. Figure 1.5 illustrates the relational aspects of various planning activities.

An inventory of existing land uses must be undertaken for the planning area. General types of uses are single-family residential, multifamily residential, apartments, commercial, light industrial, heavy industrial, floodway, public, semiprivate (churches, etc.), streets, and vacant. Aerial photographs, Sanborn Insurance Maps, and tax maps are useful in compiling this initial data, which then must be verified in the field.

A map of existing land uses can then be made, similar to the one illustrated in Figure 1.6. Various uses can be indicated with different colors on the map. Other factors to be considered include topography and ridge line locations for drainage basin boundary delineation, flood areas, and a soils inventory.

An important aspect of urban planning includes transportation activities. During the 1920's, increased automobile ownership allowed greater mobility. This increased the movements of persons, which allowed the day-to-day range of daily activities to expand greatly. Movement to the suburbs was experienced, and residential developments followed. As vehicle travel expanded, new types of streets allowing higher capacities were needed. Rail, truck, water, air, and mass transit all have been impacted by growth.

Due to this migration away from the cities, central business districts (CBDs), which are the heart of the city, began to deteriorate. Inadequate parking, traffic congestion, crime, blight, outdated utilities, and depressed markets were some of the problems facing the declining CBDs. As a result of urban planning, renewal efforts have helped bring about revitalization of some CBDs.

1.2.5   Plan Implementation

Upon completion of the economic, transportation, population, housing inventory, community facilities, and environmental studies, along with the preparation of various base maps showing existing land use, the planning agency should integrate into its strategy not only the goals and objectives of the planning area, but also those requirements set forth by governmental bodies having broader and more encompassing authority. Future requirements are formulated by forecasting and projecting trends. Needs associated with land use, community facilities, transportation, and capital improvements can be established that will guide future growth in an orderly manner. Plan implementation utilizes such devices as generalized land use maps, Figure 1.7, zoning maps, zoning ordinances,

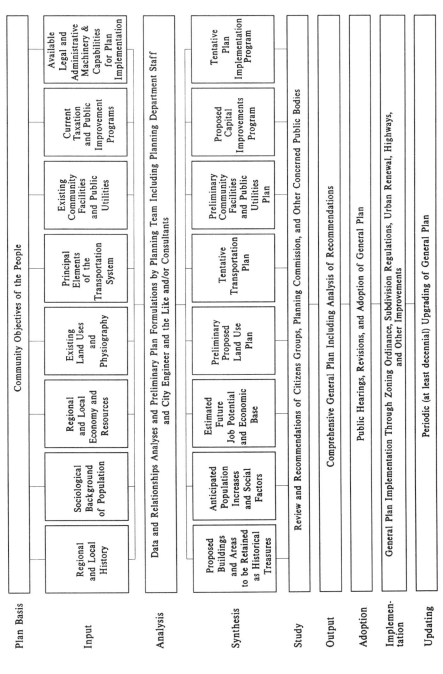

Figure 1.5   Simplified diagrammatic outline of general planning process. Source: William H. Claire, Ed. *Urban Planning Guide—ASCE Manual No. 49*. New York: ASCE, 1969. Reprinted by permission of the American Society of Civil Engineers.

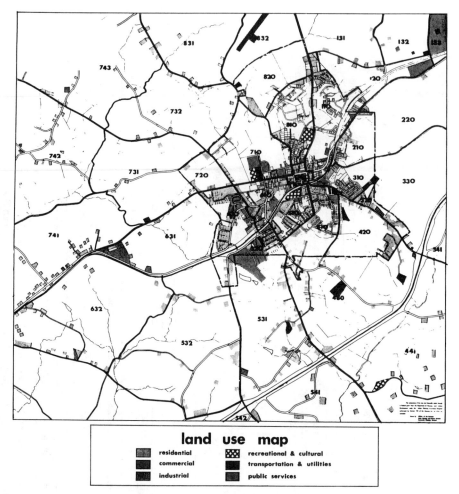

Figure 1.6 Existing land use map for Liberty, S.C. Source: E. L. Falk, Planner in Charge. *Liberty, S.C. Land Use Plan.* Clemson University, 1972. Reprinted by permission of Clemson University.

subdivision regulations, urban renewal, and the adoption of codes for buildings, electrical, and plumbing activities.

The general land use plan or master plan is not a zoning map, as it is used only as a guide for controlling future growth patterns. A zoning map, on the other hand, is an integral part of the zoning ordinance. It is site specific and cannot be altered without legal process.

1.2.6   Plan Maintenance

Maintaining the currency of these devices is vital to insure that the plan remains responsive throughout the entire planning period. Periodic jurisdictional modifications may be necessary to update the plan. As the future unfolds, various "planned" elements based on

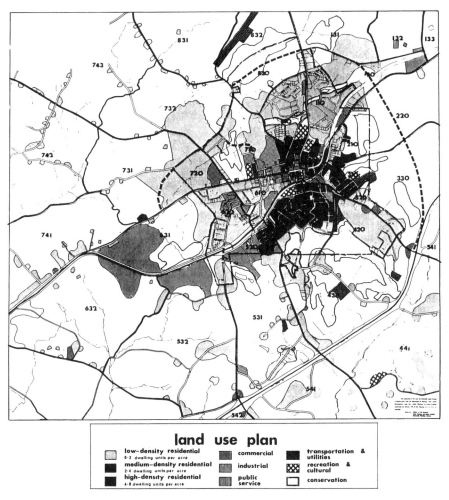

Figure 1.7 20-year land use plan for Liberty, S.C. Source: E. L. Falk, Planner in Charge. *Liberty, S.C. Land Use Plan.* Clemson University, 1972. Reprinted by permission of Clemson University.

previous uncertainties give rise to a variable regulatory condition that requires the successful developer to be well informed.

1.3 THE DEVELOPER'S ROLE

The person or organization that is the catalyst for land development is called a developer, or owner. This may be an individual, partnership, corporation, or trustee. The developer is usually credited with the conception of the project. The preliminary market potential must be assessed by the developer so that an initial risk assessment can be made. If the developer believes the project has potential and has reason to believe that project funding can be obtained, several preliminary activities must be undertaken. First, an attorney may be retained to handle legal matters, including title searching and agreement drafting, which are required when developing land. Secondly, the developer may schedule a formal market analysis to obtain an indication of potential success while investigating and soliciting property availability for consideration as the potential site.

1.3.1 Design Professionals

The developer would then solicit the services of a design professional to investigate, plan, design, and complete the desired project. Professionals normally associated with land developments include registered civil engineers, land surveyors, architects, land planners, and/or landscape architects. These professionals play a vital role in shaping the project's cost when performing their services. Combined with the cost of land, the cost of site improvements and structures cannot be so high as to make the final product unmarketable. Figure 1.8 illustrates the cost components of a new single-family home in a residential land development.

It is the responsibility of the developer to establish the project parameters and goals that are used to guide the design professional. If the proposed development is to include commercial areas, the developer usually contacts key regulatory personnel to inspect and approve the completed project. Involved in this process are zoning, building, electrical, fire, and plumbing inspectors, along with local police agencies and jurisdictional highway, health, and utility organizations.

During the design process, the design professional will produce data, studies, alternatives, and recommendations that must be analyzed by the developer for decision making activities. These decisions form the basis for final project design and include preliminary studies, cost estimates, and a reanalysis of the project's risk by the developer.

1.3.2 Project Approval Activities

The developer can then seek preliminary project approval from regulatory agencies through the assistance of the design professional, and at the same time solicit tentative financial commitments to fund the project. Once notified of preliminary plan approval, and having received a tentative commitment for project funding, the developer will engage the design professional to complete work on final construction plans, specifications, and cost estimates. Review of these documents allows the developer to evaluate the

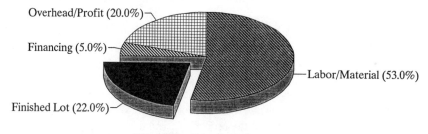

1991 - Sales Price $120,000

Figure 1.8 Cost components of a new single-family home. Source: National Association of Home Builders (NAHB). *1992 Housing Backgrounder.* Washington: NAHB, 1992. Reprinted by permission of the National Association of Home Builders.

project's pro forma (estimate of project costs and income to establish the rate of return) and to make yet another final risk analysis. Assuming the developer is still comfortable with the project's scope and potential, design professionals will seek approval for the developer from regulatory authorities for permission to construct the project. The financial commitment for property acquisition and construction improvements is formalized at this time by the developer. Any temporary utility service hook-ups required during construction can be arranged by the developer; however, some contractors may have this burden placed upon them in the construction specifications.

1.3.3   Activities During Construction

At this stage, a construction contractor can be engaged through the bid process or through direct negotiations. The design professional's assistance is normally utilized by the developer in this process.

If premarketing activities are to be undertaken by the developer prior to the satisfactory completion of all required improvements, many jurisdictional authorities will accept a bond or other posted security to insure the project's completion.

Safety is another protection the developer must provide, using reasonable care to those who pass by or trespass on the development. Those persons, especially children, must not be exposed to unnecessary risk or harm. Prudent developers take precautions to protect the public during construction and also require the contractor to assume a similar role (Sweet, 1985).*

Progress by the contractor will be monitored by the design professional or construction manager and the developer. Upon completion, the project will be inspected by the design professional and all appropriate regulatory review agencies having jurisdiction over the development to ascertain whether the finished site improvements meet the scope and intent of the approved plans and specifications. Certificates for occupancy or system operational approval will be issued when construction work is substantially completed and satisfactory. Final marketing arrangements which may include sales or even lease agreements, can be undertaken once all constraints have been met by the developer. Compliance with all rules and regulations, such as the Federal Interstate Land Sales Full Disclosure Act (15 USC 1701), can allow the developer access to the funds necessary to undertake and complete the project.

REFERENCES

Barnhart, C. L. and Jess Stein, Eds. *The American College Dictionary.* New York: Random House, 1963, p. 242. Copyright © 1963 by Random House, Inc. Reprinted by permission of Random House, Inc.

Claire, William H., Ed. *Urban Planning Guide—ASCE Manual No. 49.* New York: American Society of Civil Engineers, 1969, pp. 27, 82–83, 149. Referenced material is reproduced by permission of The American Society of Civil Engineers.

Falk, E. L., Planner in Charge. *Liberty, S.C. Land Use Plan.* Unpublished planning report. Clemson, S.C.: Clemson University, 1972, pp. 26, 35, 45, 85.

National Association of Home Builders (NAHB). *1992 Housing Backgrounder.* Washington, NAHB, 1992.

*See references at the end of this chapter.

So, Frank S., Israel Stollman, Frank Beal, and David S. Arnold, eds. *The Practice of Local Government Planning*. Washington: International City Management Association, 1979, pp. 262–272, 389–415.

Sweet, Justin. *Legal Aspects of Architecture, Engineering, and the Construction Process,* 3rd ed. St. Paul, Minn.: West, 1985, pp. 129–134.

2 Developer's Constraints

A major problem affecting a developer is the time required to complete a project, during which interest payments accumulate and consumer markets, along with construction costs, fluctuate. The engagement level a developer seeks in a project directly relates to the demand for committed personal time. There are several approaches a developer might use to undertake a project as illustrated in Figure 2.1. If the project is small and all work will be accomplished in a single phase, the project would reflect a consummate approach. If the project is large, and will be undertaken in incremental sections through phased construction, a staged approach would be reflected. Within each of these approaches the developer must further establish a role of either providing only improved land activities or total development activities. Improved land activities normally include the procurement and transfer of raw property into improved building sites. These improved sites, having transportation access, utilities, and storm drainage facilities, could be sold to others, including potential individual homeowners or professional builders, who would then undertake the completion of all remaining site improvements. Possible site improvements include erection of structures, landscaping, driveways, electrical, water, and sanitary sewer service ties. This type of sale generally contains restrictions on the minimum cost of the proposed dwelling unit, along with its type of construction, size, and possibly other elements. Total development activities include those undertaken in improved land developments and also all other activities required to produce a complete project. A professional builder could be engaged in a total development situation to build on an improved site owned by the developer. The finished product might then be jointly sold. Another approach might be for the developer to engage a single firm to both design and build the project.

2.1 MARKET ANALYSIS

An evaluation of the project area is necessary to determine whether:

(a) There exists a need for development.
(b) The resultant product will possess utility when completed.
(c) There is a scarcity of residential units, due to a lack of inventory or created by filtration.
(d) There is a capability of a buyer to purchase the product.
(e) There is reasonable expectation that the project will be absorbed by the target market and that the project is economically feasible.
(f) A profit potential exists.

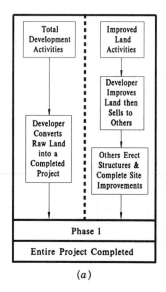

(a)

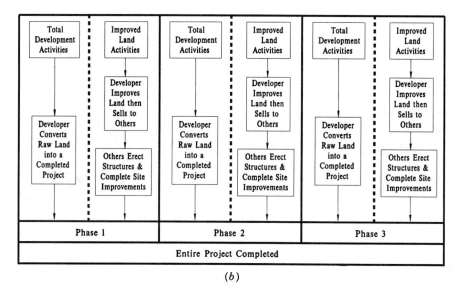

(b)

Figure 2.1   Examples of two approaches used to develop property. (a) Consummate approach. (b) Staged approach.

These essential conditions, which are normally investigated by a marketing research consulting firm for the developer, are evaluated using a procedure known as market analysis (Ashbury, 1970; Casey, 1968; FHA, *Techniques of Housing Market Analysis*, 1970). The results of the analysis are used to determine the highest and best use of a particular site. The analysis must be subject to certain boundary conditions within which the market area exists.

2.1.1 Market Area

It is not an easy task to select a market area. Generally, the approximate realm of influence exerted by a development can only be guessed. This may appear easy, but when one begins to draw lines of delineation on a map, many problems arise. Questions confronting the analyst include whether to follow natural features (ridge lines or drainage basins), or political subdivisions (county, city, or state boundaries), or census tracts, or municipal districts (school, water, fire, or sewer), or even road systems for defining the market area. Information from various sources, which was gathered according to specific originator needs, may be required to aid in the conduct of an analysis. Initial study immediately shows that overlapping or incomplete information will probably be present, since resource dividing lines do not always coincide. The best approach is to pick a combination of sources (e.g., census information and utility company records) that produces the most complete area coverage, along with estimated partial values obtained from other self-serving sources that have data gathered within unique differing confines. An alternate, more accurate, method is to disregard existing sources and conduct a first-hand study of the area for market analysis information, although time and expense usually preclude this approach.

Once a market area has been determined a two-part analysis can commence. The macro analysis will yield a telescopic view of the area. The micro analysis will zoom in on specific entities. Together they should indicate development marketability.

2.1.2 Macro Analysis

By placing the market area in broad perspective, the macro analysis will give an indication of prosperity, viability, and durability of existing conditions. Some of the topics to be addressed are demographic analysis of population characteristics and trends, income distribution, employment trends and levels, existing housing stock, potential and firm housing starts, military facilities (if applicable), and retiree population. Short term population trends, calculated using linear regression or other curve fitting methods, will normally yield reasonable projections from which family numbers can be estimated, based on a locally determined factor that represents the number of persons per household. Future income distributions can also be projected, enabling the developer to determine the project's target price range.

Projecting age distributions allows the analyst to determine future numbers of people within certain age brackets; these projections are required to establish bedroom demands.

Most families spend about 35 percent of their gross income on housing. This knowledge enables one to determine the magnitude of housing indebtedness that can be supported by funds, from which future mortgage sizes can be estimated for economic levels within the population.

Combining this information yields the housing demand, which includes the number of dwelling units, bedrooms, and price range at a given future time. Existing housing numbers, reduced by a factor that reflects dilapidation, can be combined with firm and potential housing starts in the market area (determined by interviews, building permit issues, or governmental sources) to produce an estimate of the housing supply. The numeric difference between the housing supply and the housing demand yields the housing need. If factors such as new industry and low unemployment simultaneously indicate economic stability and growth, the projected annual housing need is considered reasonably valid.

2.1.3 Micro Analysis

A micro analysis of possible development sites can be undertaken once the housing need has been established. The fair market value of the land and the effective drawing power of the proposed facility in relation to existing competition are worthy of consideration. A professional market analyst should be consulted for a more thorough assessment of the problem.

Sometimes area market analyses are prepared by governmental bodies such as the U.S. Department of Housing and Urban Development (HUD) and local planning agencies. Inspection of these analyses may reveal that an area is nearly in equilibrium; however, some private sector studies might indicate that the same market reflects differing conditions. To prevent market saturation a more conservative approach may have been taken by the government as compared to some private sector analyses. If equilibrium were the actual situation, one might wonder why vacancy rates do not rise appreciably. The reason might be that new industry seeks geographic areas where housing is already available; consequently the supply may precede the actual need.

If the developer plans to include some commercial area as part of the project's scope, the market analysis must include the influence, magnitude, location, and distribution of competitive facilities, as well as the income, social groups, family size, ages, and life-styles of people in the market area.

2.2 SITE LOCATION

Having selected a handful of possible sites within the market area as a result of the micro analysis, one must choose which is the most suitable for the particular development—be it a manufactured home park, manufactured home subdivision, single-family development, multifamily development, or apartments. Consideration should be given to any effect the potential development might have on the adjoining properties, and vice versa, relating to light, air, view access, and traffic generation.

2.2.1 Municipal Services

A residential site should provide the occupants with municipal services necessary for everyday family needs. These services include the availability of schools, religious facilities, places of employment, shopping centers, and medical facilities. Police and fire protection availability will make the development a safer and more appealing place to live. To mobilize the area, transportation links are required, both for private conveyance and for public modes. Arrangements must be made for solid waste disposal. Finally, if the development is to attract a particular consumer market, appropriate accommodating facilities should be included as part of the development plan. Figure 2.2 graphically illustrates several of these desired relationships (APHA, 1960; DeChiara and Koppelman, 1969).

Of significance to the developer is the relationship of the site to existing municipal utility lines. If a site is chosen that is not close to existing sewer and water service, additional developmental costs will accrue in providing these necessities. Additionally,

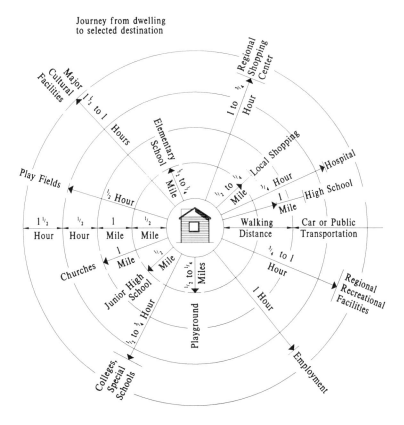

Journey from dwelling
to selected destination

Walking distance is measured in miles.
Car or public transportation is measured in time.

Figure 2.2   Maximum distance for community facilities. Source: Joseph DeChiara and Lee Koppelman. *Planning Design Criteria.* Van Nostrand Reinhold, 1969. Reprinted with permission.

depending on the size of the development, the governing body may require certain areas to be reserved for future locations of some of the above listed municipal services.

2.3   SITE GEOMETRY

Property parcels comprised of boundary lines that avoid acute convergence angles do not contribute to spurious land slivers or extremely disjointed shapes, and hence offer the greatest ease to the developer in providing site improvements. Figure 2.3*a* illustrates an attractive site, while Figure 2.3*b* represents an undesirable one.

2.4   SITE CONDITIONS

When a site is being considered for a certain proposed type of development, several physical features must be investigated to determine its suitability. The developer normally utilizes the services of design professionals who strive to minimize construction and

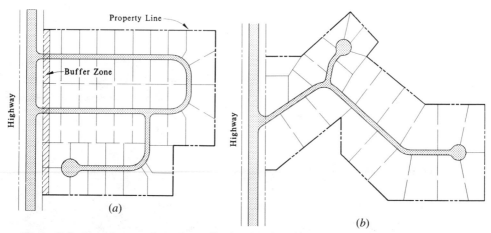

Figure 2.3   Boundary configurations affecting development planning. (*a*) Site conducive to promoting economical development; (*b*) site having an irregular shape where development costs are increased.

improvement costs, while providing a safe and workable project which, with normal maintenance, will endure.

Some potential project locations require additional considerations. Site habitation by endangered species, marine mammals, or migratory birds can appear to hamper the activities of an aggressive developer; however, if preservation and conservation are practiced in concert with established regulations, natural features supporting such habitats can be used as beneficial assets to enhance the project's attractiveness.

2.4.1   Wetlands

Potential development sites must be investigated for the presence of wetlands. A wetland has been defined as an area that contains hydric soils, hydrophile plants, and periodic flood water. According to the Conservation Foundation, *Protecting America's Wetlands: An Action Agenda. The Final Report of the National Wetlands Forum*, 1988, (Train, 1988), wetlands "provide habitat indispensable to a great and varied array of aquatic, avian, and terrestrial wildlife." Salt and fresh water wetlands can be categorized into five different system types: marine, estuarine, riverine, lacustrine, and palustrine systems (Cowardin et al., 1979). Figure 2.4 depicts these classifications of wetlands. Although the marine system is not shown, Figure 2.5 provides important details on the estuarine and riverine systems, while Figure 2.6 illustrates the lacustrine and palustrine systems. Should wetlands occur on site, an investigation and evaluation must be undertaken to determine whether the wetlands fall under the jurisdictional control of regulatory agencies. Regulated wetlands come under the jurisdiction of the U.S. Army Corps of Engineers (COE) and the Environmental Protection Agency (EPA), in concert with state and local units of government (Dunkle et al., 1989). These organizations are responsible for maintaining the chemical, biological, and physical well-being of the country's waters. Potential discharge hazards resulting from dredging or filling activities must be minimized and therefore cannot be undertaken without permit approval as provided under Section 404 of the Federal Clean Water Act (P.L. 92-500), as amended. The U.S. Fish and Wildlife Service and the National Marine Fisheries Service play important advisory roles in this permitting process (DeVereaux, 1982; U.S. Army Corps of Engineers, 1991).

2.4.2 Environmental Assessment

Another condition that might constrain a developer's activity is the receipt of an adverse environmental report on a potential site. Under the Comprehensive Environmental Response, Compensation, and Liability Act of 1980 (CERCLA) owners of property contaminated with hazardous substances are strictly, jointly, and severally liable for conducting or financing cleanup of such substances under provisions contained under Section 107. In 1986 the Superfund Amendment and Reauthorization Act (SARA) allowed, "innocent purchasers" a defense to this liability if, prior to property acquisition, "all appropriate inquiry into the previous ownership and uses of the property" along with an "appropriate inspection" of the property was undertaken. A developer who acquires property today with even the slightest indication of possible hazardous substances being present might not enjoy the benefits embodied in SARA and could be liable for all future cleanup costs. In an effort to avoid potential cleanup liability resulting from foreclosure action, where remediation costs may exceed the property's market value, financial institutions and prudent developers are undertaking "due diligence" investigations to protect themselves and the end user. Should contamination be discovered later, developers will bear the burden of proof to demonstrate that "due diligence" was in fact exercised in order to acquire "innocent purchaser" status (Gustin and Neal, 1990; Leifer, 1989; Moskowitz, 1989).

Another type of environmental investigation, precipitated by the National Environmental Policy Act of 1969, may be required if the federal government is involved in the development. HUD requires local environmental reviews for community block grants, rental rehabilitation, and housing development grant programs, under the format of an Environmental Assessment. This process will indicate the project's compliance with applicable rules and regulations while determining whether an Environmental Impact Statement (EIS) or a Finding of No Significant Impact (FONSI) is appropriate (U.S. HUD, undated).

2.4.3 Coastal Zone Regulations

If the potential project is located in an area that comes under state jurisdiction being regulated through the Federal Coastal Zone Management Act (P.L. 92-583), the developer may be required to limit development in certain areas by conforming to established critical zones, setback lines, and pollution control activities that conform to the regulatory coastal zone requirements (S.C. Coastal Council, 1979, 1988). These regulations help protect coastal land and waterways, while providing a coordinated approach to minimizing the effects of erosion related property losses and flood damage. Structural improvements, such as the installation of docks, bulkheads, or revetments within the coastal zone, may require COE approval under Section 10 of the Federal Rivers and Harbors Act of 1899.

The terrain controls much of the development's layout and makes certain parcels unbuildable due to their location within a jurisdictional wetland, floodplain, or drainage basin. Flood plain management practices require the developer to provide minimum grades on roads, establish minimum finished floor elevations on dwellings, and add structural integrity to dwellings also located within floodway "V zones," or velocity zones. As most developers approach the economics of a project from the standpoint of minimizing costs while maximizing profits, the cost of improvements to accommodate certain topographical features such as ravines, natural bottoms, and sloughs may preclude development of a particular site.

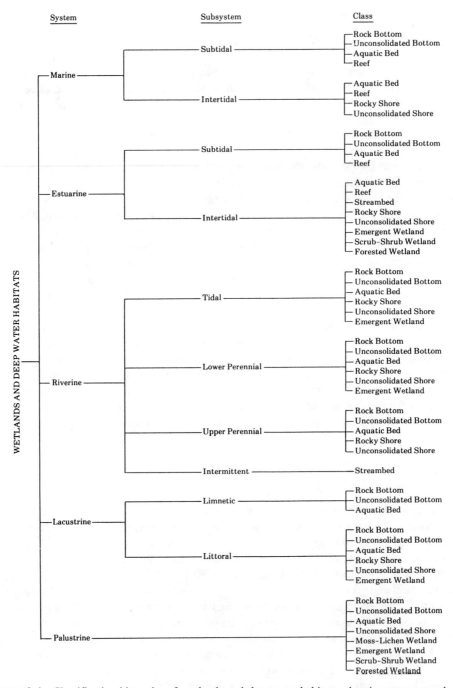

System   Subsystem   Class

Marine
- Subtidal
  - Rock Bottom
  - Unconsolidated Bottom
  - Aquatic Bed
  - Reef
- Intertidal
  - Aquatic Bed
  - Reef
  - Rocky Shore
  - Unconsolidated Shore

Estuarine
- Subtidal
  - Rock Bottom
  - Unconsolidated Bottom
  - Aquatic Bed
  - Reef
- Intertidal
  - Aquatic Bed
  - Reef
  - Streambed
  - Rocky Shore
  - Unconsolidated Shore
  - Emergent Wetland
  - Scrub-Shrub Wetland
  - Forested Wetland

Riverine
- Tidal
  - Rock Bottom
  - Unconsolidated Bottom
  - Aquatic Bed
  - Rocky Shore
  - Unconsolidated Shore
  - Emergent Wetland
- Lower Perennial
  - Rock Bottom
  - Unconsolidated Bottom
  - Aquatic Bed
  - Rocky Shore
  - Unconsolidated Shore
  - Emergent Wetland
- Upper Perennial
  - Rock Bottom
  - Unconsolidated Bottom
  - Aquatic Bed
  - Rocky Shore
  - Unconsolidated Shore
- Intermittent
  - Streambed

Lacustrine
- Limnetic
  - Rock Bottom
  - Unconsolidated Bottom
  - Aquatic Bed
- Littoral
  - Rock Bottom
  - Unconsolidated Bottom
  - Aquatic Bed
  - Rocky Shore
  - Unconsolidated Shore
  - Emergent Wetland

Palustrine
- Rock Bottom
- Unconsolidated Bottom
- Aquatic Bed
- Unconsolidated Shore
- Moss-Lichen Wetland
- Emergent Wetland
- Scrub-Shrub Wetland
- Forested Wetland

(Left margin, vertical text: WETLANDS AND DEEP WATER HABITATS)

Figure 2.4 Classification hierarchy of wetlands and deepwater habitats, showing systems, subsystems, and classes. The palustrine system does not include deepwater habitats. Source: L. M. Cowardin, Virginia Carter, F. C. Golet, and E. T. La Roe. *Classification of Wetlands and Deepwater Habitats of the United States.* U.S. Department of the Interior, 1979.

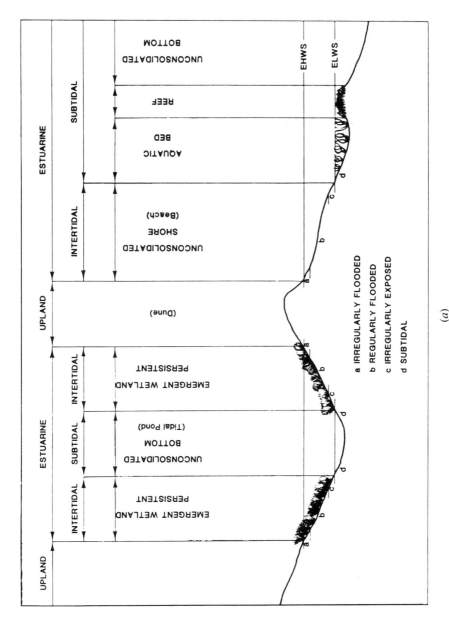

Figure 2.5 Distinguishing features and examples of habitats. (*a*) In the estuarine system. (*b*) In the riverine system. EHWS = extreme high water of spring tides; ELWS = extreme low water of spring tides. Source: L. M. Cowardin, Virginia Carter, F. C. Golet, and E. T. La Roe. *Classification of Wetlands and Deepwater Habitats of the United States.* U.S. Department of the Interior, 1979.

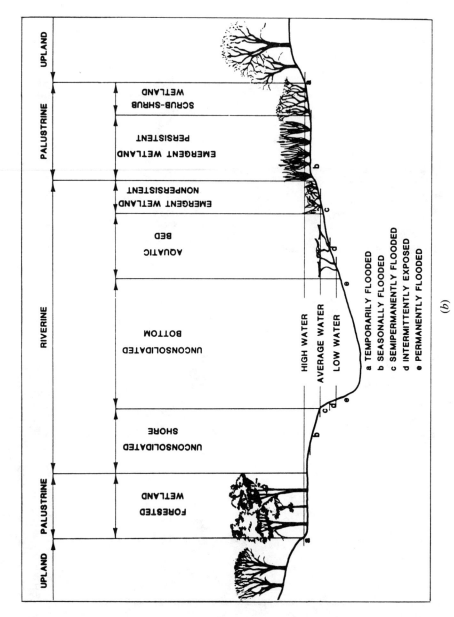

(b)

Figure 2.5 continued.

24

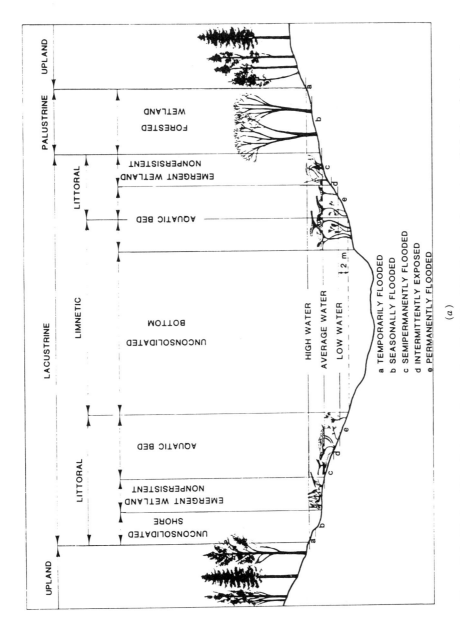

Figure 2.6 Distinguishing features and examples of habitats. (*a*) In the lacustrine system. (*b*) In the palustrine system.
Source: L. M. Cowardin, Virginia Carter, F. C. Golet, and E. T. La Roe. *Classification of Wetlands and Deepwater Habitats of the United States.* U.S. Department of the Interior, 1979.

(*a*)

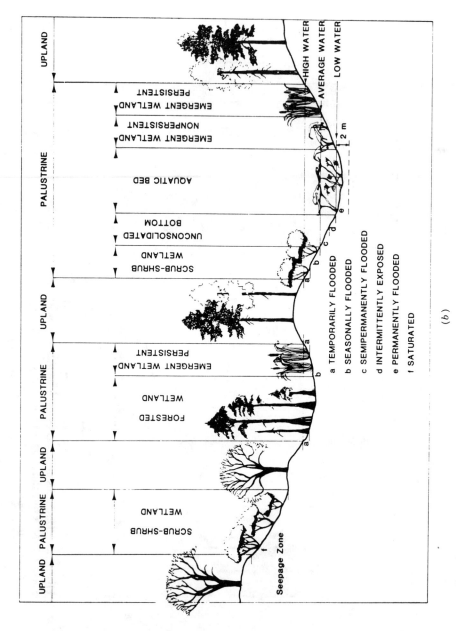

Figure 2.6 continued.

2.4.4 Existing Site Terrain

Attributes of a well developed project often preexist with the developmental plans, merely magnifying them. For example, if a knoll is present with a deep ravine running through the property, the knoll should be considered a good location to place structures due to view, drainage, and maybe even exposure to prevailing favorable winds. The ravine, a natural drainage feature, should be maintained if possible.

When considering drainage, a gradual and uniformly sloped parcel, which has a natural drainage basin into which storm water can be diverted, is the most desirable since this represents site conditions that would require minimal improvement costs. Should the site be level, higher development costs will be incurred because of the increased excavation costs required for gravity drainage line installations. Land that possesses little topographic relief normally suffers from a high groundwater table, and as a result requires additional earthwork activities to provide positive site drainage features. If the site is located where the closest direct access to a natural drainage basin exists across another's property, then an easement for a drainage outfall may be required to provide proper discharge. Should this easement be granted, potential flood waters can be properly channeled, and better drainage throughout the area will result. Where natural drainage basin access is unavailable, provisions must be made for holding storm water runoff until it can evaporate, percolate, or slowly be piped away. This holding process is known as ponding, and can be either in the form of retention (runoff is retained without discharge) or detention (runoff is held for a period of time while being slowly discharged).

It would be foolish for a developer to confine all interests in topography strictly to the proposed site just because it has boundaries that limit ownership and development interests. Topographic boundaries (e.g., ridge lines, valleys, and edges of plateaus) usually are not confined to the site boundaries, as the site may be located somewhere between a ridge and a valley. If this is the case, any alteration to the area within the natural drainage boundaries may affect the physical characteristics of interior areas. Should a natural drainage system be realigned to provide better road access, the altered course may result in flooding of adjacent land if there is no place for the water to be channeled to allow continued flow downstream. Developing land also causes runoff to increase considerably, which results from several factors. One factor that accounts for more runoff volume is the increase in impervious areas. The coefficient of runoff associated with development related impervious surfaces, such as asphalt pavements, concrete sidewalks, driveways, shingle roofs, grass lawns, and others, is much higher than that of forest land with its prolific foliage and ground cover. This additional runoff must be controlled and channeled through existing outfall drainage facilities. An investigation will determine whether there will be an overload placed on these facilities, resulting in flooding.

Road systems also have a relationship to the existing terrain. Most developers are aware that earthwork is expensive and that it is desirable to minimize cutting and filling operations as far as it is consistent with overall improvement requirements. The better a road adapts to a site's topography, the more construction costs are reduced. Figure 2.7 illustrates three situations reflecting placement of development streets as they relate to site topography.

Terrain directly affects utility improvement costs. Depths of embedment necessary for proper covering and for attaining the required installation slopes are dependent on hilly ground and variation in slope.

Topography also plays an important role in the development of the sanitary sewer

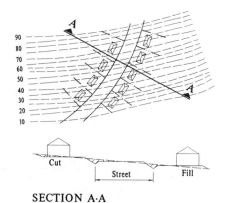

SECTION A·A

A. Design Road to Diagonally
Cross Steep Terrain

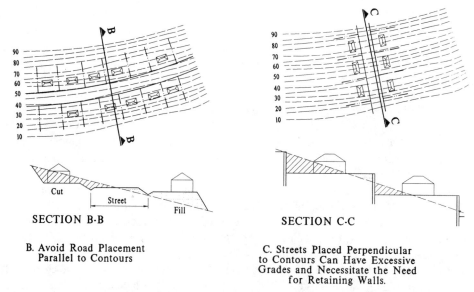

SECTION B·B

B. Avoid Road Placement
Parallel to Contours

SECTION C·C

C. Streets Placed Perpendicular
to Contours Can Have Excessive
Grades and Necessitate the Need
for Retaining Walls.

Figure 2.7 Placement of development streets. Source: U.S. Department of Housing and Urban Development.

system plan. If municipal service is not available adjacent to the site, then an on-site centralized treatment facility might be considered. The project site layout may dictate building the plant in an area that conflicts with the unifying theme of the development, thus causing relational problems. If municipal services are available some distance from the site, the topography of the site and that of the land separating the site from the municipal line are determining factors when contemplating utility line extensions to join existing off-site facilities. These extensions may require the installation of a sanitary lift or pumping station with an emergency generator when gravity flow is not feasible for sanitary sewer lines.

2.4.5 Subsurface Site Conditions

An important concern is the investigation of the underlying soil at the site. Results of this investigation have an effect on mucking and excavation costs, fill material needs, pavement thickness requirements, and foundation requirements. Soil permeability is a parameter that relates to the soil's percolation capability necessary for drain field sewage effluent dissipation associated with septic tank applications.

An attractive site would have fairly stable soils free from rock outcrops and organic content; thus a soil exploration program is necessary to insure satisfactory results. If the proposed site contains unsuitable soil types, the soil exploration program should yield useful information for correcting the problem. For example, if an area drains poorly as a result of underlying impervious material, subsurface drains may be used to improve these conditions. Data on the soil's permeability would be helpful in this analysis, despite the fact there was no initial requirement to determine the soil's permeability for purposes of evaluating drain field effluent dissipation needs because municipal sewage facilities were available.

Certain basic properties of the soils should be determined for all projects. These are usually confined to the liquid limit, plastic limit, grain size distribution, and possibly the unconfined compression tests, along with the optimum moisture content (OMC) determination for soil located in subgrade and foundation areas. If a multifamily development has sizable structures, additional soil testing may be required.

Geology is another consideration of subsurface investigation. Investigation of a site may reveal fault lines or the potential for landslides or mudslides, which would deter locating the project there. Sinkholes in limestone areas also could pose a problem and therefore should be investigated (U.S. Department of Interior and U.S. HUD, 1970). In addition, the presence of expansive soils on site can pose a potential problem to the integrity of completed site improvements and structures (Jones and Jones, 1987).

Although groundwater table elevations fluctuate, their range is important when considering a community potable water system. Removal of groundwater from a particular unconfined aquifer may result in lowering the stabilized groundwater table. This results in a cone of depression, called drawdown, which may be of significance. If groundwater is used as the supply, this information must be available to determine a suitable design. On sites having rolling terrain it is necessary to investigate all potential locations where groundwater might seep to the surface through hillsides, resulting in piping, which is a phenomenon that can occur in an earth dam. Should piping exist, it would be hazardous to build in that location, as the ground is saturated, and generally unstable. Incorporation of a subsurface filter blanket may minimize this problem.

2.4.6 Climate

Climatic activities, including temperature, humidity, wind, and precipitation, can affect a potential development site, and thus give rise to various conditions the developer must address. The area's temperature range should be considered as some design parameters are temperature dependent. Subgrade depths must be of sufficient magnitude to withstand frost action and heave. Also, building foundations and the depth of utility line burials are dependent on frost penetration depth. All of these items reflect costs to the developer to improve the land. The range of temperature also plays an important part in the amenities to be provided in the development. For example, if the area is subject to warm summers, a community swimming pool may be included as part of the recreational amenities afforded the occupants. Air conditioning and heating demands are directly related to the ambient air temperature and relative humidity, resulting in costs to the developer in providing these services. Use of active and passive solar heating and cooling systems has proven efficient in providing comfort, with consideration being given to properly positioning structures to provide comfort control and energy conservation.

Precipitation is another factor that affects a developer's final product. Areas subject to snowfall require increased structural strength and abatement provisions for removal. Rainfall affects the scope and magnitude of storm water collection, delivery, and containment facilities and the frequency and magnitude of flooding. Precipitation also affects the availability of water for consumption by recharging ground aquifers or by contributing to the resupply of surface water sources.

2.4.7   Blight

Unappealing visual, environmental, or auditory conditions can affect the developer's assessment of a potential project's location. Scenes of unpleasant neighboring land use activities may render the site unappealing to the passerby, who might be a potential renter or buyer. Adverse environmental conditions, such as hazardous wastes or contaminated groundwater, on or close to the site, pose serious and significant development potential liability. Loud noises from heavy traffic, manufacturing, aircraft operations, or industrial related activities are considered auditory blight and pose potential problems that can undermine the peace and tranquility of the development.

2.5   SITE LAYOUT

Once the proposed location is chosen, a development plan must be drawn up that economic analysis can be undertaken to insure a feasible and workable project. These include a preliminary cost estimate, preliminary financing negotiations, and preliminary negotiations with regulatory officials. The site layout on the preliminary plat is usually one of many that have been generated by a design professional during the project's planning process. Many site plans can be fitted to a particular land plot; however, there is usually one plan that suits best by fulfilling requirements of workability, safety, economy, personal edification on the part of the developer, and finally, compliance with developmental regulatory guidelines.

2.5.1   Development Theme

Site configuration is largely influenced by the theme of the development. If the developer's goal is strictly to provide a neighborhood atmosphere, similar to that proposed by Perry in 1929 (Perry, 1929), it would include all necessary services, facilities, and activities for a coordinated and connected setting that would accommodate the everyday needs of the occupants.

Should the developer attempt to attract a certain clientele using a particular marketing theme for appeal, amenities such as golf courses, marinas, and airports have potential for project success.

On a large project, the developer may wish to include some flexibility in his marketing plan by sequentially phasing the development. This employs the staged approach in combination with flexible or multiple land use zoning (if available) to compensate for future market fluctuations and trends.

2.5.2 Allocation of Nonresidential Areas

Depending on the size and location of the development, the developer may have to consider allocating some area of the site to nonresidential uses such as commercial, recreational, educational, and open space areas (FHA, *Techniques of Housing Market Analysis,* 1970).

Commercial facilities such as shopping centers have estimated drawing radii depending on their classification. In addition there is an estimated minimum population required to support such facilities (Casey, 1968). A neighborhood shopping center of about 50,000 square feet of floor space and about 150,000 square feet of parking can be supported by about 1000 families. If the development's scope and magnitude warrants consideration of including commercial areas, it would be advantageous to do so, as high rates of return may be obtained if the commercial area prospers. Normally, the developer undertakes the development of a commercial area as a total development activity where all improvements including buildings, parking, and appurtenant work are installed. Figure 2.8 illustrates a typical neighborhood shopping facility.

Theme enhancement can be promoted with the inclusion of recreational areas that involve active or passive participation. Examples of active areas include swimming pools, tennis courts, basketball courts, and tot lots. Passive areas encompass neighborhood and community parks. User safety is paramount; therefore high visibility from motorists must be provided or adequate spacial separation included. Figure 2.9 shows various types of recreational facilities and their nomenclature.

Nature and conservation areas, which would include fresh and salt water wetlands, create passive recreation areas and provide natural beauty while also protecting habitat and timber. Barrier protection and flood routing benefits also can be attained.

The developer of a large project may be confronted with issues that center on allocation of areas for educational facilities. Figure 2.10 exhibits one set of guidelines that could vary based on jurisdictional requirements.

2.5.3 Configurations for Residential Developments

A development can take form in many ways; however, there are five basic configurations. The first utilizes a series of orthogonal streets, named gridiron. This configuration is appealing because it is usually the most economical to design and lay out. Also, construction costs are lower when considering form work for curbing and sidewalks. The gridiron configuration is applicable to communities that have experienced little growth and do not want to alter street patterns. Certain multifamily developments are conducive to this type of configuration because cost reductions may be possible.

Gridiron development, illustrated in Figure 2.11, can produce hazards that are minimal in other configuration types. Long straight streets can induce high vehicular speeds that should be lowered within the confines of a residential area. Through traffic can intrude into the area because of the accessibility provided. One may find the gridiron arrangement presents a monotonous appearance. Drainage and construction costs may be appreciably higher with this type of layout because it may not conform naturally with the land. If the

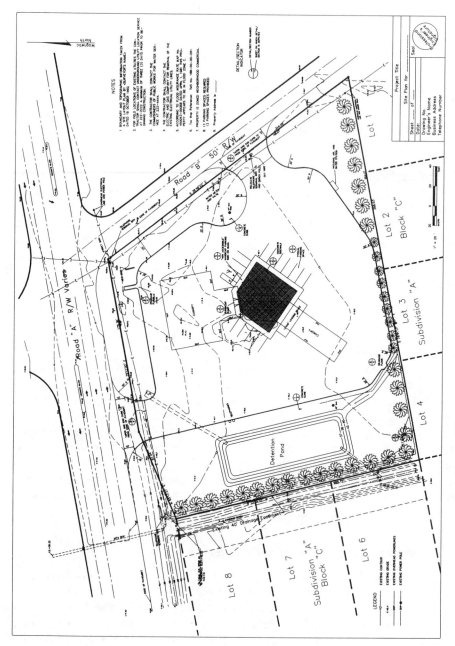

Figure 2.8 An example of a neighborhood shopping center.

32

FACILITY TYPE	SIZE	ORGANIZATIONAL LEVEL (POPULATION)	TARGET AGE GROUP	STANDARD[1]	SERVICE AREA	RECOMMENDED ACTIVITIES AND REMARKS
tot lot	2000-2500 sq. ft.	dwelling group (40)	5 years & under	20f/1000	300 ft. radius	defined areas for individual creative play and areas which encourage social order (stand in line, take turns, share, etc.) remark: this facility primarily intended for use in urban residential areas; may not be feasible in suburban and rural residential areas.
sand lot	½-1 a	small neighborhood (250)	6 - 12 years	4f/1000	½-½ mile radius	multiuse open area (sports field, social activities, etc.), some play equipment, plus areas like those in tot lot above, but more advanced. remark: this facility may not prove suitable in rural residential areas.
neighborhood park	3-6 a	neighborhood (1500)	all ages	3a/1000	½-1 mile radius	multiuse fields, paved sports facilities (tennis, basketball, etc.), shelters, picnic facilities and areas designed to stimulate individual and social growth.
community park	10-30 a	community (9000)	all ages	2a/1000	3-5 mile radius	recreation center building (possibly with gymnasium), natural and/or nature study area, possibly a swimming facility, plus the same areas which are found in the neighborhood park.
county park	100-300 a	county (54,000)	all ages	3a/1000	10-20 m radius	organized and individual camping facilities (possibly cabins), possibly fishing and boating (especially applicable in Pickens County), plus the same areas found in the community park, natural environmental areas, recreation building (including hardcourt sports facilities) multiuse fields, daycamp facilities, paved sports facilities, could also include golf and large scale spectator sports facilities.
regional park	500-2000a	region (300,000)	all ages	5a/1000	1-1½ hr. drive	wide range of active and passive recreation including day use and overnight use facilities, should provide much the same facilities and services as the county park but with more natural areas and more extensive hiking, camping, nature trail and overnight facilities.

[1]The following abbreviations are used: f–facility; a–acre; m–mile. STANDARDS are expressed in facilities (f) or acres (a) per 1000 population.

Figure 2.9 Recreational facility standards. Source: E. L. Falk, Planner in Charge. *Pickens County Recreation and Open Space Plan.* Clemson University. 1975. Reprinted by permission of Clemson University.

Type	Area Required
Elementary school	10 acres + 1 acre per 100 students
Middle and junior high school	20 acres + 1 acre per 100 students
Senior high school	30 acres + 1 acre per 100 students

Figure 2.10 Educational facility standards. Source: S.C. Department of Education. *South Carolina Facilities Planning and Construction Guide,* 1983.

design is found forced, then a curvilinear design might suit more pleasingly and efficiently the terrain and boundary configuration.

The curvilinear layout, illustrated in Figure 2.12, consists of roads that follow the terrain in an unrestricted manner while adhering to regulations. This results in winding road patterns that extend through the project and present an interesting feeling to the traveler because of the variations in road alignment. The curvaceous road system helps reduce traffic speeds within the development. The length of curvilinear roads, compared to an equivalent gridiron design, is usually less, resulting in reduced improvement costs. This concept provides little protection to residents from offsite traffic. Vacant land and open space are not usually accommodated when either of the aforementioned types of land developments are undertaken. However, the curvilinear design does allow for preservation of some natural terrain features, which may result in theme improvement while reducing construction costs attributed to drainage, earthwork, and accessibility requirements.

The curvilinear method allows greater freedom for the developer to work with the land. Another type of development, known as the density zoned development, at times called a cluster development, is shown in Figure 2.13. This approach allows even more flexibility in preserving natural topographical features and vegetation. This type of development is controlled by zoning that restricts the number of dwelling units per gross acre. Some regulations require the developer to allocate a minimum percentage to open space and

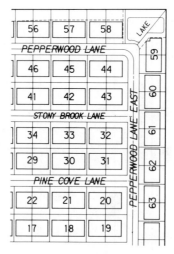

Figure 2.11 An example of the gridiron development configuration. Courtesy of Post, Buckley, Schuh, and Jernigan, Inc., Tampa, Fla., 1992.

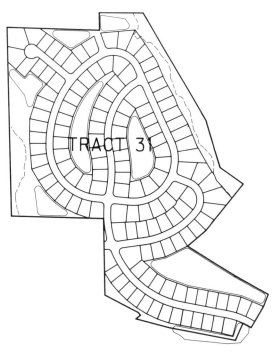

Figure 2.12 An example of the curvilinear development configuration. Courtesy of Post, Buckley, Schuh, and Jernigan, Inc., Tampa, Fla., 1992.

public facilities, while maintaining a maximum allowable percentage for commercial and residential areas. This configuration results in groups of dwelling units being surrounded by green belts and open space corridors.

To assist in high density areas the Federal Housing Administration (FHA) has devised a scheme called land use intensity (LUI), which describes numerically the ratio of the relationships of gross land area, building area, and open space area allowed for a given parcel. Most zoning ordinances, however, do not refer to LUI.

Density zoned developments provide many things missing from rectilinear and curvilinear developments. They allow the developer to create a project of any size, incorporating any dwelling type or size, provided it conforms with the intent of the zoning laws. Unbuildable areas, such as wetlands, are utilized as open space, while more suitable areas are used for dwelling unit locations. Improvement costs are thus reduced. Increased access and serviceability are obtained through the use of density zoned developments. Amenities such as variations in parking, interesting courtyard layouts, and sculptured features can be provided by development. Lot sizes are small, providing reduced maintenance obligations to the occupant. Density zoned developments also provide for seclusion from nonessential traffic, thus affording greater safety for residents. For the developer, this type of development allows for reduced improvement costs, and often eliminates the need for curb and gutter and sidewalks. This concept accommodates amenities such as golf courses, picnic areas, playgrounds, conservation, and historic preservation because adequate land area is available with no sacrifice of density.

There are, however, additional problems associated with density zoned developments. One is maintaining and controlling open and vacant land available for use by residents.

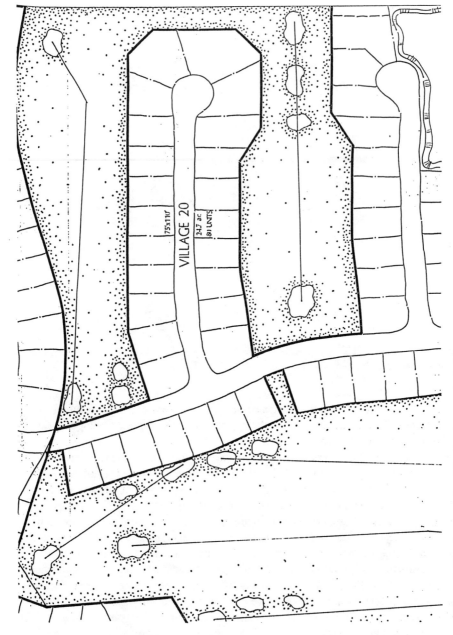

VILLAGE 20
75'x110'
24.7 ac
80 UNITS

Figure 2.13 An example of the density zoned configuration. Courtesy of Post, Buckley, Schuh, and Jernigan, Inc., Tampa, Fla., 1992.

There are questions about whether the common land should be dedicated to the public, maintained by the dwelling unit owners through a homeowners' association, or maintained by the developer. Another problem might be that local ordinances do not permit density zoned developments. Legal action may be required to amend the local ordinances.

A common misconception about a density zoned development is that inadequate recreational areas are better than none. This may seem true until a golf ball shatters the picture window of a dwelling fronting a cramped golf course. Amenities and basic essentials must be adequate for their intended usage.

Figure 2.14 depicts a cluster development, which is a particular kind of density zoned development. What distinguishes the cluster type development is that dwelling unit sites are located in clusters centering on cul de sacs or common courts and each of these clusters is separated from the other by open space. In density zoned developments the dwelling unit sites are not necessarily located in scattered groups or clusters, but may all be centrally positioned. Both, however, allow for buildings to be placed for maximum effectiveness while minimizing construction improvement costs.

Planned unit developments (PUD), which evolved during the 1960's and 1970's, are similar to density zoned developments. They contain both residential areas and recreational open space, and may include commercial sites that can be centrally and conveniently located. A PUD is designed to self-regulate common areas using a homeowners' association that is legally organized and chartered. Such an organization, which can be formed for other types of developments as well, is composed of members of the planned unit. The homeowners' association governs, through authority embodied in its charter, such regulatory items as new development conformance, restrictive covenant enforcement, grounds maintenance, operation, and the general character of the area. Most PUDs have a central theme such as a golf course, swimming pool, marina, beach, or air facility. Other amenities might include parks, boat docks, jogging trails, club houses, tennis courts, and bicycle facilities (Tomioka and Tomioka, 1984). Owners of dwelling units within the PUD have membership privileges to any or all of the facilities that are provided. The homeowners' association usually assesses each dwelling unit owner a flat fee depending on the situation, to support and maintain all common facilities. A developer should recognize that a sizable amount of legal work is necessary to set up and protect such a development. In addition, the FHA has ruled that any development that has mandatory homeowner association fees (even if it is only one dollar) qualifies as a PUD; therefore, if potential purchasers are to have the opportunity of utilizing FHA insured financing, the developer must submit specific information to the FHA for approval.

Condominium developments, which can include high-rise, townhouse, and detached units, follow the general pattern of the PUD, except that the individual units may have a specific amount of land sold with each unit designated for yard space.

As a result of urban sprawl, development related problems, and a reduction in sites suitable for development, land planners and architects during the 1980's formulated an innovative approach to urban-suburban designs called "Traditional New Towns." They can vary from several hundred acres in size to tens of acres and are usually not incorporated. According to Lewis (1990), Miami architects Andres Duany and Elizabeth Plater-Zyberk have pioneered the use of:

> Traditional townscape imagery, streetscapes, and civic spaces, street grid patterns, mixed uses and higher densities, new approaches to zoning, multi-modal transportation and "pedestrianization."

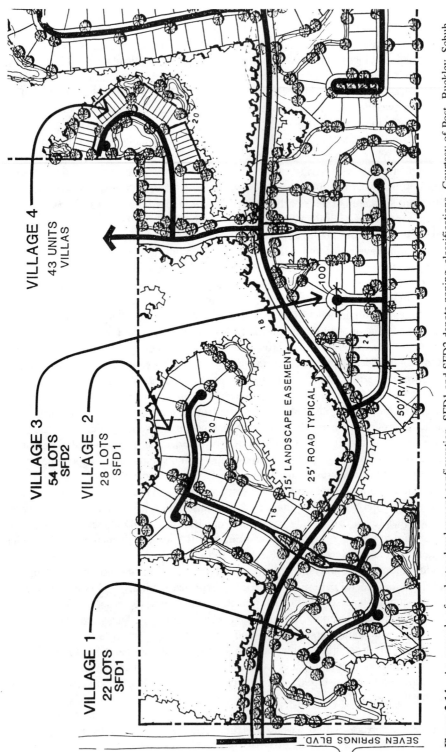

VILLAGE 1
22 LOTS
SFD1

VILLAGE 2
28 LOTS
SFD1

VILLAGE 3
54 LOTS
SFD2

VILLAGE 4
43 UNITS
VILLAS

15' LANDSCAPE EASEMENT

25' ROAD TYPICAL

50' R/W

SEVEN SPRINGS BLVD.

Figure 2.14 An example of a cluster development configuration. SFD1 and SFD2 denote zoning classifications. Courtesy of Post, Buckley, Schuh, and Jernigan, Inc., Tampa, Fla., 1992.

38

Traditional new towns have a town center with spatial identity, a system of tree lined narrow streets that integrate with site conditions to produce rectilinear or curvilinear grids that range from alleyways to avenues, and pedestrian facilities. Short blocks promote traffic dispersion, while parking is provided within blocks at building sides or to the rear, and pedestrian paths are networked with public parks and places.

Civic buildings are located to provide landmarks. They form part of the streetscape system, along with a variety of dwelling types, mixed with compatible nonresidential areas and buildings. As a result, the concept of the traditional neighborhood development (TND) evolved as successor to the PUD.

According to Spielberg (1989), the TND has regulated maximum and minimum proportions of land allocated to various uses (housing, retail, commercial, services, and light industry). In addition other requirements, such as pedestrian accommodations, lot sizes, layout, and street frontages have been determined.

The TND is contrasted in Figure 2.15 with typical development patterns that have prevailed since the 1950's.

Changes in local development regulations may be required to accommodate a TND because streets are scaled to slow traffic and promote pedestrian movements while street intersections are more prevalent to promote the dispersion of vehicular traffic. In addition, some communities have promoted "affordable housing" developments, where development regulations have been modified to accommodate small lot developments with a reduction in lot area, frontage, and setbacks—including provisions for zero lot lines (ZLL). Small lot developments require a more integrated use of the indoor and outdoor areas while providing adequate parking and privacy.

2.5.4 Aesthetics

To promote a project's beauty and harmony, the developer usually relies on architects and landscape architects. As part of the professional team they work to refine spacial relationships that yield pleasant views and vistas, resulting in an aesthetically pleasing experience (Rubenstein, 1987). Development planting schemes can enrich aesthetic response by including trees that are suitable to the area's climate, exposure, and soil conditions. Some communities have tree ordinances that help to protect mature on-site trees and provide for plant renewal as part of the construction activity. Deciduous and evergreen shrubs, hedges, and ground cover can control erosion and also provide beauty. Natural open spaces and curvaceous lake perimeters can help promote aesthetic continuity. Planting associated with parking lots helps provide visual screens and sight continuity.

The Historical Preservation Act of 1966 may affect the aesthetics of a development if federal participation is involved. If the site includes areas of historic significance similar to those in the National Register of Historic Places or if it has natural resource features that warrant preservation, the developer should consider the inclusion of conservation easements to protect these valuable assets.

2.5.5 Transportation Considerations

Transportation systems within a development are major concerns of the developer. Facilities to allow the safe and efficient movement of people and goods are essential and should be made within the framework of the existing urban transportation plan. Vehicular driver

Suburban Sprawl

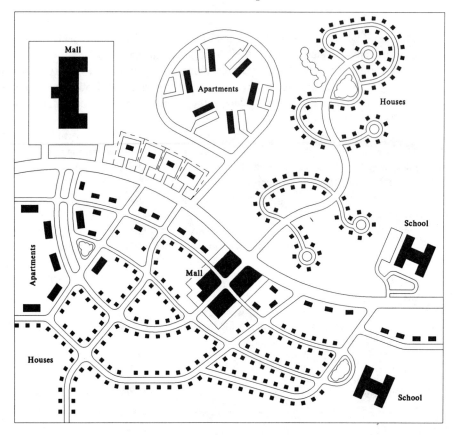

Traditional Neighborhood

Figure 2.15   An example of a community developed as a traditional neighborhood development (TND) (bottom) versus that developed under existing development guidelines (top). Taken from *ITE Journal,* September 1989, "The Traditional Neighborhood Development: How Will Traffic Engineers Respond?" by Frank Spielberg. Copyright © 1989 by the Institute of Transportation Engineers. Reprinted with permission.

requirements must be coordinated with pedestrian and bicycle activities to promote the development's transportation network (ASCE et al., 1974).

Initial concerns to the developer are the proximity of adequate highway and street accesses to allow service to the site. Off-site studies of adjacent land can provide data that will be useful in determining the adequacy of existing access facilities. Streets within the development can augment and reinforce the jurisdictional area's transportation plan.

Generally, road system requirements are prescribed in local regulations that attempt to create networks that provide ease in vehicular accessibility and require minimal maintenance. These regulations exist in the expectation that sound engineering practice and principles, as well as quality construction, will yield a workable, efficient, and safe development that, during the life of the improvement, will minimize cost to the taxpayer by minimizing maintenance activities.

Public residential streets are peculiar in that they produce no tax revenue but require tax dollars for maintenance. In addition, the developer receives no direct monetary compensation from the sale of streets but is obligated to pay for their construction. To minimize construction costs for the developer and maintenance costs for the public body assuming custody, the roads should be as narrow as possible per dwelling unit while still yielding a safe and workable system. Utility costs generally are closely related to the length of road per dwelling unit in a project; thus a reduction in road length usually results in reduced utility construction costs.

When at least two entrances into a site are included, emergency access can still be provided even if one of the entrances becomes blocked. Another benefit of providing multiple entrances is to reduce the concentration of traffic by allowing the dispersion of vehicular movements.

Figure 2.16 shows the general types of publicly dedicated streets based on their use and function. Arterial streets handle large traffic volumes, resulting in heavy usage. Driveway entrances should be avoided along these streets for safety purposes. Collector streets act as feeder elements to convey traffic from local to arterial streets. The main entrance to a residential development can act as a collector street if the development is large enough. Local streets are usually designed to prohibit through traffic and to provide vehicular access to collector streets from interior property locations. A local street that allows frontage property protection from heavy arterial traffic is called a frontage street. Figure 2.17 illustrates two methods of protecting property contiguous to arterial streets by using a frontage street access approach, or reverse frontage lots with protective plant barriers, fences, or walls that are located in a nonaccess easement to prevent double frontage access. Lots with access to more than one street, other than those located at street intersections called corner lots, have double frontage, which is undesirable unless precautions are taken similar to the use of a nonaccess easement (U.S. HUD 4140.1, 1973).

Narrow streets, called alleys, usually have a minimum width of 20 feet, which allows on and off loading activities in commercial and industrial areas. Alleys can be used in small lot residential developments to improve the streetscape by eliminating views of

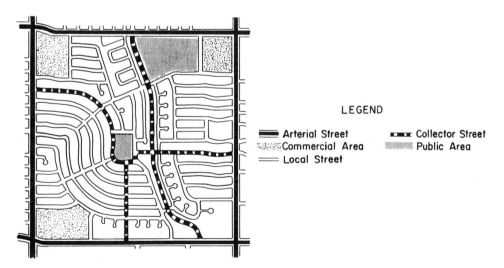

LEGEND

━━━ Arterial Street   ▭▬▭ Collector Street
▒▒▒ Commercial Area   ▓▓▓ Public Area
═══ Local Street

Figure 2.16 Types of streets found in land developments. Taken from *A Policy on Geometric Design of Highways and Streets*. Copyright © 1990 by the American Association of State Highway and Transportation Officials. Reproduced with permission.

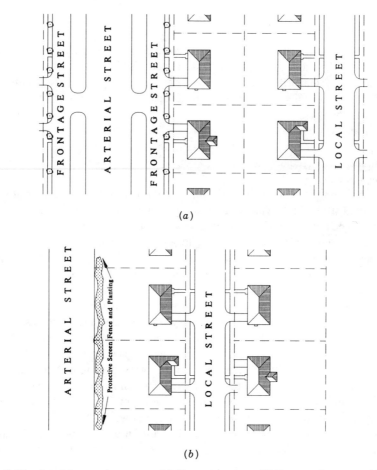

Figure 2.17 Arterial street lot access. (*a*) Frontage streets. (*b*) Reverse frontage with protective screen easement. Source: U.S. Department of Housing and Urban Development.

garages and eliminating curb cuts for driveways. Should the developer require access to a residential development through adjacent off-site property, an ingress-egress easement, if granted by the adjoining landowner, would accommodate such access.

A cul de sac is another street type that permits flexibility in utilizing various difficult areas within the site. Normally cul de sacs are limited to a maximum length by local regulations. Cul de sacs assume different configurations that allow flexible applications including circular, "T", and "Y" shapes. Figure 2.18 illustrates several circular applications.

Reserve strips can cause problems for the developer. Reserve strips occur at the end of a terminal road between a road right of way and the property boundary for the purpose of limiting road access and thus prohibiting adjacent landowners' use of the road. If off-site street access is needed to service an interior portion of the site, and a public road exists close to the boundary but is separated by a very small strip of land called a reserve strip, the chances are very good that access through the strip will be granted by the jurisdictional authorities. In the case of a private road, a reserve strip is legal; however, a knowledge-able landowner who seeks safety and orderly traffic patterns may allow access to promote

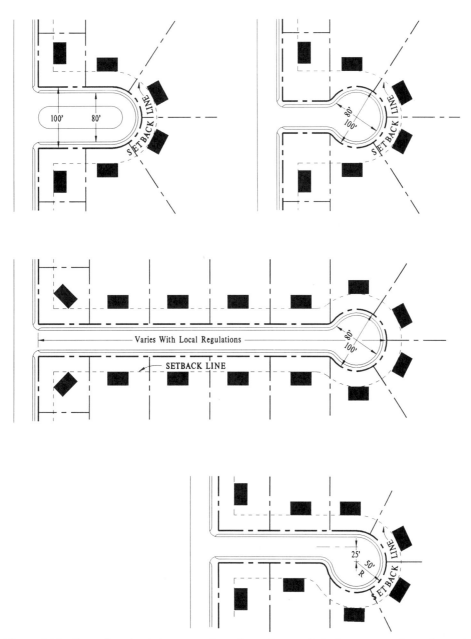

Figure 2.18   Examples of cul de sac configurations. Source: U.S. Department of Housing and Urban Development.

good planning if the imposed traffic does not cause problems. An exception occurs when a development site is landlocked. Then the matter is usually decided in court if the adjacent landowner does not concede an access.

Finally, the name of streets is usually a requirement for all developers to undertake as part of the regulatory process. Figure 2.19 illustrates a method used by the Montgomery

County (Ohio) Planning Commission to preserve continuity; however, care must be exercised to insure that existing names are not duplicated, resulting in confusion.

The developer must consider the right location within the project for signs that provide both information and traffic control. Motorists expect traffic signs and pavement markings to conform to the standards established in the *Manual on Uniform Traffic Control Devices* (MUTCD) (U.S. Department of Transportation, 1988), and the developer should insist that these standards be maintained if traffic control devices are required. Should a traffic accident occur as a result of a nonstandard traffic control device being installed, the developer may be liable as a contributory party.

Street intersection design is an integral element of the road system. No more than a maximum of four streets should be allowed to intersect at any one point. Three-way intersections reduce the number of conflicts and are preferred if centerline offsets of 125 feet or more are available to provide adequate site distances, as shown in Figure 2.20.

The angles used in intersections should be as close to right angles as possible, as indicated in Figure 2.21. If large volumes of traffic pass through an intersection, consideration should be given to including turning lanes.

Parking requirements vary depending on the type of development and zoning requirements. Most single-family developments utilize off street parking; therefore, no special provisions are usually required. For multifamily and commercial developments, adequate parking facilities must be available to provide convenience and safety to the user. Parking areas must be proportioned for ease in vehicle maneuvering and proper drainage.

2.5.6 Pedestrian Considerations

In large scale developments there is an acute need for included pedestrian facilities to augment and complement circulation patterns formulated by local authorities. Large developments may contain open space or recreational areas where safe pedestrian access can enhance the overall project cohesiveness. This will simultaneously allow for such activities as lingering and browsing, or an opportunity for a user to socialize with neighbors or perhaps enjoy the scenery. Other pedestrian activities might include trips related to work, shopping, school, and a means to interconnect with public transportation facilities.

Pedestrian facilities should be fairly level. Obstacles, obstructions, lack of maneuvering room to avoid others, and congestion resulting in an inability to maintain reasonable walking rates are potential problems. System continuity that incorporates a minimum number of potentially conflicting situations with vehicles and other pedestrian movements enhances user efficiency and comfort.

User safety not only entails a consideration of vehicle and pedestrian crossings, but also protection from pollution associated with noise, refuse, and odor. Amenities such as street furniture, fountains, and other aesthetic improvements are beneficial and can help to convey the underlying project theme. A developer's use of adequately designed ramps, steps, guards, and safety barriers can provide pedestrian safety, as can the use of night or security lighting to enhance visual safety.

2.5.7 Handicap Access

Physically handicapped access should be of importance to a developer. The Americans with Disabilities Act (ADA), PL 101-336, provides that all private establishments/facilities considered "public," such as those provided in some commercial developments, must provide "reasonable accommodations" to disabled persons. Also, the 1988 Fair

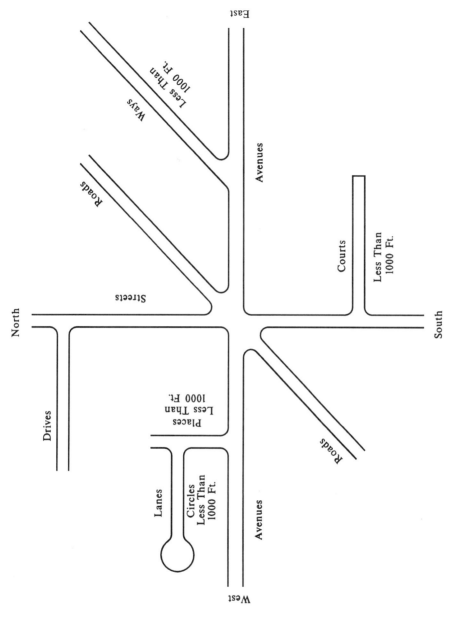

Figure 2.19 Method for naming streets. Source: Montgomery County Planning Commission, Dayton, Ohio. Reproduced with permission.

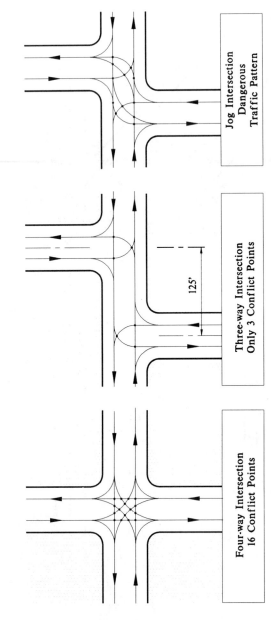

Figure 2.20 Street intersection types and resulting traffic patterns. Source: U.S. Department of Housing and Urban Development.

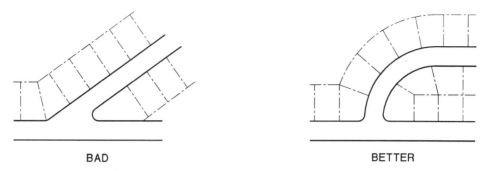

BAD   BETTER

Figure 2.21 Examples of desirable and undesirable street intersection configurations. Source: U.S. Department of Housing and Urban Development.

Housing Act amendments address handicap access in multifamily developments. The Architectural Barriers Act (P.L. 90-480) requires all developments that contain public facilities, such as retail stores, or that rely on federal grants, loans, or utilization to provide ample accessibility to all pertinent buildings and site facilities, including those for trash disposal and mail pickup. This encompasses not only federally assisted rental and home ownership units but also federally owned properties and qualified assembly areas.

Initially, the American National Standards Institute (ANSI) approved the first national standard for access design in 1961, which was designated as ANSI A117.1. In 1980 modifications were made and incorporated as part of the Minimum Guidelines and Requirements for Accessible Design (MGRAD) that was mandated by Congress. In 1984 the General Services Administration, the Department of Defense, the Postal Service, and the Department of Housing and Urban Development promulgated the Uniform Federal Accessibility Standards (UFAS), which were consistent with MGRAD.

According to UFAS, a number of aspects must be addressed, including provisions for adequate passage and maneuvering space to accommodate handicap transit from public streets, sidewalks, and passenger loading zones to the closest serviceable building access. Handicap parking spaces, bearing the recognized symbol, and loading zones must be contiguous to a five foot free access zone, as illustrated in Figure 2.22. The minimum number of designated spaces should be in accordance with Figure 2.23. Provisions promoting handicap access are illustrated in Figure 2.24. If the access grade is greater than 20:1, then it is treated as a ramp, which may require landings and handrails, and cannot exceed a 1:12 gradient (General Services Administration et al., 1988).

2.5.8 Bicycle Considerations

Some projects allow developers to provide special facilities for bicyclists, called bikeways. Bikeways should be smooth riding surfaces free from construction joints, holes, and differential settlement. They should be located in areas away from congestion with motor vehicles, pedestrians, and other opposing cyclist movements. They also should incorporate adequate horizontal and vertical design controls to allow proper maintenance of sight distances. Bike/vehicular/pedestrian crossings should be kept to a minimum for improved safety. Bikeways can be bike routes, lanes, paths, and trails, as illustrated in Figure 2.25. Bike routes normally utilize the same roadway riding surface as motor vehicles, whereas bike paths parallel existing roadways while being offset by a small

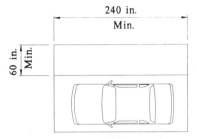

Access Aisle at Passenger Loading Zone

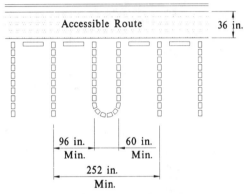

Dimensions of Parking Spaces

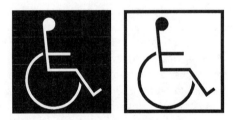

International Symbol of Accessibility

Figure 2.22 Handicap parking details. Source: General Services Administration, U.S. Department of Defense, U.S. Department of Housing and Urban Development, and U.S. Postal Service. *Uniform Federal Accessibility Standard*, 1988.

shoulder. Bike lanes can be either unprotected (where only a paint strip separates the motor vehicle from the cyclist) or protected (where a physical barrier is in place). Bike trails normally are located independent of other transportation facilities (Daecher, 1975).

2.5.9 Storm Water Management

In addition to local regulations, storm drainage facilities installed by the developer are subject to provisions of an area wide water quality plan implemented as a result of

Total Parking in Lot	Required Minimum Number of Accessible Spaces
1 to 25	1
26 to 50	2
51 to 75	3
76 to 100	4
101 to 150	5
151 to 200	6
201 to 300	7
301 to 400	8
401 to 500	9

Figure 2.23   Handicap parking requirements. Source: General Services Administration, U.S. Department of Defense, U.S. Department of Housing and Urban Development, and U.S. Postal Service. *Uniform Federal Accessibility Standard*, 1988.

provisions contained within Section 208 of the Clean Water Act, as amended. This plan coordinates the overall surface water quality within the area. Point sources, such as culvert discharges, and nonpoint sources, such as overland sheet flows, are included in the analysis. These waters contain settled air hydrocarbons, metals, eroded soils, salts, chemicals, litter, and animal refuse, with the highest concentration of pollution being contained within the first one inch of runoff. Best management practices (BMPs) are encouraged to reduce adverse clean water conditions. Wetlands can be used to remove nutrients, soils, and other pollutants, as can retention/detention ponds with filtration. Local authorities may not issue building or grading permits prior to the developer's compliance with the overall management plan.

Under the Environmental Protection Agency's (EPA's) new General Storm Water Permits Program, which became effective October 1, 1992, development sites comprised of five or more acres of disturbed land are affected and require a National Pollution Discharge Elimination System (NPDES) General Permit. To qualify under the General Permit, a "Notice of Intent" (NOI) must be filed at least 48 hours prior to construction ground-breaking, and a storm water pollution prevention plan must be available on site for review by regulatory authorities if requested.

This plan must include a description of the development site and a map showing areas that will be disturbed during construction. The plan must also show the location of stabilization activities and control measures. The location of surface waters at discharge points, as well as the location of slopes and drainage patterns after development, must be included. This plan must also contain the sequencing of construction activities (supplied by the contractor). Other requirements of the plan include:

(a) An estimate of the coefficient of runoff from the site after construction.

(b) Data on the quality of the storm water draining from the site.

(c) An inventory of the types of soils located on site.

Provisions for sediment basins are also required for development sites where 10 or more acres of land are to be disturbed. These basins must provide at least 3600 cubic feet of storage for each acre drained. In addition, the plan must include controls for vehicle soil tracking, and disposal of construction solid wastes, while providing for site inspec-

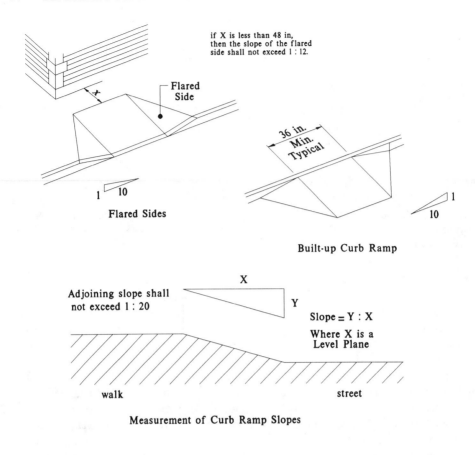

if X is less than 48 in,
then the slope of the flared
side shall not exceed 1 : 12.

Flared
Side

36 in.
Min.
Typical

1 10

Flared Sides

1
10

Built-up Curb Ramp

X

Adjoining slope shall
not exceed 1 : 20

Y

Slope = Y : X

Where X is a
Level Plane

walk

street

Measurement of Curb Ramp Slopes

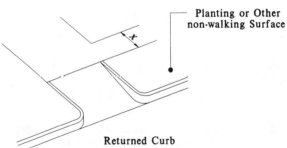

Planting or Other
non-walking Surface

Returned Curb

Figure 2.24 Handicap access requirements. Source: General Services Administration, U.S. Department of Defense, U.S. Department of Housing and Urban Development, and U.S. Postal Service. *Uniform Federal Accessibility Standard,* 1988.

tions every seven days, or within 24 hours after a rainfall of 0.5 inches or more has occurred.

Should the development be located within a designated urban area or within the jurisdiction of a medium or large municipality where the population exceeds 100,000, point source discharges will be subject to provisions of an NPDES permit, as provided under Section 402 of the Clean Water Act, as amended. Discharge data collection,

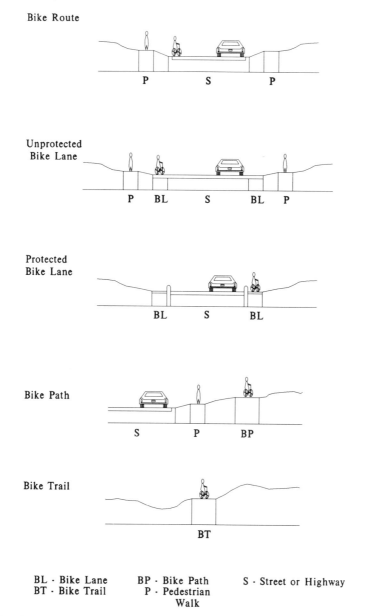

Bike Route

P   S   P

Unprotected
Bike Lane

P   BL   S   BL   P

Protected
Bike Lane

BL   S   BL

Bike Path

S   P   BP

Bike Trail

BT

BL · Bike Lane   BP · Bike Path   S · Street or Highway
BT · Bike Trail   P · Pedestrian
                         Walk

Figure 2.25 Bikeway nomenclature. Source: C. W. Daecher. *Proceedings of the 4th National Seminar on Planning, Design, and Implementation of Bicycle and Pedestrian Facilities.* New York: The American Society of Civil Engineers, 1975. Reprinted by permission of the American Society of Civil Engineers.

sampling, and analysis are required with remedial treatment action being prescribed if necessary (U.S. EPA, 1990).

Diffused surface storm water has been considered a "nuisance" (Rice and White, 1987) as each landowner can dispose of it as long as it does not affect another property owner in a detrimental manner. Installed drainage facilities therefore should be placed to protect other site improvements from flooding while minimizing erosion and excessive ponding in streets, which would hamper vehicle travel.

Operation, maintenance, control, and ownership of the completed facility is normally undertaken by the jurisdictional governing body as public works activities. However, with increased emphasis being placed on managing storm water facilities such as detention ponds, the developer may find some municipalities have formed separate storm and surface water utilities. These storm and surface water utilities have been vested authority to generate operational revenues by assessing monthly charges, similar to those for water and sewer, from landowners based on individual runoff contributions (Bissonnette, 1985).

2.5.10   Wastewater Management

Proper wastewater disposal is an essential element of a successful project. The developer may be fortunate enough to find an existing wastewater main, which was installed in concert with an area wide wastewater plan, provided by Section 201 of the Clean Water Act, as amended, that is close enough to the project that feasible tie-in arrangements can be made. Improvements would include collection lines, with appurtenances, and, if required, a pumping facility with force main to allow a means to transport wastewater to the connection point.

Should the development be distantly located from an existing system, an alternate means of handling the wastewater might include on-site ground disposal through the use of properly designed septic tank and drain field systems that will neither contaminate groundwater nor leach into waterways. Another means might be the installation of a central wastewater treatment plant facility.

In compliance with requirements of the Clean Water Act, as amended, a central facility would have to play an integral role within the area wide 208 plan. The developer could obtain a Section 401 water quality certificate by utilizing spray irrigation on suitable land, such as a golf course, to dispose of the treated wastewater. Should a point discharge be proposed, the developer would need an NPDES permit for treated effluent release based on dissolved oxygen sag curve depletion analysis and other parameters.

2.5.11   Potable Water Considerations

Every effort should be made by the developer to provide potable water from an established municipal system. Should this not be available, then a suitable supply from a reliable groundwater aquifer or surface impoundment must be located. The U.S. Bureau of Reclamation and the U.S. Soil Conservation Service (SCS) are federal organizations involved with water supply activities. Treatment and storage facilities must be provided so that safe drinking water can comply with the National Primary Drinking Water Regulations and the Safety of Public Water Systems Act (P.L. 93-523). Should conditions warrant the withdrawal of water from one basin with ultimate discharge being made into another, a regulatory interbasin transfer permit likely will be required.

Availability of water for a development's supply from groundwater aquifers and surface water impoundments involves water rights doctrines that vary throughout the country. Surface waters can be classified as following riparian or appropriation doctrines. The riparian doctrine allows a contiguous landowner the right to "continual natural flow of the stream" (Rice and White, 1987) with reasonable use. The appropriation doctrine, which prevails where water is scarce, promotes the taking of water for beneficial use based on "first in time is first in right" (Rice and White, 1987). Control of groundwater falls under several doctrines. One is the absolute privilege doctrine, which basically provides overlying rights, without taking into account groundwater movement. Another is the reasonable use doctrine, which controls extracted groundwater strictly to use on the overlying property. Correlative rights define shared usage based on overlying ownership, whereas the prior appropriation doctrine examines percolating groundwater as underground streams. Permitting encompasses the remaining control system, which is normally handled on a state and regional basis, along with Federal Reserved Water Rights. Water lawyers and engineers proficient in water rights allocations will be required by the developer if a water supply is to be perfected for a development. For further details, consult specialized references on this important topic (Rice and White, 1987).

2.5.12   Electrical Considerations

Most power company policies vary with respect to developer responsibilities (SCE & G, 1967). To qualify for underground service the developer may be required to have at least 24 contiguous subscribers situated in a configuration acceptable to the power company. A minimum of 15 feet dispersion may be required between the dwelling unit and the sewage system (either centrally located system and appurtenances or individual septic systems). A manufactured housing development (mobile home type), which can take the form of a subdivision or park, must have a minimum lot width of 40 feet. The terrain must not be excessively rocky or entail unusually nonuniform slopes. Many companies prefer rear lot electric distribution configurations (a must for manufactured housing developments), as most water and sewer services are along the dwelling unit front. In the case of the manufactured housing (mobile home) park, a five year contract between the power company and the developer could allow the power company to amortize its installation investment. The preliminary plat generally contains the minimum information needed by the power company for initial design purposes. In addition, the developer must indicate the planned project construction time and phasing, which must be consistent and logical. To assist the power company in the system design, an estimated electrical load for each dwelling unit must be provided by the developer. Once these requirements have been satisfied, a wiring diagram for the project can be formulated by the power company. The developer is then able to review the proposed wiring diagram and grant utility easements necessary for plan implementation. If the developer expresses dissatisfaction with the plan, which is most likely to occur in street lighting schemes and transformer pad locations, the power company will normally make revisions with the understanding that any additional construction costs will be paid by the developer.

Electric lights are necessary along streets, parking lots, walkways, and areas requiring nighttime illumination. They are available in several styles, with the developer usually paying for nonstandard types, if selected. A contractual agreement that stipulates how revenue will be raised by the power company to support lighting services centers either on the developer or dwelling unit occupant and is normally undertaken as part of the project planning process.

Wiring configurations vary according to development type and are based on where the power company terminates its facilities and where the developer must make connection. In the case of apartments and mobile homes, the power company may supply all material and labor through the installation of meter boxes. The service entrance must be provided by the developer, and, if materials of a certain established quality are provided by the developer, power company reimbursement may be available. In most single-family developments the developer must include materials and labor from the secondary connection unit to the meter box and then within the dwelling unit. If materials meet certain standards, the developer may receive reimbursement for some incurred costs.

The power company would prefer a firm commitment from regulatory agencies prior to commencement of distribution system design. Once an agreement has been reached on the configuration, the developer must grant easements, usually ten feet wide, for installation and maintenance purposes. The developer normally is required to clear the easements and remove existing stumps and debris. Often the telephone company will utilize one side of the easement while the power company uses the other.

Construction of the power distribution system will not start until all lots are staked and the road system has been established in the field. Care must be taken to insure close coordination between the power, telephone, gas, and construction companies installing the sewer, storm drainage, and water distribution systems, as carelessness and lack of coordination often result in cut power lines and broken utilities.

2.5.13 Gas Distribution Considerations

Major gas companies have policies that establish relationships between developers and the gas company. Gas service is normally provided by the company to the dwelling unit meter. Piping from the meter through the dwelling unit is the dwelling unit owner's responsibility.

As in the case of the electrical distribution system, a gas line configuration usually will be provided by the gas company for review by the developer. Necessary easements must be granted, similar to those required for electrical distribution. If gas service cannot be provided by a public utility company, the developer then must decide whether to provide a centrally located master tank and distribution system or individual l.p. tanks.

2.5.14 Telephone Considerations

Telephone companies have requirements that should be satisfied to simplify service availability. Developers often do not consult initially with the phone company because installation and services are usually provided at no cost. The need for on-site electric power occurs early because construction tools must be energized; however, telephone service need is not usually felt until the dwelling unit occupants arrive. Below are several reasons why the telephone company should be consulted punctually:

(a) It alerts the telephone company that installation cables must be ordered from the supplier as materials generally are not locally stocked.

(b) It allows time to study the relationship of present and future demands to trunk line capacities to determine whether adequate capacities exist.

(c) It allows the telephone company to avoid piecemeal planning by addressing area wide service needs.

(d) It allows time for the telephone company to obtain any additional easements or encroachment permits (e.g., rail, highway, and others) necessary to locate service cables within controlled rights of way.

The telephone company will provide the developer a proposed distribution plan for review. Conflicting interests can be discussed and the plan modified if necessary. Ten foot minimum easements must be provided. Construction procedures are much the same as for electric and gas companies, except that during the interior construction of a structure the telephone company should have access to install or inspect interior wiring prior to wall finishing. The telephone company, as in the case of the other utility companies, experiences cut cables during the construction phases. Maximum coordination and respect for one another's services must command prime consideration.

2.5.15 Cable Television Considerations

Cables for the distribution of television broadcasts should be installed much the same as telephones. The utility that has the franchise to service a development area should be contacted early to provide efficient service.

2.5.16 Adjacent Utility Usage

On occasion utility systems installed by a developer are replaced by larger capacity lines, even though the original lines would accommodate the development's needs. This may be necessary because of new demands that have been placed on the system by the development of adjacent land. A prudent developer, who knows adjacent property will be developed in the future, should make arrangements initially with regulatory officials to install facilities compatible with all anticipated future loadings. This will reduce future public expenses, assist in area wide planning, and prevent unnecessary future burdens on the completed development. Incurred expenses over and above that required for the developer's installation are usually reimbursable.

2.5.17 Solid Waste Collection

Solid waste collection services must be available to occupants of the development. The service may be provided by a public body, such as the city or county, or by a private firm acting on an agreement with the developer or the homeowners' association. Depending on the development, certain types of pickups are available. For single-family developments generally house to house service is standard. Manufactured housing communities and multifamily developments utilize centrally located dumpsters for point pickups. The developer must provide easy access to these collection points to insure adequate receptacle capacity. High subgrade compaction and sufficient pavement depth are necessary to accommodate the applied concentrated tire loads that are nearly static and the additional loads of the container and waste.

   If the project includes total development activities, provisions for trash compactors can be included to help reduce solid waste volumes and the separation of waste materials (plastics–newspapers–metals–glass) for recycling. This will help to prolong the life of the receiving sanitary landfill.

2.5.18 Municipal Services

Many projects are regulated to incorporate various user service facilities within the development plans. For example, postal delivery facilities must be available and must meet requirements of the U.S. Postal Service. If public transportation or schools have routes that will provide service to the development, then access stops must be investigated. In addition, fire alarms and police call facilities may be required to enhance neighborhood protection and provide increased security.

2.5.19 Property Ownership

A developer may dispose or depart from property located within the development site by deed or gift conveyance, forced sale for back taxes, foreclosure sale, bankruptcy, adverse possession, eminent domain, or dedication. Planned activities would focus on the conveyance of improved property by sale; however, land dedication may be considered appropriate for certain parcels. During the planning process, the developer should avoid creating random residual parcels that could be left over as a result of a boundary that has an irregular configuration. If such parcels are created, the developer may be unable to market them, resulting in continued ownership, liability, tax obligations, and maintenance requirements.

2.5.20 Dedication of Land

Many times a site plan has provisions for pedestrian easements, natural conservation areas or buffer zones, tot lots, storm drainage retention or detention ponds, and recreational areas. In the case of the PUD the homeowners' association usually assumes maintenance and control of such areas. When such an organization does not exist, a decision must be made by the developer to either maintain these areas or turn them over to a public body (So et al., 1979). The latter is known as dedication. For land to be legally dedicated there must be:

(a) An offer by the developer to turn the land over to the public body.
(b) An acceptance by the public body once the offer has been formally made, usually done by the adoption process constituting legal acceptance of the offer.

If either element is not present, the act of dedication has not transpired, and the public body may rightfully assume no responsibility. Therefore, intent of dedication by a developer is not a valid reason for a public body to assume responsibility.

The petitioned public body customarily investigates to determine whether the proposed land offered for dedication will serve the public or only a chosen few, whether the area can be easily accessed by service vehicles, and whether any sizable expenses for maintenance are foreseeable. Such factors may result in a "white elephant" gift.

2.6 REGULATORY REQUIREMENTS

A number of previously discussed issues confronting a developer are related to federal involvement in protecting the country's environment. Figure 2.26 illustrates a typical regulatory network that includes some interrelationships that could affect project permitting activities. Sizable developments having the potential to create impacts over large areas relative to infrastructure capacities, environmental quality, or economic activities may require special regional review.

2.6.1 Regulations

Developments are controlled by local jurisdictional requirements, which include, but are not limited to, subdivision regulations, zoning, land use maps, environmental regulations, and requirements of the municipal engineer.

The guidelines referred to most often by a developer are the subdivision regulations adopted by the unit of government having regulatory power over the area. The local planning organization, or board, oversees the enforcement of these regulations through project reviews. Extraterritorial police power may be included in the organizations mandate, which would allow the regulation of areas that could be annexed in the future into the jurisdictional area. A subdivision is usually defined as any division of a tract or parcel of land into two or more lots, building sites, or other divisions for purposes of sale, legacy, or building. Included are divisions of land involving new or altered streets, and parcel redivisions when the following occur:

(a) The combination or recombination of portions of previously platted lots where the total number of lots is not increased and the resultant lots are equal to the standards of the governing authority.

(b) The subdivision of land into parcels of five acres or more where no new street is involved.

This definition may vary from regulation to regulation, depending on the scope and wording; however, the basic idea should remain constant. A typical review procedure for developments is shown in Figure 2.27. The purpose of subdivision regulations are:

> To provide for the harmonious development of the [area] . . . to secure a coordinated street and road layout with relation to major streets and adjoining subdivisions and adequate provisions for traffic, to protect residential areas from through traffic and other traffic hazards, to insure proper street intersection design, to achieve individual property lots of maximum utility, to secure adequate provisions for light, air, water supply, drainage, and sanitary sewer facilities and other public health requirements, to secure adequate provisions for transportation, recreational areas, school sites, and other public services and facilities, to provide accurate land records for the convenience and protection of the public and for adequate identification and permanent location of real estate boundaries, and to ensure the recording of necessary survey data prior to the selling of land.*

Because subdivision regulations vary from locale to locale, no attempt will be made to discuss detailed information contained within a set of regulations.

2.6.2 Zoning

A study of the legally adopted zoning ordinance for the market area, if one exists, must be undertaken. This legal device is supplemented by the zoning map, which physically ties the zoning ordinance to the land it polices. Zoning provides compatibility of land use and insures availability of adequate space for a particular land use. By controlling the use of land in areas of future growth, the local governing body maintains a reasonable expectation of providing adequate streets, schools, refuse collection, police and fire protection, utility systems, and recreational areas. Additionally, zoning protects existing property by causing developments to conform to certain established standards that allow for adequate air, light, and privacy for persons who live and work in the area. Land use, setbacks, and parking requirements are some of the elements regulated by zoning (So et al., 1979). Figure 2.28 contains various relationships that are often regulated through zoning.

Zoning is an exercise of police power on future situations, rather than past, and

---

*Taken from Charleston County Planning Board (1955).

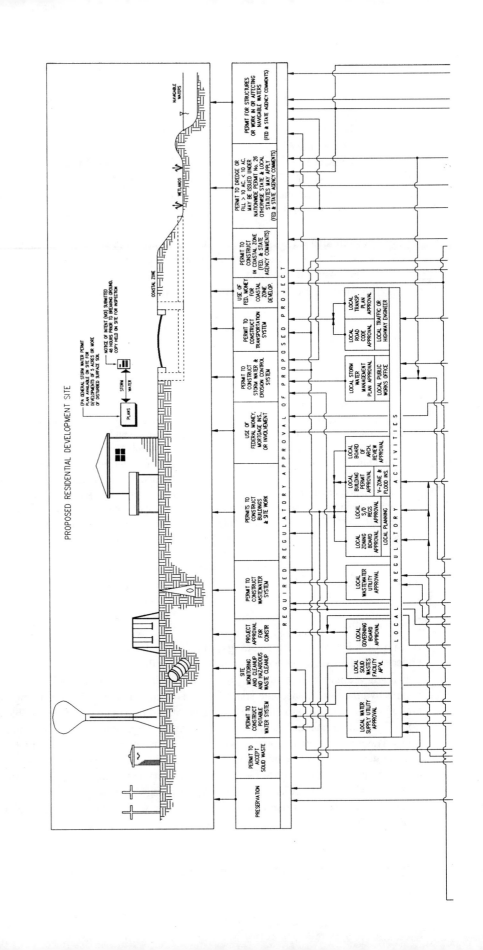

PROPOSED RESIDENTIAL DEVELOPMENT SITE

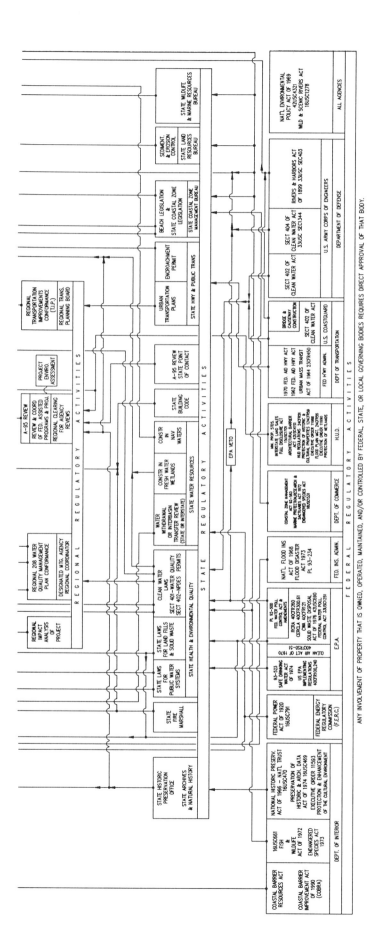

ANY INVOLVEMENT OF PROPERTY THAT IS OWNED, OPERATED, MAINTAINED, AND/OR CONTROLLED BY FEDERAL, STATE, OR LOCAL GOVERNING BODIES REQUIRES DIRECT APPROVAL OF THAT BODY.

Figure 2.26 Regulatory permitting and commenting agencies that influence land developments.

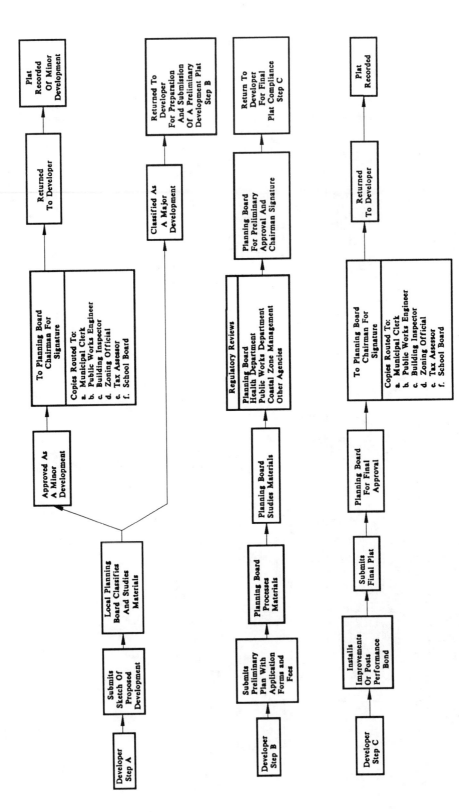

Figure 2.27 Typical procedure for local regulatory approval of land developments.

INTENSITY STANDARDS FOR RESIDENTIAL DEVELOPMENT

DWELLING UNIT TYPE	DWELLING UNITS PER ACRE	COMMON OPEN SPACE AS PERCENTAGE OF TOTAL SITE	PARKING PER UNIT (1)	TREES PER ACRE OF TOTAL SITE AREA (2)	PRIVATE OPEN SPACE (3)
Single family estate		Usually not provided	3+	15+	Depends on lot size
Single family	3 to 5 depending on lot size	Usually not provided	2.5+	15+	Depends on lot size
Duplex	5 to 10 depending on lot size	Usually not provided	2+ (4+ for each structure)	15+	Depends on lot size
0-lot line (4)	4 to 7 units per acre	Usually not provided	2.25+	15+	Depends on lot size
Single family cluster (5)	4 to 7	25 to 50%	2.25+	15 to 20± for first 3 acres, 10± for remaining acreage	Depends on lot size
Atrium (6)	4 to 8 units per acre	Usually not provided	2.25+	15+	25% of total square footage
Suburban townhouse (7)	6 to 9	25% to 40%	2.25+	15 to 20± for first 3 acres, 10± for remaining acreage	500 to 700 sq ft
Urban townhouse (8)	8 to 16	15% to 25% if provided	2.0+	15 to 20± for first 3 acres, 10± for remaining acreage	400 to 600 sq ft
Walk-up apartments	10 to 25	25% to 40%	2.0± depending on number of bedrooms	15 to 20± for first 3 acres, 10± for remaining acreage	Usually not provided except on ground floor apartments
Midrise apartment up to 6 stories	15 to 35	30% to 40%	2.0± depending on number of bedrooms	15 to 20± for first 3 acres, 10± for remaining acreage	Usually not provided except on ground floor apartments
Highrise apartments	30 to 75	35% to 60%	2.0± depending on number of bedrooms	15 to 20± for first 3 acres, 10± for remaining acreage	Usually not provided except on ground floor apartments
Planned unit development (9)	10 to 25 depending on type of planned unit development	25% to 60%	2.0± depending on unit type	15 to 20± for first 3 acres, 10± for remaining acreage	Depends on individual unit type

Gary Greenan, AIP, ASLA, Miami, Florida

NOTES

1. PARKING PER UNIT: In determining the parking per unit, the site design should consider the influence of public transportation and location of the development in relation to employment centers and supporting facilities. An excessive amount of parking spaces to accommodate infrequent special activities could result in excess pavement; depending on soil conditions, it is sometimes better to accommodate infrequent overflow parking on grassed areas or, where conditions permit, on commercial parking areas when not in use or at community centers.

2. TREES: Trees provide one of the major unifying design elements and act as climate modifiers; where a substantial number of trees do not already exist, extensive tree planting should be undertaken as part of most planning programs. The determination of the number of trees per acre should be based more on the particular locale than on a uniform standard.

3. PRIVATE OPEN SPACE: Private open space should be private to the unit concerned and may be provided in the form of courtyards, entrance courts, and rear, side, and front yards. Both visual and aural privacy is important in the design of these spaces.

4. 0-LOT LINE: A single family unit located directly on one side property line and possibly on the rear property line; the emphasis is on eliminating small side yards for larger, more usable open spaces.

5. SINGLE FAMILY CLUSTER: Single family units grouped in clusters in order to maximize open space; units may or may not be attached.

6. ATRIUM: A single family unit related to the early Greek and Roman prototypes which incorporate interior living spaces fronting on an interior court.

7. SUBURBAN TOWNHOUSE: A single family unit attached to other single family units with a common party wall; open space is usually a major element in the suburban townhouse development.

8. URBAN TOWNHOUSE: Early prototypes appear in the urban areas of many older cities; similar to suburban townhouse, although densities are usually higher and common open space is usually provided in the form of public open space.

9. PLANNED UNIT DEVELOPMENT: A mixture of housing types with emphasis on total community design; all the housing types included in the chart, plus associated retail support facilities and community amenities, may be found in a PUD development.

Figure 2.28 Intensity standards for residential developments. Source: John Hoke, Ed. *Architectural Graphic Standards*, 8th ed. Copyright © 1988 by John Wiley & Sons, Inc. Reprinted by permission of John Wiley & Sons, Inc.

therefore cannot be made retroactive to correct undesirable existing conditions other than through natural attrition. A physical investigation of a possible development site may uncover blighted areas, public nuisances, or nonconforming land uses. These undesirable land uses may be undergoing elimination by such actions as urban renewal, health department actions, or building code enforcement. Nonconformances may be the result of a variance from the zoning ordinance granted by zoning board of appeals. These items must be investigated carefully as the land use of surrounding areas will influence the development.

Zoning is undoubtedly the most commonly used legal device for implementing land use plans. Zoning can regulate the size of buildings, parcel areas, density, and use. The landmark court case that upheld the constitutionality of zoning was *Euclid versus Ambler Realty Co.* in 1926 (Yokley, 1978).

What happens if a site is chosen in an unzoned area and:

(a) The surrounding area is undeveloped?

(b) The surrounding area has undesirable uses present?

Two courses of action the developer might follow entail either an effort to enact a zoning ordinance that would support the development or an effort to purchase the problem land. The latter method is an expensive but effective undertaking and, by the use of restrictive deed covenants, the property could be resold as an asset to the development.

Zoning also can regulate the location and size of signs. If the development includes a neighborhood shopping center, these regulatory requirements may be of concern.

Additionally, certain provisions are sometimes utilized by municipalities to control activity outside their jurisdiction. Such display of extraterritorial regulation of unincorporated land adjacent to a municipality's boundary insures compatibility and coordination with the existing municipality plan should future site annexation be instituted.

2.6.3 Restrictive Covenants

Another land use restraint is a restrictive deed covenant, which may designate such items as allowable land uses, minimum lot size if the land is subdividable, regulations regarding construction of substructures, minimum house size or relative value, and numerous other possibilities. Covenants are placed on the land by the developer. They remain in force no matter who owns the property and run with the land. During formulation of the municipality's land use plan, the restrictive deed covenants of all properties under the jurisdiction of the municipality should be investigated and incorporated provided they are reasonable and compatible to the zoning ordinance. Sometimes a conflict exists between zoning and the restrictive deed covenant for a particular parcel. In practice, when the restrictive deed covenant is more stringent than the zoning ordinance, then the zoning ordinance must be satisfied along with the more stringent requirements set forth in the covenant. Should covenants exist, they could further limit the usability of the land under investigation.

2.6.4 Permit Approvals

Some state and local permits that may be required by the developer are:

(a) Permits to construct and operate potable water systems.

(b) Permits to construct and operate wastewater systems.

(c) Permits to construct and operate storm drainage systems.

(d) Permits to construct across navigable streams.

(e) Permits to encroach on public highways, railroads, and public utility rights of way.

(f) Permits to construct within the coastal zone.

(g) Permits to withdraw surface and subsurface water.

(h) Permits for the interbasin transfer of water.

(i) Permits to construct public facilities such as swimming pools.

(j) Permits for building construction.

(k) Permits to grade sites and construct streets and roads.

(l) Permits for landscaping and tree removal.

(m) Permits to dredge and fill wetlands.

The developer also must adhere to various rules and regulations pertaining to employment conditions if federal money is used under various laws such as the Davis Bacon Act and the Civil Rights Act of 1964. Federal and state labor laws, tax regulations, business licenses, unemployment insurance requirements, and workman's compensation are examples of other constraints.

2.7 ECONOMIC CONSIDERATIONS

A potential project must be economically feasible for a developer and financial institution to desire undertaking its completion. This feasibility is based not only on the market's supply and demand but also on the anticipated absorption rate of the new dwelling units that will be produced. Developers undertaking total development activities may establish a two year product inventory as a maximum that will govern the rate of project phasing.

2.7.1 Initial Costs

The developer is faced early with development related expenditures resulting from initial studies and evaluations conducted to establish a basis for financial lending and project potential. Some of the initial expenditures are costs associated with preliminary engineering studies, land surveying activities, architectural/landscape architectural services, environmental assessment, archeological studies, legal costs associated with title and abstracts, interim interest, permit application review fees, and other miscellaneous costs.

2.7.2 Impact and Linkage

Impact fees and linkage are two variable items of great economic concern to a developer as they can amount to appreciable expenditures. Impact fees are used to create jurisdictional capital funds based on units, structures, or development density for water, sewer, roads, fire protection, or recreational improvements necessary to respond to new demands created by the development. Linkage or inclusionary zoning is used to create some housing units that will be sold at below market values in an effort to increase supplies of affordable housing in exchange for permitting the developer to increase the usual density allocations. Linkage also can be accomplished through incentive zoning, where the developer receives increased density allocations in exchange for paying mitigating fees or providing public amenities to offset development related strains being placed on urban systems (Nelson, 1988).

Other regulatory fees might include posting a performance bond or negotiable instrument with the appropriate jurisdictional body. This will provide project completion security, allowing the developer to premarket the project without having to wait until all improvements have been installed and approved by regulatory authorities.

2.7.3 Project Cost Estimate

The estimated cost to install development related improvements is provided by the design professional. This estimate includes the costs for constructing all transportation, drainage, wastewater, and potable water systems if it is an improved land development. Care must be taken by the developer when undertaking total development activities to include not only the costs to erect structures, but also the costs associated with service entrances to the buildings, driveways, parking, and other improvements. This is true for most single-family and multifamily developments. Manufactured housing parks and subdivisions usually require the developer to include the costs of utility service entrances through the concrete pad.

A complete cost estimate includes all anticipated nonconstruction developmental costs. These costs usually include applicable engineering, architectural, surveying, interest, taxes, and legal fees. Also, water supply and treatment facility surety bond costs, mortgage loan insurance, filing fees, and mortgage application fees are included, and vary according to the company to which applications are submitted.

Architectural fees vary according to the type of development. For a multifamily development 12 percent of the construction costs are generally allowed for the architectural firm. This should include building plans and any engineering fees involved with the structure. Single-family-unit architectural fees can be as high as 15 or 20 percent of the value of the house.

Engineering services for the design of the sewer system, water distribution system, storm drainage system, and road network are negotiable and generally range from 6 to 10 percent of the cost of construction. Costs associated with soils investigations can vary depending on the project size, terrain, equipment accessibility, and prevailing subsurface conditions. In addition, surveying costs for site boundary work are variable, based on the property complexity, the number of contiguous landowners and their properties' relationships along with topographic conditions. Topographic survey costs vary greatly depending on:

(a) The amount of site vegetation.
(b) The ground surface uniformity.
(c) The amount of detail required.
(d) The maximum allowable distance between rod shots.
(e) The desired contour interval.
(f) The datum requirements.

For a single-family development lot staking generally is negotiated on a per lot basis, whereas multifamily developments normally use a per unit basis. Unforeseeable problems in the field could increase surveying costs.

Legal fees are negotiable for title research and deed transactions, with the developer usually purchasing land title insurance for protection. Once the initial land transaction has taken place, legal fees are charged for such things as drafting partnership or corporate declarations, negotiations with governmental agencies, obtaining rights of way for connection to existing services if required, drafting protective deed covenants, and drafting individual lot deeds. PUDs require much legal work, and as in the case of surveying, many uncertainties exist. Provisions are normally made to provide for compensation for professional services arising from unforeseeable problems. Although there are no hard and fast rules for such fee determination, one method used is to base additional professional fees on the additional time expended.

Mortgage loan insurance is available to mortgage lenders with the fee being paid by the developer upon loan closing. Most institutions limit their maximum loan to 75 percent of the raw land fair market value, or 75 percent of the raw land plus improvement value, regardless of the insurance provisions. Mortgage application fees generally range from 1 to 2 percent of the mortgage value. Some lending institutions account for this fee in their first payment disbursement.

2.7.4 Recurring Costs

A developer must determine, for feasibility analysis, costs of a recurring nature that can be expected during project development. Recurring costs might include mortgage retirement payments, interest on the mortgage, taxes, insurance, promotion, advertising, real estate agent fees, management fees, utilities, depreciation, solid waste collection, fire protection, power, and maintenance/repair costs.

The life of a mortgage on a land improvement loan can range from three to five years. Retirement payments are flexible and can be arranged to be paid weekly, biweekly, monthly, quarterly, semiannually, or annually. A mortgage life of 25 years may be available on a manufactured home park. Mortgage interests are variable and can range from 9 to 15 percent, depending on the market.

The valuation of the land before and after the development varies, as does the corresponding *ad valorem* tax liability imposed on the developer. Taxes must be paid as long as the developer owns the property, and should be included in the analysis.

Promotion, advertising, and real estate agent fees vary according to the type of development. For an apartment complex or a manufactured home park, continuous advertising may be necessary throughout the life of the development. Contrarily, a single-family or multifamily development may require a short term, but intense, advertising campaign because the need for advertising decreases as dwelling units are sold. Real estate agents normally charge commissions of 6 percent of the sale price of a dwelling unit. Agent's fees can approach 10 percent of a month's rent in apartment complexes or manufactured home parks. Fees may vary and must be negotiated for a particular situation.

Development management costs vary with the type of development and the amount of management required. Management costs of a mobile home park or an apartment complex would be appreciably more than those of a multifamily development. These costs must be estimated on an individual basis.

Utility costs are variable according to development type. Electric power costs should be included for the construction period, while street light expenditures will be absorbed by the developer, the renter, or the dwelling unit purchaser.

Development depreciation must be investigated based on the remaining economic life of the facility. An economic life can be assigned to buildings, privately owned utility systems, and amenities. As the development depreciates, it may allow the developer to decrease tax payments, because of decreased valuation, and allow incremental savings to replace the facility in the future.

Where solid waste collection services are contracted, a variety of costs can result, depending the service provider. Single-family and some multifamily developments utilize individualized pick-up contractual operations, whereas apartment facilities may require point pick-up contracts.

When municipal fire coverage is not available, negotiations with the nearest municipality may be necessary. Such charges for fire protection are generally on an annual basis, either in the form of a bill or bond payment.

Maintenance and repair incurred by the developer vary with the type of development and construction. Weather conditions play an important part in the amount of maintenance required, as does the unit's quality of workmanship and materials. Estimation of such costs must be made on an individual basis.

2.7.5 Revenues

Revenue is money earned. Direct revenue results from consideration paid by the renter or buyer during the course of renting or purchasing a lot and dwelling unit. Indirect revenue results from money generated by vending machines, telephones, laundry facilities, and other such operations that are provided in conjunction with the development. Their combination represents the earning potential of the development. The amount of anticipated revenue can be estimated from the preliminary site plan and market analysis. The number of dwelling units to be sold or rented within a certain price range determines the gross direct revenue. This should be adjusted downward to indicate something less than 100 percent occupancy or immediate sellout. When the adjusted value is added to the indirect revenue, the result yields the gross adjusted income. Once the gross adjusted income is known, the feasibility of the preliminary site plan can be determined.

2.7.6 Feasibility

The feasibility of a project, as indicated in a *pro forma* analysis, is simply an indicator that the project will be a success or a potential failure. The difference between the gross adjusted income, the developmental costs, and operating expenses represents the developer's profit. This is usually expressed as a percentage of the developer's equity in the project. A feasible development could produce from 10 to several hundred percent return on an initial investment. However, some developers expect to produce gross sales that approach 200 percent of the acquisition and development costs from which 100 percent of the acquisition and development costs can be satisfied, leaving the developer approximately a 40 percent profit. The other 60 percent is accounted for by legal fees, interest, real estate agent fees, and other costs of doing business. Once the developer has determined that the project will be feasible, arrangements for a mortgage loan must be made.

2.7.7 Financing

Agencies that finance land development loans are banks, savings and loan associations, mortgage investment companies, insurance companies, and others. All have policies and directives that limit their amount of project financing. This protects their investment. The developer strives to obtain as much financial leverage as possible to accomplish the development. The developer uses the least amount of personal capital possible, and secures the maximum loan available to accomplish project financing. There are two basic reasons for this. First, it allows the developer to maintain diversity in investing capital that is not committed to one development. Second, the developer can obtain a larger rate of investment return, which means additional earning power.

The lending institution is very interested in who is borrowing the money for a development, be it an individual, partnership, or corporation. Certified financial statements concerning the development organization and persons involved are required by the lending agencies to determine financial stability. Past experience in developing and competent members in the development organization are definite assets in obtaining a loan. Financial institutions also undertake an independent assessment of those same items the developer investigated during site selection and site planning to insure:

(a) That the proposed development is competitive in costs and amenities.

(b) That harmony is maintained throughout the interior and surrounding area of the development.

(c) That there are no local hazards adjacent to the development and that the development does not create any hazards.

(d) That the contractors engaged for the work are competent.

Financial institutions require a written appraisal for all real estate related financial transactions normally handled by savings institutions, in accordance with the Financial Institutions Reform, Recovery, and Enforcement Act of 1989 (FIRREA). This would encumber all sales, leases, purchases, interest in, or the exchange of real property, or property financing.

The appraisal must be independently and impartially prepared by a licensed or certified appraiser. The appraiser must set forth an opinion of value, which is described as of a particular date. All materials used as a basis for value, including market data, must be included, along with a legal description of the appraised property that exactly mirrors conveyance documents contents.

Upon establishment of a favorable developer profile, completion of a formal supporting appraisal, and submission of a feasible plan, the lending institution will process the mortgage request, pending regulatory compliance certification. If the applicant is acceptable to the lending institution's system of checks and balances, the loan is granted.

2.7.8 Mortgages

The mortgage loan is normally issued to the developer in increments, called disbursements, for work completed. For example, if 20 percent of the water distribution system were completed and the developer received a bill from the contractor covering this phase of the construction, the mortgage company would disburse to the developer the percentage of the loan allocated to cover that percent of the water system costs. The developer would in turn pay the contractor. The total amount of money disbursed on a percentage basis never exceeds the percent financing for the entire project.

Mortgage retirement, or repayment, is accomplished after all improvements have been installed and approved, by releasing land according at a predetermined amount, which might be 120 percent of the improved property value. In addition to retiring the mortgage, the property release provides a way to offer clear title on a lot or dwelling unit to a potential buyer, even though the lending institution still holds the development mortgage.

In summary, most developers, while being confronted by numerous constraints, concentrate their efforts in seeking project success and completion. A high degree of dependence therefore is placed on the various professionals engaged by the developer to provide safe, economical, and acceptable solutions to the many problems that arise out of land development. The civil engineer plays a key role in the total process.

REFERENCES

American Association of State Highway And Transportation Officials (AASHTO). *A Policy on Geometric Design of Highways and Streets—1990*. Washington: AASHTO, 1990, p. 8.

American Public Health Association (APHA), Committee on the Hygiene of Housing. *Planning the Neighborhood*. Chicago: Public Administration Services, 1960.

American Society of Civil Engineers (ASCE), National Association of Home Builders, and Urban Land Institute. *Residential Streets*. New York: ASCE, 1974.

American Society of Civil Engineers (ASCE), National Association of Home Builders, and Urban Land Institute. *Residential Streets*, 2nd ed. New York: ASCE, 1990.

Ashbury, N. G. *A Formula for Financing Mobile Home Housing Developments*. An unpublished report. Chicago: The Mobile Home Manufacturers Association, 1970.

Bissonnette, Pam. "Bellevue Experiences with Urban Runoff Quality Control Strategies," *Perspectives on Nonpoint Source Pollution*. Proceedings of a National Conference. Washington: U.S. Environmental Protection Agency, 1985, pp. 279–280.

Casey, William J. *Real Estate Investments and How to Make Them,* 3rd ed. New York: Institute for Business Planning, Inc., 1968.

Charleston County Planning Board (CCPB). *Subdivision Regulations: Charleston County*. An unpublished regulation. Charleston, S.C.: Charleston County Government, 1955.

Cowardin, L. M., Virginia Carter, F. C. Golet, and E. T. La Roe. *Classification of Wetlands and Deepwater Habitats of the United States*. Washington: U.S. Department of the Interior, 1979, pp. 3–14.

Daecher, C. W. "A Guide for Municipalities for Bicycle Planning," *Proceedings of the 4th National Seminar on Planning, Design, and Implementation of Bicycle and Pedestrian Facilities*. New York: American Society of Civil Engineers, 1975, pp. 117–131.

De Chiara, Joseph and Lee Koppelman. *Planning Design Criteria*. New York: Van Nostrand Reinhold, 1969, pp. 180–190.

DeVereaux, A. B., Victoria J. Tschinkel, Elton Gissendanner, and State of Florida. *Joint Application for Permit Dredge-Fill-Structures*. An unpublished report. Jacksonville, Fla.: U.S. Army Corps of Engineers, 1982.

Dunkle, F., Rebecca Hanmer, R. W. Page, and Wilson Scaling. *Federal Manual for Identifying and Delineating Jurisdictional Wetlands*. Washington: U.S. Government Printing Office, 1989.

Falk, E. L., Planner in Charge. *Pickens County Recreation and Open Space Plan*. An unpublished report. Clemson, S.C.: Clemson University, 1975, p. 27.

Federal Housing Administration (FHA). *FHA Techniques of Housing Market Analysis*. Washington: U.S. Department of Housing and Urban Development, 1970.

Federal Housing Administration (FHA). *Mobile Home Park Development Standard G 4200.7a*. Washington: U.S. Department of Housing and Urban Development, 1970.

General Services Administration, U.S. Department of Defense, U.S. Department of Housing and Urban Development, and U.S. Postal Service. *Uniform Federal Accessibility Standard.* Washington: U.S. Government Printing Office, 1988, pp. 5, 23–25.

Goodman, William I. and E. C. Freund, Eds. *Principles and Practice of Urban Planning.* Washington: The International City Managers' Association, 1968, pp. 441–484.

Gustin, James D. and Larry A. Neal. "Site Assessments," *Civil Engineering.* New York: American Society of Civil Engineers, August 1990, pp. 53–55. Referenced material is reproduced by permission of the American Society of Civil Engineers.

Hoke, John, Ed. *Architectural Graphic Standards,* 8th ed. New York: Wiley, 1988, p. 148.

Jones, D. Earl, Jr., and Karen A. Jones. "Treating Expansive Soils," *Civil Engineering.* New York: American Society of Civil Engineers, August 1987, pp. 62–65. Referenced material is reproduced by permission of the American Society of Civil Engineers.

Leifer, S. L. *EPA's Innocent Landowner Policy: A Practical Approach to Liability Under Superfund.* Washington: Bureau of National Affairs, 1989, pp. 646–649. Reprinted with permission from *Environmental Reporter,* Vol. 20, pp. 646–649 (August 4, 1989). Copyright 1989 by The Bureau of National Affairs, Inc. (800-372-1033).

Lewis, Roger K. "Shaping the City: Designers Are Going Back to Basics with the 'Traditional New Towns.'" *The Washington Post.* June 13, 1990. Referenced material is reproduced by permission of Roger K. Lewis, 5034½ Dana Place, N.W., Washington, D.C. 20016.

Montgomery County Planning Commission. A method of naming streets. Dayton, Ohio, 1992.

Moskowitz, Joel S. *Environmental Liability and Real Property Transactions: Law and Practice.* New York: Wiley, 1989.

Nelson, Arthur C., Ed. *Development Impact Fees: Policy Rationale, Practice, Theory, and Issues.* Chicago: Planners Press—The American Planning Association, 1988, pp. 219–277.

Perry, Clarence A. *The Neighborhood Unit, A Scheme of Arrangement for the Family Life Community.* New York: Russell Sage Foundation, 1929.

Rice, Leonard and M. D. White. *Engineering Aspects of Water Law.* New York: Wiley, pp. 13–19. Copyright © 1987 by John Wiley & Sons, Inc. Reprinted by permission of John Wiley and Sons, Inc.

Rubenstein, Harvey M. *A Guide to Site and Environmental Planning,* 3rd ed. New York: Wiley, 1987.

Salvesen, David. *Wetlands—Mitigating and Regulating Development Impacts.* Washington: Urban Land Institute, 1990.

Sanders, Welford, Judith Getzels, David Mosena, and Jo Ann Butler. *Affordable Single-Family Housing—A Review of Development Standards—Planning Advisory Service Report No. 385.* Chicago: The American Planning Association, 1984.

So, Frank S., Israel Stollman, Frank Beal, and David S. Arnold, Eds. *The Practice of Local Government Planning.* Washington: International City Managers' Association, 1979, pp. 389–415.

South Carolina Coastal Council. *Guidelines and Policies of the South Carolina Coastal Management Program.* Washington: U.S. Department of Commerce, 1979.

South Carolina Coastal Council. *Storm Water Management Guidelines.* An unpublished report. Charleston, S.C.: S.C. Coastal Council, 1988.

South Carolina Electric and Gas Co. (SCE & G). *Electrical Underground Distribution.* An unpublished manual. Columbia, S.C.: SCE & G, 1967.

South Florida Water Management District. *Guidance for Preparing an Application for Surface Water Management Permit.* An unpublished report. West Palm Beach, Fla.: South Florida Water Management District, 1990.

Spielberg, Frank. "The Traditional Neighborhood Development: How Will Traffic Engineers Respond?" *ITE Journal.* Washington: ITE, September 1989, pp. 17–18.

Tomioka, Seishiro and E. M. Tomioka. *Planned Unit Developments—Design and Regional Impact.* New York: Wiley, 1984.

Train, R. E., Chairman. *Protecting America Wetlands: An Action Agenda. The Final Report of the National Wetlands Policy Forum.* Washington: The Conservation Foundation, 1988, p. 1.

U.S. Army Corps of Engineers and the South Carolina Coastal Council. *Developer's Handbook for Freshwater Wetlands.* An unpublished handbook. Charleston, S.C.: S.C. Coastal Council, 1991.

U.S. Department of Housing and Urban Development (U.S. HUD). *Land Planning Principles for Home Mortgage Insurance, A HUD Handbook, 4140.1.* Washington: HUD, 1973, pp. 5-9–5-18.

U.S. Department of Housing and Urban Development (U.S. HUD). *Land Planning Data Sheet Handbook, Circular 4140.3.* Washington: HUD, 1973, Data Sheet 6.

U.S. Department of Housing and Urban Development (U.S. HUD). *Manual of Acceptable Practices, 1973 ed., 4930.1 Vol. IV.* Washington: HUD, 1973.

U.S. Department of Housing and Urban Development (U.S. HUD). *Environmental Reviews at the Community Level.* Washington: HUD, undated.

U.S. Department of Housing and Urban Development (U.S. HUD) and the NAHB Research Foundation, Inc. *The Affordable Housing Demonstration—A Case Study: Phoenix, Arizona, HUD-PDR-761-1* Washington: HUD, 1984.

U.S. Department of Housing and Urban Development (U.S. HUD) and the NAHB Research Foundation, Inc. *The Affordable Housing Demonstration—Two Case Studies: Charlotte, N.C. and Greensboro, N.C., HUD-1073-PDR.* Washington: HUD, 1987.

U.S. Department of Housing and Urban Development (U.S. HUD) and the NAHB Research Foundation, Inc. *The Affordable Residential Land Development, HUD-1128-PDR (vi).* Washington: HUD, 1987.

U.S. Department of the Interior and U.S. Department of Housing and Urban Development (U.S. HUD). *Environmental Planning and Geology.* Washington: U.S. Government Printing Office, 1970.

U.S. Department of Transportation. *Manual on Uniform Traffic Control Devices for Streets and Highways.* Washington: U.S. Government Printing Office, 1988.

U.S. Environmental Protection Agency. *Release of Final Storm Water Regulations.* An unpublished memorandum. Washington: EPA, 1990.

U.S. League of Savings Institutions. *Special Management Bulletin.* Chicago: U.S. League of Savings Institutions, 1990.

Yokley, E. C. *Zoning Law and Practice,* 4th ed. Charlottesville, Va.: Michie Company, 1978, pp. 10–11.

3 Engineering Administration

Civil engineering focuses on solving problems that pertain to the health, safety, welfare, and economic well-being of the populace. This involves the principle that the engineer is an independent thinker who makes decisions based on factual information and not as a result of outside influence. Classical undergraduate civil engineering programs provide students with special knowledge in mathematical, physical, and engineering sciences along with studies in land surveying, transportation, structures, geotechnical, and environmental engineering. As a result of their education, training, and experience, these students are generally well suited for participating in land development activities. Civil engineers who have met various criteria set forth by state law are qualified to offer professional services to the public as registered professional engineers.

These activities center on problem solving within the constraints of various regulations, economics, and project goals set forth by the developer. The design professional is therefore not involved in the valuation, potential marketability, or financeability of the project, and would not normally be in a position to provide the developer with any financial recommendations other than those pertaining to costs and operation of improvements.

3.1. ORGANIZATION

A civil engineer who provides consulting engineering services can offer professional services as a sole proprietor, as a partnership or professional association member, or even under the umbrella of a corporation. Traditionally, sole proprietorships and partnerships are the means most utilized by design professionals. As a sole proprietor, the professional engineer is personally responsible for all actions that arise from the conduct of business. As a result the engineer reaps all profits from produced work, while at the same time being responsible for any financial losses.

A partnership allows for two or more professional engineers to conduct business activities as co-owners, with each partner acting as an agent for the organization. Each partner therefore shares in the management and operations of the business, generally using the premise of majority rule unless otherwise agreed. Unanimous consent is needed for a change in business activities. Obligations made by individual partners during the course of ordinary partnership business transactions are binding on all partners, just as tort liability (Section 3.3) created by one of the partners is answerable by all. Likewise, partnership assets are subject to co-ownership status while each partner shares in any profits and losses. It is advisable to form a partnership by written agreement to establish a clear understanding of the rights and obligations of the members. The Uniform Partnership Act can provide an easy form to follow.

A corporation for profit may allow design professionals to organize effectively, especially if large numbers of engineers are involved. The corporation acts as a legal entity, with the shareholders or stockholders possessing undivided ownership of assets. Stockholders are only liable to pay for stock shares purchased in the firm, and not for organizational indebtedness or business liability. Normally the articles of incorporation set forth authorized activities of engagement, shareholder rights, and the organizational structure and governance. Should the corporation enter into a contract that falls outside the purview of its authorized activity, the contract is unenforceable at law, and termed *ultra vires* (Abbett, 1963). Corporate profits can be distributed by the board of directors in accordance with the articles of incorporation or reinvested into the organization. Many design professionals have adopted the professional corporation form of organization to conduct business, which allows various tax advantages.

Other forms of organizations include associations and joint ventures. Regardless of the organizational forum within which the design professional operates, all engineering work is considered personal service, performed by a professional. Therefore, each professional is individually responsible for the work's adequacy, including work performed in a design-build situation.

3.2 CONDITIONS OF EMPLOYMENT

The developer can engage a design professional to perform various services using one of several procedures, which include the customary approach, the customary approach with project management, and services rendered in conjunction with a professional construction manager (Fisk, 1988).

3.2.1 Customary Approach

The most common situation, which historically has been the method of choice, is for the developer to solicit the design services from the engineer with an arrangement for intermittent site visitations being performed by the engineer during construction. Site visitations are for the purposes of monitoring the general compliance of material and accomplished work with the construction documents. Intermittent observation made by the design professional in an attempt to prevent project deficiencies generally does not imply a guarantee of the contractor's work on the part of the engineer, nor does it relieve the contractor of contractual responsibility. The engineer therefore would act as the developer's agent in all activities associated with the project except when performing design activities that require the design professional to act as an independent contractor. Other duties would entail the review of shop drawings, approving pay requests for the contractor, activating change orders, and final job inspection. Figure 3.1 illustrates this relationship.

3.2.2 Customary Approach with Project Management

Should the developer request a full-time job inspector, or should a jurisdictional regulatory agency require a full-time inspector, the design professional could provide in addition to those customary services previously described, various management services as shown in Figure 3.2. Again, the presence of a full-time inspector does not relieve the contractor

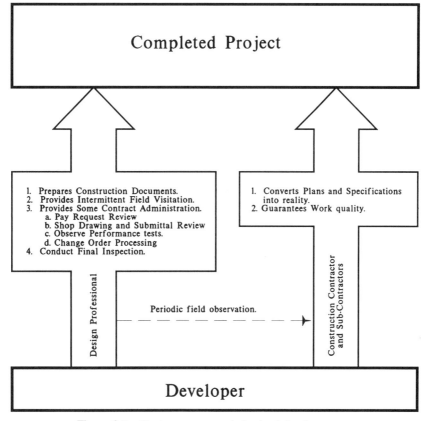

Figure 3.1 Customary approach for land developments.

of any obligation to properly complete all contracted work, nor does it place any conditions on the design professional to assume responsibility for consequences arising from methods, procedures, or construction sequencing utilized by the contractor, nor is there any obligation for the engineer to guarantee any of the completed work other than for adequacy of design.

3.2.3 Professional Construction Manager Approach

Should the project be large or complex, requiring either the services of several design professional firms or the use of several prime construction contractors, a professional construction manager may be procured by the developer for overall project coordination. This manager would in turn prosecute for the owner the engagement of the design professionals and contractors required to complete the project, resulting in the developer entering into individual contracts with the professional construction manager, the design professional(s), and the construction contractor(s). The construction manager would provide guidance and make recommendations relating to design reviews, bid packaging, scheduling, coordination, budgeting, quality control, progress analysis, construction methods, and claims management, as illustrated in Figure 3.3.

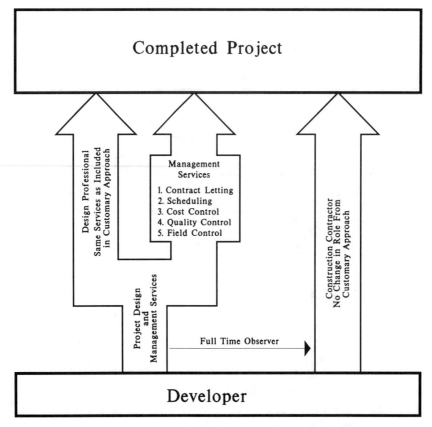

Figure 3.2   Customary approach for land developments that incorporate project management.

3.2.4   Registration Law Compliance

On occasion, developers may seek engineering type services from nonlicensed sources, as a result of personal friendship or otherwise, which would in effect bypass state registration laws that are there to protect the public. This practice should be discouraged; however, the nonlicensed source can only persist as a project participant by working under the supervision of a licensed engineer. The registered supervising party must be completely responsible as the primary design consultant to prevent circumventing the law.

3.2.5   Engineering Services: Private and Public

Work associated with consulting engineering can deal with either public or private interests. Public interest work involves public funding and is associated with the needs of, or in conjunction with, a governmental agency. Private work is conducted using private business relationships and private funds. This distinction is important because competitive activities are required when public involvement exists.

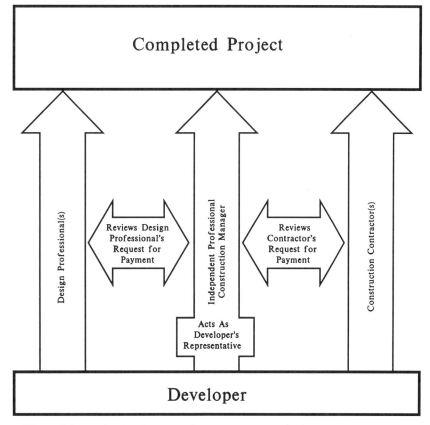

Figure 3.3 Professional construction manager approach for land developments.

3.3 CONDUCT OF AFFAIRS

The engineer who undertakes a project must possess the developer's good faith and loyalty in order to foster the mutual trust and confidence required for successful project completion. The engineer's position is one of great responsibility; therefore, any engineer who accepts a job implies that adequate competence is possessed to undertake such a job, and that diligence will be exerted to complete the project professionally and satisfactorily.

3.3.1 Engineer's Liability

The engineer's responsibility has been recognized since the early code of Hammurabi. Under this code, if the owner of a dwelling was killed due to structural failure, the builder would suffer death (Nischwitz, 1984). Professional services normally are offered with the understanding that the engineer will exercise ordinary professional skill and diligence while conforming to acceptable standards of practice. These standards include building and design codes, recognized textbook procedures, and standard practices in the industry.

Possible sources of liability could center on defective plans and specifications where the measurement of damage may range from repair costs for rectification of the resulting deficiency to the resulting loss in value of the finished product. Other areas of possible liability stem from specifying inferior materials, delaying the project, or even significantly underestimating the project's cost. The engineer can be held liable, as defined by applicable state law, if the resulting plans are grossly defective or for failure to observe obvious construction activity deficiencies during site visits.

3.3.2 Governing Laws

Two forms of law that regulate a consulting engineer's activities are statutory law and common law. Statutory law consists of rules of conduct enacted by duly authorized legislative authority; it includes federal, state, and local statutes. Common law consists of various doctrines and maxims that originated in court decisions and generally are not found in the statutes.

Torts are civil wrongs caused either by intentional actions or through negligence. Common torts are: assault and battery; defamation of character, which is verbal slander, or libel, which results from disreputable written or pictorial pronouncements; infringements of copyrights, trademarks, or patents; trespassing; conversion, or seizing of other's property for personal use; creating a nuisance; and riparian rights violations.

3.3.3 Negligence

If an engineer is found negligent as a result of some less than desirable condition being associated with the project, several conditions must be met to show that the engineer had a duty to conform to various standards of conduct to protect against unreasonable danger and that the engineer did not conform to those standards. In addition, a "legally protected interest" must have been infringed upon as a result of a "reasonably close causal connection" that existed (Sweet, 1985).*

Many states are making efforts to alter tort related liability by abolishing or limiting the use of joint and several liability, which provides for damages to be assessed on ability to make restitution rather than on the degree of contribution when several parties are at fault. The engineer may find relief from a negligence action if the damaged party in fact contributed to the negligent condition or risk in use was assumed by the damaged party.

3.3.4 Fraud

In addition to confronting liability resulting from the nonperformance of contractual duties, the engineer can also be subject to liability resulting from fraudulent activity—intentionally misrepresenting facts that were reasonably relied upon by others who, as a result, suffered damage. An example would be an engineer who had knowledge of design errors on a set of plans but took no action to correct them prior to job bidding, causing the bids to come in too low.

---

*Taken from Sweet, Justin. *Legal Aspects of Architecture, Engineering, and the Construction Process,* 3rd ed. St. Paul, MN: West, 1985.

3.3.5 Equity

Good practice dictates that a practicing engineer should therefore use no short cuts and should heed standards that could be applied to scrutinize professional performance. One set of standards used in settling matters concerning questions in equity are the maxims of equity. Some of these maxims are included in Figure 3.4.

3.3.6 Insurance

Professional liability insurance can be purchased to offset risks that endanger the practicing engineer during the performance of professional services. The cost of coverage may be five percent of the engineering firm's gross income.

3.3.7 Ethics

The American Society of Civil Engineers (ASCE) instills professionalism within its membership by upholding a professional code of ethics (ASCE, 1987). Figure 3.5 includes the fundamental principles and canons that are included in the code.

3.4 ENGINEERING COMMUNICATION

As professionals, civil engineers must be able to conduct effectively all duties required for the successful execution of the project, which includes possessing adequate skills to communicate through verbal, written, and graphical means. Public hearings and regulatory review proceedings often require engineers to make verbal presentations where effective public speaking ability is essential. Similarly, report preparation and correspondence require the engineer to clearly and adequately express intended issues. Finally, well organized drawings that contain adequate information will minimize confusion, and it is therefore essential that graphical communication skills are mastered by the design professional.

Advances in communication technology have produced innovative devices that allow

> He who seeks equity must do equity.
> He who comes into equity must come with clean hands.
> Equity aids the vigilant, not those who slumber on their rights.
> Equity acts specifically.
> Equity follows the law.
> Equity delights to do justice and not by halves.
> Equity will not suffer a wrong to be without a remedy.
> Equity regards that as done which ought to be done.
> Equity regards substance rather than form.
> Equity imputes an intention to fulfill an obligation.
> Equity is equality.
> Between equal equities the first in order of time shall prevail.

Figure 3.4 Maxims of equity. Source: Robert W. Abbett. *Engineering Contracts and Specifications,* 4th ed. Copyright © 1963 by John Wiley & Sons, Inc. Reprinted by permission of John Wiley & Sons, Inc.

CODE OF ETHICS
FUNDAMENTAL PRINCIPLES

Engineers uphold and advance the integrity, honor, and dignity of the engineering profession by:
  1. *Using their knowledge and skill for the enhancement of human welfare.*
  2. *Being honest and impartial and serving with fidelity the public, their employers and clients.*
  3. *Striving to increase the competence and prestige of the engineering profession; and*
  4. *Supporting the professional and technical societies of their disciplines.*

ASCE GUIDELINES TO PRACTICE UNDER THE FUNDAMENTAL CANNONS OF ETHICS

Canon 1. Engineers shall hold paramount the safety, health, and welfare of the public in the performance of their professional duties.

Canon 2. Engineers shall perform services only in areas of their competence.

Canon 3. Engineers shall issue public statements only in an objective and truthful manner.

Canon 4. Engineers shall act in professional matters for each employer or client as faithful agents or trustees, and shall avoid conflicts of interest.

Canon 5. Engineers shall build their professional reputation on the merit of their services and shall not compete unfairly with others.

Canon 6. Engineers shall act in such a manner as to uphold and enhance the honor, integrity, or dignity of the engineering profession and shall not knowingly engage in business or professional practices that are of a fraudulent, dishonest or unethical nature.

Canon 7. Engineers shall continue their professional development throughout their careers, and shall provide opportunities for the professional development of those engineers under their supervision.

Figure 3.5 A civil engineer's code of ethics. Source: The American Society of Civil Engineers (ASCE). *ASCE Official Register—1987.* Reprinted by permission of the American Society of Civil Engineers.

the design professional to utilize time more effectively. Electronic personal pagers, telephone answering machines, and electronic mailboxes, along with cellular and fax telephone services, allow avenues of communication for engineers to conduct everyday business. In addition, digital personal computers, including laptop models or main frame networked computer systems, provide the engineer means to send or receive data including reports, calculations, or even files of drawings.

3.5   TYPES OF ENGINEERING SERVICES

Most civil engineering services are categorized into two general areas. The first involves either performing investigations or making consultations, with the final product being the formation of a report. The second involves the provision of services associated with the construction of a project.

  The engineer requires knowledge of the existing site conditions to form a basis for design; this requires a certain amount of investigative work. A detailed discussion is included in Chapter 4 concerning most required activities associated with land development; however, the services requested may include not only feasibility, life cycle, planning, and rate studies, but also specialty services such as materials evaluation, pilot and equipment evaluations, site evaluations, engineering related surveying services, and others.

Professional services that pertain to land development construction projects are discussed in detail throughout the remainder of this book and include the following:

(a) Background investigation.
(b) Preliminary design.
(c) Final project design.
(d) Job bidding activities.
(e) Construction related activities.
(f) Project completion.

3.6 SELECTION OF A DESIGN PROFESSIONAL

The developer should have a procedure established for selecting a consulting engineer prior to soliciting proposals from consulting firms interested in performing services connected with a projected job. Several things that will probably be evaluated by the developer are the consulting firm's reputation, the breadth and depth of talent within the organization, information pertaining to completed projects, and evidence of financial stability. Typical information that might be requested is shown in Figure 3.6.

3.6.1 Selection Process

The developer should extend notice to potential consultants by making available details that depict the nature of the solicited work and should request interested parties to respond by providing submitted prerequisite information and a statement of interest.

The developer would select possibly no more than five of the most promising firms, based on review of the requested material, to continue further negotiations and evaluation. The preferred firms would be given a more complete description of the anticipated scope of work, and the developer would request these firms to prepare a proposal that would outline their anticipated conduct and scope of proposed services for completing the associated work if awarded the job.

Upon receipt of these proposals, the developer normally conducts individual interviews with each candidate firm where all shared information would be discussed. After completing inquiries into past performance by contacting supplied references, the developer would be able to rank order in decreasing preference the remaining firms. Following the order of preference, each consultant is afforded an opportunity to negotiate compensation and other matters for contractual agreement. If agreement cannot be reached with the highest ranked firm, that firm is eliminated and the process is repeated.

3.6.2 Bidding for Professional Service

Good business procedures would dictate that a developer follow the process just outlined in order to procure the best possible service for each particular job; however, many developers operating with private funds will engage a design professional based on favorable past work experiences, using an informal selection process. In addition, developers who rely too heavily on the bidding process to solicit professional services often receive unsatisfactory results because professional abilities vary among firms, as does the quality of work. In much the same manner, a client creates the potential of being ill served if

CONSULTING ENGINEERING FIRM'S VITAE

Firm Name
Year established
Former firm names
Business address
Telephone number and name of person to contact
Type of service particularly qualified to perform
Names of principals of firm and where registered
Names of key personnel, with experience of each and years of member-
  ship in the firm, and identified under special headings such as:

Civil Engineers	Architects
Structural Engineers	Transportation Engineers
Mechanical Engineers	Environmental Engineers
HVAC Engineers	Geotechnical Engineers
Electrical Engineers	Surveying & Mapping Engineers
Industrial Engineers	Computer Specialists
Planners—site, city, community, other	

Number of staff at present time
Approximate square feet of office space
Outside consultants and associates usually employed
Completed similar work on which the firm was the Engineer of Record
Present activities:
  Number and identity of projects
  Estimated total construction cost
Completed similar work on which the firm was associated with others
Present activities on which the firm is associated with others
Estimated annual work capacity (revenue) in dollars
Average annual volume of work (revenue) in last five years in dollars
Largest project in last five years (construction cost)
Largest current project (construction cost)
Financial capability
Banking references
Date of submission

Figure 3.6 Consulting engineering firm's vitae. Source: The American Society of Civil Engineers (ASCE) Committee on Standards of Practice. *ASCE Manuals and Reports on Engineering Practice No. 45—Consulting Engineering*–a guide for the engagement of engineering services, 1988. Reprinted by permission of the American Society of Civil Engineers.

solicitation for professional services is made on a contingency basis. According to ASCE Manual No. 45 (ASCE, 1988):

> Such practice, which requires a favorable recommendation for the proposed project as a condition of receiving subsequent compensation, could project an undesirable appearance of tending to influence the engineer's objectivity.*

3.7 ENGINEERING SERVICES CONTRACT

A contract is an agreement to exchange things of value—an agreement that can be enforced between two or more parties bound by specific terms contained therein. Engi-

---

*Taken from ASCE (1988) by permission.

neering contracts normally provide for documents resulting from services rendered to remain the design professional's property unless otherwise specified. The developer can obtain reproduced copies of drawings and job related documents at cost. Contracts for auxiliary professional services are made either with the primary design professional or developer. The ensuing discussion is focused on a primary design professional's contract.

3.7.1 Elements of a Contract

If it is to be enforceable, the contract must include a clear description of an agreement to which all parties concur. Should the agreement include a mistake in the terms or conduct of the contract, the contract may void utilizing a doctrine called unreality of consent. This basis can also be used to void a contract if fraud, duress, or undue influence was used (Abbett, 1963). Contracts can be expressed or implied. Express contracts involve a definite offer by one party and an unqualified acceptance by the other. Most contracts may be assigned to a third party unless the terms of the contract specifically prohibit such activity. Normally the assignor is not relieved of obligations unless all original parties agree to such action. Contracts also can be changed, modified, or terminated by mutual agreement.

An enforceable contract must contain material that is lawful, and there must be an obligation or something must be accomplished. Should the contractual matter violate federal, state, or local law, or should it be contradictory to acceptable public processes (if, e.g., bid rigging were involved), the contract would be unenforceable. In addition, should the contract not include real and valuable consideration as part of the contractual elements, that too would cause the contract to be unenforceable.

All parties to a contract, acting in good faith, must be competent to produce a binding contract. An incompetent person, such as a minor, an insane person, or a substance abuse victim, in addition to a corporation operating outside of its articles of incorporation, or even a governmental agency operating outside of its relegated authority, cannot produce a binding contract.

Finally, certain contracts by law must be reduced to writing before they can become enforceable. Included are conveyances of interests in real property, agreements that extend more than a year, contracts for sale of goods over a minimum value, and suretyship and guarantees. Forms of agreement typically used by civil engineers for various professional services have been prepared by the Engineers Joint Contract Documents Committee (EJCDC), the National Society of Professional Engineers, the American Consulting Engineers Council, the American Society of Civil Engineers, and the Construction Specification Institute, as indicated in Figure 3.7.

3.7.2 Contract Termination

Contract termination can be attained by several means. The most common is the satisfactory completion of all contractual provisions in the form of specific performance, while substantial performance indicates substantial completion of the contractual elements excepting some minor conditions that by uncontrollable conditions were unattainable. When a party refuses or fails to perform contractual obligations, a termination by breach occurs, with the resulting damages being assessed for nonperformance, while the damaged party is relieved of contractual liability obligations. Finally, if unforeseeable conditions should make it impossible or radically more expensive to carry out the terms of the contract, such as, in the case of personal services, the death of one of the parties, violation of a recently

The Engineers Joint Contract Documents Committee
Typical Standard Forms

(a) Standard Form of Agreement Between Owner and Engineer for Study and Report Professional Services (EJCDC No. 1910-19)

(b) Standard Form of Agreement Between Owner and Geotechnical Engineer for Professional Services (EJCDC No. 1910-27A)

(c) Standard Form of Agreement Between Engineer and Geotechnical Engineer for Professional Services (EJCDC No. 1910-27B)

(d) Standard Form of Agreement Between Owner and Engineer for Professional Services (EJCDC No. 1910-1)

(e) Standard Form of Letter Agreement Between Owner and Engineer for Professional Services (EJCDC No. 1910-2)

(f) Guide Sheet for Including Limitation of Liability in the Standard Form of Agreement Between Owner and Engineer for Professional Services (EJCDC No. 1910-9F)

(g) Standard Form of Agreement Between Engineer and Architect for Professional Services (EJCDC No. 1910-10)

(h) Standard Form of Agreement Between Engineer and Associate Engineer for Professional Services (EJCDC No. 1910-13)

(i) Standard Form of Agreement Between Engineer and Consultant for Professional Services (EJCDC No. 1910-14)

Figure 3.7 Typical standard forms of agreement for various professional services. Source: The Engineers Joint Contracts Documents Committee (EJCDC): National Society of Professional Engineers, American Consulting Engineers Council, American Society of Civil Engineers, The Construction Specifications Institute.

passed law, or destruction of the subject matter of the contract, the contract can be terminated by impossibility or impracticability of performance.

3.8 MOBILIZATION PLAN

The design professionals on which the remainder of this book focuses are consulting civil engineers who have a primary responsibility to their clients to undertake and see projects through to completion. Details of responsibilities outlined in the engineering services contract relegate activities that the consulting engineer must undertake. Work assignments based on available staffing with proper planning will provide systematic efforts to insure timely completion of the project. Good practice includes the assignment of a permanent project coordinator who will provide continuity of services. Various memoranda of understanding are helpful, especially when coordinating the efforts associated with the procurement of auxiliary services, such as subsurface site investigations. Of utmost importance is the necessity for all parties involved to maintain clear and factual records that will minimize confusion and hopefully prevent the occurrence of major problems.

3.9 REMUNERATION FOR ENGINEERING SERVICES

The consulting engineer can receive compensation for services rendered in any number of ways. Most land developments entail the design consultant being paid a percentage of the costs of construction. Figure 3.8 illustrates possible relationships between engineering services costs and construction costs for projects similar to land developments. Although this graph is not included in the current edition of ASCE Manual No. 45, it is useful to illustrate general cost trends. Negotiated engineering percentage costs associated with an improved land development could range from 6 to 10 percent of the cost of construction, with provisions of on-site observation ranging from 3 to 5 percent. Other services, such as engineering surveying for construction control, may range between 2 and 4 percent of the construction cost.

Some projects warrant payments to the consulting engineer on a lump sum basis, especially if the job is clearly defined and small. Should this method be inappropriate, a cost plus fixed fee approach could be employed, where the engineer would collect reimbursements for all costs associated with salaries, overhead costs, auxiliary expenses, and a fixed fee for profit and intangible costs. A final method of compensation uses a cost multiplier approach, where all costs associated with salaries are increased by a factor that may range from two to three to provide income for profit, contingencies, and overhead costs. The cost multiplier approach usually provides that incidental costs associated with the necessary conduct of work be reimbursed, including an administrative charge of up to 15 percent being assessed. Other methods of payment include retainage and *per diem* fees.

Although various methods of allocating partial payments to the consulting engineer can be used, one effective method is the use of monthly billings. These may be based either on

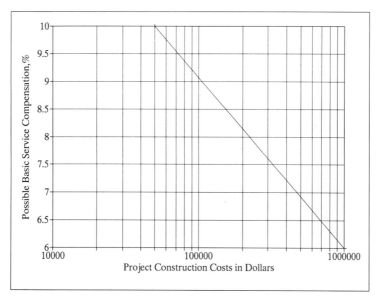

Figure 3.8 Possible costs of engineering services for land developments. Source: The American Society of Civil Engineers (ASCE) Committee on Standards of Practice. *ASCE Manuals and Reports on Engineering Practice No. 45—Consulting Engineering*–a guide for the engagement of engineering services, 1972. Reproduced by permission of the American Society of Civil Engineers.

percentage of work completed, if the method of payment is related to a percent of construction costs, or on the number of actual accrued hours and expenses. Another schedule of payment might be:

Completion through preliminary phase: 40 percent.

Completion through final design phase: 90 percent.

Completion of the construction phase: 100 percent.

It is important for the engineer to determine prior to entering into an agreement with a developer that the project is financially sound. This will help to insure that the engineer will receive payment for professional services rendered.

When there is no express agreement as to the compensation for services rendered, the law implies that an engineer is worthy of the services provided. The doctrine of *quantum meruit* provides that a party who accepts professional services is obligated to pay reasonable and commensurate remuneration in return.

3.10   ENGINEER'S ROLE

A civil engineer who has been engaged to perform professional services must function within the terms of the contractual agreement. Simultaneously the engineer must assist the developer in addressing the various constraints associated with the development. An initial step in this process requires the engineer to begin with an investigation into existing conditions that affect the project. This step is normally associated with what is called preliminary engineering activities, and is further described in Part II.

REFERENCES

Abbett, Robert W. *Engineering Contracts and Specifications,* 4th ed. New York: Wiley, 1963, pp. 23–24, 85–88, 112–113. Copyright © 1963 by John Wiley & Sons, Inc. Reprinted by permission of John Wiley & Sons, Inc.

American Society of Civil Engineers (ASCE). *ASCE Manuals and Reports on Engineering Practice—No. 45 A Guide for the Engagement of Engineering Services.* New York: ASCE, 1972, p. 29 and 1988, pp. 4, 42–43. Referenced material is reproduced by permission of The American Society of Civil Engineers.

American Society of Civil Engineers (ASCE). *ASCE Official Register—1987.* New York: ASCE; 1987, pp. 313–316.

Fisk, E. R. *Construction Project Administration,* 3rd ed. New York: Wiley, 1988.

National Society of Professional Engineers (NSPE), American Consulting Engineers Council, and American Society of Civil Engineers (ASCE), *Standard Form of Agreement Between Owner and Engineer for Professional Services (EJCDC No. 1910-1).* New York: ASCE, 1984.

National Society of Professional Engineers (NSPE), American Consulting Engineers Council, and American Society of Civil Engineers (ASCE), *Standard Form of Letter Agreement Between Owner and Engineer for Professional Services (EJCDC No. 1910-2).* New York: ASCE, 1985.

National Society of Professional Engineers (NSPE), American Consulting Engineers Council, and American Society of Civil Engineers (ASCE). *Guide Sheet for Including Limitation of Liability in the Standard Form of Agreement Between Owner and Engineer for Professional Services (EJCDC No. 1910-9F).* New York: ASCE, 1988.

National Society of Professional Engineers (NSPE), American Consulting Engineers Council, and American Society of Civil Engineers (ASCE), *Standard Form of Agreement Between Engineer and Architect for Professional Services (EJCDC No. 1910-10)*. New York: ASCE, 1985.

National Society of Professional Engineers (NSPE), American Consulting Engineers Council, and American Society of Civil Engineers (ASCE). *Standard Form of Agreement Between Engineer and Associate Engineer for Professional Services (EJCDC No. 1910-13)*. New York: ASCE, 1985.

National Society of Professional Engineers (NSPE), American Consulting Engineers Council, and American Society of Civil Engineers (ASCE). *Standard Form of Agreement Between Engineer and Consultant for Professional Services (EJCDC No. 1910-14)*. New York: ASCE, 1985.

National Society of Professional Engineers (NSPE), American Consulting Engineers Council, and American Society of Civil Engineers (ASCE). *Standard Form of Agreement Between Owner and Engineer for Study and Report Professional Services (EJCDC No. 1910-19)*. New York: ASCE, 1985.

National Society of Professional Engineers (NSPE), American Consulting Engineers Council, American Society of Civil Engineers (ASCE), and Construction Specifications Institute (CSI). *Standard Form of Agreement Between Owner and Geotechnical Engineer for Professional Services (EJCDC No. 1910-27A)*. New York: ASCE, 1989.

National Society of Professional Engineers (NSPE), American Consulting Engineers Council, and American Society of Civil Engineers (ASCE), *Standard Form of Agreement Between Engineer and Goetechnical Engineer for Professional Services (EJCDC No. 1910-27B)*. New York: ASCE, 1986.

Nischwitz, J. L. "The Crumbling Tower of Architectural Immunity Evolution and Expansion of the Liability to Third Parties," *Ohio State Law Journal*, Volume 45, 1984, p. 217.

Sweet, Justin. *Legal Aspects of Architecture, Engineering, and the Construction Process*, 3rd ed. St. Paul, MN.: West, 1985, p. 105.

PART II
Preliminary Engineering Activities

4 Inventory, Analysis, and Reporting

A consulting engineer's initial activity would be to hold informal detailed discussions with the developer to review pertinent factors pertaining to the project. Topics of discussion might include potential phasing requirements and development theme guidelines that are necessary for project design. Economic and technical aspects also must be discussed. The developer may be considering several sites for the project, requiring the consulting engineer to investigate the conditions of each site. The investigation would include analysis of how well the site potentially supports the development's theme as well as the impact on surrounding areas. The design professional can ascertain during the initial discussions whether services of other professionals are required and mutually agree with the developer on the scope of these auxiliary services. Of keen interest to the developer is the potential cost of the development, which the design professional will have to estimate.

It is imperative that the consulting engineer make initial contact with jurisdictional regulatory authorities on local, state, and federal levels (Figure 2.26) to establish communications that will aid in the progression of subsequent review activities. Hopefully, the consulting engineer will glean through these early contacts important guidance that otherwise might not surface until later, during the review process. The engineer can take proper action while planning the project to prepare for the project's review. If regulatory concerns have been satisfactorily addressed in the preliminary development plan, the project will receive approval. The consultant's role is to seek regulatory approval on the developer's behalf while incorporating into the project plans and specifications provisions that address concerns raised by regulatory agencies. The developer, however, is obligated to perfect and obtain regulatory approval (or the approval of others in the case of an easement) since many approvals require negotiation and fee payments.

4.1 DESK TOP STUDY AND SITE INSPECTION

Initially, the engineer must make every reasonable effort to obtain existing information pertaining to the potential site that might reflect material useful to the evaluation process.

4.1.1 Reference Sources

An excellent source of material is a U.S. Geological Survey (USGS) map on a scale of 1 : 24,000, similar to that shown in Figure 4.1. This map shows general locations, transportation systems, and topography. Information pertaining to waterways can be obtained from the U.S. Department of Commerce hydrographic charts. Local tax maps show general property boundaries and streets, including serialized identification of individual parcels for specific property information such as current landowner names. Figure 4.2 shows details associated with a typical tax map.

Figure 4.1 An example of a USGS 1:24,000 quad map—Summerville, S.C. Source: U.S. Geological Survey, 1990.

Local authorities have available zoning and/or land use maps, similar to Figure 4.3, which are useful in determining how the proposed site fits within an area's overall land use plan. If the proposed development site would create an incongruent situation, the developer may have to consider petitioning local authorities to rezone the property.

Flood insurance rate maps (FIRM) contain drainage basin information that could affect

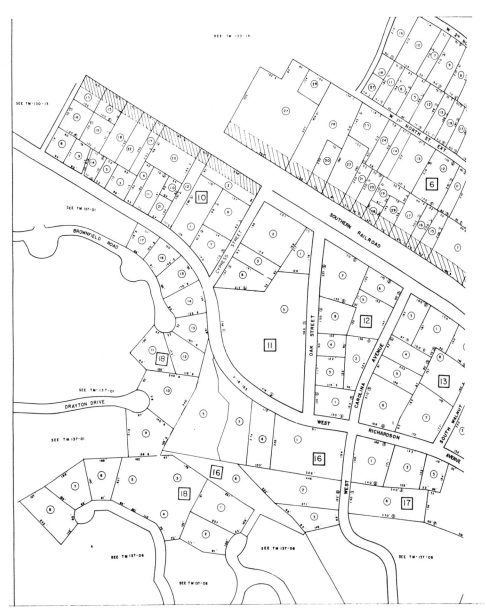

Figure 4.2 A sample section of a tax map—town of Summerville, Dorchester County, S.C. Source: Dorchester County Tax Assessor's Office, St. George, S.C., 1980.

a potential development by delineating areas of potential inundation. Figure 4.4 shows a portion of a typical flood map.

Another important reference is a U.S. Department of the Interior national wetlands inventory (NWI) map, which indicates possible jurisdictional wetlands being present. Figure 4.5 shows one of these maps.

Other sources for potential information include documents contained within local public record offices such as recorded deeds, plats, easements, and other legal instru-

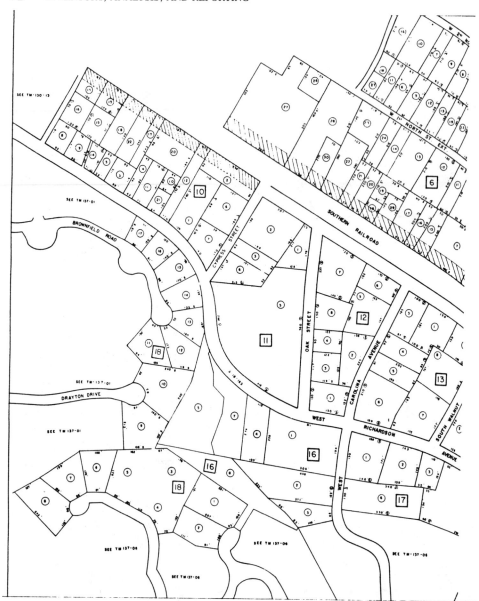

Figure 4.3   A sample section of a zoning map—town of Summerville, Dorchester County, S.C. *Note:* R-1 and R-2 are single family residential, R-3 is single family residential (attached), B-3 is general business, R-6 is multifamily, and PL is public land. Source: Planning Office for the town of Summerville, S.C.

ments. Local inquiries can also include interviews with residents who are familiar with the area. Transportation maps, water resources management maps, soils inventory maps, stream gauge data, historical and archaeological maps, aerial photographs, satellite imagery, and utility system maps similar to Figure 4.6 can also provide information that documents prevailing conditions.

An emerging reference system, called a geographic information system (GIS), is

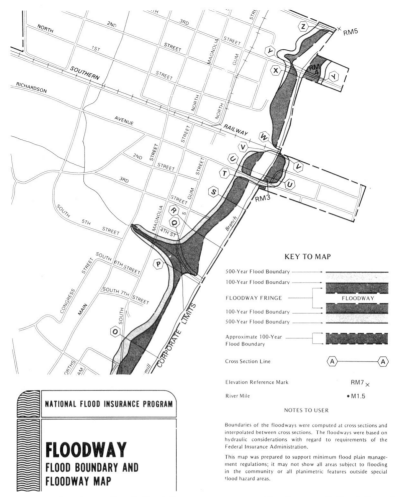

Figure 4.4 An example of a national flood insurance program floodway and flood boundary map—town of Summerville, S.C. Source: Federal Emergency Management Agency. *Flood Insurance Rate Map 450073—Town of Summerville, South Carolina,* 1980.

comprised of computer based spacial data and offers benefits to a developer and design professional. The general concept of a GIS system is illustrated in Figure 4.7. Various stored geographical data overlays can be combined to show relational attributes of prescribed areas that incorporate previously mentioned resource data. Agencies creating a GIS facility have encountered a herculean task of inputting and updating system data; nevertheless, the system does provide enormous benefits to many users.

4.1.2 Base Map Construction

After compiling and assembling existing site boundary data, the design professional can sketch a draft copy of a base map, similar to the one illustrated in Figure 4.8. The map would include preliminary information that pertains to on-site and significant off-site areas, including transportation networks, adjoining properties and owners, existing util-

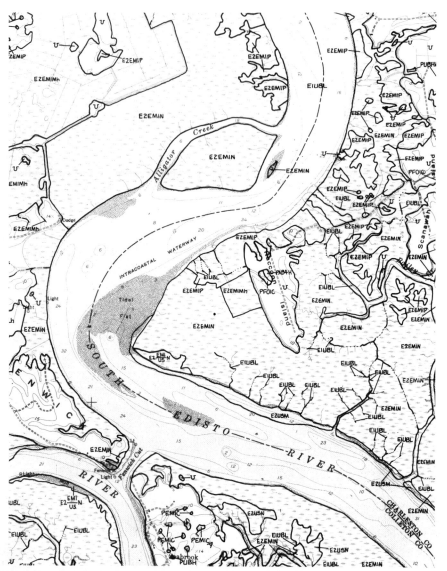

Figure 4.5 A portion of a national wetlands inventory (NWI) map—Raccoon Island, Charleston County, S.C. Source: U.S. Department of Interior. National Wetlands Inventory Map (NWI)— *Bennetts Point, S.C., 1989.*

ities and storm drainage facilities, easements, rights of way, and critical geologic and topographical features such as wetlands.

4.1.3 Site Inspection

The base map is useful when conducting an on-site inspection since it provides a background where observed field conditions can be recorded for later office use. Photographs, sketches, and significant findings should be included in the field investigation notations.

Figure 4.6 Inventory of existing water mains in Liberty, S.C. Source: E. L. Falk, Planner in Charge. *Liberty, S.C. Land Use Plan*. Clemson University, 1972. Reprinted by permission of Clemson University.

Apparent property corner monuments should be flagged using plastic surveying ribbon; these corners would subsequently be verified during the boundary survey. Obvious encumbrances, such as contaminated waste materials or property line encroachments, should be noted for further investigation, although appropriate auxiliary professionals who conduct subsequent evaluative work should investigate them. The resulting site sketch, similar to the one illustrated in Figure 4.9, would include base map information and field notations. On completion of the site inspection, the engineer may begin planning conceptual alternatives from which the developer could later choose a preference. These alternatives are subject to revision since a number of formal studies must be undertaken prior to forming final concepts. Some of these studies follow.

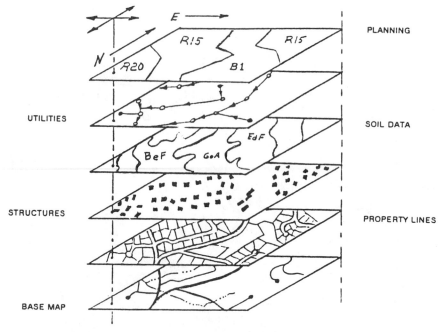

Figure 4.7   An example of a geographic information system (GIS). Source: Graphic Mapping Group, Inc., Hendersonville, N.C. Reprinted with permission.

4.2   WETLANDS DETERMINATION STUDY

Under Section 404 of the Clean Water Act, the COE, in conjunction with the EPA, issues permits for dredge and fill activity that occurs within jurisdictional wetlands. In addition, the COE has, under the Rivers and Harbors Act, similar authority, while state and local authorities may have additional control over wetlands especially in coastal zone areas. The presence of hydrophytic vegetation, hydric soils, and wetland hydrology constitutes a jurisdictional wetland. Private consultants are usually employed to perform extensive on-site studies to identify and delineate jurisdictional saltwater and freshwater wetlands (Dunkle et al., 1989; U.S. Army COE and S.C. Coastal Council, 1991).

4.2.1   Defining Wetlands

Hydrophytic vegetation is defined as macrophytic plant life growing in water, soil, or on a substrate that is at least periodically deficient in oxygen, as a result of excessive water content* (Dunkle et al., 1989).

As part of an on-site inventory, dominant plant genus and species must be inventoried to determine the existence of significant occurrences in obligate, or various facultative plants, which would establish one criterion for jurisdictional wetland declaration. Significant occurrences could include having obligate and/or facultative species comprising

*Taken from Dunkle et al. (1989).

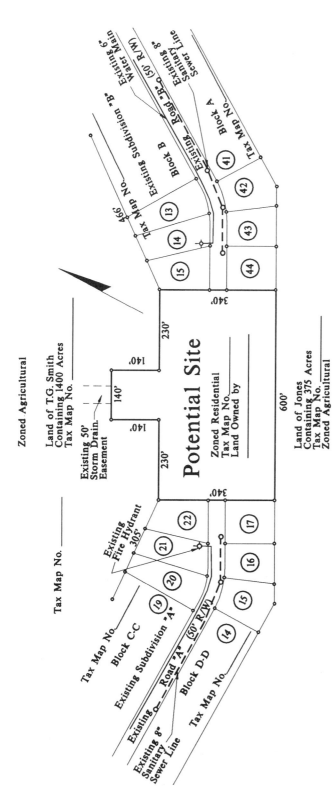

Figure 4.8 An example of a base map for a land development.

97

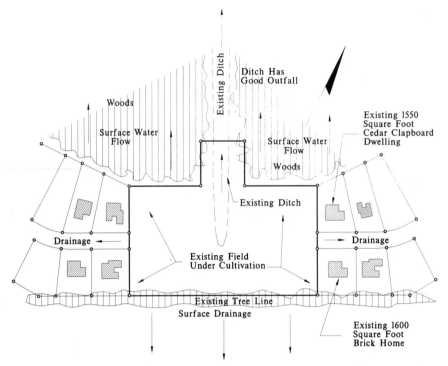

Figure 4.9 An example of a site sketch showing existing features.

more than 50% of the dominant types or a prevalence index of less than three based on specific criteria.

Hydric soils have been defined for jurisdictional wetland identification purposes as "soils that are saturated, flooded, or ponded long enough during the growing season to develop anaerobic conditions in the upper part" (Dunkle et al., 1989). This would include soils that are poorly drained, inundated for long durations, or frequently flooded during the growing season. They are normally categorized into organic muck, organic peat, or a combination thereof. The Soil Conservation Service (SCS) and the National Technical Committee for Hydric Soils (NTCHS) have compiled a listing of hydric soils. Local soil surveys published by SCS can be used to determine initially whether prevalent soil types in the vicinity of the site fall within hydric soil classifications. If so, soil samples will be required for further investigation.

Wetland hydrology has been defined for jurisdictional wetland determination purposes as the "permanent or periodic inundation, or soil saturation to the surface, at least seasonally, [which] are the driving forces behind wetland formation" (Dunkle et al., 1989). Water from precipitation, runoff, groundwater, or tidal activity may be contributory sources. The local offices of COE, USGS, SCS, the U.S. National Oceanographic and Atmospheric Administration (NOAA), along with state and local units of government, can provide information relating to site hydrology. Aerial photographs and field inspections provide amplifying data that can reveal hydrologic activities affecting a site.

The definition of jurisdictional wetlands has undergone an evolutionary change in the past and is still being debated by both the governmental and private sectors. As a result of

(1) Locate and delineate the development site on a USGS 1:24,000 map and note the topographic features that are encompassed by the site that connote wetlands (e.g., lakes, marshes, and swamps).

(2) Study the appropriate National Wetlands Inventory map to determine the potential wetland areas on site.

(3) Study the soil survey map by SCS to determine if hydric soils exist on site.

(4) Study any available aerial photographs to seek wetland indications.

(5) Review any amplifying information on the area to include other wetland studies.

(6) Make a determination if wetlands exist based on the reviewed information. If an indication of wetlands is present, an on-site inspection would establish areas directly on the ground that could be surveyed and platted.

Figure 4.10   Off-site wetlands inventory procedure.

these debates, modifications to current regulations defining wetlands could be forthcoming.

4.2.2   Jurisdictional Wetlands Determination

Professional consultants can choose from four recognized methods a means to conduct a wetlands inventory. If sufficient information is available, an off-site procedure can be used when field inspection is either unwarranted or unattainable. Figure 4.10 indicates a method of conducting an off-site analysis. On-site physical investigations consist of routine, intermediate, or comprehensive assessments. Routine procedures are used primarily on areas of five acres or less, while intermediate and comprehensive methods can apply to any site size including small ones with intensive wetland activities.

Figure 4.11 illustrates a process for undertaking an on-site jurisdictional wetland inventory that a professional consultant would follow, with routine results being recorded in a style similar to Figure 4.12.

Should jurisdictional wetlands comprising 10 acres or less be identified on site, then nationwide permit No. 26 (33 CFR 330.5) provides a means to authorize dredge or fill activities that affect nontidal waterways or isolated wetlands. Other regulations included in most coastal zone management programs may require strict permitting provisions for any size jurisdictional wetland, which in effect would cover areas exempt under nationwide permit No. 26. Jurisdictional authorities may allow mitigation, or the creation of new wetlands, in exchange for approving wetland alterations within critical project areas. Without mitigation an otherwise satisfactory and beneficial project might be prohibited.

4.3   ENVIRONMENTAL EVALUATION

Provisions contained within federal regulations such as CERCLA and SARA, or other parallel state requirements such as New Jersey's Environmental Cleanup Responsibility Act (ECRA), place a high priority on the investigation for hazardous wastes.

4.3.1   Innocent Landowners

Relying on the provisions for an innocent landowner embodied in SARA, a developer requires professional services of a environmental consultant to undertake and perfect "due

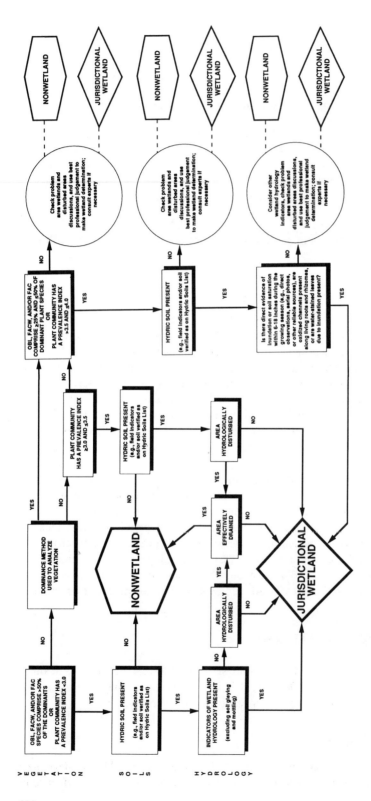

Figure 4.11 On-site jurisdictional wetland determination showing conceptual (not decision making) approaches. OBL = obligate wetland plants, or plants that almost always occur in wetlands; FACW = faculative wetland plants, or plants that usually occur in wetlands; FAC = faculative plants, or plants that are equally likely to occur in wetlands or nonwetlands. Source: F. Dunkle, Rebecca Hanmer, R. W. Page, and Wilson Scaling. *Federal Manual for Identifying and Delineating Jurisdictional Wetlands*, 1989.

DATA FORM
ROUTINE ONSITE DETERMINATION METHOD

Field Investigator(s): _____ Date: _____
Project/Site: _____ State: _____ County: _____
Applicant/Owner: _____ Plant Community #/Name: _____
Note: If a more detailed site description is necessary, use the back of data form or a field notebook.

- -

Do normal environmental conditions exist at the plant community?
Yes _____ No _____ (If no, explain on back)
Has the vegetation, soils, and/or hydrology been significantly disturbed?
Yes _____ No _____ (If yes, explain on back)

- -

VEGETATION

Dominant Plant Species	Indicator Status	Stratum	Dominant Plant Species	Indicator Status	Stratum
1.			11.		
2.			12.		
3.			13.		
4.			14.		
5.			15.		
6.			16.		
7.			17.		
8.			18.		
9.			19.		
10.			20.		

Percent of dominant species that are OBL, FACW, and/or FAC _____
Is the hydrophytic vegetation criterion met?   Yes _____ No _____
Rationale: _____

SOILS

Series/phase: _____ Subgroup: _____
Is the soil on the hydric soils list?   Yes _____ No _____   Undetermined _____
Is the soil a Histosol? Yes _____ No _____ Histic epipedon present? Yes _____ No _____
Is the soil: Mottled? Yes _____ No _____ Gleyed? Yes _____ No _____
Matrix Color: _____ Mottle Colors: _____
Other hydric soil indicators: _____
Is the hydric soil criterion met?   Yes _____ No _____
Rationale: _____

HYDROLOGY

Is the ground surface inundated?   Yes _____ No _____ Surface water depth: _____
Is the soil saturated?   Yes _____ No _____
Depth to free-standing water in pit/soil probe hole: _____
List other field evidence of surface inundation or soil saturation.

Is the wetland hydrology criterion met?   Yes _____ No _____
Rationale: _____

JURISDICTIONAL DETERMINATION AND RATIONALE

Is the plant community a wetland?   Yes _____ No _____
Rationale for jurisdictional decision: _____

Figure 4.12 Routine wetlands analysis report. OBL = obligate wetland plants, or plants that almost always occur in wetlands; FACW = faculative wetlands plants, or plants that usually occur in wetlands; FAC = faculative plants, or plants that are equally likely to occur in wetlands or nonwetlands. Source: F. Dunkle, Rebecca Hanmer, R. W. Page, and Wilson Scaling. *Federal Manual for Identifying and Delineating Jurisdictional Wetlands,* 1989.

diligence" activities. For legal and ethical purposes the intent of this action would be to establish that the developer did not know contamination existed and that there was no reason to know of such occurrence. Considerable burden is placed on the developer to obtain adequate information to insure the production of safe and usable dwelling units. Environmental consultants, in conjunction with the developer and legal council, may determine the need for investigative activities that will satisfy the developer that due diligence has been met. The American Society for Testing and Materials (ASTM) is developing a voluntary standard on "due diligence" performance.

4.3.2 Phase I Environmental Audit

Due diligence activities are used to determine whether hazardous substances are present or threaten a site. A Phase I Environmental Audit includes the following courses of action:

(a) A check of the ownership title chain to include wills, deeds, easements, leases, restrictions, and covenants for at least 50 years to determine whether any previous owners appear capable of potentially contaminating the site.

(b) An inspection of any available aerial photography or topographic maps that might reflect prior property usage.

(c) An investigation of the existence of any recorded environmental clean-up liens placed against the property as a result of federal, state, or local action within three miles of the site.

(d) A check of all federal, state, and local governmental regulatory records that would indicate a release of hazardous substances including:

(1) Landfills and other disposal records.

(2) Underground storage tank records.

(3) Hazardous waste handling and generation records.

(4) Accidents and spill reporting records including crash and rescue, police, fire fighting records and any local planning board records.

(e) A visual inspection of the property that includes looking for distressed vegetation, above and underground storage tanks, trash, rubbish, fertilizers, old electrical transformers, junk cars, batteries, and tires. In addition, this inspection should include investigating past potential on-site chemical use, storage, or treatment, and disposal practices. Surface water wells present should be inventoried to determine whether driller's logs are available. Also, information from adjoining landowners about the site's past usage can be helpful, along with information pertaining to the use of their contiguous property (Weldon et al., 1989).

A number of naturally occurring environmental hazards might be present at the site. These could detrimentally affect the completed project's quality, and include any release of subsurface methane or radon gas. Special testing requirements to identify the presence of hazards similar to these must be discussed by the consultant and developer.

When information and site reconnaissance results indicate contamination is unlikely, no further activity may be required. Although the environmental consultant has the obligation to investigate contamination potential within the bounds of SARA and the scope of engagement services, the decision to conclude the investigation is solely the developer's. Some may wish to undertake some type of subsurface investigation even if a favorable Phase I Environmental Audit is received, especially if the site was previously used.

Prior to commencing a Phase I audit, most environmental consultants require the current landowner or developer to complete a questionnaire pertaining to the site. Figure 4.13 shows typical questions that might be included.

  I. Existing and Previous Site Conditions and Use
    (a) Primary use including agricultural, commercial, or industrial.
    (b) Do above- or underground storage tanks exist?
    (c) Are buildings on-site that contain asbestos?
    (d) Is this site an EPA identified waste generating site?
    (e) Has this site been identified as a hazardous waste clean-up site?
    (f) What permits have been issued pertaining to this site?
    (g) What zoning classification has been assigned to this parcel in the past?
    (h) Are there any wells present on site?
    (i) Has any metal plating, chemical, or petroleum processing been conducted on site?
    (j) Has electrical equipment been manufactured, repaired, or stored on site?
 II. Future Site Conditions
    (a) Will this parcel be used as a residential area?
    (b) Will child care facilities be located on-site?
    (c) Will commercial facilities be located on-site?
III. Adjacent Property Use Both Past and Present
 IV. Past and Present Site Owners
  V. Special Sampling Requirements
    (a) Owner specified for radon, methane, or other hazards.
 VI. Special Requirements
    (a) Confidentiality of Report
    (b) Was the site ever a location for an accident?
    (c) Does the report require special certifications?
    (d) Do you have access to reference plats, soil borings, or site drawings?

Figure 4.13 Pre-audit Phase I questionnaire for conducting an environmental site assessment.

4.3.3 Phase II and Phase III Activities

A Phase II activity reflects a discovery stage that includes some type of site exploration and testing. This action would be necessary if the Phase I assessment indicates a possibility of hazardous wastes on site. Should no contamination be found during the Phase II activities, due diligence requirements most likely would have been met. However, if contamination is confirmed during Phase II, then a Phase III site clean-up program must be undertaken by the current landowner, unless the developer elects to pursue the property's purchase. The developer would assume at purchase the obligation for full site remediation.

Installing groundwater monitoring wells and collecting subsurface soil samples for chemical testing can be expensive in terms of time and money; nevertheless, adequate coverage is necessary. When a Phase I audit does not clearly indicate the site is contamination free or when land use activities on adjacent property are questionable, a preliminary exploration program should be undertaken to obtain initial data. The environmental consultant would have to make a determination of methods and procedures to use, which could range from surface soil sampling to subsurface investigations. Care must be exercised to account for migratory patterns associated with certain types of contaminants, such as polychlorinated biphenyls (PCBs) or metallic compounds. These substances should be sampled close to the suspected source.

When a Phase I assessment indicates the likelihood of site contamination, it is essential to conduct a comprehensive exploratory program that includes groundwater sampling.

Test borings, soil gas analysis, groundwater monitoring wells, and test pits are possible ways to investigate a site. Other methods could include ground penetrating radar, or magnetometer, seismic, gravimetric, or electromagnetic conductivity surveys. Sample testing might utilize a variety of means including mass spectrometry, atomic absorption, or gas chromatography.

Environmental consultants that perform Phase I audits and Phase II site evaluation activities should consider potential conflicts of interest, as results of the investigation could affect many personal contacts. These consultants also must maintain knowledge of rapidly changing events that occur in this rapidly evolving field (Moskowitz, 1989).

4.3.4   Water and Wastewater Service

Other environmental issues that the design professional must consider include the investigation of a potable water supply source and an assessment of wastewater disposal availabilities. This would necessitate checking with local public service utilities to determine whether service is available, as these organizations will be required to approve and accept the proposed system once installed and placed into service for operation, maintenance, control, and ownership. If potable water service is available, the engineer will have to gather existing flow data to determine whether excess system capacity is sufficient to accommodate the new demands. Should capacity not be available, the developer may be required to provide various capital improvements in the project's development, such as the inclusion of a booster pumping station to maintain adequate operating pressure. Similarly, the availability of wastewater treatment and disposal must be investigated to determine if the capacity on hand is sufficient to accommodate new development loads. The systems gravity and pressure mains, and the wastewater treatment plant must be capable of accommodating all loads. Various improvements might be required to accommodate the new project, such as the installation of an improved sanitary sewer pumping station for increased wastewater flows. The construction costs associated with these types of improvements are usually shared on a *pro rata* basis with the developer's financial obligation being linked to the amount of utilization. Remaining costs may be borne by the public service utility until other users request service.

If utility service is not available, an alternate source such as a surface water impoundment or a subsurface water well for water supply must then be investigated. Alternate wastewater disposal methods include on-site subsurface disposal facilities or a wastewater treatment plant.

4.3.5   Other Environmental Concerns

A final environmental concern is the proximity of the proposed development to areas that produce noise and polluted air. Some noise-producing facilities, such as airports, have maps that show noise level intensities as emanating contours. Potential development problems could be eliminated by examining and using the information in these documents while planning the project. Likewise, air quality must be investigated if a potential problem exists close to the site.

4.4   HISTORICAL AND NATURAL RESOURCE SURVEY

Any development that affects federal property, is subject to federal licensing, receives federal funds, or requires federal approval must include an investigation to determine whether valuable natural or historical resources exist on site. If so, a suitable inventory must be made. Some states require similar undertakings, especially if marked or unmarked burials are present. A natural resource survey should be conducted prior to undertaking any physical land survey work so that potential "finding" locations can accurately be tied to the property boundary.

4.4.1   Historical Resource Survey Procedure

Historical resource surveys are conducted by certified professional archaeologists, who normally use a two-step approach. The first step is to check all pertinent records including historical maps, state and local archives, and the register of potential historical locations to obtain direct and corollary information. Upon completion, a site inspection is conducted to determine visually whether any artifacts or features of cultural importance are present. This process usually entails a visual inspection of the property in conjunction with spot shovel testing. Should the first step indicate the possibility that noteworthy features exist, additional and more intense field inspection procedures may be required in concentrated locations to ascertain whether the findings are significant. Investigators conclude the inspection process by preparing a written report that justifies the site's potential eligibility for inclusion in the National Register.

Should significant findings occur on private property, a memorandum of agreement is normally drafted. Either a protective easement is instituted to insure the area's preservation, or the area is set aside to be sold to a preservation type organization. A last resort would be to approve the site for excavation so that information can be recorded prior to the removal of important features, which will result from the development construction.

4.5   LAND AND PHOTOGRAMMETRIC SURVEYS

Project planning must begin with a knowledge of the metes and bounds (bearings and distances) of the prospective property boundary. This information precisely defines the development's area and provides physical site orientation with respect to geographical references. These are important for the placement of certain recreational facilities and for proper residential unit orientation.

Some regulatory agencies require a project's horizontal and vertical control to be based on adopted systems of reference. Within the United States most past vertical control has been linked to the National Geodetic Vertical Datum of 1929 (NGVD 29), which is the result of a national vertical adjustment performed by constraining the heights of 26 tidal stations in the United States and Canada. One geographical area, relying on geopotential surfaces for hydraulic and engineering projects, encountered problems using NGVD 29. As a result it developed the International Great Lakes Datum of 1955 (IGLD 55), based on mean sea level at Father's Point, Quebec, Canada. While NGVD 29 has served the engineering, surveying, and mapping community well, it contains distortions not understood when the adjustment was made. This, coupled with the dynamic nature of the earth's crust in certain parts of the country, resulted in problems with the overall integrity of the NGVD 29. As a consequence the National Geodetic Survey (NGS), a component of the Coast and Geodetic Survey, in conjunction with international parties, redefined the vertical reference surface to better service user needs. This resulted in the North American Vertical Datum of 1988 (NAVD 88), publication of which began in 1991. This datum holds the local mean sea level value (1970–88 tidal epoch) for the tidal bench mark at

Father's Point/Rimouski fixed, thus making IGLD 85 and NAVD 88 "one and the same," according to Zilkoski (1992). Differences in elevation between NGVD 29 and NAVD 88 will vary up to approximately 1.5 meters, as indicated in Figure 4.14 (where the longitude values are shown in degrees east). NAVD 88 is not designed to coincide with local mean sea level at any point in the United States. Local mean sea level and NAVD 88 zero geodetic elevation values will differ by large enough amounts to be of much concern, especially on the west coast.

NAVD 88 will replace NGVD 29 over a five to seven year period. Appropriate NAVD 88 elevations pertaining to NGVD 29 marks not yet published can be obtained from NGS or converted for mapping purposes using NGS prepared computer software under the name "VERTCON."

Likewise, the reference surface for horizontal control used for plane coordinate systems has undergone recent revisions through the redefinition and readjustment of the North American Datum of 1927 (NAD 27). NAD 27 was established using the previous coordinates of horizontal control station Meades Ranch in Kansas as the origin and the Clarke Spheroid of 1866 to define the earth's shape. In order to remove distortions that developed in the NAD 1927 coordinates as new surveys were forced to fit the existing horizontal network and to provide for a system that is compatible with satellite geodesy, the ellipsoid was changed to the Geodetic Reference System of 1980 (GRS 80), the origin was moved to the earth's center, and the entire U.S. horizontal control network of approximately 250,000 control points was simultaneously recomputed or "readjusted," resulting in a more accurate set of coordinates. The result is the current North American Datum of 1983 (NAD 83). Each state's plane coordinate system (SPCS) has been redefined by the states (in conjunction with NGS) to conform to NAD 83. Figure 4.15 shows current SPCS 83 zones for which unique zoned project constants apply to define either Lambert or Transverse Mercator state plane coordinate projection surfaces (oblique Mercator for southeast Alaska) similar to those shown in Figure 4.16 (NOAA, 1990). Jurisdictional authorities are placing increased emphasis on the use of the SPCS to control data uniformity necessary for compatibility between varying GIS based activities.

4.5.1 State Plane Coordinate Systems

When the project's boundary is tied to an SPCS, all measured horizontal distances must be reduced to an ellipsoid to obtain the equivalent geodetic length using an elevation factor which takes into account the geoid (sea level) and ellipsoid separation called the geoid height. This value must further be reduced to its equivalent grid length by using a grid scale factor. Grid scale factors increase in magnitude as the distance of a point increases from the central parallel in a Lambert projection (scale varies with latitude), while in a Transverse Mercator projection, scale increases as the distance of a point increases from the central meridian. Grid scale factors can be different at each end of a surveyed line. The greatest difference occurs for a north-south line in a Lambert projection, and an east-west line in a Transverse Mercator projection. Several methods to obtain a grid scale factor, depending on the accuracy desired according to NOAA Manual NOS NGS 5 (1990) include:

(a) For high accuracy the grid scale factor can be obtained for each line in the survey, using

$$k_{12} = (k_1 + 4k_m + k_2)/6 \qquad (4\text{-}1)$$

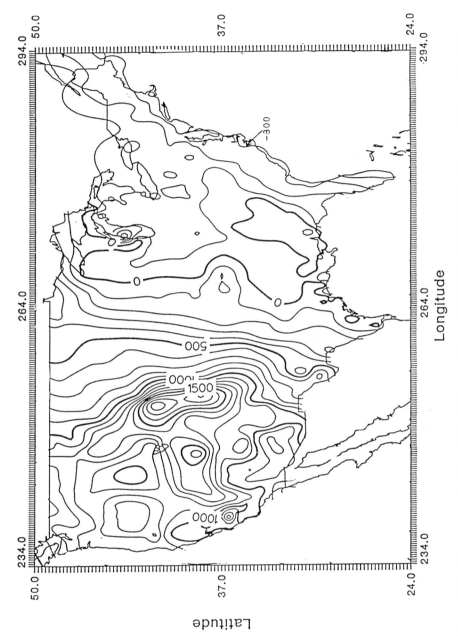

Figure 4.14 Contour map depicting height differences between NAVD 88 and NGVD 29 (units = mm). Source: David B. Zilkoski. *U.S. Hydrographic Conference*, 1992. Reprinted with permission.

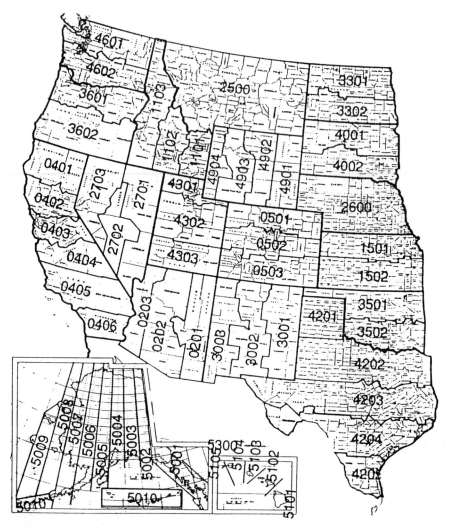

Figure 4.15   Index of state plane coordinate (SPC) zone codes. Source: National Oceanographic and Atmospheric Administration. *State Plane Coordinate Systems of 1983: NOAA Manual NOS—NGS 5,* 1990.

where:

k_1 and k_2 = the grid scale factors at each end of the line.
k_m = the grid scale factor at the midpoint of the line.

(b) A less accurate method includes using the grid scale factor of the line's midpoint or the mean of the line's endpoint grid scale factors.

(c) For lower accuracy, the grid scale factor for a project can be obtained by using the grid scale factor that corresponds to the project's center point.

Figure 4.15   (continued)

Figure 4.17 graphically depicts the relationship of various distances as they apply to state plane coordinate calculations. Elevation factor relationships are shown in Figure 4.17*a, c,* and *d,* while grid scale factor relationships are depicted in Figure 4.17*b.* The product of these two factors is known as the combined factor and its application is demonstrated in Section 4.5.5.

Some states have legislation that defines one of the two available standards for converting meters to feet, which are:

(a) International foot: 1 inch = 2.54 centimeters (exactly).
(b) U.S. Survey Foot: 1 meter = 39.37 inches (exactly).

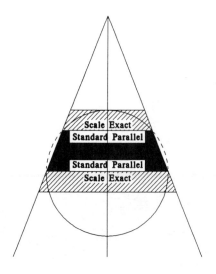

Lambert Projection - Cone Secant to Sphere
Defined by Two Standard Parallels and the Origin

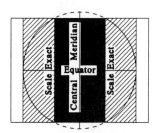

Transverse Mercator Projection -
Cylinder Secant to Sphere
Defined by Central Meridian and Its Scale
Factor, and the Origin

Figure 4.16 Projections for state plane systems. Source: National Oceanographic and Atmospheric Administration. *State Plane Coordinate Systems of 1983: NOAA Manual NOS—NGS 5,* 1990.

It is noted that the difference in these two conversion factors is only 1 part in 500,000. While the difference is insignificant for a survey measurement, it is very important when converting large numbers, such as state plane coordinates, which are usually in the millions. For instance, using the wrong conversion on a state plane coordinate of 2,000,000 would result in a four foot error. This is especially important because NGS publishes state plane coordinates in meters.

Also, observed azimuths require reduction to grid equivalents. This is done by first

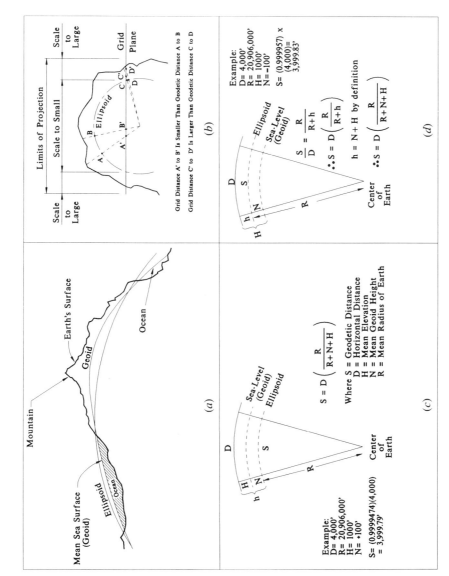

Figure 4.17 State plane system distance relationships. (*a*) Geoid-ellipsoid surface relationships. (*b*) Geodetic versus grid distance. (*c*) Reduction to the ellipsoid. (*d*) Reduction to the ellipsoid (shown with a negative geoid height). Source: National Oceanographic and Atmospheric Administration. *State Plane Coordinate Systems of 1983: NOAA Manual NOS—NGS 5*, 1990.

111

applying a La Place correction to the astronomic observations, which yields an equivalent geodetic value. La Place corrections can be obtained from NGS or by using NGS prepared computer software titled "DEFLECT90." (La Place corrections may be small in most parts of the United States, while the largest corrections can occur in the Rocky Mountains). Obtaining a reduced grid equivalent for this equivalent geodetic value requires corrections for mapping angle convergence and arc-to-chord corrections $(t - T)$ to obtain a reduced grid equivalent. Mapping angle convergence is the relationship between the projected geodetic azimuth and grid azimuth.

The convergence angle (or mapping angle) depends on the position of the azimuth's origin in relation to the projection's central meridian and therefore is a function of longitude. In the Lambert system the convergence angle can be obtained using:

$$\text{convergence angle} = (L_0 - L_P) \sin(B_0) \tag{4-2}$$

where:

L_0 = Central meridian—Longitude of the true and grid origin.
L_p = Meridian of geodetic longitude, positive west.
B_0 = Central parallel, the latitude of the true projection origin.

For example, the convergence angle at a point where the longitude is 82° 01′ 05″ W can be determined using Equation 4-2. If the central meridian for this example is taken as 81° W and B_0 as 33.6693534716, then:

$$
\begin{aligned}
\text{convergence angle} &= (81°\ 00′\ 00″ - 82°\ 01′\ 05″)\ (0.554399350) \\
&= (-3665\ \text{seconds})\ (0.554399350) \\
&= -33′\ 52″
\end{aligned}
$$

NOAA Manual NOS—NGS 5 (1990) contains a series of equations for determining convergence angles in Transverse Mercator projections, while NGS prepared computer software titled "SPCS83" can provide both the convergence angle and scale factor at a point in either projection (in addition to converting latitude and longitude to state plane coordinate northing and easting values, and vice versa).

The $(t - T)$ correction is usually small and not needed for ordinary surveys associated with land developments except for extreme cases or where high precision is required. These angular relationships are illustrated in Figure 4.18.

Corrections for $(t - T)$ can be determined using the following equation:

$$(t - T) = 2.36\ (\Delta N)\ (\Delta E) \times 10^{-10} \tag{4-3}$$

where:

$(t - T)$ is the correction in seconds of arc.
N and E are in feet.

For a Lambert projection, ΔE is the departure of the line and ΔN is the difference between the mean north coordinate value of the line and the projection's central axis. In a

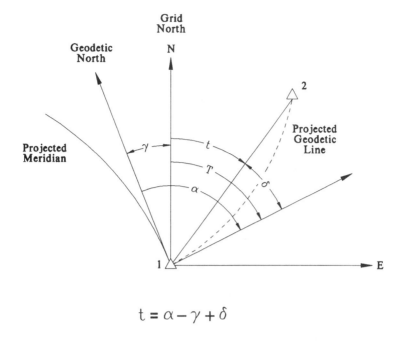

$$t = \alpha - \gamma + \delta$$

α = **Geodetic Azimuth Reckoned From North**

T = **Projected Geodetic Azimuth**

t = **Grid Azimuth Reckoned From North**

γ = **Mapping Angle = Convergence Angle**

δ = $t - T$ = **Second Term Correction = Arc-to-Chord Correction**

Figure 4.18 State plane system azimuth relationships. Source: National Oceanographic and Atmospheric Administration. *State Plane Coordinate Systems of 1983: NOAA Manual NOS—NGS 5*, 1990.

Transverse Mercator projection, ΔE is the difference between the east value of the projection's central meridian and the mean east coordinate value of the line. ΔN is the line's latitude, as shown in Figure 4.19. Section 4.5.5 includes a procedure for determining a line's latitude and departure.

Other horizontal control systems that the development might reference are the public land survey system (based on township and range control) and local urban control networks resulting from city surveys.

4.5.2 Property Boundary Determination

A property line survey is required either to establish the true location of the site extremities or to verify current recorded information along with assurance that no encroachments exist. Adjusted results of this survey will yield a mathematically closed figure, called a closed traverse, whereby a person walking the property lines, while following the defining boundary data, would travel as far to the north as to the south, and as far to the east as to

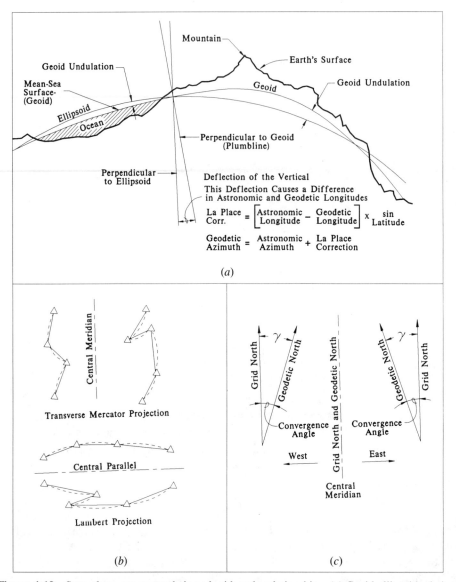

Figure 4.19   State plane system geodetic and grid angle relationships. (*a*) Geoid-ellipsoid relationships: deflection of the vertical. (*b*) Projected geodetic versus grid angles: arc to chord (*t* − *T*) relationships. (*c*) Lambert and Mercator convergence angles. Source: National Oceanographic and Atmospheric Administration. *State Plane Coordinate Systems of 1983: NOAA Manual NOS—NGS 5,* 1990.

the west, such that the ending point is the same as the point of beginning (POB). To obtain a legal boundary survey requires the services of a registered land surveyor who possesses educational and practical experience along with having successfully completed licensing examinations. Arrangements to procure these services would have to be made, unless the project's consulting engineer holds dual registration that would allow these activities to be provided as part of the rendered services.

4.5.3 Background Investigation

The land surveyor (or dual registrant) begins a survey by obtaining from public records repositories (possibly the county records office or register mesne conveyances (RMC) office) information pertaining to deeds, mortgages, easements, and rights of way of record that are either on or affect the subject property. The tax assessor's office, in conjunction with city or county engineering offices, can provide additional data, while the highway and public transportation departments, along with utility companies, can further augment background information. Once this material has been studied, a base map, similar to Figure 4.8, is assembled. This map includes additional details pertaining to property line definitions. These are helpful for measuring property lines on the ground.

If the survey is to be referenced to a state plane system, then the surveyor must locate two separate control marks from which to reference the survey (or a verified mark) where the coordinate values are known. NGS prepared computer software that can convert NAD 27 to NAD 83 datums, and vice versa, is available under the name "CORPSCON."

Analysis of this preliminary information on the base map may indicate potential problems such as discrepancies between measurements recorded for contiguous property lines. This knowledge will alert the field party to anticipate possible problems in questionable areas. The surveyor is cautioned, however, that no judicial authority has been vested through licensing that provides a platform from which to settle boundary disputes. All a surveyor is entitled to do after following such doctrines as senior rights is either to offer suggestions for a solution based on available information, or, that failing, to record the conditions that produce the conflict on a plat to show the basis of conflict that can be used for legal proceedings.

Prior to conducting field work, the surveyor must establish the minimum acceptable precision that the work must meet for both angular and linear measurements. Some states have established minimum standards that regulate the conduct of boundary surveys; however, other standards exist, including one prepared by the American Land Title Association (ALTA), in conjunction with the American Congress on Surveying and Mapping (ACSM) which is entitled "Minimum Standard Detail Requirements for ALTA/ACSM Land Title Surveys" (1988). Figure 4.20 contains some of the information included in these requirements.

4.5.4 Field Work

Activities undertaken in the field commence by initially walking the site with the current landowner, if possible, to obtain assistance in the initial spotting of property corners. This will reduce the amount of time spent in subsequent recovery efforts required to locate hidden monuments. Additional time spent in interviewing adjacent landowners is helpful, especially in establishing a rapport that could aid in verifying measurements of their intersecting property lines, which, if measured, would help eliminate the possibility of boundary conflicts. Once initial site reconnaissance has been completed, the surveyor should have a fuller understanding of the impending field work, along with the approximate limits of the work area. All information gathered thus far is only preliminary and will be subject to verification with instrument measurements and mathematical calculations. Rules of conduct require the person in charge of the field work to pursue the work in a professional manner. Of primary importance is that a surveyor's services are procured but for one reason, and that is to report through measurements what is seen in the field, be it

Survey Class	Dir. Reading Of Instrument 2)	Instrument Reading Estimated 3)	Number of Observations Per Station 4)	Spread From Mean Of D&R Not To Exceed 5)
A	20" <1'> [10"]	5" <0.1'> N.A.	2 D & R	5" <0.1'> [5"]
B	20" <1'> [10"]	10" <0.1'> N.A.	2 D & R	10" <0.2'> [10"]
C	(20') <1'> [20"]	N.A.	1 D & R	(20") <0.3'> [20"]
D	(1') <1'> [1']	N.A.	1 D & R	(30") <0.5'> [30"]

	Angle Closure Where N = No. Of Stations Not To Exceed	Linear Closure 6)	Distance Measurement 7)	Minimum Length Of Measurements 8), 9), 10)
A	10"$\sqrt{N}$	1:15,000	EDM or Double steel tape	8) 81m, 9) 153m, 10)20m
B	15"$\sqrt{N}$	1:10,000	EDM or Steel Tape	8) 54m, 9) 102 m, 10) 14m
C	20"$\sqrt{N}$	1:7,500	EDM or Steel Tape	8)40m, 9) 76m, 10) 10m
D	30"$\sqrt{N}$	1:5,000	EDM or Steel Tape	8) 27m, 9) 51m, 10) 7m

Figure 4.20   ALTA/ACSM survey specifications. Source: American Land Title Association and the American Congress on Surveying and Mapping. *Minimum Standard Detail Requirements for ALTA/ACSM Land Title Surveys,* 1988. Reprinted by permission of the American Congress on Surveying and Mapping, Bethesda, Md.

good or bad. In addition, care must be exerted to avoid damage to landowner's shrubs, fences, and property, while at the same time observing applicable laws of trespass, or even fence laws that require gates to be secured.

4.5.5   SPC Example Survey

Figure 4.21 illustrates reduced observed field measurements on a potential development's boundary. Local regulatory authorities, for this example, have adopted utilization of the established SPC Lambert projection system as a basis for horizontal control that will provide consistent data necessary for incorporation into the local GIS data base. A Transverse Mercator projection would follow this same procedure when using published coordinates and convergence angles, except that for Transverse Mercator projections, the arc-to-chord relationships would reflect Transverse Mercator configurations illustrated in Figure 4.19 and $(t - T)$ values would be determined using Equation 4-3 as applied to Transverse Mercator projections. The regulations also are assumed to require that the final plat will show:

(a) A bearing and distance tie between an SPC monument and the property boundary using grid values.

(b) State plane coordinates for the property corner located by the tie obtained above.

(c) Grid bearings denoting each property line's direction.

(d) Project (or ground) distances for each property line.

(e) A combined factor for the survey that will allow property line distances (shown as project distances) to be reduced to equivalent grid distances. The end user can

SURVEY CLASSES BY LAND USE

Class A—Urban Surveys

Surveys of land lying within or adjoining a City or Town. This would also include the surveys of Commercial and Industrial properties. Condominiums, Townhouses, Apartments, and other multi-unit developments, regardless of geographic location.

Class B—Suburban Surveys

Surveys of land lying outside urban areas. This land is used almost exclusively for single family residential use or residential subdivisions.

Class C—Rural Surveys

Surveys of land such as farms and other undeveloped land outside the suburban areas that may have a potential for future development.

Class D—Mountain and Marshland Surveys

Surveys of land that normally lie in remote areas with difficult terrain and usually have limited potential for development.

Notes:
 (1) All requirements of each class must be satisfied in order to qualify for that particular class of survey. The use of a more precise instrument does not change the other requirements such as number of angles turned, etc.
 (2) Instrument must have a direct reading of at least the amount specified (not an estimated reading), i.e.: 10″ = Micrometer reading theodolite, ⟨1′⟩ = Scale reading theodolite, [10″] = Electronic reading theodolite, (20″) = Micrometer reading theodolite, or a vernier reading transit.
 (3) Instrument must have the capability of allowing an estimated reading below the direct reading to the specified reading.
 (4) D & R means the Direct and Reverse positions of the instrument telescope, i.e., Class A requires that two angles in the direct and two angles in the reverse position be measured and meaned.
 (5) Any angle measured that exceeds the specified amount from the mean must be rejected and the set of angles remeasured.
 (6) Ratio of closure after angles are balanced and closure calculated.
 (7) All distance measurements must be made with a properly calibrated EDM or steel tape, applying atmospheric, temperature, sag, tension, slope, scale factor, and sea level corrections as necessary.
 (8) EDM having an error of 5 millimeters, independent of distance measured (Manufacturer's specification).
 (9) EDM having an error of 10 millimeters, independent of distance measured (Manufacturer's specification).
(10) Calibrated steel tape.

Figure 4.20   (continued)

determine SPC values for each property corner by using the recorded grid bearings and reduced grid distances.

It is emphasized that this example is intended only to demonstrate activity process. Actual procurement of field data would vary to accommodate the presence of several short lines, along with the included limitation that the assumed next closest SPC monument exists several miles away, and the one utilized has been recently verified by another survey. Also, a Global Positioning System (GPS) similar to the one described at the end of this section is assumed not to be available. Sequential lettering indicates the order in which points were occupied during traverse operations, beginning with point "*A*" being closest upon site arrival. Precision requirements for conducting this survey are assumed to be those specified for ALTA/ACSM Class B surveys.

Initially, astronomic observations were made by occupying both the existing SPC

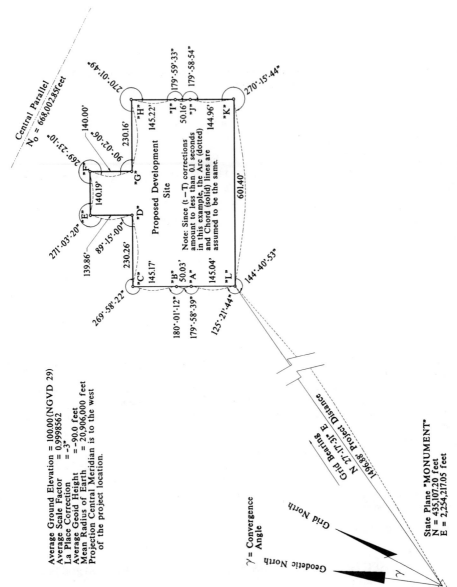

Figure 4.21 An example of field angles and distances for a boundary survey of a land development site.

118

monument (MONUMENT)—a fictitious monument for illustration only—and point "L," which produced correlative data that differed insignificantly after application of La Place corrections. Concurrent with each astronomic observation, a series of horizontal line measurements were made between these points, resulting in the most probable value of the observed horizontal length of line MONUMENT–L as being 1496.88 feet. Calculations for determining the grid bearing of line MONUMENT–L using "published" data pertaining to MONUMENT are shown in Figure 4.22.

Field angles can now be reduced to grid by first determining the $(t − T)$ corrections. On a traverse the size of this example, all $(t − T)$ corrections are negligible; however, $(t − T)$ corrections should be considered on larger surveys. The dotted lines in Figure 4.22 represent observed geodetic lines, while the straight lines represent equivalent grid counterparts. In this example the difference between them is less than 0.1 second and therefore ignored for the rest of this illustration.

The angular error of closure is determined using equivalent grid values, and checked against the appropriate summation for either interior angles $(N − 2) 180°$, exterior angles

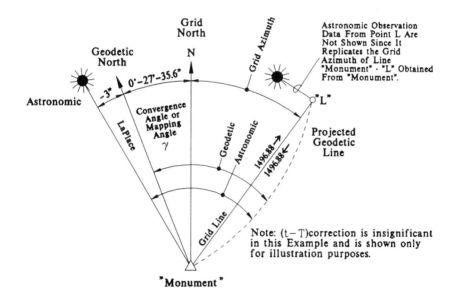

Astronomic Az.	=	27° − 45' − 09.64"
LaPlace Corr.	=	00° − 00' − 03.00"
Geodetic Az.	=	27° − 45' − 06.64"
Convergence Angle	=	00° − 27' − 35.60"
Convergence Adjustment	=	27° − 17' − 31.04"
(t− T) Corr.	=	00° − 00' − 00.04"
Grid Az.	=	27° − 17' − 31.00"

Figure 4.22 An example of a grid bearing determination for a boundary survey of a land development site.

(N + 2) 180°, or deflection angles (summation angles equal 360° while accounting for directional signs). In this case, the sum of the exterior angles should be (12 + 2) × 180° = 2520°; therefore, the angular error of closure is 26″, which is within the allowable ALTA/ACSM Class B survey limits. For this example, the angles are adjusted by assuming equal weight and reliability given that field conditions remain fairly constant. Figure 4.23 shows these steps.

The observed horizontal project distances can be reduced to grid by applying both an elevation factor and a grid scale factor. When combined, these form what is known as the combined grid factor. Since the geoid height for the example's location is −90.00 feet, and the average project height is approximately 100 feet (NGVD 29), the elevation factor can be computed using information contained in Figure 4.17d as:

$$\text{elevation factor} = \frac{20,906,000}{20,906,000 + 100 - 90} = 0.9999995$$

The combined factor is then calculated (using an average grid scale factor of 0.9998562 for this example) as:

$$(0.9999995)\,(0.9998562) = 0.9998557$$

This correction will be shown on the final plat and is the amount each measured horizontal distance must be corrected to obtain equivalent grid distances. Since it is assumed that local regulations require the plat to include the SPC values of only one of the property corners (along with a tie bearing and distance), the equivalent grid distance for line MONUMENT–L must be determined by 1496.88 × 0.9998557, or 1496.66 feet.

Grid Angle Adjustment Calculations					
Point	Obs. Angle	Angle Corr'n.	Adjusted Grid Angle	Line	Grid Bearing
L	125-21-44	-2"	125-21-42	MON - L	N 27-17-31 E
A	179-58-39	-2"	179-58-37	L - A	N 27-20-47 W
B	180-01-12	-2"	180-01-10	A - B	N 27-22-10 W
C	269-58-22	-2"	269-58-20	B - C	N 27-21-00 W
D	089-15-00	-2"	089-14-58	C - D	N 62-37-20 E
E	271-03-20	-2"	271-03-18	D - E	N 28-07-42 W
F	269-23-10	-2"	269-23-08	E - F	N 62-55-36 E
G	090-02-06	-2"	090-02-04	F - G	S 27-41-16 E
H	270-01-49	-2"	270-01-47	G - H	N 62-20-48 E
I	179-59-33	-2"	179-59-31	H - I	S 27-37-25 E
J	179-58-54	-2"	179-58-52	I - J	S 27-37-54 E
K	270-15-44	-2"	270-15-42	J - K	S 27-39-02 E
L	144-40-53	-2"	144-40-51	K - L	S 62-36-40 W
Sum	2520-00-26	-26"	2520-00-00		

Note: 1. All angles are shown as degrees, minutes, and seconds unless otherwise noted.

2. The observed angles are taken as grid in this example because (t-T) corrections are less than 0.1 second.

Figure 4.23   An example of angular adjustment to grid.

Using the verified grid bearing and distance of line MONUMENT–*L*, the SPCs of point *L* can be determined by first obtaining the latitude and departure of line MONUMENT–*L* as shown in Figure 4.24.

Point *L* now serves as the beginning for traverse closure calculations. Figure 4.25 contains the closing calculations and demonstrates that the actual precision obtained meets minimum specifications set forth under Class B surveys of 1 : 10,000. If the calculated precision was out of compliance, remedial field work would be required.

Although adjustment of traverse closure errors can be accomplished by several methods, the Compass or Bowditch Rule is utilized in this example. The Compass Rule states that the magnitude of correction applied to any traverse leg shall be allocated as part of the

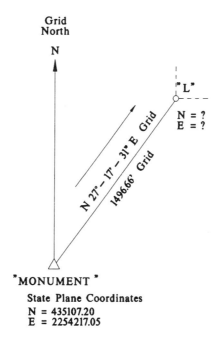

The latitude from "MONUMENT" to point "L" calculation :

$$1496.66 \times \text{Cos} \, (27° - 17' - 31") \, = \, +1330.05'$$

The departure from "MONUMENT" to point "L" calculation :

$$1496.66 \times \text{Sin} \, (27° - 17' - 31") \, = \, +686.26'$$

The northing of "L" = 1330.05 + 435107.20 = 436,437.25 feet

The easting of "L" = 686.26 + 2254217.05 = 2,254,903.31 feet

Figure 4.24 An example of calculations for coordinate determination from latitudes and departures.

Error of Closure Calculations					
Side	Quadrant	Bearing	Distance	Latitude	Departure
L - A	N W	27.34639	145.04	128.83114183	-66.62686016
A - B	N W	27.36944	50.03	44.42967606	-23.00010403
B - C	N W	27.35000	145.17	128.94241066	-66.69470469
C - D	N E	62.62222	230.26	105.88630753	204.46945365
D - E	N W	28.12833	139.86	123.34167258	-65.93672273
E - F	N E	62.92667	140.19	63.80474884	124.82864305
F - G	S E	27.68778	140.00	-123.96898682	65.05144354
G - H	N E	62.34667	230.16	106.82203064	203.86927029
H - I	S E	27.62361	145.22	-128.66674752	67.33287818
I - J	S E	27.63167	50.16	-44.43912074	23.26349388
J - K	S E	27.65056	144.96	-128.40476150	67.27271976
K - L	S W	62.61111	601.40	-276.66060335	-533.98583366
Summation			2162.45	-0.08223178	-0.15632292
Linear Error of Closure:			0.176632		
Precision:		1 to	12242.67		

Note: 1. Grid bearings are shown in degrees and decimal degrees.
　　　2. Distances are horizontal project distances in feet.

Figure 4.25　An example of traverse error of closure calculations.

total error proportioned at a ratio of each side's length to the sum of all sides; more specifically:

$$\text{the correction in latitude per side} = (\text{sum of all latitudes}) \, (L)/(P) \qquad (4\text{-}4)$$

$$\text{the correction in departure per side} = (\text{sum of all departures}) \, (L)/(P) \qquad (4\text{-}5)$$

where
　L is the length of each traverse side.
　P is the traverse perimeter.

Figure 4.26 shows the necessary calculations to produce adjusted distances and directions pertaining to each property line that will be included on the final plat.

Final area calculations are made by utilizing either the coordinate or the double meridian distance (DMD) method, both of which are illustrated in Figure 4.27 using a fictitious 10 unit by 10 unit boundary. Figure 4.28 contains a tabulation of the area calculations for this example using the coordinate procedure.

If SPC values are desired for each property corner, then each adjusted boundary distance must first be reduced to grid using the combined factor. The adjusted boundary grid bearing and this adjusted grid distance will provide grid latitudes and grid departures necessary to calculate SPCs, as indicated in Figure 4.29.

This example utilizes a field procedure that consists of occupying each sequential point that in fact is a true property corner; therefore, no ancillary calculations are necessary to solve for unknown sides using a process called inversing. Tie line MONUMENT–L can be the tie reference on the plat; however, in order to provide a direct reference tie to MONUMENT from point "A," which is the most obvious property corner spotted upon first arrival, an inverse grid bearing and grid distance must be computed as shown in Figure 4.30. This is the tie that will be included on the plat.

			Traverse Adjustments Using The Compass Rule					
Side	Length	Correction for Latitude	Correction for Departure	Adjusted Latitude	Adjusted Departure	Adjusted Distance	Adjusted Direction	Bearing
L - A	145.04	0.00551546	0.01048490	128.83665729	-66.61637525	145.04	-27.341709	N 27-20-30 W
A - B	50.03	0.00190250	0.00361666	44.43157855	-22.99648738	50.03	-27.364764	N 27-21-53 W
B - C	145.17	0.00552040	0.01049430	128.94793106	-66.68421039	145.17	-27.345320	N 27-20-43 W
C - D	230.26	0.00875613	0.01664543	105.89506366	204.48609909	230.28	62.622192	N 62-37-20 E
D - E	139.86	0.00531848	0.01011044	123.3469106	-65.92661229	139.86	-28.123653	N 28-07-25 W
E - F	140.19	0.00533102	0.01013430	63.81007987	124.83877734	140.20	62.926612	N 62-55-36 E
F - G	140.00	0.00532380	0.01012056	-123.96366302	65.06156411	140.00	-27.692458	S 27-41-33 E
G - H	230.16	0.00875233	0.01663820	106.83078297	203.88590850	230.18	62.346659	N 62-20-48 E
H - I	145.22	0.00552230	0.01049791	-128.66122522	67.34337609	145.22	-27.628291	S 27-37-42 E
I - J	50.16	0.00190744	0.00362605	-44.43721330	23.26711994	50.16	-27.636347	S 27-38-11 E
J - K	144.96	0.00551241	0.01047912	-128.39924909	67.28319888	144.96	-27.655236	S 27-39-19 E
K - L	601.40	0.02286952	0.04347504	-276.63773383	-533.94235862	601.35	62.611140	S 62-36-40 W
Sum	2162.45	0.08223178	0.15632292	-0.00000000	0.00000000			

Note: 1. Distances are horizontal project distances in feet.
2. Adjusted directions are shown in degrees and decimal degrees.
3. Bearings are shown in degrees, minutes, and seconds.

Figure 4.26 An example of traverse adjustment using the Compass Rule.

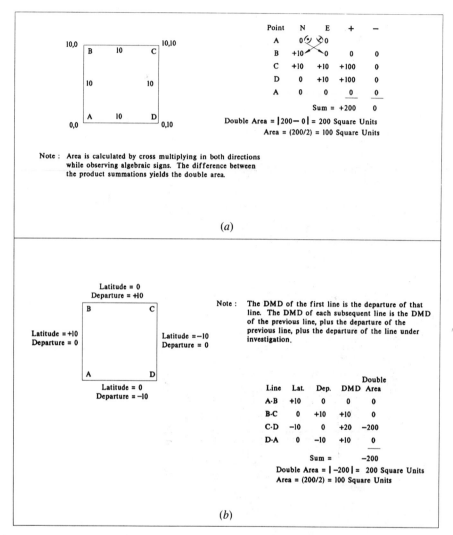

Figure 4.27   Examples of area calculations. (*a*) By coordinates using a square traverse with sides north, south, east, and west and lengths of ten units. (*b*) By double meridian distances (DMDs) using a square traverse with sides north, south, east, and west and lengths of ten units.

After the plat has been drafted into its final form, as shown in Figure 4.31, either by manual or computer aided drafting (CAD) methods, and after all remedial field stake out data have been determined if resetting obliterated or lost monuments is required, then a return trip to the site is necessary. The plat should be field checked by rewalking the boundaries to insure all pertinent features, including encroachments, are properly reflected on the plat. In addition, the plat's final boundary information should be verified independently in the office for notation errors. This is done by double checking each line's bearing and distance to insure that in fact a close traverse is represented by the shown data. This must be done prior to disseminating copies of the plat.

Modern surveying systems allow field data collection systems using total surveying

Area Calculations by Double Meridian Distances				
Side	Adjusted Latitude	Adjusted Departure	D M D	Double Area
L - A	128.83665729	-66.61637525	-66.616	-8582.631
A - B	44.43157855	-22.99648738	-156.229	-6941.512
B - C	128.94793106	-66.68421039	-245.910	-31709.577
C - D	105.89506366	204.48609909	-108.108	-11448.109
D - E	123.34699106	-65.92661229	30.451	3756.093
E - F	63.81007987	124.83877734	89.364	5702.299
F - G	-123.96366302	65.06156411	279.264	-34618.582
G - H	106.83078297	203.88590850	548.211	58565.855
H - I	-128.66122522	67.34337609	819.441	-105430.245
I - J	-44.43721330	23.26711994	910.051	-40440.139
J - K	-128.39924909	67.28319888	1000.602	-128476.484
K - L	-276.63773383	-533.94235862	533.942	-147708.604
			Sum	-447331.635
			Area, s.f.:	-223665.817
			Area,Ac.:	-5.135

Note: The absolute value of the area is used.

Figure 4.28 An example of area determination by double meridian distances (DMDs) using adjusted latitudes and departures.

stations and electronic notebooks, which allow field data to be directly entered into a digital computer for data reduction and processing. A finished plat is plotter drawn using the corrected measurements. A typical system is shown in Figure 4.32.

Another modern surveying system, GPS, is an emerging technology that provides surveyors with a new approach for locating the position of points on or near the earth's surface. A total of 24 satellites will orbit the earth by the mid 1990's to provide an

State Plane Coordinate Calculations								
Side	Adjusted Bearing	Adjusted Project Distance	Grid Distance	Grid Latitude	Grid Departure	Point	Grid North	Grid East
L - A	-27.34171	145.04008	145.01915	128.81806616	-66.60676251	L	436437.25	2254903.31
A - B	-27.36476	50.03003	50.02281	44.42516708	-22.99316899	A	436566.07	2254836.70
B - C	-27.34532	145.17008	145.14913	128.92932387	-66.67458786	B	436610.49	2254813.71
C - D	62.62219	230.27881	230.24558	105.87978300	204.45659174	C	436739.42	2254747.04
D - E	-28.12365	139.85992	139.83974	123.32919209	-65.91709908	D	436845.30	2254951.49
E - F	62.92661	140.20145	140.18122	63.80087207	124.82076311	E	436968.63	2254885.57
F - G	-27.69246	139.99999	139.97979	-123.94577506	65.05217572	F	437032.43	2255010.40
G - H	62.34666	230.17880	230.14558	106.81536729	203.85648776	G	436908.49	2255075.45
H - I	-27.62829	145.21998	145.19902	-128.64265940	67.33365844	H	437015.30	2255279.30
I - J	-27.63635	50.15999	50.15275	-44.43080101	23.26376249	I	436886.66	2255346.64
J - K	-27.65524	144.95998	144.93906	-128.38072107	67.27348991	J	436842.23	2255369.90
K - L	62.61114	601.35088	601.26410	-276.59781501	-533.86531074	K	436713.85	2255437.18
				-0.00000000	0.00000000	L	436437.25	2254903.31

Note: 1. Adjusted grid bearings are shown in degrees and decimal degrees.
2. Distances are shown in feet.
3. The combined grid factor used is 0.999856 .

Figure 4.29 An example of state plane coordinate (SPC) calculations.

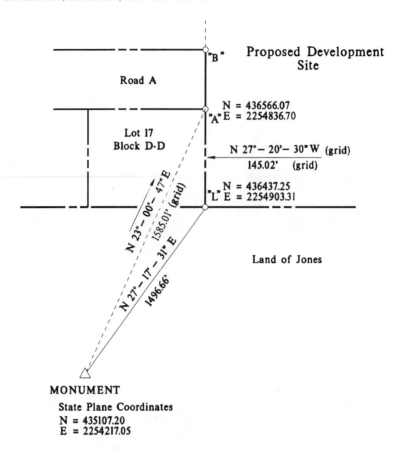

Inverse bearing determination for line "MONUMENT" to "A" = :

$$\mathrm{Tan}^{-1}\!\left(\frac{E_2-E_1}{N_2-N_1}\right) = \left[\frac{2254836.70 - 2254217.05}{436566.07 - 435107.20}\right]^{\mathrm{Tan}^{-1}} = \left[0.4247466\right]^{\mathrm{Tan}^{-1}}$$

Bearing from "MONUMENT" to "A" = N 23° − 00' − 47" E

Inverse distance determination for line "MONUMENT" to "A" =

$$\sqrt{(E_2-E_1)^2 + (N_2-N_1)^2} = \sqrt{(619.65)^2 + (1458.87)^2} = 1585.01 \text{ feet}$$

Figure 4.30   An example of inverse calculations or solving for an unknown side.

accurate navigational system for the U.S. Department of Defense; however, surveyors can utilize this constellation of satellites to pinpoint a location within centimeters, if certain equipment and procedures are employed. Four satellites are required to remove system timing inaccuracies when locating a point in three dimensions. Initially GPS was utilized by surveyors for locating primary control points; however, receiver costs and improved

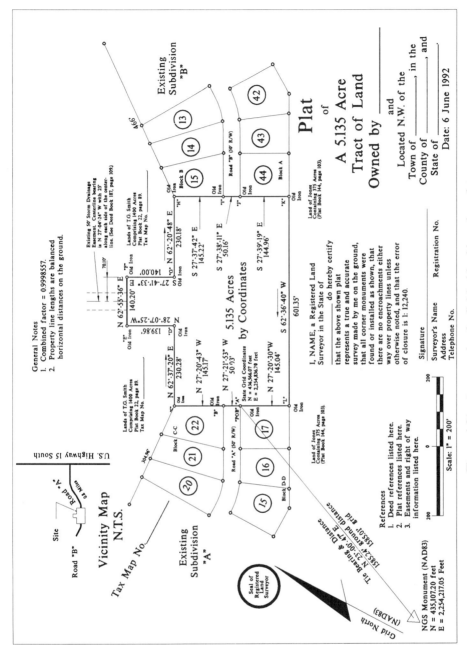

Figure 4.31 An example of a finished land boundary plat.

127

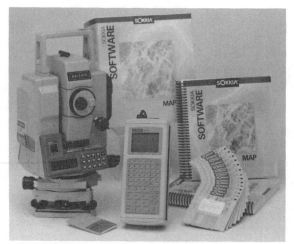

Figure 4.32   An example of a modern surveying system for integrated office and field activities. Courtesy of SOKKIA Corporation, Overland Park, Kan., 1992.

equipment availability have allowed surveyors an opportunity to obtain increased use of GPS equipment. This increased equipment use has resulted in more localized GPS applications. Figure 4.33 illustrates a GPS system.

Although the subject of property surveying is immense, this presentation has been focused on some of the required activities that are necessary to produce a final recordable plat. Topics involving field procedures, rules of evidence, adjoining landowner calls, and much more are included in various recognized texts on the subject (Davis et al., 1981; McCormac, 1991; Moffitt and Bouchard, 1987).

4.5.6   Property Boundary Description

The surveyor, following requirements set forth in Section 2.7.7, is now able to draft a legal description of the property boundary based on final plat data. Much has been written

Figure 4.33   An example of a geographic positioning system (GPS). Courtesy of Trimble Navigation, Sunnyvale, Calif., 1992.

and said about property descriptions; nevertheless, the essence of these activities centers on one issue: Describing property with clarity. Several methods are utilized including descriptions by lot and block, SPCs, public land surveys, and metes and bounds. A metes and bounds description provides exact boundary information that replicates the plat's information using written words. References to terms such as "left" or "right" for direction can be confusing to a nonsurveyor or nonengineer and should be avoided by using substitute compass-related phrases like "northeasterly" or "southerly." The metes and bounds method provides a means to describe any shape parcel, including ones bounding on curves, by utilizing phrases such as:

Thence, along the arc of a curve, whose segment area, being defined by the arc and chord, is (*part* or *not part*) of the included tract, subtended by a radius of _____ feet, in a southwesterly direction for a distance of _____ feet and having an included chord bearing of _____ and distance of _____, to a _____ monument (*found or set*), point _____;

To illustrate the metes and bounds process, the accompanying 5.135 acre tract depicted in Figure 4.31 will be described. Each description, which includes a formal title for identification, commences by identifying and describing a beginning point, called the point of beginning (POB), from which all references are based, such that upon completion a closed traverse has been described. The POB should be one of the most accessible points that a casual site visitor would most easily locate when orienting the property on the ground. From the purest of interpretations, the POB is exactly that—the point that begins the description of the closed traverse. It is from this point, which is part of the described boundary, that back identification references are made. An easy way to facilitate clarity in writing descriptions includes the use of identification letters being placed at each property corner on the final boundary plat that is cross-referenced in the description, as shown in Figure 4.34.

Copies of the completed plat and description can now be distributed to end users such as the developer, the project attorney, the appraiser, and local jurisdictional authorities that approve plats prior to recording.

4.5.7 Site Topography

Once the boundary, or cadastral, survey has been completed, a topographic survey may be undertaken. This survey shows the natural features of the property that will influence the project planning activities. Several methods for conducting topographic surveys are used, including the grid, radial, and traverse methods. Also, aerial photographs can be utilized as alternate means to produce a topographic map.

Topographic information provides a viewer in-depth knowledge of the land's surface features. In order to visualize these features, an understanding of contour lines is required. A contour line is a line that depicts connected points of equal elevation. All contour lines sooner or later close on themselves. Natural contour lines depicting areas of natural curvature are rounded and free flowing. Man-made improvements can result in contour lines that usually appear straight with sharp bends. Contour lines are shown at some predetermined interval and must be positioned in logical sequence. The lines do not touch or cross each other unless contiguous to a wall or part of an overhanging cliff. Figure 4.35 contains various contour line characteristics.

Topographic maps normally show, in addition to contour lines, all site improvements

LEGAL PROPERTY DESCRIPTION OF A 5.135 ACRE TRACT OF LAND

OWNED BY _____, BEING LOCATED AND CONTIGUOUS

TO THE EAST OF SUBDIVISION "A" AND

TO THE WEST OF SUBDIVISION "B'

SITUATED NORTHWEST OF _____ TOWN, _____ COUNTY,

STATE OF _____

Beginning at an old iron found in the southern edge of the right of way of Road "A", Point A, the Point of Beginning (POB), from whence State Plane Monument "MONUMENT" bears S 23°-00'-47" W, at a distance of 1585.24 feet;

Thence N 27°-21'-53" W crossing the right of way of Road "A" for a distance of 50.03 feet to an old iron found, Point B;

Thence N 27°-20'-43" W along the eastern side lot line of Lot 22, Block C-C of existing Subdivision "A", for a distance of 145.17 feet, to an old iron found, Point C;

Thence N 62°-37'-20" E along the lands of Smith for a distance of 230.28 feet to an old iron found, Point D;

Thence N 28°-07'-25" W along the lands of Smith for a distance of 139.86 feet to an old iron found, Point E;

Thence N 62°-55'-36" E along the lands of Smith for a distance of 140.20 feet to an old iron found, Point F;

Thence S 27°-41'-33" E along the lands of Smith for a distance of 140.00 feet to an old iron found, Point G;

Thence N 62°-20'-48" E along the lands of Smith for a distance of 230.18 feet to an old iron found, Point H;

Thence S 27°-37'-42" E along the western side lot line of Lot 15, Block "B" in Subdivision "B", for a distance of 145.22 feet to an old iron found in the northern edge of the right of way of Road "B", Point I;

Thence S 27°-38'-11" E across the right of way of Road "B" for a distance of 50.16 feet, to the southern edge of the right of way of Road "B" to an iron pipe found, Point J;

Figure 4.34 An example of a property boundary description of a parcel of land using metes and bounds.

Thence S 27°-39'-19" E along the western side lot line of Lot 44, Block "A" contained within Subdivision "B", for a distance of 144.96 feet to an old iron found, Point K;

Thence S 62°-36'-40" W along the lands of Jones for a distance of 601.35 feet to an old iron found, Point L;

Thence N 27°-20'-30" W along the eastern side of Lot 17 , Block D-D contained within Subdivision "A" for a distance of 145.04 feet to an iron pipe found, Point A, the Point of Beginning.

Said tract measures and contains 5.135 Acres more or less, and is more clearly shown on a Plat by _____ surveyor, dated 6 June 1992, and recorded in Plat Book _____ page _____, RMC Office, _____ County, in the State of _____-, and is subject to a contiguous 50 foot storm drainage easement whose centerline is located at midpoint on side E-F, which then proceeds from the property along a bearing of N 27°-04'-24" W with 25 feet along each side of the centerline.

Date Of Preparation: Certified to be Correct:

Seal Impression   Surveyor's Name and Signature
                                        Surveyor's Address

Figure 4.34   (continued)

including those above and below ground if definable. Trees and their canopy spreads are usually shown if of a certain type or size.

4.5.8 Topographic Survey

Prior to starting the topographic survey, a benchmark, or point of known elevation, must exist on or near the site based on the required vertical datum. Often verified benchmarks exist some distance from the site. In such cases a level loop must be run using differential leveling techniques from the known elevation point to the job site and then, reversing the process, to tie back to the starting point. To help eliminate mistakes, and thus reduce field time consumed by remedial action while searching for leveling errors, the field crew should establish along the level run turning points that can easily be identified. These same points are utilized both ways when performing differential leveling, which provides a means to check accumulated field data by verifying both differences in elevation between adjacent turning points. If the difference between the beginning elevation and the computed closing elevation for the starting point (the vertical error of closure) exceeds the allowable tolerance, a check of the turning point differences likely will indicate where the problem exists in the field, allowing efficiency in eliminating unreliable data. It is also helpful to use the three wire leveling process to conduct a control level loop that will help

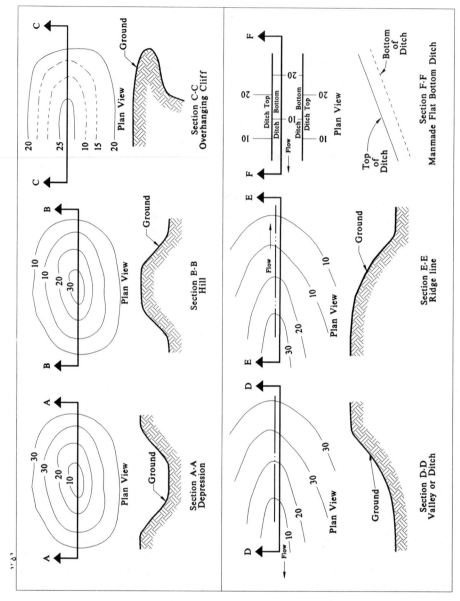

Figure 4.35 Examples of various contour line configurations.

eliminate possible data errors. Although Figures 4.36 and 4.37 illustrate these procedures, the reader is referred to other sources for leveling adjustments using least squares as an alternative method (Davis et al., 1981; Moffitt and Bouchard, 1987).

Of the various methods available to conduct a topographic survey, the one most utilized is the grid method. This process uses an imaginary grid of predetermined size on the property with spot elevations (which are the ground elevations at a particular spot) being determined at each grid intersection using differential leveling. It is assumed that a

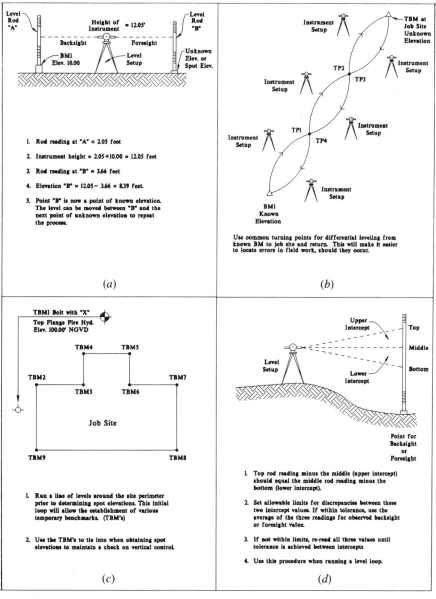

Figure 4.36 Examples of leveling procedures. (*a*) Differential leveling. (*b*) Turning point locations. (*c*) Temporary bench mark (TBM) check points. (*d*) Three-wire leveling.

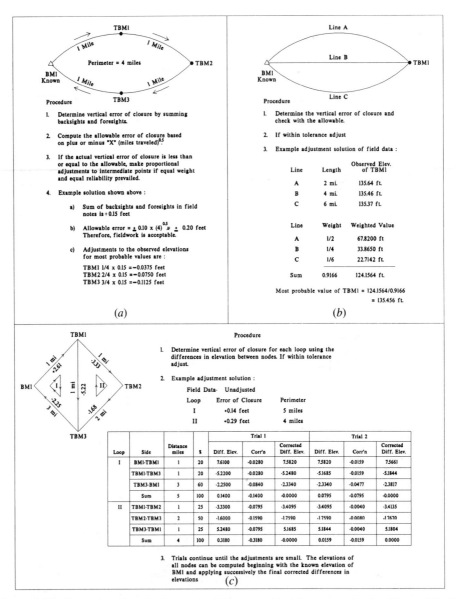

Figure 4.37 Examples of vertical control adjustments. (*a*) Adjustment of a level loop. (*b*) Adjustment of multiple level lines. (*c*) Adjustment of a level net. Source: Raymond E. Davis and Francis S. Foote. *Surveying Theory and Practice,* 4th ed. Copyright © 1962 by McGraw-Hill, Inc. Reproduced by permission of McGraw-Hill, Inc.

uniform slope exists between adjoining spot elevations; therefore, if surface irregularities occur, additional intermediate spot elevations must be obtained.

Prior to laying out a project grid in the field, it is helpful to establish a series of temporary bench marks (TBMs) along the property boundary that can be used later as references to tie each line of differential levels into when spot elevations are measured. Figure 4.38 shows an example of reduced field data, upon which contour lines must be

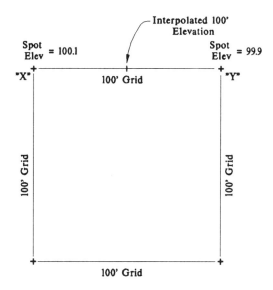

1. The reduced spot elevation of point "X" is 100.1 NGVD

2. The reduced spot elevation at point "Y" is 99.9 NGVD

3. The horizontal distance from either spot elevation must be determined for all contour lines that will be shown on the topographic map. Using a one foot contour interval, there will be only one contour line having an elevation of 100 feet between these spot elevations.

4. To mathematically proportion the distance use:

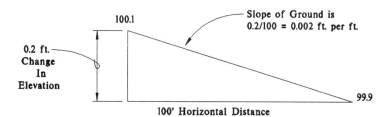

5. From the 99.9 foot spot elevation, the 100 foot contour crosses at (100−99.9)/0.002, or 50 feet.

Figure 4.38 How to determine contour line placement.

located. Digital computer software application programs can be utilized to perform this task; however, other methods are often utilized, including mathematical or graphical interpolation. The interpolation process determines intermediate positions of various contour lines that must be shown on the final topographic map based on the interval used.

If the graphical method is utilized, a transparent paper sheet with a series of parallel lines must be available. This overlay is positioned over two adjoining spot elevations within the grid. Each parallel line represents an integer value elevation with the distance between lines being proportional subdivisions. The resulting positions of intermediate contour lines are located as illustrated in Figure 4.39.

The process usually begins at the lowest or highest spot elevation in an area, and systematically progresses outward in order to maintain contour line integrity. A finished working drawing from which the final topographic map will be derived is shown in Figure 4.40. The final drawing should be field verified to insure accurate information is included and that no details are overlooked prior to design use.

4.5.9 Utility Extension Surveys

Often it is necessary to include as part of the project some utility line connections that join existing off-site facilities. These improvements could, for example, include installation of gravity or pressure sanitary sewer mains or an extension to the existing potable water system. Field data about the area between the project site and the point of connection are required for both horizontal and vertical control. These data are necessary for the preparation of a plan and profile that show the proposed extension and that are included with the other construction drawings. Best results can be obtained when a foreknowledge of the proposed utility system's general features allows positioning field control points during the survey operations on which the design is based. Most utility line extensions occur within public rights of way that require a permit to be issued prior to construction. Often other buried facilities exist in the right of way where the proposed improvement needs to be located, including possibly storm drains. It is prudent to request underground location

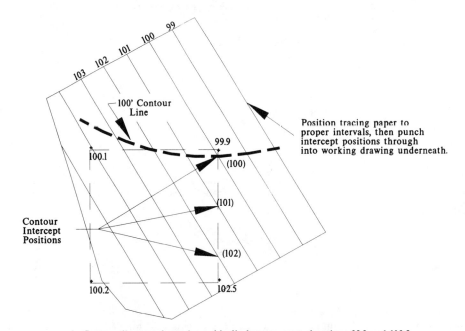

1. Contour lines are located graphically between spot elevations 99.9 and 102.5 .
2. This process is repeated between all adjacent pairs of spot elevations.
3. Contour lines are drawn between logical intercept points.

Figure 4.39   Graphical interpolation for contour line positions.

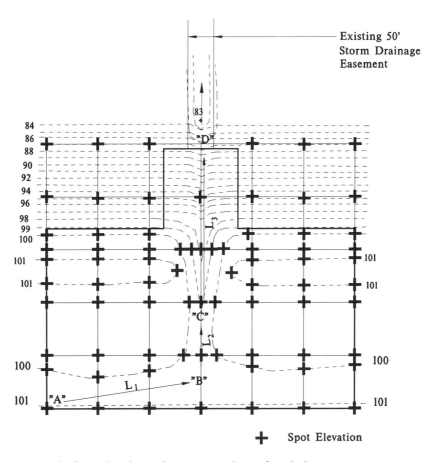

Figure 4.40 An example of a working drawing for a topographic map.

1. Spot elevation values are not shown for clarity.
2. Grid interval is 100' x 100'.
3. Points "A", "B", "C" and "D" refer to Section 7.5.2

services and to excavate test holes so control points can be properly positioned to miss these features prior to making any field measurements. The resulting survey will produce measured design stationing that will result in construction plans that contain a minimum of conflicts. This translates into less contractor change order requests being submitted during construction. Figure 4.41 provides an example of how the process works in the field when the position of a future sanitary sewer manhole is being tentatively located. In this example the existing storm drainage system is already in place. The new manhole, after being staked and measured, will produce excellent horizontal and vertical control data that will allow the engineer to show completed information on the plans, including the final elevation of the manhole's rim and cover.

4.5.10 Photogrammetric Surveys

Available aerial photographs possess a vast amount of information helpful to the design professional; however, the photograph's average scale must be known. The average scale

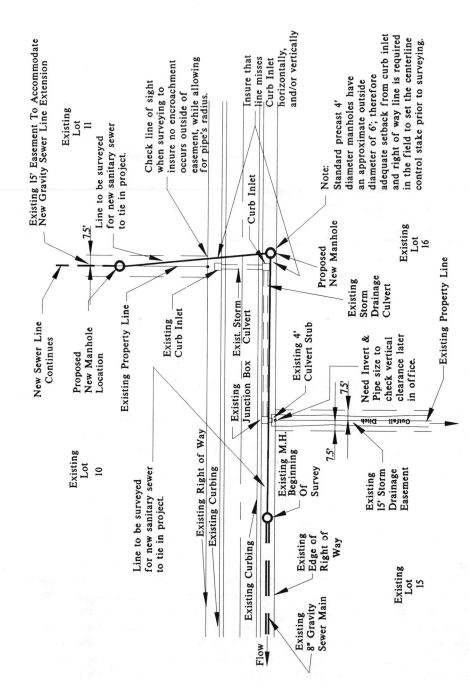

Figure 4.41 An example of the field survey techniques for an extension of an existing utility system.

can be determined knowing the focal length of the camera, the elevation of the ground, and the flying height above datum, as illustrated in Figure 4.42.

If aerial photographs are desired but not available for use, a flight plan must be designed for features desired in the finished photographs. These photographs should be taken during winter and low tide for best results. A key element in flight planning is the availability of both horizontal and vertical control points within the flight area. If ground

Given : f = 12 inches

H = 6,100 feet above sea level

h_a = 110 feet NGVD 29

h_b = 90 feet NGVD 29

Determine the average scale of the photograph.

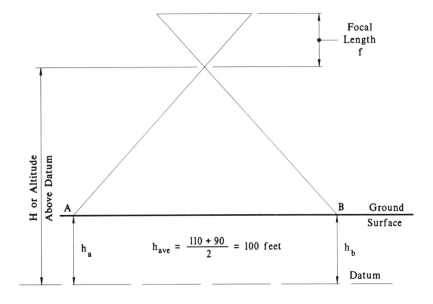

1. The average numeric scale is :

$$S_{ave} = \frac{f}{H - h_{ave}} = \frac{12"}{6,100 - 100} = \frac{1"}{500'}$$

Therefore, S_{ave} is : $1" = 500'$.

2. The average reference fraction is :

$$S_{ave} = \frac{12"}{6000'} = \frac{1'}{6000'} \text{, or } 1 : 6,000 \text{ .}$$

Figure 4.42   Average scale of a vertical photograph.

control points exist, they must be recovered and targeted. If they do not exist, then they must be established and located, using land surveying techniques, prior to being targeted. According to Lillesand and Kiefer (1979):

> Accurate ground control is essential to virtually all photogrammetric operations because photogrammetric measurements can only be as reliable as the ground control on which they are based.*

A sample flight plan design follows for a proposed development site where the average elevation is 100 feet NGVD 29. The site consists of approximately 5000 acres in a rectangular boundary of 20,800 feet by 10,400 feet. A camera with a 12 inch focal length will be used.

(a) Desired photographic scale: 1 inch = 500 feet.

(b) Size of photographs: 9 × 9 inches.

(c) Flight map scale: 1 : 24,000.

(d) Minimum sidelap: 30 percent.

(e) Minimum overlap: 65 percent.

(f) An intervalometer is to be used.

(g) The first and last flight lines will coincide with the larger boundaries.

(1) First determine the flight height above datum the aircraft must fly to produce the desired photographic scale:

$$\frac{1 \text{ inch}}{500 \text{ feet}} = \frac{f}{H - \text{have}} = \frac{12}{H - 100}$$

$$H - 100 = 6000$$

$$H = 6100 \text{ feet NGVD 29}$$

(2) Next, a determination of the ground distance between exposures must be undertaken. If the minimum overlap is given as 65 percent, then the net gain per picture is 35 percent, resulting in an effective photographic distance of 0.35 × 9 inches = 3.15 inches. This would equate to a corresponding ground distance of 3.15 inches × 500 feet per inch = 1575 feet on the ground.

(3) If the anticipated flight speed is to be maintained at 100 miles per hour, the velocity of transit can be computed as:

$$\frac{100 \text{ miles per hour} \times 5280 \text{ feet per mile}}{3600 \text{ seconds per hour}} = 146.7 \text{ feet per second}$$

Because an intervalometer will be used to activate the camera, the exposure interval can be computed by 1575 feet/146.7 feet per second = 10.7 seconds; however this must be

---

*Taken from Lillesand and Kiefer (1979), by permission.

rounded downward in integer form to 10 seconds. The adjusted ground distance would equal 10 seconds × 146.7 feet per second = 1467 feet of ground coverage between exposures, which equates to an adjusted overlap of 67.4 percent, which is greater than the allowable minimum.

(4) The total number of photographs per flight line, with the aircraft flying parallel to the long boundaries, is (20,800 feet/1467 feet) + 4 photographs (2 additional at each end for complete coverage) = 18.2, or 19 photographs per flight line.

(5) The number of flight lines can be determined by first computing the initial spacing between flight lines using the minimum allowable sidelap of 30 percent. This leaves a net gain of 70 percent between flight lines. The corresponding ground spacing is therefore 0.70 × 9 inches × 500 feet per inch = 3150 feet. The number of corresponding flight lines is:

$$\frac{10,400 \text{ feet}}{3150 \text{ feet}} + 1 = 4.3, \text{ but round to 5}$$

The adjusted net gain distance between flight lines accounting for rounding that equalizes the number of trips is:

$$\frac{10,400}{5 - 1} = 2600 \text{ feet}$$

This distance produces a 42 percent sidelap, which is in excess of the minimum allowed.

(6) The total number of pictures required for the mission would be 5 × 19 or 95 photographs.

(7) To properly draw flight lines on the flight map, the spacing must be determined using the adjusted ground distance of 2600 feet between flight lines. At a scale of 1 : 24,000 the equivalent map spacing for flight lines would be every 1.3 inches.

The completed flight plan would then be:

(a) Flight height = 6100 NGVD 29.

(b) Camera focal length = 12 inches.

(c) Speed = 100 miles per hour.

(d) Intervalometer setting = 10 seconds.

(e) Total number of photographs = 95.

(f) Map spacing for drawing flight lines on the flight map is 1.3 inches apart.

Figure 4.43 illustrates some photogrammetric relationships that were alluded to in this example.

Should the design professional require photographs that depict the area using true directions, distances, and areas, then orthophotos must be made using conventional photographs that overlap. The reader is referred elsewhere for more complete discussions on this and other important topics relating to photogrammetry (Lillesand and Kiefer, 1979; Moffitt and Mikhail, 1980; Wolf, 1983).

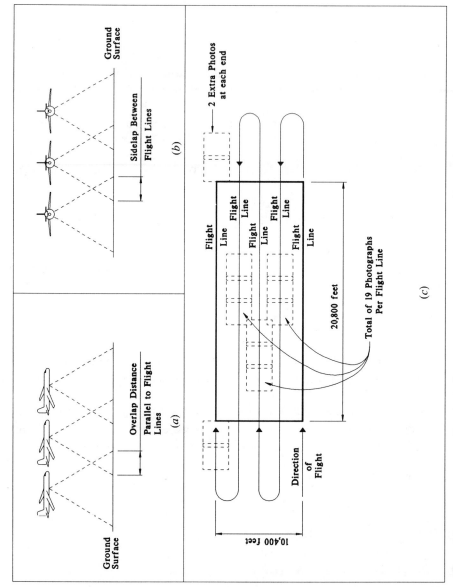

Figure 4.43 Photogrammetric relationships. (*a*) Photo overlap. (*b*) Photo sidelap. (*c*) Relationships pertaining to the sample flight plan.

4.6 GEOTECHNICAL SURVEYS

Four reasons why a subsurface exploration program is necessary:

(a) The developer must purchase property wisely.

(b) The engineer must locate improvements efficiently on the site.

(c) The engineer must design site structures properly.

(d) The engineer must design all site improvements correctly.

Subsurface investigations usually are performed by geotechnical engineering testing firms whose services generally are classified as auxiliary services. A determination of the structural loads, locations of proposed improvements, and proposed sampling sites must be available prior to commencing any subsurface investigation. Investigations can include exploratory borings, extraction of undisturbed soil samples, or employment of sounding rods for crude investigations. Hand augers are excellent for roadbed and floor slab investigations where a visual inspection of the soil can be made; however, these soil samples are totally destroyed and provide no indication of a clay's stiffness or a sand's density. The investigator is limited as to depth, especially in sandy conditions below the groundwater table.

Wash boring samples may be obtained where the sample hole is advanced by the jetting action of water, but again no idea of the soil's stiffness or density is obtained. A better method is to employ the spoon sample boring with a standard penetration test. This process is in part similar to the washed boring method except that at various intervals a standard split casing is driven into the soil using a 140 pound weight that falls 30 inches. The number of blows to drive the standard spoon one foot is the standard penetration value from which the apparent soil strength can be estimated. The hole, rather than being advanced by jetting, utilizes a rotary drill in conjunction with a clay slurry to maintain the integrity of the interior wall. If an on-site test pit is available, excellent samples can be obtained from within; nevertheless, many patented samplers including thin walled tubes are available today to extract subsurface "undisturbed" samples for laboratory testing without requiring a pit facility.

4.6.1 Soils Exploration Program

Because the complexity and proposed uses of sites vary, as does the uniformity of the underlying soil, exploratory activities cannot be based strictly on economics (although economics always play a part). If the soil if fairly uniform and the anticipated structural loads are light, a minimum program that encompasses at least one deep boring every 10 acres will probably provide sufficient preliminary investigative data. Also required are supplemental hand auger borings in locations where improvements such as roads are proposed, or where soil problems may exist. Proposed boring locations should be shown with distances on a site plan to allow accurate field positioning. Each boring site will yield a soil log showing the extents of the underlying strata of soil. In addition, soil samples and standard penetration values can be obtained if spoon or thin walled tube sampling is employed. Finally, the stabilized groundwater table and possibly an indication of the seasonal table fluctuations can be determined. Should results of this exploratory program warrant additional investigative work, arrangements with the developer must be made.

Other procedures used for subsurface exploration may incorporate geophysical methods, which are good for locating bedrock or gravel deposits. One method utilizes seismic shocks that are measured on a calibrated recorder. Another uses electrical resistivity by measuring electrical currents through electrodes driven into the soil. Finally, air photo interpretation can be utilized to identify soils on the potential site if such photos are available.

4.6.2 Soil Classification

A soil's composition varies relative to particle size distribution and the amount of plasticity. Two primary systems used for classifying soils for purposes of construction are the American Association of State Highway and Transportation Officials (AASHTO) and the Unified Classification Systems (UCS). These systems require the soil's grain size distribution and Atterberg limit values, which indicate the amount of cohesive material present.

The AASHTO method (M 145-87) classifies soils into seven groups based on grain size distribution and Atterberg limits. Evaluation of soils within each group is made using a "group index" that is obtained with an empirical formula. The group classification and group index can indicate the relative quality of the soil; however, additional soil testing is usually required for design purposes.

Figure 4.44 can be used to classify soils, while Figure 4.45 can be used to further define soils by subgroups. Liquid limit and plasticity index ranges for A-4 through A-7 soils are shown in Figure 4.46.

AASHTO soil classification is accomplished by a process of elimination beginning with entering Figure 4.44 or 4.45 at the left, and then proceeding toward the right. The first group that the grain size distribution and Atterberg limits satisfy is the correct classification. Group index values are shown in parentheses after the group symbol. The group index can be obtained using Equation 4-6 where the nearest computed integer value is used.

$$\text{group index} = (F - 35) [0.2 + 0.005 (LL - 40)] + \\ 0.01 (F - 15) (PI - 10) \qquad (4\text{-}6)$$

where:

F = percentage passing 0.075 mm (No. 200 sieve), expressed as a whole number. This percentage is based only on the material passing the 75 mm (3 inch) sieve.
LL = liquid limit.
PI = plasticity index.

If the calculated group index is negative, then zero is used for the index value.

The Unified Classification System (UCS) uses Figure 4.47 to base the classification of soils into one of 15 major soil groups. Soils are classified using:

(a) Percentage of gravel, sand, and fines (fraction passing the No. 200 sieve).

(b) Shape of the grain size distribution curve using C_u (coefficient of uniformity) and C_g (coefficient of gradation). D_{10} is the grain size diameter that corresponds to 10 percent on the grain size distribution curve.

(c) Plasticity and compressibility characteristics.

General Classification	Granular Materials (35% or Less Passing 0.075 mm)			Silt-Clay Materials (More than 35% Passing 0.075 mm)			
Group Classification	A-1	A-3[a]	A-2	A-4	A-5	A-6	A-7
Sieve analysis, percent passing:							
2.00 mm (No. 10)	—	—	—	—	—	—	—
0.425 mm (No. 40)	50 max.	51 min.	—	—	—	—	—
0.075 mm (No. 200)	25 max.	10 max.	35 max.	36 min.	36 min.	36 min.	36 min.
Characteristics of fraction passing 0.425 mm (No. 40)							
Liquid limit	—	—	b	40 max.	41 min.	40 max.	41 min.
Plasticity index	6 max.	N.P.	b	10 max.	10 max.	11 min.	11 min.
General rating as subgrade	Excellent to good			Fair to poor			

Figure 4.44 Classification of soils and aggregate mixtures. Taken from *Standard Specifications for Transportation Materials and Methods of Sampling and Testing* AASHTO M145-87, "Recommended Practice for the Classification of Soils and Soil-Aggregate Mixtures for Highway Construction Purposes." Copyright © 1990 by the American Association of State Highway and Transportation Officials. Reproduced with permission.

[a]The placing of A-3 before A-2 is necessary in the "left to right elimination process" and does not indicate superiority of A-3 over A-2.
[b]See Figure 4.45 for values.

General Classification

	Granular Materials (35% or Less Passing 0.075 mm)							Silt-Clay Materials (More than 35% Passing 0.075 mm)			
	A-1		A-3	A-2				A-4	A-5	A-6	A-7
Group Classification	A-1-a	A-1-b		A-2-4	A-2-5	A-2-6	A-2-7				A-7-5, A-7-6
Sieve analysis, percent passing:											
2.00 mm (No. 10)	50 max.	—	—								
0.425 mm (No. 40)	30 max.	50 max.	51 min.								
0.075 mm (No. 200)	15 max.	25 max.	10 max.	35 max.	35 max.	35 max.	35 max.	36 min.	36 min.	36 min.	36 min.
Characteristics of fraction passing 0.425 mm (No. 40)											
Liquid limit			—	40 max.	41 min.	40 max.	41 min.	40 max.	41 min.	40 max.	41 min.
Plasticity index	6 max.		N.P.	10 max.	10 max.	11 min.	11 min.	10 max.	10 max.	11 min.	11 min.[a]
Usual types of significant constituent materials	Stone fragments, gravel and sand		Fine sand	Silty or clayey gravel and sand				Silty soils		Clayey soils	
General Ratings as Subgrade	Excellent to Good							Fair to poor			

Figure 4.45 Classification of soils and aggregate mixtures, continued. Taken from *Standard Specifications for Transportation Materials and Methods of Sampling and Testing* AASHTO M145-87, "Recommended Practice for the Classification of Soils and Soil-Aggregate Mixtures for Highway Construction Purposes." Copyright © 1990 by the American Association of State Highway and Transportation Officials. Reproduced with permission.

[a]Plasticity index of A-7-5 subgroup is equal to or less than LL minus 30. Plasticity index of A-7-6 subgroup is greater than LL minus 30 (see Figure 4.46).

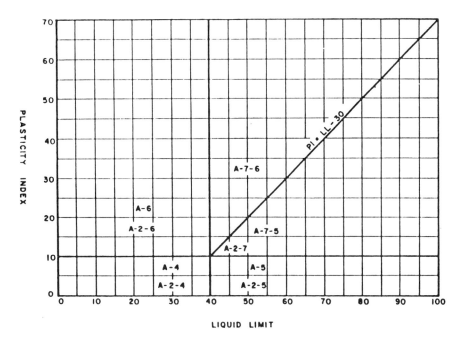

NOTE—A-2 soils contain less than 35 % finer than 200 sieve.

Figure 4.46 Liquid limit and plasticity index ranges for silt-clay materials. Taken from *Standard Specifications for Transportation Materials and Methods of Sampling and Testing* AASHTO M145-87, "Recommended Practice for the Classification of Soils and Soil-Aggregate Mixtures for Highway Construction Purposes." Copyright © 1990 by the American Association of State Highway and Transportation Officials. Reproduced with permission.

Generally clay (*C*) type materials are located above the "*A*" line of the plasticity chart, while silts (*M*) are below. Silts and clays are further classified based on high (*H*) or low (*L*) liquid limit values. The reader is referred elsewhere for a more complete presentation of this soil classification system (Joint Publication of the U.S. Army, Navy, and Air Force, 1971).

A third method of soil classification utilized by the U.S. Department of Agriculture is illustrated in Figure 4.48.

4.6.3 Combined Mechanical Analysis of a Soil

When a soil contains particle sizes ranging from clay to gravel, it is necessary to run both the dry (sieve) and wet (hydrometer) analyses to determine the grain size distribution curve. Sieve analysis is suitable for particle sizes down to the No. 200 sieve, or 0.074 mm, whereas hydrometer analysis is suitable for particles ranging from 0.2 mm down to 0.2 micron, which provides some redundant information in the range between 0.2 mm and 0.07 mm.

A dry representative soil sample is carefully weighed and then washed (without screening) with wash water containing a dispersing agent. The muddy wash water is retained for hydrometer analysis. The washed soil is dried and then used in the sieve analysis. The mechanical analysis is conducted by placing the dry soil sample in the top of a nest of

Major Divisions			Group Symbols	Typical Name	Field Identification Procedures (Excluding Particles Larger Than 3 Inches and Basing Fractions on Estimated Weights)
1	*2*		*3*	*4*	*5*
Coarse-grained Soils. More than half of material is larger than No. 200 sieve size. The No. 200 sieve size is about the smallest particle visible to the naked eye.	Gravels. More than half of coarse fraction is larger than No. 4 sieve size. (For visual classification, the 1/4-inch size may be used as equivalent to the No. 4 sieve size.)	Clean gravels (little or no fines)	GW	Well-graded gravels, gravel-sand mixtures, little or no fines	Wide range in grain sizes and substantial amounts of all intermediate particle sizes
			GP	Poorly-graded gravels, gravel-sand mixtures, little or no fines	Predominantly one size or a range of sizes with some intermediate sizes missing
		Gravels with fines (appreciable amount of fines)	GM	Silty gravels, gravel-sand-silt mixtures	Nonplastic fines or fines with low plasticity (for identification procedures see ML below)
			GC	Clayey gravels, gravel-sand-clay mixtures	Plastic fines (for identification procedures see CL below)
	Sands. More than half of coarse fraction is smaller than No. 4 sieve size.	Clean sands (little or no fines)	SW	Well-graded sands, gravelly sands, little or no fines	Wide range in grain size and substantial amounts of all intermediate particle sizes
			SP	Poorly-graded sands, gravelly sands, little or no fines	Predominantly one size or a range of sizes with some intermediate sizes missing
		Sands with fines (appreciable amount of fines)	SM	Silty sands, sand-silt mixtures	Nonplastic fines or fines with low plasticity (for identification procedures see ML below)
			SC	Clayey sands, sand-clay mixtures	Plastic fines (for identification procedures see CL below)

Major Divisions			Group Symbols	Typical Name	Identification procedures on fraction smaller than No. 40 sieve size		
					Dry strength (crushing characteristics)	Dilatancy (reaction to shaking)	Toughness (consistency near PL)
Fine-grained Soils. More than half of material is smaller than No. 200 sieve size.	Silts and Clays. Liquid limit less than 50		ML	Inorganic silts and very fine sands, rock flour, silty or clayey fine sands with slight plasticity	None to slight	Quick to slow	None
			CL	Inorganic clays of low to medium plasticity, gravelly clays, sandy clays, silty clays, lean clays	Medium to high	None to very slow	Medium
			OL	Organic silts and organic silty clays of low plasticity	Slight to medium	Slow	Slight
	Silts and Clays. Liquid limit greater than 50		MH	Inorganic silts, micaceous or diatomaceous fine sandy or silty soils, elastic silts	Slight to medium	Slow to none	Slight to medium
			CH	Inorganic clays of high plasticity, fat clays	High to very high	None	High
			OH	Organic clays of medium to high plasticity, organic silts	Medium to high	None to very slow	Slight to medium
Highly organic soils			Pt	Peat and other highly organic soils	Readily identified by color, odor, spongy feel and frequently by fibrous texture		

(1) **Boundary classifications** Soils possessing characteristics of two groups are designated by combinations of group symbols. For example
Adopted by Corps of Engineers and Bureau of Reclamation, January, 1952.

Figure 4.47   The Unified Classification System (UCS) for soils. Source: Joint publication of the U.S. Army, Navy, and Air Force. *Materials Testing—TM 5-530/NAVFAC MO-330/AFM 89-3.* Philadelphia: Dept. of U.S. Navy, 1971.

sieves, which range usually from 1.5 inches through the No. 200 sieve, for the purpose of separating the soil particles to determine the percent weight retained on and passing through each sieve. The hydrometer analysis, based on Stokes law where "Particles of equal specific gravity settle in water at a rate which is in proportion to the size of the particle" (Seelye, 1960), is conducted on the wash water using a hydrometer for sequentially timed readings (Lambe, 1951).

Figure 4.49 indicates a graph of the combined analysis results from which the D_{60} and D_{10} UCS values can be obtained.

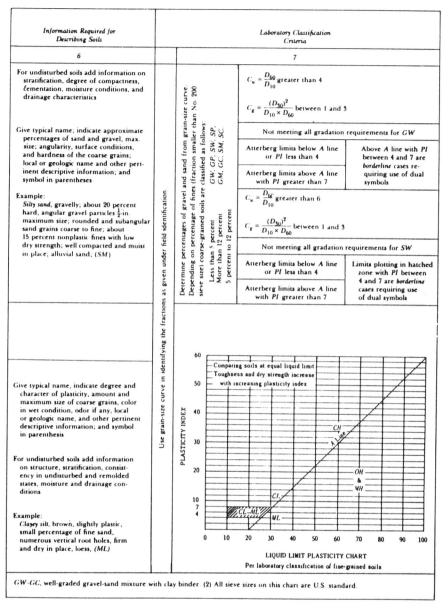

Figure 4.47 (continued)

4.6.4 Atterberg Limits Testing

The Atterberg limits indicate the amount of cohesive material present in the soil and consist of the liquid limit, plastic limit, and the shrinkage limit (Lambe, 1951).

The liquid limit is the minimum moisture content, expressed as a percentage of the oven dry soil weight, at which the soil will begin to flow when subjected to a small shearing force. This test is conducted using a standard liquid limit device, which consists

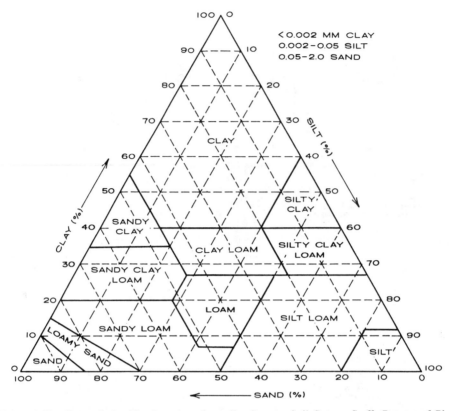

Figure 4.48   Textural classification chart for soils. Source: Soil Survey Staff, Bureau of Plant Industry, Soils, and Agricultural Engineering. *Soil Survey Manual—USDA Agricultural Handbook No. 18*. Washington: U.S. Govt. Printing Office, 1951.

of a small cup fastened on a pivot that can be lifted and dropped. The soil sample is placed in the shallow cup and grooved with a special tool. The cup and sample are then dropped under the action of a cam that is turned at a rate of two rotations per second until the groove closes for a distance of 0.5 inch. This process is repeated several times while altering the soil's moisture content. A plot of the data is then made to determine the moisture content that would allow groove closure using exactly 25 blows, as illustrated in Figure 4.50.

The plastic limit is the minimum moisture content at which the soil can be rolled into a thread one-eighth of an inch in diameter without crumbling. It is determined by trial and error. Finally, the shrinkage limit is the moisture at which a reduction in moisture will not cause a volume decrease in the soil mass, but at which an increase in the moisture will result in a corresponding increase in soil mass. The plasticity index is the difference between the liquid limit and the plastic limit.

4.6.5   Compaction Testing

The requirements of constructing excavations or placing controlled fills within a development to reduce settlement and other considerations necessitate adequate assessment of a

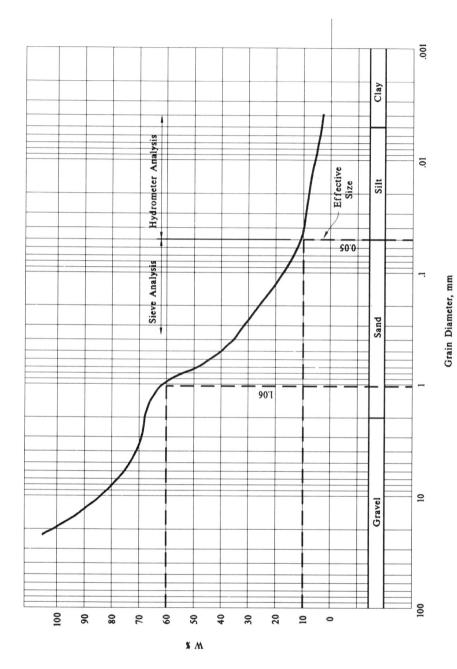

Figure 4.49 An example of a grain size distribution curve.

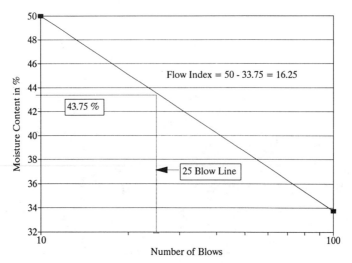

Figure 4.50   A sample graph for liquid limit determination.

soil's compaction capability. Granular soils usually rely on vibration, water, and rolling for best results. Soils containing more cohesive materials rely on pneumatic rubber-tired rollers for silty soils or sheeps foot rollers for plastic soils. Factors affecting a soil's capability of being densified include moisture content, compactive effort exerted, and soil types. Standard and modified Proctor testing will indicate the optimum moisture content (OMC) of a soil passing the No. 10 sieve using a given compactive effort. Using either designated test procedure, samples are compacted at various moisture contents with the resulting equivalent dry unit weights being plotted. The OMC then can be determined at the point of maximum dry unit weight, as shown in Figure 4.51.

Results of the compaction test provide the engineer with information for proper field control during construction.

4.6.6   Other Soil Tests

Although there are many additional tests available to analyze soils, four of them are applicable to land developments that do not include structures requiring deep foundations. These tests are:

(a) The field density test, which is used to determine a soil's density either in its natural state or in a fill section. The test provides an estimate of compaction strength or checks results of field compaction. This test typically is conducted using a nuclear density meter or a sand cone apparatus.

(b) The California Bearing Ratio (CBR) test can be conducted either in the field or in the laboratory. In laboratory tests the soil is compacted to anticipated field conditions and then soaked to duplicate field moisture. A surcharge weight is placed on top of the sample to simulate overburden pressures; then a 1.95 inch diameter piston is forced into the prepared soil sample at a rate of 0.05 inch per minute. Corresponding loads for 0.1, 0.2, 0.3, 0.4, and 0.5 inch penetrations are recorded and plotted. From this information a

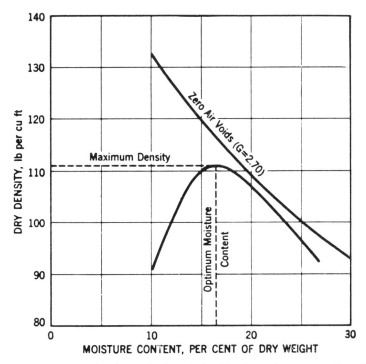

Figure 4.51 An example of moisture density relationships for a given compaction effort. Source: Paul H. Wright and Radnor J. Paquette. *Highway Engineering,* 5th ed. Copyright © 1987 by John Wiley & Sons, Inc. Reprinted by permission of John Wiley & Sons, Inc.

corrected load value representing the 0.1 inch penetration is obtained. The CBR value is then defined as:

$$\text{CBR}(\%) = \frac{\text{unit load for 0.1 inch penetration}(100)}{1000} \qquad (4\text{-}7)$$

Thus the relative shear strength of a soil can be compared to a standard material's strength, which in this case is crushed stone with a corresponding bearing strength of 1000 pounds at 0.1 inch penetration (Wright and Paquette, 1987).

(c) Resistance "*R*"-value of compacted soils can be determined in the laboratory using procedures contained in AASHTO T 190-90.* Four soil specimens 2.5 inches high by 4 inches in diameter are formed by compacting them at various moisture contents to anticipated field conditions using a mechanical kneading compactor that simulates the kneading action soil is subjected to during field compaction. Each sample is then subjected to a uniformly increasing pressure using a compression testing machine. The pressure required to cause exudation of water from the sample is the "exudation pressure." The moisture content used to prepare the specimens should yield exudation pressures that bracket a value of 300 psi. Each sample is allowed to rebound and then is placed in an expansion pressure device for 16 to 24 hours. This device measures the expansive force exerted by

*Taken from AASHTO T 190-90 by permission.

the swelling soil. The resistance "R"-value can then be obtained for each specimen by placing it in an Hveem stabilometer where lateral specimen deformation caused by vertical loadings results in compressive lateral pressures being developed. Materials can be rated from 0 to 100 based on their ability to limit lateral pressure from developing. For example, an R-value of 0 indicates a liquid. [The R-value used for pavement thickness design can be estimated from either the R-value that corresponds to an exudation pressure of 300 psi or the R-value based on the thickness of pavement required to maintain the soil's compaction as described in MS-10(1991)].

(d) Permeability testing can yield an indication of how quickly water is capable of flowing through the soil structure.

4.6.7 Additional Considerations

The subsurface site investigation should include an examination of soil conditions that will allow the determination of safe slope angles for stability, if applicable. Volume swelling and shrinkage factors also are required for earthwork mass diagram balances. Some sites contain expansive clays that could be detrimental to foundations and road subgrades, while other sites may have sinkholes. Other concerns might focus on the investigation of previously low areas where logs and stumps are covered with soil, which could cause problems. The potential for soil liquefaction resulting in mudslides must also be investigated, along with earthquake fault data.

4.7 TRAFFIC STUDIES

Most transportation systems are studied on a local or regional basis, as indicated in Chapter 1, consistent with federal and state planning. Included is a classification inventory of existing streets and highways into usage categories such as arterial, collector, or local streets. Existing traffic studies include traffic volume counts, travel time studies, street capacities, accident studies, parking studies, and traffic control device studies. Land use changes resulting from development cause a change in traffic activity in terms of trip generating and reduced levels of service on existing facilities. Figure 4.52 illustrates the interrelationship between land use and traffic activity. One of the most important goals of a designer is to minimize the degradation of service levels on transportation networks.

4.7.1 Traffic Associated with Development

Land developments include improvements that act as traffic generators affecting servicing traffic facilities. Multifamily developments containing commercial areas are of concern and may require the anticipated traffic demands to be evaluated to determine how the facilities will perform.

Proper evaluation techniques require the determination of existing and future traffic conditions, including existing peak hour levels of service within the affected area. A development's anticipated contribution would be included when analyzing the effect of traffic on the system's carrying capability. Should the existing system indicate inadequate future capacity, various improvement alternatives must be investigated to determine the best course of action.

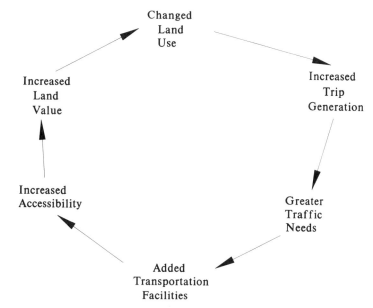

Figure 4.52 Land use transportation cycle. Source: Paul H. Wright and Norman J. Ashford. *Transportation Engineering, Planning, and Design*, 3rd ed. Copyright © 1989 by John Wiley & Sons, Inc. Reprinted by permission of John Wiley & Sons, Inc.

Some information required to undertake an analysis such as this includes:

(a) Proposed development site plan showing the proposed system and internal street layouts and the anticipated average daily traffic (ADT) volumes obtained from the size and type of development and trip generation factors.

(b) Location and characteristics of all traffic accesses to the proposed site including intersections used currently and those planned.

(c) An analysis of service levels using conditions given to various system inputs to accommodate future requirements.

4.7.2 Highway Capacity Studies

Highway capacity is defined as the maximum hourly rate at which vehicles reasonably can be expected to pass through a section of roadway during a given time period under prevailing roadway, traffic, and control conditions (TRB, 1985). Roadway conditions usually refer to street geometry, including alignment, geometric cross section, and geographic location, whereas traffic conditions refer to the mixture and characteristics of the flow stream. Control conditions refer to traffic regulatory devices. These conditions affect the level of service of a given roadway ranging from "A," which consists of free flow conditions, to "F," which denotes congestion. The service flow rate is the facility's capacity when a designated level of service is maintained in hourly flow measured at 15 minute intervals. Service flow rates that have no traffic signals or intersections operate as uninterrupted flows, whereas the presence of signals and intersections results in interrupted flows. Uninterrupted flows are affected by various roadway factors such as align-

Level Service	Characteristics of Traffic
A	Average travel speed of about 90 percent of free flow speed. Stopped delay at signalized intersections is minimal.
B	Average travel speeds drop due to intersection delay and intervehicular conflicts, but remain at 70 percent of free flow speed. Delay is not un-reasonable.
C	Stable operations. Longer queues at signals result in average travel speeds of about 50 percent of free flow speeds. Motorists will experience appreciable tension.
D	Approaching unstable flow. Average speeds down to 40 percent of free flow speed. Delays at intersections may become extensive.
E	Average travel speeds 33 percent of free flow speed. Unstable flow. Continuous backup on approaches to intersections.
F	Average travel speed between 25 and 33 percent of free flow speed. Vehicular backups, and high approach delays at signalized intersections.

Figure 4.53 Level-of-service characteristics for urban and suburban arteries. Taken from *A Policy on Geometric Design of Highways and Streets*. Copyright © 1990 by the American Association of State Highway and Transportation Officials. Reproduced with permission.

ment, grades, surface riding quality, shoulder extent, and lane widths. Uninterrupted flows also are affected by various traffic factors including the mixture and direction of trucks, buses, and passenger vehicular volumes, including driver characteristics. Urban streets similar to those associated with land developments would operate under interrupted flow conditions and would be affected by the surrounding land use. The number of available traffic lanes and their widths, street alignment, and the existence of turning lanes or parking characteristics affect the flow. Interrupted flow traffic factors could be dependent on traffic volumes and peaking factors, pedestrian interfaces, public transportation facilities, and parking activities. Control factors usually include signal timing (TRB, 1985).

A Policy on Geometric Design of Highways and Streets—1990 (AASHTO) provides characteristics associated with the five levels of service associated with several highway types, of which the urban and suburban arterial characteristics are shown in Figure 4.53.

During the preliminary stages of project planning, only an indication of potential traffic problems can be ascertained since the project's scope has not been fully established. Once the preliminary design has been refined, a definitive impact analysis of the traffic can be made as described in Chapter 6.

4.8 MATERIALS EVALUATION

On occasion, a development might include the use of novelty or specialty materials that may require testing and evaluation to insure adequate performance. Various materials testing firms are available to undertake such evaluations as auxiliary project consultants.

4.9 ANALYSIS AND REPORT ON PROSPECTIVE SITES

Background data pertaining to all possible sites must be gathered and assembled in a manner conducive to thorough analysis. Information from site inspections and other background data that could be in various stages of completion are useful to the design professional when generating a proposed site plan that satisfies the developer's needs and regulatory requirements.

4.9.1 Developing a Schematic Site Plan

The design professional starts by taking a piece of tracing paper and placing it over a scaled display of the site boundary that also contains the topography. Significant areas are delineated to indicate wetlands, drainage facilities, positions for improved views, areas for potential open space, and so on. The site sketch is useful to assist the designer in this endeavor, along with other information.

Once controlling factors have been indicated on this initial overlay, another tracing paper sheet is placed as before and the design professional sketches to approximate scale a layout as specified by the developer. There are probably many things that appear on the sketch that indicate a poor design. If so, another fresh piece of tracing paper must be substituted to allow sketching yet another configuration. In the resketching process insufficiencies may again develop, necessitating even yet another approach. This process requires skill and knowledge, and it may take 10 to 20 tries before a product of quality is obtained. Figure 4.54 demonstrates two plans for a particular site. A precise drawing can be made once a desirable solution is attained. During this planning process colored pencils are used to great advantage to indicate various utilities or areas of different land use. They add perception to the drawing and often, because of the resulting contrasting effect, flaws are noticed.

Utilization of CAD systems to generate schematic plans can parallel the manual approach just described. The major difference is that the tracing paper is replaced by an electronic drawing "layer."

The need for drawing these preliminary plans to scale is apparent when one counts the number of lots or dwelling units contained in a resulting configuration. If no scale were used, a piece of leftover land thought to be a great recreational area may not even exist because the actual lot sizes or road rights of way might result in no excess area. At true scale, such fallacies in the plan are revealed. Additional complexities resulting from interrelationships of roads to car storage to dwelling unit to recreational areas necessitate the use of specialists such as architects and land planners. In group developments or multifamily developments problems may arise concerning space relationships, solar lighting, and view. These are prime considerations a specialist must address. By doing so, the specialist will upgrade the project from mediocre to excellent.

Each potential site's conceptual plan will be assessed by the engineer regarding how well it meets assigned criteria. All beneficial features as well as potential problems are summarized in a comprehensive report to the developer that includes rough cost estimates. The developer can study the various alternatives to establish a site of choice. Once the project's site is chosen, the design engineer can finalize a preliminary development plan by improving the schematics that pertain to that site. This plan will be submitted to the jurisdictional regulatory review agencies in order to seek permission to proceed with the

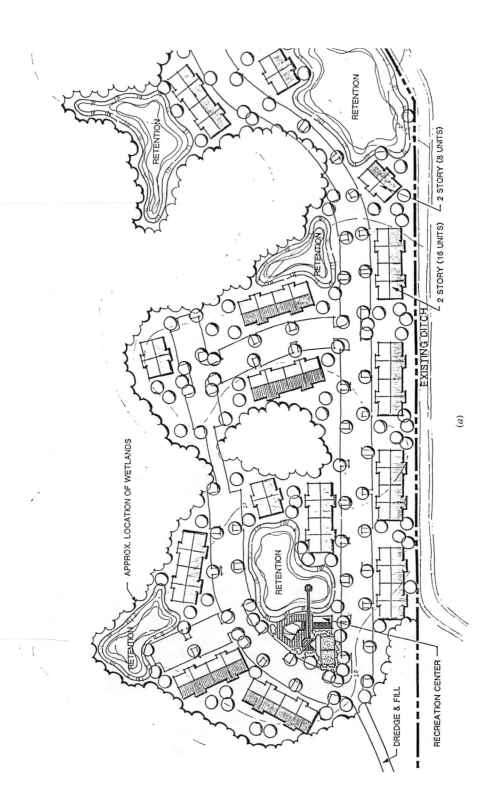

RETENTION

RETENTION

RETENTION

RETENTION

RETENTION

RETENTION

2 STORY (8 UNITS)

2 STORY (16 UNITS)

EXISTING DITCH

APPROX. LOCATION OF WETLANDS

DREDGE & FILL

RECREATION CENTER

(a)

158

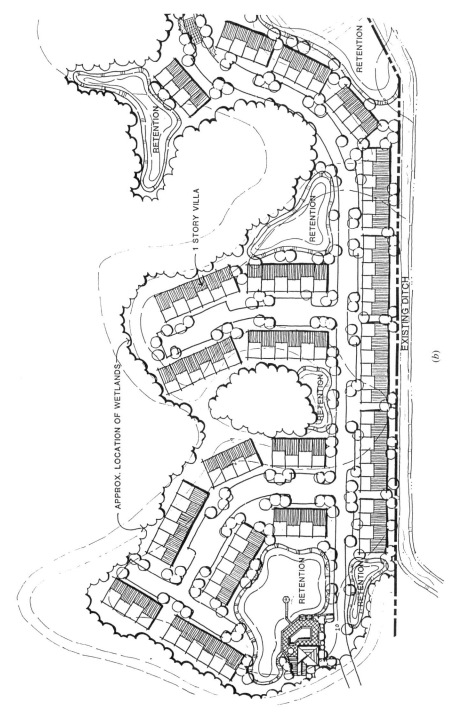

Figure 4.54 Examples of several site layout studies for a single-family development. (*a*) Condominium concept plan. (*b*) Villa concept plan. Courtesy of Post, Buckley, Schuh, and Jernigan, Inc., Tampa, Fla., 1992.

(*b*)

design of the project. The estimated construction cost is prepared from the preliminary plan and used as a basis for financial feasibility analysis by the developer.

REFERENCES

American Association of State Highway and Transportation Officials (AASHTO). *Standard Specifications for Transportation Materials and Methods of Sampling and Testing–AASHTO M 145-87, Recommended Practice for the Classification of Soils and Soil-Aggregate Mixtures for Highway Construction Purposes.* Washington: AASHTO, 1990, pp. 152, 154.

American Association of State Highway and Transportation Officials (AASHTO). *Standard Specifications for Transportation Materials and Methods of Sampling and Testing–AASHTO T 190-90, Standard Method of Test for Resistance R-Value and Pressure Expansion of Compacted Soils.* Washington: AASHTO, 1990, pp. 472–476. Copyright © 1990 by the American Association of State Highway and Transportation Officials. Referenced material reproduced by permission of the American Association of State Highway and Transportation Officials.

American Association of State Highway and Transportation Officials (AASHTO). *A Policy on Geometric Design of Highways and Streets—1990.* Washington: AASHTO, 1990, pp. 90–91.

American Land Title Association and American Congress on Surveying and Mapping (ALTA/ACSM). *Minimum Standard Detail Requirements for ALTA/ACSM Land Title Surveys.* An unpublished standard. Bethesda, Md.: ACSM, 1988.

Asphalt Institute, *Soils Manual* (MS-10) 5th ed. Lexington, KY.: Asphalt Institute. 1991.

Davis, Raymond E. and Francis S. Foote. *Surveying Theory and Practice,* 4th ed. New York: McGraw-Hill, 1962, pp. 198–203.

Davis, Raymond E., Francis S. Foote, James M. Anderson, and Edward M. Mikhail. *Surveying Theory and Practice,* 6th ed. New York: McGraw-Hill, 1981.

Dorchester County Tax Assessor. *Tax Map Showing a Portion of the Town of Summerville, S.C.* St. George, S.C.: Dorchester County Government, 1980.

Dunkle, F., Rebecca Hanmer, R. W. Page, and Wilson Scaling. *Federal Manual for Identifying and Delineating Jurisdictional Wetlands.* Washington: U.S. Government Printing Office, 1989, pp. 5, 23–25, B-2.

Falk, E. L., Planner in Charge. *Liberty Land Use Plan.* Unpublished planning report. Clemson, S.C.: Clemson University, 1972, p. 35.

Federal Emergency Management Agency (FEMA). *Flood Insurance Rate Map: 450073—Town of Summerville, South Carolina.* Washington: Federal Insurance Administration, 1980.

Graphic Mapping Group, Inc. *Utilimap.* An unpublished flier. 2001 Asheville Highway, Hendersonville, N.C.: Graphic Mapping Group.

Joint Publication of the U.S. Army, Navy, and Air Force. *Materials Testing—TM 5-530/NAVFAC MO-330/AFM 89-3.* Philadelphia: Dept of U.S. Navy, 1971.

Lambe, T. William. *Soil Testing for Engineers.* New York: Wiley, 1951.

Lambe, T. William and Robert V. Whitman. *Soil Mechanics.* New York: Wiley, 1969.

Lillesand, Thomas M. and Ralph W. Kiefer. *Remote Sensing and Image Interpretation.* New York: Wiley, 1979, p. 312. Copyright © 1979 by John Wiley and Sons, Inc. Reprinted by permission of John Wiley & Sons, Inc.

McCormac, Jack C. *Surveying Fundamentals,* 2nd ed. Englewood Cliffs, N.J.: Prentice-Hall, 1991.

Moffitt, Francis H. and Harry Bouchard. *Surveying,* 8th ed. New York: Harper and Row, 1987.

Moffitt, Francis H. and Edward M. Mikhail. *Photogrammetry,* 3rd ed. New York: Harper and Row, 1980.

Moskowitz, Joel S. *Environmental Liability and Real Property Transactions: Law and Practice.* New York: Wiley, 1989, pp. 313–330.

National Oceanic and Atmospheric Administration (NOAA). *State Plane Coordinate Systems of*

1983: NOAA Manual NOS NGS 5. Rockville: National Geodetic Information Center, 1990, pp. 1–7, 18–20, 46–53.

Seelye, Elwyn E. *Data Book for Civil Engineers—Design, Volume I.* New York: Wiley, 1960, pp. 9-25–9-27. Copyright © 1960 by John Wiley & Sons, Inc. Reprinted by permission of John Wiley & Sons,Inc.

SOKKIA Corporation. Overland Park, Kansas, 1992.

Spielberg, Frank. "The Traditional Neighborhood Development: How Will Traffic Engineers Respond?" *ITE Journal.* Washington: ITE, September 1989, pp. 17–18. Reference material is reprinted by permission of The Institute of Transportation Engineers.

Terzaghi, Karl and Ralph B. Peck. *Soil Mechanics in Engineering Practice,* 2nd ed. New York: Wiley, 1967.

Town of Summerville, S.C. "Official Zoning Map," 1992.

Transportation Research Board (TRB), National Research Council. *Highway Capacity Manual— Special Report 209.* Washington: TRB, 1985, pp. 1-1–1-14.

Trimble Navigation. Sunnyvale, California, 1992.

U.S. Army Corps of Engineers (COE) and South Carolina Coastal Council. *Developer's Handbook for Freshwater Wetlands.* An unpublished handbook. Charleston: South Carolina Coastal Council, 1991.

U.S. Department of the Interior. *National Wetlands Inventory Map (NWI)—Bennetts Point, S.C.* Atlanta: U.S. Fish and Wildlife Service, 1989.

U.S. Geological Survey. *7.5 Minute Series Map of Summerville, South Carolina.* Washington: U.S. Geological Survey, 1990.

Weldon, Curt and Messers Frank, Kolter, Dannemeyer, Bliley, Chapman, Lagomarsino, and James. *H.R. 2787—A Bill to Amend the Comprehensive Environmental Response, Compensation and Liability Act of 1980.* 101st Congress, 1989.

Wolf, Paul R. *Elements of Photogrammetry,* 2nd ed. New York: McGraw-Hill, 1983.

Wright, Paul H. and Norman J. Ashford. *Transportation Engineering, Planning, and Design,* 3rd ed. New York: Wiley, 1989, pp. 222, 280–286.

Wright, Paul H. and Radnor, J. Paquette. *Highway Engineering,* 5th ed. New York: Wiley, 1987, pp. 289, 433, 474–476. Copyright © 1987 by John Wiley & Sons, Inc. Reprinted by permission of John Wiley & Sons, Inc.

Zilkoski, David B. "North American Vertical Datum and International Great Lakes Datum: They Are Now One and the Same." The Hydrographic Society, Publication No. 28, U.S. Hydrographic Conference 1992. Baltimore: 1992, pp. 68–74.

Zilkoski, David B., John H. Richards, and Gary M. Young. "Results of the General Adjustment of the North American Vertical Datum of 1988," *Journal of the American Congress on Surveying and Mapping—Surveying and Land Information Systems.* Bethesda, Md.: September 1992.

5 Preliminary Design and Costs

Developers often question information in the report that summarizes various alternative conceptual plans and costs. This largely results from a lack of the details and complete information expected in final construction documents. As a result, the engineer can be expected to clarify and amplify information contained in the report, allowing the developer to analyze and consider alternatives. On occasion, preliminary inquiries made by the engineer to jurisdictional authorities, while seeking initial conceptual guidance may trigger follow-up agency clarification inquiries requiring the engineer to respond. Resulting information can affect various conceptual assumptions. The developer can choose the most appealing project conceptual plan after reviewing the report. Design professionals can then be directed by the developer to prepare all required preliminary project documents necessary for regulatory and financial institution approval.

5.1 DETERMINATION OF ADDITIONAL AUXILIARY SERVICE

During the report review, the developer and/or engineer may determine a need for additional background studies. The scope of these additional services must be defined in a manner that will result in the studies yielding adequate information. The summary report may need to be modified by the engineer as a result of these activities; nevertheless, the developer normally continues analysis until the best conceptual plan and/or site is determined.

5.2 CONCEPTUAL PLAN APPROVAL

Once the developer decides on the project's conceptual plan, a conference usually follows that culminates with the developer, engineer, land planner, architect, landscape architect, marketing specialists, financial analysts, and lenders reaching agreement on the project's preliminary design features and phasing requirements. This agreement usually is put in writing to eliminate possible misunderstandings, especially if variable conditions prevail that could alter the project's scope. If the project is small, the conceptual agreement can allow the developer to commence activities to procure title to the desired project site. However, if the project is over a certain threshold where a large portion of the locale has the potential of being affected by the proposed development, jurisdictional regulations might require the developer to supply a regional impact report for approval. This report will include information pertaining to the development such as maps, environmental factors, transportation factors, and factors pertaining to housing, education, recreation, health, police, fire, natural hazard mitigation, and preservation. Once this report is approved, the preparation of the preliminary project documents can begin.

5.3 PREPARATION OF PRELIMINARY DOCUMENTS

The engineer can now refine the applicable conceptual scheme and make additional preliminary studies or analyses necessary to formulate a preliminary development plan, or preliminary plat, suitable for regulatory authority reviews. Approval of the preliminary plan allows the developer to pursue preparation of construction plans through the action of the design professional.

5.3.1 Preliminary Development Plans

Depending on the site size, availability of financing, and market absorption potential, the approach used by developers can vary from project to project. For a small project a single preliminary plan may be sufficient for regulatory and financial review; however, on large projects that are phased developments, several plans may be required. These might include a comprehensive master development plan, neighborhood plans, or preliminary development plans for each phased section. Typical plans are illustrated in Figures 5.1 through 5.3. Supplemental plans might contain master development plans for storm drainage, wastewater, transportation, and potable water systems. These plans usually address major system components that will allow subsequent development phases. Many regulatory agencies want information indicating the interrelationship between the various proposed development improvements that are contained in the preliminary plan. Local planning agencies usually require this plan to show the property's metes and bounds along with all easements and rights of way. The property area and adjacent landowners are shown with information indicating how their boundary intersects the development. All proposed land divisions within the development must be drawn to scale to indicate new road rights of way, roadway widths, street names, locations and approximate sizes of proposed utilities, lot lines, dimensions, reserved areas, floodways, wetlands, and areas for recreational, religious, educational, or public use. Land use area summaries usually are included for land use density analysis. The plat should include a title, graphic and numeric scale, north arrow with reference meridian, date, and the names of the developer, engineer, and land surveyor.

The preliminary plan must delineate all existing physical features including contour lines referenced to an approved datum, drainage features, and wooded areas. The plan also should contain all proposed storm drainage facilities with the anticipated finished surface flows being delineated with directional arrows, in addition to flows in proposed ditches and retention/detention facilities.

5.3.2 Preliminary Cost Estimate

On completion of the preliminary plan, a more exact, but still preliminary, cost estimate can be prepared by the engineer. This estimate is intended to reasonably ascertain the cost of the development; however it is subject to several variables that make accuracy difficult. For example, larger projects can offer cost savings attributed to economy of scale resulting in the estimated construction costs being higher than actual costs. Also, the site's location and condition can affect actual costs of improvements, as can various seasonal restrictions placed on certain types of construction. The uncertainty of regulatory review comments could result in required plan revisions that might alter the cost of the project. The

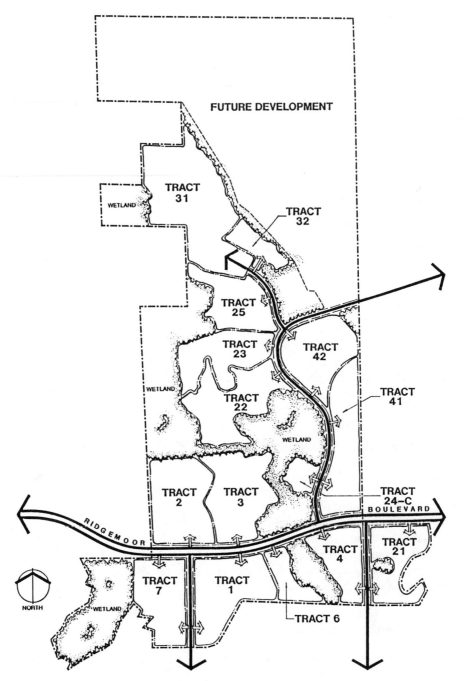

FUTURE DEVELOPMENT

TRACT
31

WETLAND

TRACT
32

TRACT
25

TRACT
23

TRACT
42

TRACT
22

WETLAND

WETLAND

TRACT
41

TRACT
2

TRACT
3

TRACT
24–C
BOULEVARD

RIDGEMOOR

TRACT
4

TRACT
21

NORTH

WETLAND

TRACT
7

TRACT
1

TRACT 6

Figure 5.1   An example of a project's conceptual master plan. Courtesy of Post, Buckley, Schuh, and Jernigan, Inc., Tampa, Fla., 1992.

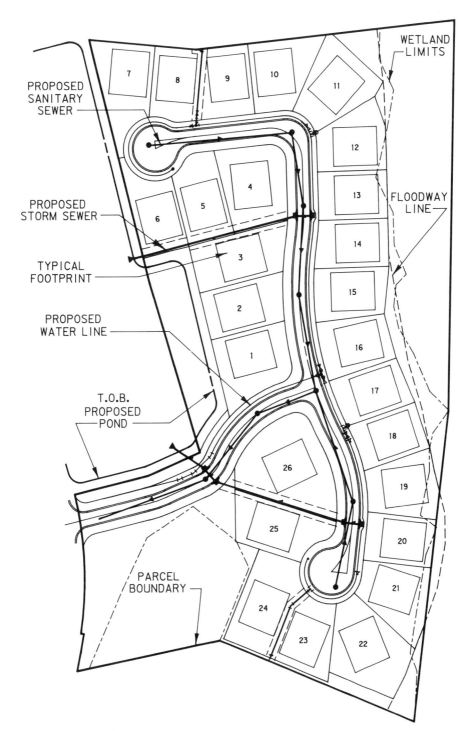

Figure 5.2 An example of a preliminary site development plan. Courtesy of Post, Buckley, Schuh, and Jernigan, Inc., Tampa, Fla., 1992.

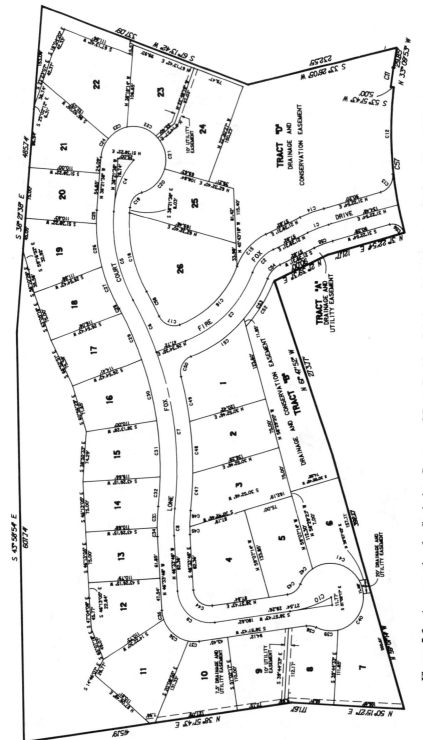

Figure 5.3 An example of a final plat. Courtesy of Post, Buckley, Schuh, and Jernigan, Inc., Tampa, Fla., 1992.

prevailing economic climate also can affect final construction costs resulting in large pricing fluctuations.

This estimate should specify any anticipated expenses not included as part of the estimate to alert the developer before they occur. Procedures for preparing a preliminary cost estimate parallel those contained in Chapter 14 for the final cost estimate.

5.3.3 Specification Outline

Preliminary activities include preparation of a specification outline for the proposed project. This outline should contain significant specification elements to be included in the final draft. Construction specifications are more fully covered in Chapter 13.

5.3.4 Project Synopsis

Most preliminary documents include a synopsis that describes various aspects of the proposed project. The synopsis is useful to the developer, regulatory review agencies, and financial lending institutions. It can include various components of the previously prepared summary report; however, more explicit detail is presented to focus on only one plan, as indicated by Figure 5.4.

5.4 PRELIMINARY PROJECT SUBMITTALS

Prior to releasing copies of the preliminary plan, construction specification outline, cost estimate, and proposed project synopsis, the engineer should double-check every document to insure completeness and to verify that no obvious errors exist.

5.4.1 Developer Approval

The first outside review of the preliminary documents is usually performed by the developer. This review is important to insure that the preliminary project appeals to the developer with regard to theme, estimated cost, and anticipated regulatory compliance. Any concerns expressed by the developer during this review are usually minor because previous communications have allowed prior concerns to be incorporated in these documents.

Once the developer has satisfactorily reviewed the documents, the engineer, acting for the developer, can petition all required regulatory authorities and request preliminary project approval. Application fees for petitions are paid by the developer.

5.4.2 Regulatory Agency Approval

Regulatory agencies that review development plans may include the local planning board, the health department, municipal engineer, building inspector, zoning officer, and jurisdictional utility companies. A typical procedure for local project approval, shown previously in Figure 2.27, varies from locale to locale; however, all regulatory review processes require time. Maintaining lines of communication is helpful to allow timely discussions for satisfactory reconciliation of problems.

PROJECT NAME

Preliminary Report
1. Property Identification
   (a) Current owner.
   (b) General tract identification (e.g. tract number, tax map number, etc.).
   (c) General location.
2. Background Information
   (a) Reduced copy of boundary plat.
   (b) Reduced copy of topographic map.
   (c) Description of the property.
   (d) Zoning classification.
   (e) Wetland findings.
   (f) Phase I Environmental Assessment results.
   (g) Historical and Natural Resource findings.
   (h) Potable water and wastewater factors.
   (i) Soil and geologic conditions.
   (j) Transportation factors.
   (k) Other factors:
       (1) Natural disasters (earthquakes, hurricanes, etc.).
       (2) Landslide and mudslide potential.
       (3) Noise.
       (4) Air pollution.
3. Proposed Development Strategy
   (a) Theme.
   (b) Project configuration and phasing.
   (c) Required improvements.
   (d) Estimated construction costs.
4. Project Summary
   (a) Potential problems.
   (b) Presubmission regulatory comments.
   (c) Estimated schedule for construction document preparation.
   (d) Suggestions.

Figure 5.4   An example of a synopsis of a land development project.

5.4.3   Financial Lending Agencies Reviews

Information in the preliminary documents is important to a financial lending institution's review process. The engineer usually provides copies of these documents to the institution while the developer makes all other arrangements. Revisions necessitated by regulatory agency comments must be reflected in an updated set of documents that would be redistributed to the lender.

5.5   PRELIMINARY PROJECT APPROVAL

Approval of the preliminary documents by regulatory agencies establishes a common understanding of the project's scope and direction, allowing the developer to secure financing. Preliminary approval enables jurisdictional regulatory authorities to anticipate improvements that are compatible with established systems. This approval also allows the engineer to move into the final project design phase when construction plans and specifications are prepared.

PART III
Final Design Related Activities

Transforming information on the preliminary plan into construction drawings is called design. Design is based on theory, technical knowledge, and practical experience. The ability to design is acquired by personal involvement and the willingness of an experienced designer to provide guidance. Many project designs can be formulated for a site; however, there are usually a limited number that best satisfy regulatory, financial, and project goal constraints.

The design process is iterative, with various elements being incorporated into a central project core, which then provides the framework for subsequent attachment of design details. Adjustments may be required to attain the proper relationship between the various designed components. This requires the designer to constantly check and reevaluate the overall project. At any stage during the project's design, other system alternatives that are subsequently designed may cause adverse impacts on previously designed components. This requires the designer either to make adjustments to the previously designed features or to modify the features currently being designed (or both). Refinements are therefore required as the process continues to completion.

Design activities also include preparation of construction specifications (Project Manual), a final cost estimate, and document review by regulatory agencies for construction approval.

6 Transportation System Design

The limiting feature of a land development is the site's property boundary. It is within this boundary that the designer must work to fabricate individual system components and appurtenances. If the boundary is not properly defined as described in Section 4.5, then the designed systems will not fit when the project is constructed, resulting in problems.

The common thread that weaves through the entire project is the transportation system. According to The Institute of Transportation Engineers (ITE) (1984), transportation considerations in land developments include the layout of streets and pedestrian systems (as related to land use) and also the engineering dimension for vehicular, pedestrian, and bicycle facilities. Both must be addressed during design. The importance of designing the transportation system first is that all building sites are contiguous to it, most utilities are located within it, and the storm drainage is usually affected by it. It is therefore necessary to initially establish horizontal and vertical control of the streets prior to undertaking the sequential design of other site improvements, including the storm drainage system, followed by the sanitary sewer system, and finally the potable water system.

6.1 TRAFFIC RELATED SITE IMPACT STUDIES AND ACCESS

Traffic related site impact studies undertaken by qualified persons can be conducted using guidelines similar to those included in ITE's *Traffic Access and Impact Studies for Site Development* (1989). These guidelines are presented in this section. (Other guidelines are available, including the U.S. Department of Transportation—Federal Highway Administration's *Site Impact Traffic Evaluation (S.I.T.E.) Handbook* (1985). Some reasons for undertaking studies such as this include safety, accessibility, and determination of street improvement needs. Increased trip generation and parking demands are additional reasons to conduct a study. Study results should help maintain satisfactory levels of transportation service; however, site related studies are not satisfactory as substitutes for area wide or regional transportation studies where many sites must be evaluated over long time periods. The overall purpose of a site study is therefore to determine what effect the proposed development will have on the transportation system in terms of site access, traffic circulation, and local transportation plans that exist to accommodate future growth. This includes determining the time when peak traffic flow is expected to occur both on and contiguous to the site, because peak site traffic can vary from that of the adjoining road. According to ITE guidelines, one reason a traffic impact study may be required occurs when the site's change in land use results in at least 100 additional peak direction trips to or from the site during the abutting street's peak hour or the proposed development's peak hour (ITE, 1989). Other reasons for conducting the study may be associated with applications for zoning changes, driveway encroachment permits, and building permits. Such accumulated traffic data are necessary not only for geometric design but also for structural design of the pavement. Safety, traffic flow, and structural integrity are basic objectives of a comprehensive traffic analysis program.

Section 4.7 indicates that a definitive traffic impact study can be undertaken once the preliminary plan is formulated. This is because the various on-site land uses are delineated. The traffic engineer should begin the actual study process, however, during the inventory, analysis, and reporting phase (Chapter 4) when other factors associated with the proposed development are being investigated. The scope of the traffic impact study should be established and agreed to by the jurisdictional regulatory review agency because not all studies require the same magnitude of detail. Initial study activities nevertheless should include the establishment of study area limits based on abutting streets, major intersections, and street segments up to the first signalized intersection in all directions from the site, if that distance is not unreasonable (possibly up to one half mile). Other areas can be included for study depending on actual study area conditions. Figure 6.1 illustrates a typical study area.

Evaluation of an appropriate project study horizon (time frame for analyzing traffic conditions) must also be undertaken. This horizon can consist of development phasing periods, or it can be synchronized with local planning horizons, area wide transportation planning horizons, or even with governmental capital improvements plans. The development shown in Figure 6.1 consists of three developmental phases.

The site impact study should include not only the effects of changing land use attributed to the study site, but also any effects attributed to other sites experiencing land use changes within the study area. In addition, the effects of through traffic growth (both short and long range) and proposed off-site traffic system improvements must be included. This

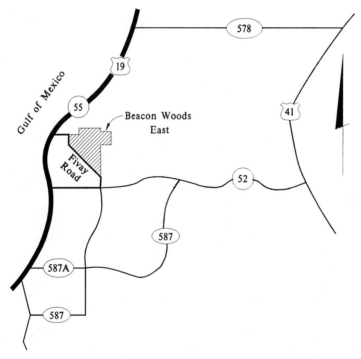

Figure 6.1 An example of a traffic study area for a typical land development. Courtesy of Post, Buckley, Schuh, and Jernigan, Inc., Tampa, Fla., 1992.

information can be requested from local regulatory authorities having jurisdiction over the system located in the study area. The engineer must conduct a field investigation of the study area and especially the proposed site to inventory traffic control devices, signal phasing and timing, roadway configurations, parking, street lighting, right of way widths, speed limits, areas involving accidents at high incident rates, travel lanes, driveway access close to the site, public transportation facilities, adjacent land uses, and turning movements at or near the proposed site. Field studies should be conducted using methods contained in ITE's *Manual of Traffic Engineering Studies* (1976).

6.1.1 Nonsite Traffic

Nonsite traffic can include traffic moving through the study area that originates and terminates outside of the study area. This is sometimes called "background" traffic. Nonsite traffic also includes traffic generated within the study area but not as a result of contributions made by the site under study. This can include present and future traffic expected in the study area during the established project study horizon that is independent of traffic generated by the study site. Figure 6.2 illustrates current daily volumes while Figure 6.3 shows initial peak hour volumes for this example.

According to ITE (1989) nonsite traffic can be estimated using either the build-up method, an existing transportation plan, the projection of historical trends, or possibly a combination of these methods.

The build-up method is usually used in areas experiencing moderate growth rates and

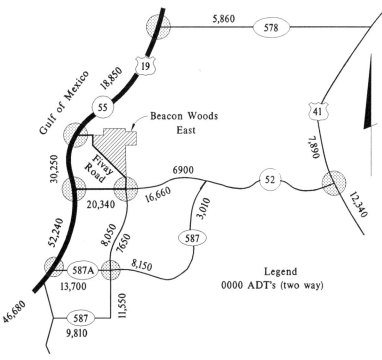

Figure 6.2 An example of daily traffic volumes for the sample traffic study area. Courtesy of Post, Buckley, Schuh, and Jernigan, Inc., Tampa, Fla., 1992.

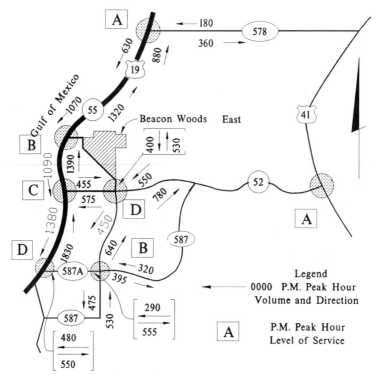

Figure 6.3   Existing P.M. peak hour traffic conditions for the sample traffic study area. Courtesy of Post, Buckley, Schuh, and Jernigan, Inc., Tampa, Fla., 1992.

in conjunction with a study site that has a study horizon of usually no more than 10 years. This method is used when several major projects are being developed over the same time period where traffic information exists on approved (but yet to be constructed developments) and other anticipated developments. The build-up method also includes investigating the impact of proposed traffic system improvements that are scheduled for implementation during the project study horizon. These factors, coupled with estimated new traffic volumes associated with other off-site land use changes, must be included.

The use of an existing transportation plan requires that the plan include the area under study. Most area wide transportation plans reflect average daily traffic (ADT): however, some include peak hour values. Use of this method often occurs in areas experiencing rapid growth where regional type project sites having phased construction are being studied. The local transportation plan must be adaptable to the project's study horizon. This method requires refinement and verification of results. Most area forecasts are focused on future traffic volumes with emphasis being placed on rights of way and roadway lane requirements. Care must be exercised to obtain the desired results relating to intersection service levels and the impact of site traffic.

The use of projected historical trends occurs when short study horizons exist. This method is used on smaller projects when locally collected data over at least a five year period are available. This method may not provide accurate results if the data are of poor quality. Average daily traffic is used to develop growth rates that can project nonsite traffic. This method, however, does not necessarily reflect localized changes or peak hour traffic trends.

Whichever method of determining nonsite traffic is chosen, it should be acceptable to the reviewing jurisdictional authority and should provide reasonable results. Projected nonsite daily traffic volume for Phase I is shown in Figure 6.4.

6.1.2 Site-Generated Traffic

Another important element of the traffic related site impact analysis is the estimation of the amount of site-generated traffic (new traffic). ITE defines a trip as a single or one directional movement that either originates or terminates within a designated area (ITE, 1989). In residential land developments the number of trips can be estimated from the number of included dwelling units or from other parameters. Estimates of trip generation can be obtained by using either trip generation references or results of local study data on comparable sites, or by conducting special field studies. Often many state and local transportation planning agencies have available trip generation data collected from within their jurisdictional area that can be used. The engineer must also determine what design level of traffic to use in the analysis, whether it is peak hour, peak season, or something else. Figure 6.5 illustrates typical peak traffic flow times for various land uses.

One technical reference, ITE's *Trip Generation* (1987), contains information obtained primarily from suburban sites. A reference such as this provides the engineer with a

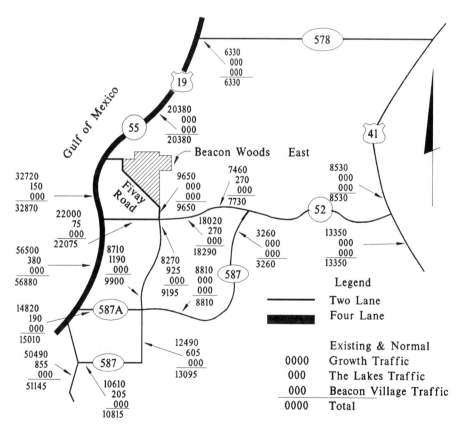

Figure 6.4   Projected nondevelopment daily traffic during Phase I for the sample traffic study area. Courtesy of Post, Buckley, Schuh, and Jernigan, Inc., Tampa, Fla., 1992.

Land Use	Typical Peak Hours[a]	Peak Direction
Residential	7:00–9:00 A.M. weekdays	Outbound
	4:00–6:00 P.M. weekdays	Inbound
Regional shopping	5:00–6:00 P.M. weekdays	Total[b]
	12:30–1:30 P.M. Saturdays	Inbound
	2:30–3:30 P.M. Saturdays	Outbound
Office	7:00–9:00 A.M. weekdays	Inbound
	4:00–6:00 P.M. weekdays	Outbound
Industrial	Varies with employee shift schedule	
Recreational	Varies with type of activity	

Notes:

[a]Hours may vary based on local conditions.

[b]Period of maximum weekday traffic impact.

Figure 6.5 Typical peak traffic flow hours for selected land uses. Taken from *A Recommended Practice—Traffic Access and Impact Studies for Site Development.* Copyright © 1989 by the Institute of Transportation Engineers. Reproduced with permission.

starting point to begin trip generation analysis; good judgment must also be used, since average rates, best fit equations, or a combination of these confronts the user. The best estimate is obtained by comparing average trip rates and the results from the equations, while adjusting the results to represent local conditions. An adjustment to this estimate of generated traffic may be required to account for pass-by trips, transit trips, internal site trips resulting from the presence of mixed land use on site, and other factors.

The study can include a summary table where different land uses within the proposed development are classified along with the chosen method of estimating trip generation for each classification and any associated adjustments.

6.1.3   Site Related Traffic Distribution and Assignment

Estimated new traffic generated by the development must be distributed and assigned along with any pass-by (captured) trips to the various study area roadways for analysis. This establishes the amount of traffic that is expected on each route and at intersections where system capacities, signalization times and needs, circulation patterns, parking needs, and safety can be analyzed. Traffic distribution is a function of the type of proposed development and its size. Variable land uses within the study area and population distributions are also factors affecting traffic distribution, as well as the condition of the roadway system. Before distributing the traffic, areas contributing to large percentages of the generated trips should be identified and their areas of influence should be delineated within the study boundary.

According to ITE (1989), two of the most common methods to establish influence areas are either to use a reasonable maximum travel time to the site or to delineate the boundary based on the proximity of competing developments. Once the influence area is defined, the distribution of trips can be determined by analogy, modeling, or by the use of other reference data.

The analogy method utilizes trip distributions that exist at similar developments near the study site. These data can include turning movements, origin-destination surveys, and driver response surveys.

A trip distribution model, such as the gravity model, can be used assuming that potential trips between two areas are proportional to the magnitude of each area and simultaneously inversely proportional to the separating distance or travel time.

Should the engineer elect to use other reference data, which can take the form of demographic, socioeconomic, or other related census survey data, trip distributions can also be estimated. Development sites connected with multiple land use areas can require traffic to be distributed and assigned several times.

Trip assignment is focused on the amount of traffic that will travel over specific routes within the study area's roadway network. Assigning trips is dependent upon street capacity, intersection capacity, and travel times. Commonly, traffic is assigned over parallel routes even though some routes require less travel time. Traffic assignment also includes pass-by trips. Pass-by trips result when traffic initially having trip destinations other than to the study site temporarily use the study site as an unplanned intermediate destination. Traffic generated from pass-by trips temporarily exits adjacent street traffic until being returned to continue travel on the adjacent street to the original destination. This results in additional trips being generated by the study site in addition to the new trips that originate with the study site being the intended destination.

Should anticipated pass-by trips be a significant factor, then the percent of pass-by trips must be estimated as a part of the total trips. This can be determined from driver response surveys. This will separate the traffic into pass-by trips and new trips. Both categories of trips can be distributed and assigned, with caution being applied to account for pass-by departure routes that may not coincide with the route used for entrance.

Traffic assignment reflects peak morning and afternoon volumes where two-way trips are usually used. Travel times are important and are included as a "function factor" in gravity model analysis. If a gravity model is not used, trip length frequency distribution curves obtained from local trends are needed for traffic assignment. Results from analyzing trip generation include the number of generated trips that are expected at each of the site's access locations. Assigned traffic volumes should be checked to insure that logical assignment has been made. Figure 6.6 illustrates projected P.M. peak hour traffic for the study in the example.

6.1.4 Impact Analysis and Alternatives

An analysis of system performance can be made after the traffic has been estimated, distributed, and assigned. This usually includes evaluation of performance levels for each major roadway and intersection. Figure 6.7 illustrates typical intersection turning movements. Other analysis considerations can include parking, pedestrian and bicycle activities, safety, and traffic control considerations. The results of this analysis should provide an indication of potential problem areas and possible ways to eliminate or mitigate them.

The most common method to evaluate transportation system performance uses the level of service concept (Section 4.7.2). Levels of service for signalized intersections are shown in Figure 6.8a and for unsignalized intersections in Figure 6.8b.

The engineer can use the *Highway Capacity Manual* (TRB, 1985), the intersection capacity analysis, the critical movement analysis, or other methods to evaluate the roadway segments. Safety analysis includes evaluating sight distances, areas experiencing high accident rates, and driveway turning speeds.

Providing access to a site requires the traffic engineer to consider providing access availability to adjoining properties, preserving roadway capacities, and maintaining safety

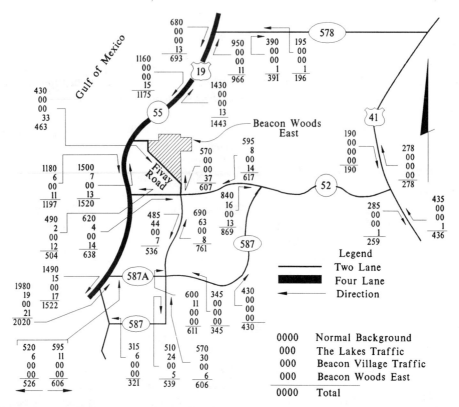

Figure 6.6   Projected P.M. peak hour traffic for Phase I for the sample traffic study area. Courtesy of Post, Buckley, Schuh, and Jernigan, Inc., Tampa, Fla., 1992.

and traffic flows. Most residential areas front on local streets (Section 2.5.5), while nonresidential developments usually access major streets that are contiguous to the development site. A limited number of driveway openings are usually provided to allow controlled and efficient ingress and egress.

The location and spacing of ingress-egress access depends on corner clearances, property clearances, median openings, traffic controls, sight distances, adjacent property access, traffic queues, traffic gaps, weaving distances, and internal traffic circulation patterns. Off-site improvements necessary to accommodate anticipated traffic demands include possible improvements to intersections, roadway segments, interchanges, and site access. If off-site improvements prove inadequate, then the scope of the proposed project may have to be altered.

Impact analysis also should include the effects of future traffic passing through roadway areas that have experienced high accident rates. Improved safety conditions should be considered as part of the overall improvement plan.

Should the impact analysis indicate that adequate levels of service are attainable, then no additional improvement need to be made. However, should potential traffic problems manifest during the analysis process, an assessment of various on-site and off-site improvements must be undertaken to identify cost effective means of providing adequate levels of service. This usually requires coordination with the jurisdictional regulatory reviewing authority.

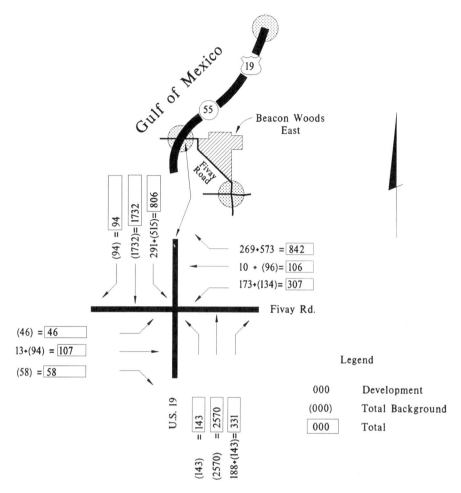

Figure 6.7   Typical intersection P.M. peak hour turning movements for the sample traffic study area—Phase III. Courtesy of Post, Buckley, Schuh, and Jernigan, Inc., Tampa, Fla., 1992.

The minimal acceptable level of service could be established at "D" during the design peak hour traffic. The alternatives can include making improvements adjacent to or close to the study site by improving site access for ingress-egress, parking, and traffic control. The proposed improvements can be scheduled for implementation either over a short or long period of time, depending on traffic growth and system needs.

6.1.5   On-Site Traffic Planning

Interior traffic patterns will affect ingress-egress access locations and their performance. According to ITE, accesses should be treated as intersections and designed accordingly, with the location being determined by:

(a) Allowing adequate distance between the abutting street and interior driveway intersections to provide space where the intersection or traffic lanes do not become blocked.

Level of Service	Average Stopped Delay Per Vehicle (seconds)	Qualitative Description
A	≤5.1	Good progression, few stops, and short cycle lengths
B	5.1–15.0	Good progress and/or short cycle lengths; more vehicle stops
C	15.1–25.0	Fair progression and/or longer cycle lengths, some cycle failures; significant portion of vehicles must stop
D	25.1–40.0	Congestion becomes noticeable; high volume-to-capacity ratio, longer delays, noticeable cycle failures
E	40.1–60.0	At or beyond limit of acceptable delay; poor progression, long cycles, high volumes, long queues
F	>60.0	Unacceptable to drivers. Arrival volumes greater than discharge capacity; long cycle lengths, unstable and unpredictable flows

(*a*)

Level of Service	Reserve Capacity (pcph)[a]	Impact on Minor Street Traffic
A	≥400	Little or no delay
B	300–399	Short traffic delays
C	200–299	Average traffic delays
D	100–199	Long traffic delays
E	0–99	Very long traffic delays
F		When demand exceeds the capacity of the lane, extreme delays will be encountered with queuing, which may cause severe congestion affecting other traffic movements in the intersection; this condition usually warrants improvement to the intersection

[a]Passenger cars per hour.

(*b*)

Figure 6.8 Levels of service criteria for intersections. (*a*) Signalized intersections. (*b*) Unsignalized intersections. Source: Transportation Research Board, National Research Council, Washington, D.C. *Transportation Research Board—Highway Capacity Manual—Special Report 209,* 1985. Reprinted with permission. Taken from *A Recommended Practice—Traffic Access and Impact Studies for Site Development.* Copyright © 1989 by the Institute of Transportation Engineers. Reprinted with permission.

    (b) Providing an adequate number of lanes and associated traffic controls to reduce conflicts in traffic flow patterns.

    (c) Providing facilities to accommodate handicapped persons and public transit users.

Interior traffic movements should be accomplished safely and efficiently. This includes the movements of automobiles, service and emergency vehicles, pedestrians, and bicy-

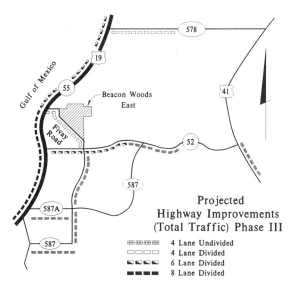

Figure 6.9.   Projected highway improvements for the sample study area. Courtesy of Post, Buckley, Schuh, and Jernigan, Inc., Tampa, Fla., 1992.

clists. Parking areas provided should be as convenient as possible to the intended point of destination, while including safety lanes for emergency vehicles, adequate space for vehicle maneuvering, and adequate space for passengers to enter and exit parked vehicles without damaging adjacent property.

6.1.6   Summary Traffic Impact Report

Usually the findings of the traffic related site impact studies and access are placed in a final report. This report can include a description of the study areas, study site, and existing conditions. Future conditions are also described along with the system's analysis. Findings and recommendations for various highway improvements resulting from the study are presented as indicated in Figure 6.9. A favorable review of this report allows the engineer to include applicable improvements in the project's construction drawings. Other improvements can be scheduled for future implementation in accordance with proposed phasing schedules.

6.2   STREET AND ROAD DESIGN CRITERIA

In designing a project's street system, the design professional faces the same constraints as the developer, the constraints described in Section 2.5. In addition, aspects other than just providing site access and conveying traffic must also be included. This is because residential streets provide movement for pedestrians, bicycles, and areas for neighborhood children to play, and at the same time give the development continuity and neighborhood cohesiveness. Streets associated with land developments are primarily classified as local and collector streets, while arterial types are not usually included.

According to AASHTO (1990),* the urban principal arterial system accommodates most trips entering and leaving the area while serving major centers of activity and accommodates the longest trip desired. A minor arterial system interconnects and augments the principal arterial system by accommodating trips of moderate lengths. The urban collector system can go through residential areas to distribute trips from arterials, and collect traffic from urban local streets and channel it to the arterial system. Urban local streets permit access to individual parcels of land (land service) and to higher order street systems.

The design of local urban streets is governed more by practical limitations than that of local rural streets. Dominant design controls include the type and extent of urban development and zoning (or regulatory) restrictions. In residential areas, many of the streets are "land service streets" where the primary consideration is to promote a safe and pleasant environment while convenience of the motorist is secondary. The other streets in residential areas can provide some "land service," while "traffic service" dominates.

Local urban land service streets in residential areas do not usually depend on traffic volumes as a criterion for design. Usually design speeds are low (ranging from 20 to 30 miles per hour) for local urban streets in residential areas, and passing sight distances are not usually a consideration because of the street's character and function. Because low speeds are promoted in residential areas, superelevation is not usually provided on local urban streets.

A primary concern is to provide the motor vehicle operator with adequate sight distance to allow safe vehicle stops under inclement conditions regardless of the street classification. On a level roadway, the estimated braking distance can be calculated using:

$$d = \frac{V^2}{30f} \tag{6-1}$$

where:

d = The braking distance in feet.
V = The initial vehicle speed in miles per hour.
f = The coefficient of friction between the tires and the roadway.

If the road is not level, the braking distance must account for the road's gradient by:

$$d = \frac{V^2}{30(f \pm G)} \tag{6-2}$$

where:

G = The road's slope in feet per foot.

Braking distances adjusted for uphill slopes are shorter than those required when braking downhill.

Stopping distances include braking distance and the distance traveled while the operator reacts to commence application of the brakes. AASHTO (1990) proposes that a

---

*Taken from AASHTO (1990), with permission.

reasonable reaction time for the driver is 2.5 seconds. The distance traveled during this time to perceive and react is:

$$d_r = V \times 2.5 \times 1.47 \qquad (6\text{-}3)$$

where:

d_r = The distance traveled in feet while the driver is perceiving and reacting.
V = The vehicle speed during the reaction time in miles per hour.
2.5 = The recommended reaction time in seconds.
1.47 = A constant to convert miles per hour to feet per second.

The sum of the braking distance plus d_r yields the required safe stopping distance. Figure 6.10 shows various recommended stopping sight distances for various design speeds on level wet pavements. Sight distances required for safe stopping along horizontal curves can be determined using the geometric relationships contained in Figure 6.11.

Most streets are designed to meet standards that will allow their dedication to the jurisdictional governing body once the streets are installed and approved. Streets that are not dedicated will remain private for the developer or homeowners' association to own, operate, maintain, and control. Regardless of the street's final disposition, it must be safe, serviceable (including provisions for the operation of refuse and fire trucks, moving vans, and snow removal vehicles), economical, and attractive. The design professional must consider criteria set forth in this section while formulating the preliminary development plan (Chapter 5) and during the design process discussed in Section 6.4.

6.2.1 Road Cross Sections

Local streets in residential areas can permit parking along one or both sides of the street, resulting in possibly one usable lane for passage. Collector roads, however, usually provide two or more active traffic lanes to function properly. Moving lanes should be designed for a minimum width of 10 feet (12 feet if cars will be allowed to park along both sides), while parking lanes should not be less than eight feet wide.

The typical road cross section includes all features contained within the street right of way. It is within this strip of property that not only the driving and parking lanes are located, but also utilities, storm drainage facilities, sidewalks, bicycle facilities, and road

Design Speed (mph)	Assumed Speed for Condition (mph)	Brake Reaction Time (sec)	Brake Reaction Distance (ft)	Coefficient of Friction f	Braking Distance on Level (ft)	Stopping Sight Distance Computed (ft)	Stopping Sight Distance Rounded for Design (ft)
20	20-20	2.5	73.3-73.3	0.40	33.3-33.3	106.7-106.7	125-125
25	24-25	2.5	88.0-91.7	0.38	50.5-54.8	138.5-146.5	150-150
30	28-30	2.5	102.7-110.0	0.35	74.7-85.7	177.3-195.7	200-200
35	32-35	2.5	117.3-128.3	0.34	100.4-120.1	217.7-248.4	225-250
40	36-40	2.5	132.0-146.7	0.32	135.0-166.7	267.0-313.3	275-325

Figure 6.10 Stopping sight distances for wet pavements. Taken from *A Policy on Geometric Design of Highways and Streets.* Copyright © 1990 by the American Association of State Highway and Transportation Officials. Reproduced with permission.

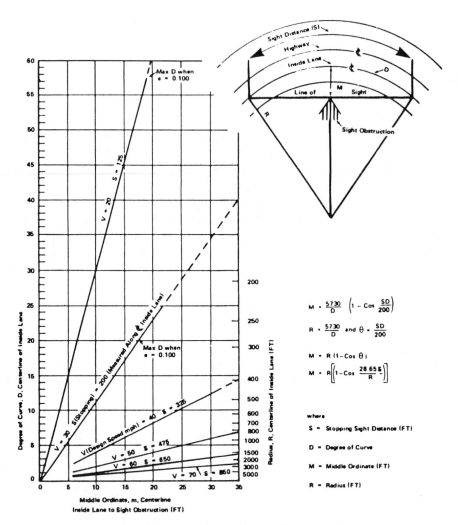

Figure 6.11 Range of upper values—relation between degree of curve and value of middle ordinate necessary to provide stopping sight distance on horizontal curves under open conditions. Taken from *A Policy on Geometric Design of Highways and Streets.* Copyright © 1990 by the American Association of State Highway and Transportation Officials. Reproduced with permission.

shoulders. Most modern developments at medium or higher densities include curb and gutter road cross sections in the design of local streets similar to those shown in Figure 6.12c. The minimum curb radius at intersections is usually 20 feet on local streets and 30 feet on collector streets. In the case of curb and gutter road sections, care must be exercised during design to incorporate the section in a "cut" environment, or a place where excavation is required, to allow positive drainage from the contiguous property. If a fill section is encountered, special provisions must be incorporated to remove otherwise trapped runoff that accumulates in abutting areas having elevations lower than the curb top.

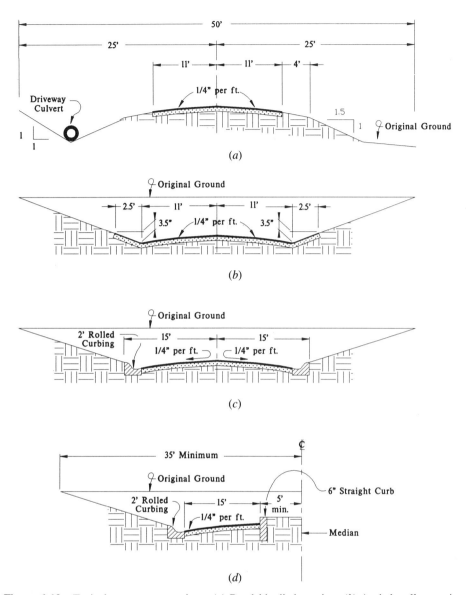

Figure 6.12 Typical street cross sections. (*a*) Roadside ditch section. (*b*) Asphalt valley section. (*c*) Rolled curb and gutter section. (*d*) Typical divided entrance.

Roadside ditch sections (Figure 6.12*a*) have been used in both cut and fill environments; however, this type of section may not be as attractive or as easy to maintain as a curb and gutter section when used with local streets. In some areas, however, a more "rural" look of no curbs may be preferred, or the design engineer may want to provide an "open" drainage system using roadside ditches to promote storm water runoff infiltration on site.

6.2.2 Local Street Design Criteria

The design of local streets located in urban settings is usually predicated by two factors. One factor is to allow street access while promoting safety, and a pleasing effect. The other factor is to promote traffic service. Some local streets primarily serve the first factor while others serve the latter. Typical guidelines for local street design contained in ITE's *Proposed Revisions to a Recommended Practice—Guidelines for Residential Subdivision Street Design* (1990) are shown in Figure 6.13.

The magnitude of traffic volumes on local streets usually does not require a capacity analysis since adequate capacity is provided by using a two lane street cross section. Other cross sections, such as divided roadway entrances, can be utilized if required by local regulations or desired by the developer for aesthetic reasons. Some types of developments, however, generate enough traffic volume to require a capacity analysis.

For land developments located in rural settings, minimum design speeds for local roads and streets that are not strictly for land service are affected by various traffic volumes as shown in Figure 6.14a. Regardless of the volume of anticipated traffic, the design engineer must provide adequate sight distance for safety. While Figure 6.10 provides stopping sight distances on level pavements under wet conditions, minimum stopping sight distances for wet pavement conditions for vertical curves are included in Figure 6.14b.

To determine the minimum length of vertical curve to allow safe stopping, the algebraic slope difference of the road on either side of the point of vertical intersection is multiplied by the tabulated K value as shown in Figure 6.14b. Current sight distance criteria used by AASHTO (1990) for stopping utilize a height of driver's eye of 3.5 feet above the pavement and a roadway object of 0.5 foot high. The minimum lengths for a crest vertical curve area:

Terrain Classification	Level			Rolling			Hilly		
Development Density	Low	Medium	High	Low	Medium	High	Low	Medium	High
Right-of Way Width (feet)	50	60	60	50	60	60	50	60	60
Pavement Width (feet)	22-28	28-34	36	22-28	28-34	36	28	28-34	36
Type of Curb V = vertical face R = roll-type O = none	O/R	V	V	V	V	V	V	V	V
Sidewalks and Bicycle Paths (feet)	0	4-6	5-6	0	4-6	5-6	0	4-6	5-6
Sidewalk Distance from Curb Face (feet)	----	5-6	5-6	----	5-6	5-6	----	5-6	5-6
Minimum Sight Distance (feet)	200	200	200	150	150	150	125	125	125
Maximum Grade	4%	4%	4%	8%	8%	8%	15%	15%	15%
Maximum Cul-de-Sac Length(feet)	1500	1000	700	1500	1000	700	1500	1000	700
Minimum Cul-de-Sac Radius (right-of-way)(feet)	60	60	60	60	60	60	60	60	60
Design Speed (mph)	30	30	30	25	25	25	20	20	20
Minimum Centerline Radius of Curves (feet)	300	300	300	175	175	175	110	110	110
Minimum Tangent Between Reverse Curves (feet)	50	50	50	50	50	50	50	50	50

Figure 6.13 Proposed local street design guidelines. Taken from *Proposed Revisions to a Recommended Practice—Guidelines for Residential Subdivision Street Design*. Copyright © 1990 by the Institute of Transportation Engineers. Reproduced with permission.

Speeds (mph) For Design Volumes Of						
Type of Terrain	Current ADT Under 50	Current ADT 50-250	Current ADT 250-400	Current ADT 400 and Over	DHV 100-200	DHV 200-400	DHV 400 and Over
Level	30	30	40	50	50	50	50
Rolling	20	30	30	40	40	40	40
Mountainous	20	20	20	30	30	30	30

(a)

Design Speed (mph)	Assumed Speed for Condition (mph)	Stopping Sight Distance (rounded for Design)(ft)	K Value For Crest Vertical Curves (rounded)	K Value For Sag Vertical Curves (rounded)
20	20-20	125-125	10-10	20-20
25	24-25	150-150	20-20	30-30
30	28-30	200-200	30-30	40-40
35	32-35	225-250	40-50	50-50
40	36-40	275-325	60-80	60-70

Note: K value is a coefficient by which the algebraic difference in the grade may be multiplied to determine the length in feet of the vertical curve which will provide minimum sight distance.

(b)

Figure 6.14 Street design standards. (a) Minimum design for local rural roads. (b) Minimum speeds stopping sight distances for local roads and collectors—wet pavements. Design hourly volume (DHV) can be the 30th highest hourly volume for peak-hour traffic determination. Taken from *A Policy on Geometric Design of Highways and Streets*. Copyright © 1990 by the American Association of State Highway and Transportation Officials. Reproduced with permission.

$$L_{\text{min.}} = 2S - \left[\frac{1329}{A}\right] \qquad \text{for } S > L \qquad (6\text{-}4)$$

$$L_{\text{min.}} = \frac{A\,(S)^2}{1329} \qquad \text{for } S < L \qquad (6\text{-}5)$$

where:

L = Length of vertical curve in feet.
S = Sight distance in feet.
A = Algebraic difference of tangent grades in percent.

For sag vertical curves, the minimum length can be obtained where the sight distance is governed by night driving conditions. If the headlight height is two feet and a one degree upward divergence of the light beam from the vehicle's longitudinal axis is used as criteria, the following equations for sag vertical curves are obtained:

$$L_{\text{min.}} = 2S - \left[\frac{400 + 3.5S}{A}\right] \qquad \text{for } S > L \qquad (6\text{-}6)$$

$$L_{min.} = \frac{A\,(S)^2}{400 + 3.5S} \qquad \text{for } S < L \qquad (6\text{-}7)$$

where:

L = Length of vertical curve in feet.
S = Light beam distance in feet.
A = Algebraic difference of tangent grades in percent.

For example, consider a design speed of 40 miles per hour for a symmetrical vertical curve, between a +1.5 percent slope and a −3.5 percent slope. The minimum downhill braking distance using an average running speed of 36 miles per hour (see Figure 6.10) and the −3.5 percent slope in conjunction with Equation 6-2 is:

$$\text{braking distance} = \frac{36^2}{30(0.31 - 0.035)} = 157 \text{ feet}$$

Although motor vehicle passing conditions do not usually prevail on local residential streets, certain situations may allow safe vehicle passing to occur. Typical minimum passing sight distances for safe maneuvering are shown in Figure 6.15a, along with K factors that produce the minimum length of vertical crest curves to allow safe passing.

Design considerations for establishing vertical control of local streets include allowable centerline grades. Local regulations usually prescribe these limiting gradients; however, typical maximum values are shown in Figure 6.15b for local rural roads, along with typical values for pavement cross slopes in Figure 6.15c. The minimum centerline slope is usually not less than 0.3 percent.

Horizontal alignment considerations include establishing minimum centerline radius values for design based on design speeds and the amount of superelevation, as shown in Figure 6.16.

Superelevation incorporates the banking of the riding surface to counteract the centrifugal force that otherwise would be counteracted only by side frictional forces produced by tire contact on the pavement. In land developments, most local streets promote low vehicle speeds and therefore are not superelevated. In some cases, however, superelevation is desired. Typical minimum values for superelevation runoff are also shown in Figure 6.17. The limiting rate of 0.08 foot per foot of road pavement width is recommended where ice and snow occur. Superelevation runoff is the horizontal distance required to restore the road cross section from its normal crown into a fully superelevated riding surface. Although there is no exact method to place superelevation runoff when transition curves (spiral) are not included, the usual procedure is to place one third of the runoff in the curve proper and two thirds outside of the curve along the tangent beginning from the point of curvature (PC) or the point of tangency (PT). A common practice in land developments is to rotate the cross section about the centerline of the street.

6.2.3 Collector Roads and Streets

Collector roads and streets in urban areas serve two functions. They collect traffic between local roads and arterial roads, and they provide access to contiguous property. These

Design Speed (mph)	Passing Sight Distance (Rounded for Design)(mph)	K Value For Crest Vertical Curves (Rounded)
20	800	210
25	950	300
30	1100	400
35	1300	550
40	1500	730

Note: K value is a coefficent by which the algebraic difference in grade may be multiplied to determine the length in feet of the vertical curve which will provide minimum sight distance.

(a)

Type of Terrain	Design Speed (mph)				
	20	30	40	50	60
Level	---	7	7	6	5
Rolling	11	10	9	8	6
Mountainous	16	14	12	10	---

(b)

Surface Type	Range In Rate Of Cross Slope (%)
High	1.5 to 2.0
Intermediate	1.5 to 3.0
Low	2.0 to 6.0

(c)

Figure 6.15 Street design standards. (a) Minimum passing sight distance for local rural roads and collector streets (rural and urban). (b) Maximum grades for local rural roads. (c) Normal cross slopes for local rural roads and local urban streets. Taken from *A Policy on Geometric Design of Highways and Streets.* Copyright © 1990 by the American Association of State Highway and Transportation Officials. Reproduced with permission.

streets accommodate higher traffic volumes than local residential streets, and therefore more emphasis is placed on traffic conveyance.

Typical collector street design guidelines taken from ITE's *Proposed Revisions to a Recommended Practice–Guidelines for Residential Subdivision Street Design* are shown in Figure 6.18.

Some design criteria for collector streets differ from local streets. Figure 6.19a shows minimum design speeds for various collector street traffic volumes.

The minimum stopping sight distance for collector roads is shown in Figure 6.14b and the minimum passing sight distance is contained in Figure 6.15a.

The maximum recommended roadway gradients for collector roads are shown in Figure 6.19b.

Required collector street roadway widths vary for various traffic volumes. Recommended widths are shown in Figure 6.20.

Design Speed (mph)	Maximum e	Maximum Degree of Curve Rounded	Minimum Radius (ft)
20	0.04	45.00	127
30	0.04	19.00	302
40	0.04	10.00	573
20	0.06	49.25	116
30	0.06	21.00	273
40	0.06	11.25	509
20	0.08	53.50	107
30	0.08	22.75	252
40	0.08	12.25	468
20	0.10	58.00	99
30	0.10	24.75	231
40	0.10	13.25	432
20	0.12	62.00	92
30	0.12	26.75	214
40	0.12	14.50	395

Note: With design speeds of 30 mph or less, conditions
may warrant elimination of superelevation.

Figure 6.16   Maximum degree of curve and minimum radius for different values of maximum superelevation for local roads and streets. Taken from *A Policy on Geometric Design of Highways and Streets*. Copyright © 1990 by the American Association of State Highway and Transportation Officials. Reproduced with permission.

Superelevation Rate, e	Length Of Runoff, L (ft) For Design Speed (mph)		
	20	30	40
Reverse Crown	50	100	125
0.04	60	100	125
0.06	95	110	125
0.08	125	145	170
0.10	160	180	210
0.12	190	215	250

Note: Length of runoff on 10- and 11-ft lanes may
be reduced proportionately but no shorter
than the minimum length shown for reverse
crown at the respective speed.

Figure 6.17.   Minimum length for superelevation runoff for two 12 foot lanes of pavement on local streets. Taken from *A Policy on Geometric Design of Highways and Streets*. Copyright © 1990 by the American Association of State Highway and Transportation Officials. Reproduced with permission.

Terrain Classification	Level			Rolling			Hilly		
Development Density	Low	Medium	High	Low	Medium	High	Low	Medium	High
Right-of Way Width (feet	70	70	80	70	70	80	70	70	80
Pavement Width (feet)	24-36	24-36	40	24-36	24-36	40	24-36	24-36	40
Type of Curb, Vertical Face (VF)	VF	VF	VF	VF	VF	VF	VF	VF	VF
Sidewalk Width (feet)	4-6	4-6	4-6	4-6	4-6	4-6	4-6	4-6	4-6
Sidewalk Distance from									
Curb Face (feet)	10	10	10	10	10	10	10	10	10
Minimum Sight Distance (feet)	250	250	250	200	200	200	150	150	150
Maximum Grade	4%	4%	4%	8%	8%	8%	12%	12%	12%
Minimum Spacing Along Major									
Traffic Route (feet)	1300	1300	1300	1300	1300	1300	1300	1300	1300
Design Speed (mph)	35	35	35	30	30	30	25	25	25
Minimum Centerline Radius									
of Curves (feet), w/ superelev.	350	350	350	250	250	250	150	150	150
Minimum Tangent Between									
Reverse Curves (feet)	100	100	100	100	100	100	100	100	100

Figure 6.18. Proposed collector street design guidelines. Taken from *Proposed Revisions to a Recommended Practice—Guidelines for Residential Subdivision Street Design*. Copyright © 1990 by the Institute of Transportation Engineers. Reproduced with permission.

6.2.4 Intersection Design

Typical intersection design guidelines for residential areas included in ITE's *Proposed Revisions to a Recommended Practice—Guidelines for Residential Subdivision Street Design* (1990) are shown in Figure 6.21.

Figure 6.22*a* and Figure 6.22*b* show examples of site distances necessary for safety.

6.2.5 Driveways

Driveways provide vehicular access to street riding surfaces, as well as allowing vehicular departure from the riding surface. Driveways must provide for certain rates of travel when exiting the street's traffic pattern and also for smooth ingress-egress. Typical guidelines for driveway design are illustrated in Figure 6.23. Typical residential driveway details are shown in Figure 6.24. Proper clearance must be maintained between the vehicle's underside and the pavement, especially where driveway slopes are steep. Provisions must be made to keep storm runoff from being encumbered by the driveway. This includes not blocking the throat of the curb, not restraining a roadside ditch's capacity with fill, and installing any driveway culvert properly.

6.3 PARKING CONSIDERATIONS

Most residential developments provide for off street parking where residential owners can place their vehicles out of the street right of way. Often in single-family detached developments, the parking area is the driveway. In multifamily developments designated parking areas are normally provided. These areas must be convenient to the user, especially when considering the need to manually transport purchased goods from the parked vehicle to the dwelling.

Type of Terrain	Current ADT 0-400	Current ADT Over 400	DHV 100-200	DHV 200-400	DHV Over 400
Level	40	50	50	60	60
Rolling	30	40	40	50	50
Mountainous	20	30	30	40	40

(a)

Rural Collectors Design Speeds (mph)			
Type of Terrain	20	30	40
Grades (Percent)			
Level	7	7	7
Rolling	10	9	8
Mountainous	12	10	10

Urban Collectors Design Speeds (mph)			
Type of Terrain	20	30	40
Grades (Percent)			
Level	9	9	9
Rolling	12	11	10
Mountainous	14	12	12

Note: Maximum grades shown for rural and urban
conditions of short lengths, (less than
500 ft), on one-way down grades and on
low-volume rural collectors may be 2%
steeper.

(b)

Figure 6.19   Collector street design standards. (a) Minimum design speeds (miles per hour) for various traffic volumes. (b) Maximum grades. Taken from *A Policy on Geometric Design of Highways and Streets*. Copyright © 1990 by the American Association of State Highway and Transportation Officials. Reproduced with permission.

Width (ft) For Design Traffic Volumes Of:					
Design Speed (mph)	Current ADT Under 400	Current ADT 400 and Over	DHV 100-200	DHV 200-400	DHV Over 400
20	20	20	20	22	24
30	20	20	20	22	24
40	20	22	22	22	24
Width Of Grade Shoulder - Each Side Of Pavement					
All Speeds *	2	4	6	8	8

* Note: Minimum width is 4 ft. if roadside barrier is utilized.

Figure 6.20   Minimum width of traveled way and graded shoulder for rural collector streets. Taken from *A Policy on Geometric Design of Highways and Streets*. Copyright © 1990 by the American Association of State Highway and Transportation Officials. Reproduced with permission.

Terrain Classification	Level	Rolling	Hilly
Development Density	All	All	All
Approach Speed (MPH)	25	25	20
Clear Sight Distance (length along each approach leg)(feet)*	90	90	70
Vertical Alignment Within Intersection Area	flat	2%	4%
Minimum Angle of Intersection in Deg. (Rt. Angle Preferred)	70	70	70
Minimum Curb Radius (feet):			
a. Local-local	20	20	20
b. Local-collector	25	25	25
c. Collector-major	30	30	30
Minimum Centerline Offset of Adjacent Intersection (feet):			
a. Local-local	125	125	125
b. Local-collector	150	150	150
c. Collector-collector	200	200	200
Minimum Tangent Length Approaching Intersection (each leg)	50	30	20

Note: * At an alley intersection with a street (or with another alley) a 15-foot minimum clear sight distance leg is recommended along each intersecting property line.

Figure 6.21 Proposed street intersection design guidelines. Taken from *Proposed Revisions to a Recommended Practice—Guidelines for Residential Subdivision Street Design.* Copyright © 1990 by the Institute of Transportation Engineers. Reproduced with permission.

Parking stalls should provide adequate space to allow vehicle doors to open when embarking or disembarking a vehicle parked adjacent to other vehicles. The parking facility should provide an adequate number of stalls for occupants, visitors, and possibly space for some recreational vehicles such as boats and their trailers or campers. Parking for handicapped persons requires additional provisions (see Section 2.5.7).

The geometric layout of all parking areas (residential and commercial) should allow the vehicle operator ease in positioning the vehicle in the parking stall by providing adequate radii on curb turnouts and islands. Drainage is also important to prevent standing surface water and degradation of the parking lot pavement and subgrade from subsurface water. Arrangements also must be made for security, night illumination, and landscaping.

6.3.1 Parking Generation

Estimates of the volume of parked vehicles in various settings can be made using references such as ITE's *Parking Generation*. As in the case of traffic generation, the designer must use this source as a beginning point because local adjustments are required. Should the parking facility accommodate "shared parking" (uses that have different peak parking periods), the total number of needed parking spaces is not necessarily the sum of the individual needs attributed to each use.

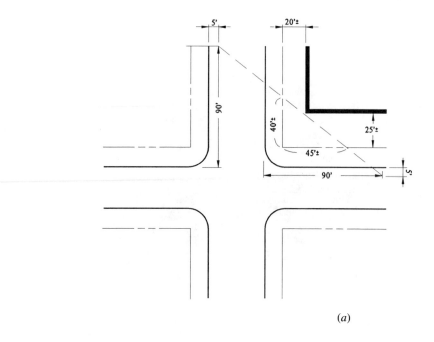

(a)

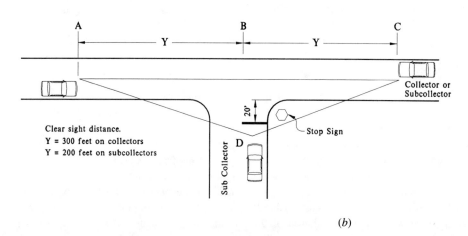

(b)

Figure 6.22   Clear sight distances at typical intersections found in land developments. (a) Local-local intersection. Taken from *Proposed Revisions to a Recommended Practice—Guidelines for Residential Subdivision Street Design.* Copyright © 1990 by the Institute of Transportation Engineers. Reproduced with permission. (b) Higher order street intersection. Source: The American Society of Civil Engineers, the National Association of Home Builders, and the Urban Land Institute. *Residential Streets,* 2nd ed., 1990. Reproduced by permission of the American Society of Civil Engineers.

Recommended Basic Driveway Dimension Guidelines				
	Dimension Reference	Residential	Commercial	Industrial
Nominal Width (See Note 1)	W			
One-way		10	15	20
Two-way		10	30	40
Right turn radius or flare	R			
Minimum (See Note 2)		5	15	20
Minimum Spacing (See Note 3)				
From property line	P	0	0	-R
From street corner	C	5	10	10
Between driveways	S	3	3	10
Minimum Angle (See Note 4)	A	45 ^{0}	45 ^{0}	30 ^{0}

Notes:

1. Residential driveway widths typically should not exceed about 24 feet. Commercial driveway widths may vary from about 24 feet for low volume activity (providing that 20 foot radii are used) to a maximum of 36 feet for undivided design, higher volume activity. A 36 foot driveway is usually marked with two exit lanes of 10 to 11 foot width, with the balance used for a single, wide entry lane. Industrial driveway widths should not exceed 50 feet.
2. On the side of a driveway exposed to entry or exit by right turning vehicles. The radii for major generator driveways should be much higher than the values shown.
3. Measured along the curb or edge of pavement from the roadway end of the curb radius or flare. For individual properties, a suggested limitation on the number of driveways is: 1 for 0-50 foot frontage, 2 for 51-150 foot frontage 3 for 151-500 foot frontage, and 4 for over 500 foot frontage.
4. Minimum acute angle measured from edge of pavement, and generally based on one-way operation. For two-way driveways and high pedestrian activity areas, the minimum angle should be 70 degrees.

Figure 6.23 Driveway layout guidelines. Taken from *Guidelines for Driveway Location and Design*. Copyright © 1987 by the Institute of Transportation Engineers. Reproduced with permission.

6.3.2 Parking Lot Configurations

Numerous configurations for parking lots are available; however, the most common relationships are shown in Figure 6.25. A percentage of the provided parking spaces can be allocated to smaller sized cars in designated areas using relationships shown in Figure 6.26.

Other forms of parking facilities include parking garages and underground parking structures. The reader is referred elsewhere for information pertaining to these alternatives (ULI and NPA, 1990).

6.3.3 Parking Aesthetics

Parking lots can be unattractive if inadequate attention is given to their appearance during design. Without attention they can be extremely hot in the summer as a result of absorbed

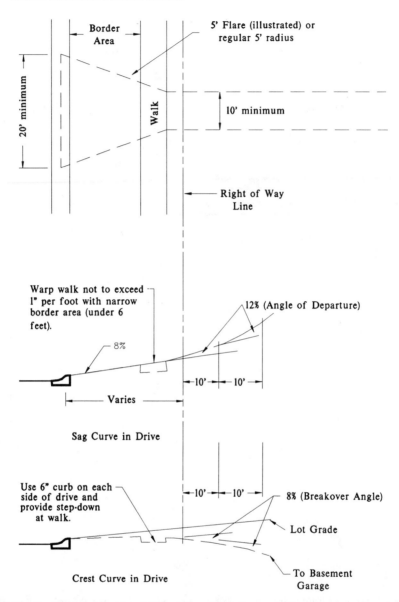

Figure 6.24 Residential driveway details. Taken from *Proposed Revisions to a Recommended Practice—Guidelines for Residential Subdivision Street Design*. Copyright © 1990 by the Institute of Transportation Engineers. Reproduced with permission.

heat and reflected energy from vehicles. In the winter they can be cold because of cross winds. Repetitive space configurations can provide mundane views for an observer, resulting in visual unattractiveness.

With landscaping, appeal can be obtained. Plants and trees can provide shade, protection from cross winds, and visual delineation to mask parked vehicles. Vegetation also helps by reducing absorbed heat and glare and filtering noises to help maintain quiet.

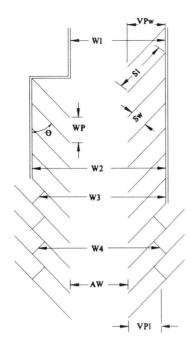

Θ = Parking Angle
W1= Parking module width (wall to wall), single loaded aisle
W2= Parking module width (wall to wall), double loaded aisle
W3= Parking module width (wall to inter-lock), double loaded
W4= Parking module width (interlock to interlock), double loaded aisle
AW= Aisle width
WP= Stall width parallel to aisle
VPl= Projected vehicle length from interlock
VPw= Projected vehicle length from wall measured perpendicular to aisle
Sl= Stall length
Sw= Stall width

Stall Width Classifications

Class	Width	Typical Uses
A	9.00	Retail customers, banks, fast foods, other very high turnover
B	8.75	Retail customers, visitors
C	8.50	Visitors, office employees, residential, airports, hospitals
D	8.25	Industrial, commuter, univesity

Note: Width is measured for large-size vehicle, at right angles to stall

	Sw	WP Stall Width	VPw	VPl	AW	Modules	
Parking Class	Basic Stall Width,ft.	Parallel To Aisle, ft.	Stall Depth to wall, ft.	Stall Depth to interlock,ft.	Aisle Width,ft.	W2 Wall to Wall,ft.	W4 Interlock to Interlock,ft.
	Two-Way Aisle- 90 Degrees						
A	9.00	9.00					
B	8.75	8.75					
C	8.50	8.50	17.5	17.5	26.0	61.0	61.0
D	8.25	8.25					
	Two-Way Aisle- 60 Degrees						
A	9.00	10.4					
B	8.75	10.1	18.0	16.5	26.0	62.0	59.0
C	8.50	9.8					
D	8.25	9.5					
	One-Way Aisle-75 Degrees						
A	9.00	9.3					
B	8.75	9.0	18.5	17.5	22.0	59.0	57.0
C	8.50	8.8					
D	8.25	8.5					
	One-Way Aisle-60 Degrees						
A	9.00	10.4					
B	8.75	10.1	18.0	16.5	18.0	54.0	51.0
C	8.50	9.8					
D	8.25	9.5					
	One-Way Aisle- 45 Degrees						
A	9.00	12.7					
B	8.75	12.4	16.5	14.5	15.0	48.0	44.0
C	8.50	12.0					
D	8.25	11.7					

Figure 6.25. Proposed parking layout criteria for large size vehicles. Taken from *A Proposed Recommended Practice—ITE Guidelines for Parking Facility Location and Design.* Copyright © 1990 by the Institute of Transportation Engineers. Reproduced with permission.

Parking Class	Sw Basic Stall Width,ft.	WP Stall Width Parallel To Aisle, ft.	VPw Stall Depth to wall, ft.	VPl Stall Depth to interlock,ft.	AW Aisle Width,ft.	W2 Wall to Wall,ft.	W4 Interlock to Interlock,ft.
						Modules	
	Two-Way Aisle- 90 Degrees						
A/ B	8.0	8.0					
C/ D	7.5	7.5	15.0	15.0	21.0	51.0	51.0
	Two-Way Aisle- 60 Degrees						
A/ B	8.0	9.3					
C/ D	7.5	8.7	15.4	14.0	21.0	52.0	50.0
	One-Way Aisle-75 Degrees						
A/ B	8.0	8.3					
C/ D	7.5	7.8	16.0	15.1	17.0	49.0	47.0
	One-Way Aisle-60 Degrees						
A/ B	8.0	9.3					
C/ D	7.5	8.7	15.4	14.0	15.0	46.0	43.0
	One-Way Aisle- 45 Degrees						
A/ B	8.0	11.3					
C/ D	7.5	10.6	14.2	12.3	13.0	42.0	38.0

Figure 6.26. Proposed parking layout dimension guidelines for smaller sized vehicles. Taken from *A Proposed Recommended Practice—ITE Guidelines for Parking Facility Location and Design*. Copyright © 1990 by the Institute of Transportation Engineers. Reproduced with permission.

Planting screens and buffer zones can be placed around the parking lot's exterior to act as a visual screen, and interior medians can provide space for other plantings.

6.4 STREET ALIGNMENT

In addition to adhering to proper geometric relationships necessary for safe vehicular operation, street and roadway design also includes the establishment of horizontal control using bearings and distances, and vertical control with stations, elevations, and grades. The design procedure presented in this section includes the following sequential steps:

(a) Determining centerline bearings and distances from point of centerline intersection (PI) to PI for the entire road system.

(b) Calculating the horizontal curve nomenclature for each included curve.

(c) Calculating the road's stations using the PI to PI distances and each horizontal curve's nomenclature.

(d) Plotting the plan view of the centerline and features that are related to centerline controls.

(e) Laying out the corresponding profile plot area using previously determined stationing along the profile bottom while providing adequate space to accommodate the maximum and minimum expected elevations along the profile's sides.

(f) Plotting the existing ground profile using contour lines contained on the preliminary approved plan.

(g) Laying out the proposed road profile on a profile drawing.

(h) Calculating grades for each desired station once the storm and sewage systems are evaluated to complete the profile drawing.

6.4.1 Horizontal Control

Horizontal control for streets and roads consists of establishing the horizontal relationship of the road's centerline to the property's boundary. This can be accomplished by scaling graphical relationships from the approved preliminary plan, computing relationships using coordinate geometry, or assigning relationships using a desired distance or angle. An example of a scaled relationship is a scaled tangent (radius or chord) distance of a horizontal curve, while a computed relationship can be represented by a calculated radius for some other horizontal curve where the tangent length is scaled and the tangent bearings are assigned.

Once the proposed road's centerline bearings and distances are defined from PI to the next PI, circular horizontal curves having constant radii (and meeting design criteria similar to those contained in Sections 6.2.2 or 6.2.3) can be superimposed on the centerline's bearings and distances. Road stationing is then calculated, where a station represents (in hundreds of feet) the distance obtained by continuous measurement along the centerline from the beginning point to the point under investigation.

To further illustrate a procedure to establish alignment for horizontal control, the approved preliminary plan shown in Figure 6.27 is utilized. In this example, the designer

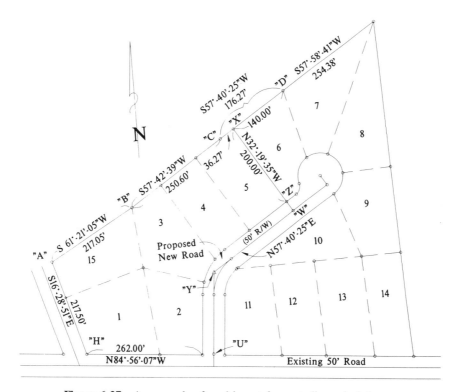

Figure 6.27 An example of road layout for centerline calculations.

decides that the property line between lots five and six will be 200 feet long and that the back lot line of lot six will be 140 feet. These two assignments can result from the designer rounding scaled distances on the approved preliminary plat. Property line C–D has a total distance of 176.27 feet. If 140.00 feet is allocated to lot six, then the remaining 36.27 feet must be included as part of lot five's back property line.

The centerline of the proposed new road leading to the cul de sac is to be made parallel to the back property line (C–D). The 200 foot lot line between lots five and six (line X–Z) is to be made perpendicular to the proposed road's alignment (by assignment). The corresponding bearing of line X–Z is obtained by subtracting the assigned bearing from 90° and giving this angular difference an adjoining bearing quadrant (northwest or southeast) by:

$$
\begin{array}{ccccc}
 & 89° & 59' & 60'' & \\
S & 57° & 40' & 25'' & W \\
\hline
N & 32° & 19' & 35'' & W
\end{array}
$$

The distance from point X to the centerline of the proposed road is 200.00 feet plus one half of the width of the road's total right of way of 50 feet, or 225.00 feet to point W. Point W is a control point used to calculate the horizontal alignment back to the existing street right of way at point U. Using the preliminary plan (Figure 6.27) for reference, the scaled distance from point H to the western edge of the right of way of the proposed road is 262 feet. If 262.00 feet plus 25.00 feet (the proposed road intersects the existing road at 90°) are used to locate point U relative to point H, coordinate geometry calculations can proceed.

The centerline alignment of the proposed road is determined by calculating the inverse bearing and distance between points U and W (with latitudes, departures, and coordinates as described in Chapter 4); then finding the point of intersection of the proposed road's tangents (point Y) based on a bearing of S 57° 40' 25'' W from point W and a perpendicular bearing to line H–U of N 05° 03' 53'' E from point U. Inverse calculations between points U and W are shown in Figure 6.28.

The inverse bearing of line U–W is determined as N 34° 20' 27'' E and the inverse distance as 364.66 feet. The proposed tangents can now be intersected mathematically to determine the distances from the PI (point Y) to both points U and W. Figure 6.29 illustrates this procedure using the law of sines.

The horizontal curve's nomenclature can now be computed based on applicable criteria. Should a particular degree of curve, or "D" (the central curve angle between two radii representing 100 feet of corresponding arc distance for arc definition), be required, the corresponding curve nomenclature can be calculated using the angular relationship between the curve's tangent bearings. In this example, however, the centerline radius (R) is assumed to be scaled from the approved preliminary plan where the approximate value of 110 feet is obtained. Since the centerline bearings have been assigned, the curve's central angle (I) can be calculated as shown in Figure 6.30.

Other centerline curve nomenclature is calculated using R and I as follows:

(a) tangent distance (T) = R tan (I/2)   (6-8)

or 110 feet $\times$ tan 26° 18' 16''. The curve's tangent is therefore 54.38 feet.

			Coordinates		
Line	Bearing	Dist,ft.	Pt.	North,ft.	East,ft.
W · X	N 32°·19'35" W	225.00	W	0.00	0.00
X · C	S 57°·40'·25" W	36.27	X	190.13	-120.32
C · B	S 57°·42'·39" W	250.60	C	170.73	-150.97
B · A	S 61°·21'·05" W	217.05	B	36.86	-362.81
A · H	S 16°·28'·51" E	217.50	A	-67.20	-553.29
H · U	S 84°·56'·07" E	287.00	H	-275.76	-491.59
			U	-301.10	-205.71

$$\text{Inverse Bearing} = \text{Tan}^{-1}\left[\frac{E_2 \cdot E_1}{N_2 \cdot N_1}\right] = \left[\frac{\cdot(205.71) \cdot 0}{\cdot(301.10) \cdot 0}\right]$$
$$(U \cdot W)$$

$$\text{Inverse Bearing} = \text{N } 34°\cdot20'\cdot27" \text{ E}$$
$$(U \cdot W)$$

$$\text{Inverse Distance} = \left[\left(E_2 \cdot E_1\right)^2 + \left(N_2 \cdot N_1\right)^2\right]^{0.5} = 364.66 \text{ ft.}$$

Figure 6.28 An example of inverse calculations to determine horizontal control for a street.

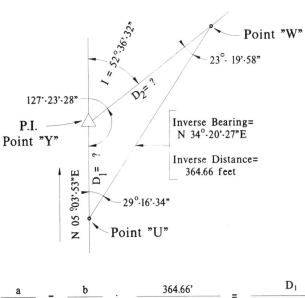

$$\frac{a}{\text{Sin A}} = \frac{b}{\text{Sin B}} \quad ; \quad \frac{364.66'}{\text{Sin } 127°\cdot23'\cdot28"} = \frac{D_1}{\text{Sin } 23°\cdot19'\cdot58"} \quad ;$$

Distance D_1 = 181.79 feet.

$$\frac{a}{\text{Sin A}} = \frac{c}{\text{Sin C}} \quad ; \quad \frac{364.66'}{\text{Sin } 127°\cdot23'\cdot28"} = \frac{D_2}{\text{Sin } 29°\cdot16'\cdot34"}$$

Distance D_2 = 224.45 feet.

Figure 6.29 Horizontal control relationships for the sample road centerline.

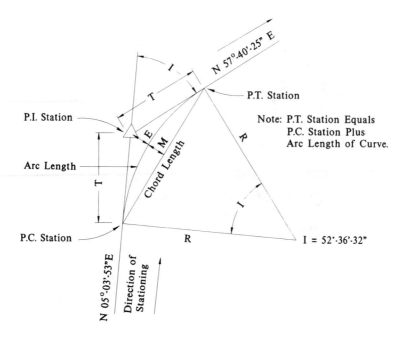

Note: P.T. Station Equals
P.C. Station Plus
Arc Length of Curve.

$I = 52°\cdot36'\cdot32"$

Curve Nomenclature

L = (l00 x I)/D , or (2 x R x π x I)/360

T = R Tan (I/2)

E = (T² + R²)⁰·⁵ R

M = R · (R²-(Chord Dist./2)²)⁰·⁵

Chord Length = 2 R Sin I/2

Where D = Degree of Curve by Arc Definition

Figure 6.30 Horizontal highway curve nomenclature.

(b) curve length–arc $(L) = 3.14 \times 2 \times R \times I/360$ (6-9)

or $3.14 \times 2 \times 110 \times 52.508889/360$. The curve's arc length is 101.00 feet. (Note: I is in decimal degrees).

(c) chord length–$(CH) = 2R \sin (I/2)$ (6-10)

or $2 \times 110 \times \sin 26° 18' 16"$. The curve's chord distance is therefore 97.49 feet.

(d) degree of curve–$(D) = (100 \times I)/L$ (6-11)

or $100 \times 52.608889/101.00$. The corresponding degree of curve is 52.088009°, or 52° 05' 17".

Station calculations for key control points along the road centerline can now begin starting with Station 0 + 00. Usually key control points consist of PIs, points of curvature (PCs), and points of tangency (PTs) (also points of vertical intersection (PVIs), points of vertical curvature (PVCs), and points of vertical tangency (PVTs) and high/low points for vertical curves described in Section 6.4.2). The beginning point (Station 0 + 00) for this example is defined by the centerline intersection of the existing and proposed roads. To establish this point relative to the proposed road's alignment, one half of the existing road's total right of way (50 feet) must be added to the distance of line U–Y. In the field, it is helpful to install a marker at this beginning point by measuring from the site's existing property lines and monuments. This installed marker locates the proposed road's beginning, which is useful for staking out the road for construction.

Point U would be assigned a station value of Station 0 + 25 and point Y would have a PI station value of Station 2 + 06.79 (181.79 feet + 25.00 feet). The corresponding PC Station for this horizontal curve is Station 1 + 52.41 (206.79 − 54.38), and the PT Station is 2 + 53.41 (PC Station plus the arc length, or 152.41 + 101.00). Point W therefore has a station value of Station 4 + 23.48 (PT Station + D_2 − T, or 253.41 + 224.45 − 54.38).

In this case the scaled distance on the preliminary plan from point W to the point where the cul de sac centerline is offset is approximately 95 feet. Therefore, the PI Station of the cul de sac is 4 + 23.48 + 95.00, or Station 5 + 18.48, and the offset station of the cul de sac's center is Station 5 + 43.48 (518.48 + 25.00). If the proposed road's geometry does not lend itself to scaling the last segment because some geometric relationship is desired, such as having the cul de sac tangent to a property line, a procedure to determine this distance is presented in Section 6.4.3.

If the road under design includes several horizontal curves, the PI to PI bearings and distances for the entire roadway should first be determined. Then each horizontal curve's nomenclature can be calculated and used to obtain road stations. Figure 6.31 illustrates this situation where the depicted road's centerline consists of four horizontal curves (one at each PI) and associated tangent segments. Should the proposed road not end with a cul de sac because it intersects another road (or itself by looping), the resulting centerline horizontal alignment (bearings and distances) must represent a closed traverse. This should be verified by the designer prior to beginning vertical control design. Once this has been verified, a plan view of the road can be plotted by using the completed horizontal alignment and by offsetting right of way lines and pavement edges (or curbing) from the centerline geometry since they are normally parallel to the centerline.

6.4.2   Vertical Control

Control for street and road vertical alignment depends on establishing the relationship of the road's grade and elevation to the site's topography. This process can begin once the road's horizontal control is established because it is necessary to horizontally locate existing contour lines where they intersect the proposed road's centerline in order to draw the ground profile.

The ground profile shows in an elevation view the surface characteristics of the existing ground along the proposed road centerline. This profile line indicates whether the ground (before construction) is level, uniformly sloped, or consists of irregular features. To construct this ground profile, the existing ground elevations are obtained where contour

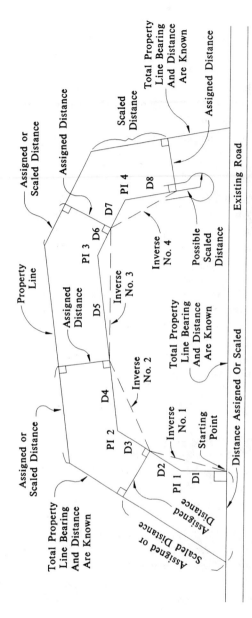

Figure 6.31 An example of a layout for horizontal control calculations using a road with four horizontal curves.

lines cross the proposed road's centerline using the approved preliminary plan. Intercept stations for each contour line must be defined by scaling the preliminary plan. On the profile drawing, road stationing is plotted along the bottom (abscissa) while elevators are plotted vertically (ordinates). Figure 6.32 illustrates a sample plan view of a portion of a proposed road and the corresponding existing ground profile.

Once the existing ground profile is plotted, a preliminary proposed road grade can be superimposed on the profile beginning at the edge of the existing pavement for uniform transition between the existing road surface and that of the proposed road. This proposed road grade usually represents the finished road surface in land developments, although the road's subgrade can be shown as an alternative. Regardless, the proposed road grade must be clearly labeled to indicate what it represents. This is important during construction because field grade stakes used for construction control must reflect intended design elevations, requiring the earthwork contractor to account for pavement thickness when grading the roadbed.

The direction of proposed roadway slope should be consistent with the approved preliminary development plan. For example, catch basins shown on the approved preliminary plan that are located in low areas must guide the designer to provide low profile elevations (low point on a sag vertical curve) there. This will allow storm water runoff to concentrate in accordance with the approved preliminary plan. For developments utilizing curb and gutter sections, the roadbed should be placed in a cut environment, if possible, to allow runoff from adjacent property to flow across the top of the curb and into the street gutter for removal. Cut sections are represented on the profile wherever the existing ground is located above the proposed road grade.

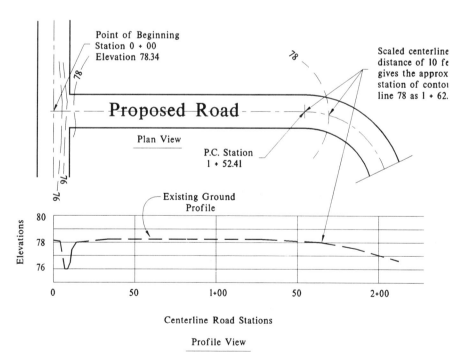

Figure 6.32 Plotting the existing ground profile on a road plan and profile sheet, using contour lines.

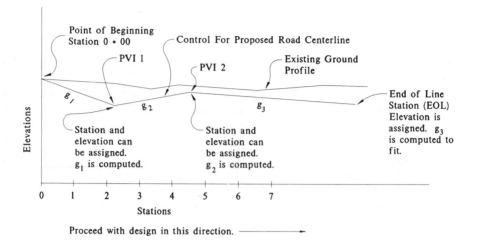

Figure 6.33   Preliminary design of a road profile.

Other criteria for laying out the preliminary road profile include minimizing earthwork (Chapter 9), providing necessary sight distances, and providing road grades that are within acceptable regulatory limits. The centerline road grade for vertical control from PVI (point of vertical curvature) to PVI is straight; therefore, initially the designer only needs to establish PVI stations and corresponding elevations to verify that the computed tangent gradients (or grades) are acceptable. At this design stage, the corresponding elevations of each station and included elevations within vertical curves are not necessarily needed (see Figure 6.33) because modification to the road's vertical alignment might be required later to accommodate design requirements for storm drains or sanitary sewers.

As in the case of horizontal control, if the road system consists of a closed centerline loop, then the vertical control network must also be a closed traverse relative to stations, elevations, and included grades. The design engineer cannot begin design of the storm drainage system if the proposed road grades are greater than the maximum allowed or less than the minimum allowed. PVI stations or elevations may have to be altered to give acceptable results. Completion of the system includes calculating vertical curve elevations and tangent elevations outside of the vertical curves once the storm and sanitary sewer systems have been evaluated.

When completing the design, symmetrical parabolic vertical curves are most commonly used to join vertical tangents in land developments, where $l_1 = l_2$. Vertical curves can either be crest types or sag types.

Figure 6.34 illustrates a typical 400 foot crest vertical curve. Intermediate elevations for stations included within the vertical curve, and the low or high point data can obtained, as shown in Figure 6.34, using the following procedure:

(a) Determine the total change of grade in the vertical curve (A) using $g_2 - g_1$, where a negative A value denotes a crest vertical curve. Data included in Figure 6.34 indicate that A equals -16 percent, -7 percent $- (+9$ percent).

(b) Determine the vertical offset (E) from the PVI to the curve (plus for sag curves, minus for crest curves) by multiplying the total change in grade (in percent) by the curve's total length (in stations) and dividing the product by the constant value eight. In this case E equals $(-16 \times 4)/8$, or -8 feet. Vertical offsets for intermediate points between the PVC station and the PVI or the PVI station and the PVT can be determined using:

$$y = E(x/l)^2 \tag{6-12}$$

where:

$x = $ The horizontal distance from the PVC or PVT to the point under investigation in feet.

$l = $ The horizontal distance from the PVC to the PVI, or the PVI to the PVT in feet.

$E = $ The vertical offset at the PVI in feet.

Figure 6.35 includes the calculations for points between Stations $28 + 00$ and $32 + 00$.

To determine the high or low point relative to the PVC or PVT station and elevation, the following can be used:

Distance to High or Low Point from PVC:

$$\frac{L\,(g_1)}{g_1 - g_2} \tag{6-13}$$

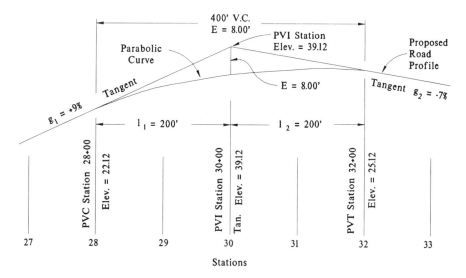

Figure 6.34 An example of a 400 foot symmetrical crest vertical highway curve.

Station	Tan. Elev.	(x/l)	(x/l)(x/l)	Vert. Off.	Curve El.
28+00 PVC	21.12	0	0	0.00	21.12
+50	25.62	1/4	1/16	0.50	25.12
29+00	30.12	1/2	1/4	2.00	28.12
+50	34.62	3/4	9/16	4.50	30.12
30+00 PVI	39.12	1	1	8.00	31.12
+50	35.62	3/4	9/16	4.50	31.12
31+00	32.12	1/2	1/4	2.00	30.12
+50	28.62	1/4	1/16	0.50	28.12
32+00 PVT	25.12	0	0	0.00	25.12

Figure 6.35   An example of vertical curve calculations.

Elevation Change to High or Low Point from PVC Elevation:

$$\frac{L\,(g_1)^2}{2(g_1 - g_2)} \tag{6-14}$$

Distance to High or Low Point from PVT:

$$\frac{L\,(g_2)}{g_1 - g_2} \tag{6-15}$$

Elevation Change to High or Low Point from PVT Elevation:

$$\frac{L\,(g_2)^2}{2(g_1 - g_2)} \tag{6-16}$$

In this example the high point from the PVT is determined to be at Station $30 + 25$ using the results of Equation 6-15. The corresponding difference in elevation between the PVT's elevation and that of the high point is found to be 6.13 feet; therefore, the elevation of the high point is 31.25 feet (6.13 feet + 25.12 feet).

Road elevations corresponding to other stations along the vertical tangent can be calculated by determining the horizontal distance between a station of known elevation and the station under investigation and multiplying this distance by the intervening slope. This product gives the difference in elevation between the point of known elevation and the one under investigation. The desired elevation is obtained by adding or subtracting this difference from the referenced elevation. For example, to determine the road elevation at Station $27 + 00$ (using the PVI at Station $30 + 00$ for reference), the horizontal distance between the two stations is 300 feet. The difference in elevation between the two stations is 27 feet (300 feet $\times$ 0.09). The required elevation at Station $27 + 00$ is 12.12 feet (39.12 − 27.00). Elevations at any station can be obtained using this same procedure.

A completed plan and profile sheet showing part of a road system is shown in Figure 6.36.

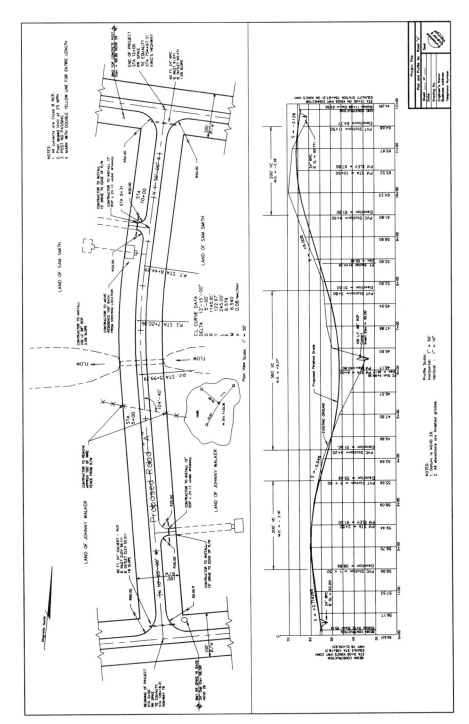

Figure 6.36 An example of a completed plan and profile for a street.

209

6.4.3 Special Geometric Conditions

Occasionally, the design professional must determine horizontal control relationships for roads, parking areas, and boundaries for parcels of land that require special techniques. This can include applications of a compound or reverse curve tangent to two lines, or a single curve tangent to a single line. Figure 6.37 contains illustrations of each of these conditions.

When a road system contains a reverse curve (Figure 6.37b) that is to be made tangent to a property line (where the road's right of way edge must not encroach on the adjoining property), a beginning point, point B (the PC), must be assigned or known, as shown in Figure 6.38, as well as the magnitude of the two radii (R_1 and R_2). Not known are the curves' remaining nomenclature and the distance from reference point D to the point of tangency along the property line (point H). Figure 6.38 shows these conditions where the right of way edge is used because it joins the property line at point H (not the road's centerline). Centerline control can be determined once I_1 and I_2 are known since I_1 and I_2 are the same for the edge of right of way and the centerline. Centerline curve data for the road can be calculated by adding or subtracting one half of the total right of way distance to R_1 and R_2.

The solution to the reverse curve problem shown in Figure 6.38 is as follows:

(a) Establish the relationship of point A, the first arc's center, to the property line $B–C$ by making the bearing of line $A–B$ perpendicular to bearing of line $B–C$ at point B (the given PC), and assign the appropriate quadrant to reflect the direction from point A to point B. The distance $A–B$ is assigned as R_1 (given). Half the problem is now solved because the location of one of the arc centers is now established.

To locate the center of the second arc (point G), two geometric conditions must prevail: First, the position of point G must be such that when an arc having a radius of R_2 is constructed, it will be tangent to line $D–H$ (the distance $D–H$ is unknown at this time). Also, the distance between the centers of the two arcs (points A and G) must be $R_1 + R_2$. These two conditions can be used to obtain the location of the second arc's center by continuing with the following solution process:

(b) Establish an imaginary line that represents a locus of points, any of which could represent the center of the second arc that would be tangent to line $D–H$. This is achieved by assigning some value to "X" (possibly 10 feet) to locate an intermediate point E. Through point E a line representing a fictitious radius is positioned, having a bearing perpendicular to side $D–H$ and a magnitude equal to R_2 (line $E–F$). A line $F–G$ passing through F that is parallel to line $D–H$ represents a locus of points, any of which, if used as a center, would produce an arc that is tangent to line $D–H$ if a radius of R_2 is used. The remaining problem is that distance $F–G$ is still unknown.

(c) The only other available relationship that can be used to locate the second arc's center is that the distance between the centers of the two arcs is equal to $R_1 + R_2$. Therefore, an additional arc equal to $R_1 + R_2$ is used with point A as its center. This additional arc intersects line $F–G$ at point G, the desired point that locates the second arc's position.

(d) The point of reverse curvature (PRC) is located along the line $A–G$ (R_1 distance from A or R_2 distance from point G). I_1 can be determined using bearings $A–B$ and $A–$

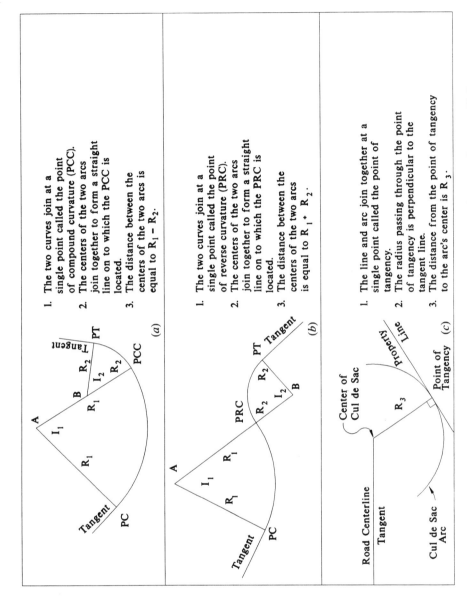

Figure 6.37 Special geometric conditions for tangent horizontal curves. (*a*) Compound curves. (*b*) Reverse curves. (*c*) Single curve tangent to a straight line.

1. The two curves join at a single point called the point of compound curvature (PCC).
2. The centers of the two arcs join together to form a straight line on to which the PCC is located.
3. The distance between the centers of the two arcs is equal to $R_1 - R_2$.

(*a*)

1. The two curves join at a single point called the point of reverse curvature (PRC).
2. The centers of the two arcs join together to form a straight line on to which the PRC is located.
3. The distance between the centers of the two arcs is equal to $R_1 + R_2$.

(*b*)

1. The line and arc join together at a single point called the point of tangency.
2. The radius passing through the point of tangency is perpendicular to the tangent line.
3. The distance from the point of tangency to the arc's center is R_3.

(*c*)

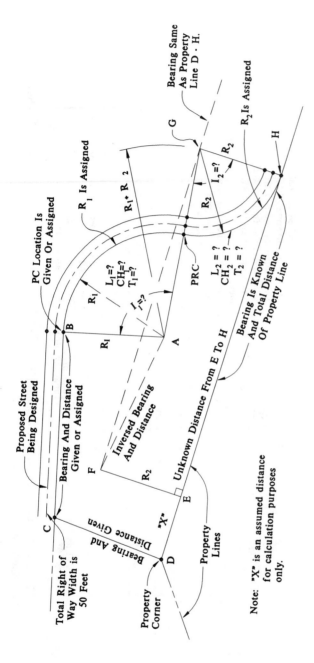

Figure 6.38 A reverse highway curve tangent to two lines.

Note: "X" is an assumed distance for calculation purposes only.

PRC. Since R_1 and I_1 are known, the remaining nomenclature for curve one can be computed.

(e) Point H (the point of tangency) is located from point G on a line perpendicular to line D–H and at a distance equal to R_2. I_2 can be determined by using bearings G–PRC and G–H. The rest of the second arc's nomenclature can now be determined.

(f) The distance D–H equals X plus the length of line F–G. The road centerline alignment can now be determined from the edge of right of way data for this portion of the road.

To solve this problem mathematically using coordinate geometry, coordinate values for points A, B, C, D, E, and F must be obtained using latitudes and departures. Line A–F is inversed to obtain its corresponding bearing and distance. The bearing of line F–G is known (same as the property line) and the distance A–G is known ($R_1 + R_2$). The unknown values are the distance of line F–G and the unknown direction of line A–G. These unknowns can be determined once triangle AFG is solved using first the law of cosines and then the law of sines. The resulting interior angles of this triangle can be used to determine the remaining bearing. From this, I_1 and I_2 can be calculated along with both curves' remaining nomenclature. The distance D–H is obtained by adding the value of X to the length of the triangle's side F–G.

If a compound curve is being designed instead of a reversed curve, each step just described is followed, except that instead of using an arc equal to $R_1 + R_2$ from point A to help find the location of the second curve as described in step c, the arc's magnitude must be $R_1 - R_2$. Figure 6.39 illustrates a compound curve tangent to a property line. In this case the outside right of way edge is used for calculations because it is this outside edge that must be tangent to the property line.

Another special problem sometimes encountered by the designer is determining the location of a cul de sac's center when a cul de sac's edge is tangent to a property line. To solve this problem, the bearing of the street's tangent where the cul de sac's center is located (line A–G) must be assigned or known (the distance is unknown). Also, the cul de sac's radius must be known. Figure 6.40 illustrates this situation.

In this example the cul de sac's center is located along bearing A–G. The center must also be located where an arc of radius R_3 will be tangent to line D–H. The distance of D–H is unknown. The solution can be obtained by intersecting two bearings formed by lines A–G and F–G. As in the case with compound or reverse curves, line F–G must be established where every point along that line could represent the cul de sac's center, such that if an arc having a radius equal to R_3 were constructed, it would be tangent to line D–H. The arc's center, however, must also be located along line A–G. In this case, coordinates for points A, B, C, D, E, and F must be obtained using latitudes and departures. Line A–F is inversed to obtain its corresponding bearing and distance. The bearing of line A–G is known, as is the bearing of line F–G (same as the back property line). The unknown distances of two sides of triangle AFG can be solved using the law of sines, as the three interior angles can be obtained from bearings and one distance is known from inversing.

6.5 PAVEMENT DESIGN

According to the Asphalt Institute's manual on *Thickness Design—Asphalt Pavements for Highways and Streets—(MS-1)* (1991), pavement design involves three basic steps: select-

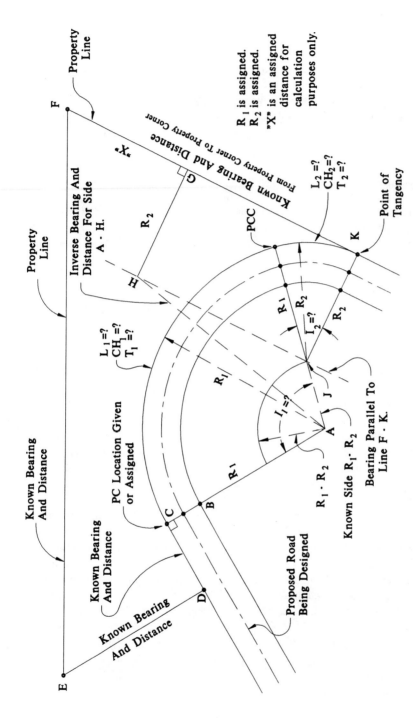

Figure 6.39 A compound highway curve tangent to two lines.

R_1 is assigned.
R_2 is assigned.
"X" is an assigned distance for calculation purposes only.

$L_2 = ?$
$CH_2 = ?$
$T_2 = ?$

Point of Tangency

PCC

$I_2 = ?$

$L_1 = ?$
$CH_1 = ?$
$T_1 = ?$

$I_1 = ?$

Known Bearing And Distance From Property Corner To Property Corner

Inverse Bearing And Distance For Side A - H.

"X"

Property Line

Property Line

Known Bearing And Distance

Known Bearing And Distance

PC Location Given or Assigned

Known Side R_1 - R_2

Bearing Parallel To Line F - K.

Proposed Road Being Designed

Known Bearing And Distance

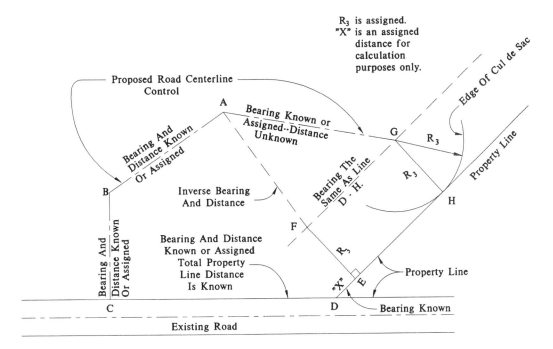

Figure 6.40 The design of a street having a cul de sac tangent to a property line.

ing construction materials, determining thickness requirements for each material used, and establishing construction related requirements (usually covered in the project's construction specifications).

Land developments require pavement as part of the incorporated site improvements to produce a finished product. Pavements can be rigid (portland cement concrete), or flexible (asphaltic concrete). The flexible pavement may be either of two types—full depth asphalt or a composite pavement. A composite pavement consists of a granular base with a hot mix asphalt surface. Parking lots and residential streets often are constructed of flexible type pavements, where maintenance crews find it relatively easy to cut pavement when necessary for effecting utility repairs. Also, resurfacing is facilitated by accommodating direct overlays, or if the existing pavement is unsound, by milling and hot mix recycling. The reader is referred to other sources for rigid pavement design and alternate methods of flexible pavement design (such as the approach used by the AASHTO). The Asphalt Institute's procedure follows.

6.5.1 The Asphalt Institute Method

The Asphalt Institute's *Thickness Design—Asphalt Pavements for Highways and Streets— (MS-1)* (1991) contains procedures based on test data, theory, and experience for structural thickness design of various asphalt pavements. Two stress strain relationships shown in Figure 6.41 are incorporated into this method as follows:

(a) Wheel loads are transferred to the pavement's structure at the tire/pavement contact interface as a uniform vertical load P_0. Stresses are spread through the pavement where a

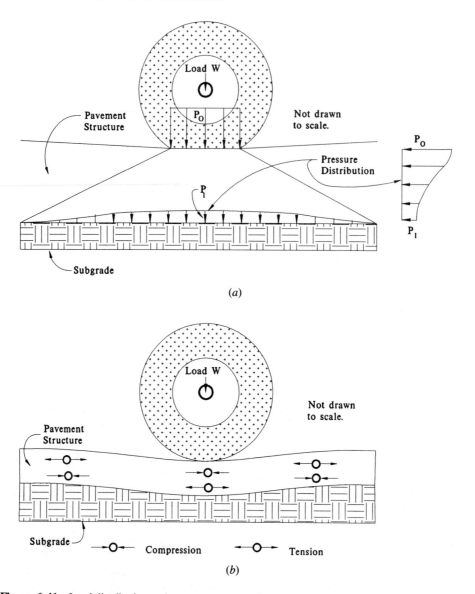

Figure 6.41 Load distribution and pavement stresses for pavement thickness design. (*a*) Spread of wheel-load pressure through pavement structure. (*b*) Pavement deflection results in tensile and compressive stresses in pavement structure. Source: The Asphalt Institute. *Thickness Design— Asphalt Pavements for Highways and Streets (MS-1)*, 1991. Reprinted by permission of the Asphalt Institute, Lexington, Ky.

reduced maximum vertical stress, P_1, results at the subgrade surface, as indicated in Figure 6.41*a*.

(b) The wheel load depicted in Figure 6.41*b* also causes pavement deflection where compressive stress occurs near the pavement's surface and tension near the pavement's bottom. The resulting strains are used to determine required pavement thicknesses for 10 different pavement structures as a result of computer analysis using a program called

DAMA (available from the Asphalt Institute). The included design nomographs address the following pavement structures:

(1) Full depth asphaltic concrete.

(2) Emulsified asphalt mix—Type I, comprised of densely graded aggregate and emulsified asphalt.

(3) Emulsified asphalt mix—Type II, comprised of semiprocessed, crusher run, bank run, or pit run aggregates and emulsified asphalt.

(4) Emulsified asphalt mix—Type III, comprised of sands or silty sands and emulsified asphalt.

(5) Untreated aggregate base thicknesses of 6 and 12 inches.

6.5.2 Traffic Analysis

Estimated initial and future traffic mixtures and volumes are required to design a pavement's thickness. Typically a pavement analysis horizon will extend to 20 years for residential areas. The analysis horizon is important for establishing the number of anticipated equivalent 18,000 pound single-axle loads (ESAL) expected to pass over the pavement. Each axle load is presumed to be applied to the pavement with two sets of dual tires that are represented by two 4.52 inch circular plates spaced 13.57 inches apart while providing a surface contact pressure of 70 pounds per square inch.

Due to the mixture of motor vehicle types, an estimate of the traffic's composition must be made to equate anticipated traffic axle loads to equivalent 18,000 pound single-axle loads. Should vehicular traffic volumes be expected to grow throughout the analysis horizon, appropriate growth factors must be applied. For local streets within residential developments, the estimated phased population or numbers of dwelling units can be used to determine incremental increases in traffic if off-site traffic is not involved with the maximum being generated at project build-out.

According to the Asphalt Institute's MS-1 (1991), national average traffic growth rates range from 3 to 5 percent compounded each year; however, rates ranging from 4 to 9 percent have been suggested for rural roads and from 8 to 10 percent for some interstate highways. Figure 6.42 can be used to obtain appropriate factors that can be used to multiply with the initial ESAL to estimate future traffic with the percentage used to reflect local growth trends. Future traffic that reflects growth should be checked against overloading the facility's capacity, thus promoting reasonable results.

ESAL determination requires knowledge of the number of various vehicle types (passenger, single-unit trucks, and multiple-unit trucks). The distribution of vehicle types can be estimated from state or local data associated with similar facilities, although Figure 6.43 can be used if local information does not exist.

According to the Asphalt Institute's MS-1 (1991), overall, heavy trucks (two-axle six-tire trucks and larger) average seven percent of the total traffic volume on all highway classes, whereas on a regional basis heavy trucks can range from 2 to 25 percent, or more, and on urban highways from 5 to 15 percent, or more. Peak hour truck traffic can be estimated at one half the daily average percentage of trucks on urban arteries, and between one half and two thirds on rural roads. Again, local traffic characteristics should be used if available.

Of primary importance is the determination of the design lane. On multilane highways the design lane is usually the outside lane, while on two lane highways and streets it

Design Period, Years (n)	No Growth	Annual Growth Rate, Percent						
		2	4	5	6	7	8	10
1	1	1.00	1.00	1.00	1.00	1.00	1.00	1.00
2	2	2.02	2.04	2.05	2.06	2.07	2.08	2.10
3	3	3.06	3.12	3.15	3.18	3.21	3.25	3.31
4	4	4.12	4.25	4.31	4.37	4.44	4.51	4.64
5	5	5.20	5.42	5.53	5.64	5.75	5.87	6.11
6	6	6.31	6.63	6.80	6.98	7.15	7.34	7.72
7	7	7.43	7.90	8.14	8.39	8.65	8.92	9.49
8	8	8.58	9.21	9.55	9.90	10.26	10.64	11.44
9	9	9.75	10.58	11.03	11.49	11.98	12.49	13.58
10	10	10.95	12.01	12.58	13.18	13.82	14.49	15.94
11	11	12.17	13.49	14.21	14.97	15.78	16.65	18.53
12	12	13.41	15.03	15.92	16.87	17.89	18.98	21.38
13	13	14.68	16.63	17.71	18.88	20.14	21.50	24.52
14	14	15.97	18.29	19.16	21.01	22.55	24.21	27.97
15	15	17.29	20.02	21.58	23.28	25.13	27.15	31.77
16	16	18.64	21.82	23.66	25.67	27.89	30.32	35.95
17	17	20.01	23.70	25.84	28.21	30.84	33.75	40.55
18	18	21.41	25.65	28.13	30.91	34.00	37.45	45.60
19	19	22.84	27.67	30.54	33.76	37.38	41.45	51.16
20	20	24.30	29.78	33.06	36.79	41.00	45.76	57.28

If Annual Growth is zero, Growth Factor = Design Period.

Figure 6.42 Traffic growth factors for pavement thickness design. Source: The Asphalt Institute. *Thickness Design—Asphalt Pavements for Highways and Streets (MS-1)*, 1991. Reprinted by permission of the Asphalt Institute, Lexington, Ky.

Truck Class	Percent Trucks *						
	Rural Systems				Urban Systems		
	Minor Arterial	Collectors		Range ***	Minor Arterial	Collectors	Range ***
		Major	Minor				
Single-unit trucks							
2-axle, 4-tire	71	73	80	43-80	84	86	52-86
2-axle, 6-tire	11	10	10	8-11	9	11	9-15
3-axle or more	4	4	2	2-4	2	<1	<1-4
All single-units	86	87	92	53-92	95	97	66-97
Multiple Unit Trucks							
4-axle or less	3	2	2	2-5	2	1	1-5
5-axle **	11	10	6	6-41	3	2	2-28
6-axle or more **	<1	1	<1	<1-1	<1	<1	<1-1
All multiple units	14	13	8	8-47	5	3	3-34
All trucks	100	100	100		100	100	

 * Compiled from data supplied by the Highway Statistics Division, U.S. Federal Highway Administration.
 ** Including full-trailer combinations in some states.
 *** These values represent truck distributions comprised of the highway classes included in this figure and for Interstate and other principal highway classes not shown.

Figure 6.43 Distribution of trucks on different classes of highways—United States. Source: The Asphalt Institute. *The Thickness Design—Asphalt Pavements for Highways and Streets (MS-1)*, 1991. Reprinted by permission of the Asphalt Institute, Lexington, Ky.

Number of Traffic Lanes (Two Directions)	Percentage of Trucks in Design Lane
2	50
4	45(35-48)*
6 or more	40(25-48)*

*Probable range.

Figure 6.44 Percentage of total truck traffic expected in the design lane. Source: The Asphalt Institute. *Thickness Design—Asphalt Pavements for Highways and Streets (MS-1),* 1991. Reprinted by permission of the Asphalt Institute, Lexington, Ky.

can be either lane. Local data can reveal the relative proportions of trucks expected to use the design lane; however, Figure 6.44 can be used if local data are not available.

To estimate ESALs from the anticipated traffic, either the number of various axle loads or the number of vehicles in each weight class must be known. Figure 6.45 can be used to convert different axle loads to equivalent 18,000 pound axle loads. When the axle loads of the vehicles are not known, Figure 6.46 can be used to estimate truck factors (TFs). Truck factors, when multiplied by the number of trucks, yield the equivalent number of 18,000 axle loads generated by those trucks. Therefore, the TF is the number of 18,000 pound single-load applications caused by a single passage of a vehicle.

Traffic on residential streets and parking lots usually is comprised of automobile traffic. Pavements designed using ESALs based on automobiles will not provide adequate pavement depths to withstand occasional traffic resulting from trash and refuse trucks, moving vans, and ready-mix concrete trucks. Estimates of use by these occasional vehicles must be included; otherwise minimum recommended pavement thicknesses must be used.

To illustrate estimating ESALs for a 330 unit single-family detached subdivision, consider these characteristics:

(a) Total number of dwelling units = 330.

(b) Average trips/day/unit = 10.

(c) Garbage pickup—two times a week.

(d) Delivery trucks—three times a week.

(e) Trash pickup—two times a week.

(f) Moving vans—in a young neighborhood, there is a four year average turnover rate.

Determination of ESALs on a weekly basis:

(1) For passenger cars, assume two each, 2000 pound axle loads with a load equivalency factor of 0.00018 (see Figure 6.45); this yields:

330 units × 10 trips per unit per day × 0.00018 × 2 axles × 7 days per week, or 8.32 ESALs per week

(2) For garbage trucks assume three axle, 8000 pound axle loads with a load equivalency factor of 0.03430), and allow for four passes per week; this yields:

3 axles × 0.03430 × 4 passes, or 0.41 ESALs per week

Gross Axle Load		Load Equivalency Factors *		
kN	lb	Single Axle	Tandem Axles	Tridem Axles
8.90	2000	0.00018		
17.80	4000	0.00209	0.00030	
26.70	6000	0.01043	0.00100	0.0003
35.60	8000	0.03430	0.00300	0.0010
44.50	10000	0.08770	0.00700	0.0020
53.40	12000	0.18900	0.01400	0.0030
62.30	14000	0.36000	0.02700	0.0060
71.20	16000	0.62300	0.04700	0.0110
80.00	18000	1.00000	0.07700	0.0170
89.00	20000	1.51000	0.12100	0.0270
97.90	22000	2.18000	0.18000	0.0400
106.80	24000	3.03000	0.26000	0.0570
115.60	26000	4.09000	0.36400	0.0800
124.50	28000	5.39000	0.49500	0.1090
133.40	30000	6.97000	0.65800	0.1450
142.30	32000	8.88000	0.85700	0.1910
151.20	34000	11.18000	1.09500	0.2460
160.10	36000	13.93000	1.38000	0.3130
169.00	38000	17.20000	1.70000	0.3930
178.00	40000	21.08000	2.08000	0.4870
222.40	50000	52.88000	4.86000	1.2200
267.00	60000		9.59000	2.5100
311.50	70000		17.19000	4.5200
356.00	80000		29.00000	7.4500

* From Appendix D of AASHTO Guide for Design
Of Pavement Structures, American Association of
State Highway And Transportation Officials,
Washington, D.C. 1986.

Figure 6.45 Axle load equivalency factors for pavement thickness design. Taken from *Guide for Design of Pavement Structures*. Copyright © 1986 by the American Association of State Highway and Transportation Officials. Reprinted with permission. Source: The Asphalt Institute. *Thickness Design—Asphalt Pavements for Highways and Streets (MS-1),* 1991. Reprinted by permission of the Asphalt Institute, Lexington, Ky.

(3) For delivery trucks assume two axle, 6000 pound axle loads with a load equivalency factor of 0.01043), and allow for six passes per week; this yields:

2 axles × 0.01043 × 6 passes per week, or 0.13 ESALs per week

(4) For trash and refuse pickup, assume: rear axle, 16,000 pounds with a load equivalency of 0.623; front axle, 6000 pounds with a load equivalency of 0.01043). Allowing four passes per week, this yields:

(0.623 + 0.01043) × 4 passes per week, or 2.53 ESALs per week

(5) For moving vans assume: rear tandem axle, 34,000 pounds with a load equivalency factor of 1.095; middle tandem axle, 34,000 pounds with a load equivalency factor of

Vehicle Type	Rural Systems *				Urban Systems *		
	Minor Arterial	Collectors		Range ***	Minor Arterial	Collectors	Range ***
		Major	Minor				
Single-unit trucks							
2-axle, 4-tire	0.003	0.017	0.003	0.003-0.017	0.006	---	0.006-0.015
2-axle, 6-tire	0.280	0.410	0.190	0.19-0.41	0.230	0.13	0.13-0.24
3-axle or more	1.060	1.260	0.450	0.45-1.26	0.760	0.72	0.61-1.02
All single-units	0.080	0.120	0.030	0.03-0.12	0.040	0.16	0.04-0.16
Tractor semi-trailers							
4-axle or less	0.620	0.370	0.910	0.37-0.91	0.460	0.40	0.04-0.98
5-axle **	1.050	1.670	1.110	1.05-1.67	0.770	0.63	0.63-1.17
6-axle or more **	1.040	2.210	1.350	1.04-2.21	0.640	---	0.64-1.19
All multiple units	0.970	1.520	1.080	0.97-1.52	0.670	0.53	0.53-1.05
All trucks	0.210	0.300	0.120	0.12-0.52	0.070	0.24	0.07-0.39

*   Compiled from data supplied by the Highway Statistics Division, U.S. Federal Highway Administration.

**   Including full-trailer combinations in some states.

*** These values represent truck factors comprised of the highway classes included in this figure and for Interstate and other principal highway classes not shown.

Figure 6.46   Distribution of truck factors (TFs) for different classes of highways and vehicles—United States. Source: The Asphalt Institute. *Thickness Design—Asphalt Pavements for Highways and Streets (MS-1)*, 1991. Reprinted by permission of the Asphalt Institute, Lexington, Ky.

1.095; front axle, 12,000 pounds with a load equivalency factor of 0.189). Allowing for occupants consisting of young adults with an average estimated turnover rate of four years, the number of estimated dwellings being vacated or occupied per week is 330 dwellings/(4 years × 52 weeks per year), or 1.6 dwellings being turned over each week. Assuming two van passes for vacating the premises and two van passes for delivery of the new occupant's goods, the total estimated passes per week are 1.6 × 4, or 6.4 passes per week. The estimated ESALs are 6.4 × (1.095 + 1.095 + 0.189), or 15.23.

Determination of ESALs for the 20 year analysis horizon:

$$\text{ESALs} = 20 \text{ years} \times 52 \text{ weeks per year} \times (8.32 + 0.41 + 0.13 + 2.53 + 15.23), \text{ or } 27,685 \text{ ESALs}$$

The percentage of traffic in the design lane is assumed to be 50 percent (see Figure 6.44); therefore, the design ESAL is 27,685 × 0.5, or 13,843 ESALs.

6.5.3   Materials Evaluation

The pavement thickness required to perform under anticipated traffic conditions is a function of the subgrade soil's strength. The strength of the subgrade is best measured by determining the resistance value of the soil (R) using AASHTO T 190-90 or by determining the resilient modulus (M_r) using the Asphalt Institute's *Soils Manual (MS-10)*. According to the Asphalt Institute's MS-1 (1991), an approximate resilient modulus can be estimated by:

$$M_r \text{ in psi} = 1500 \text{ CBR} \tag{6-17}$$

$$M_r \text{ in psi} = 1155 + 555 \, (R\text{-value}) \tag{6-18}$$

Traffic Level EAL	Design Subgrade Value, Percent
10,000 or less	60
Between 10,000 and 1,000,000	75
1,000,000 or more	87.5

Figure 6.47 Subgrade design limits. EAL = equivalent axle load. Source: The Asphalt Institute. *Thickness Design—Asphalt Pavements for Highways and Streets (MS-1),* 1991. Reprinted by permission of the Asphalt Institute, Lexington, Ky.

Equations 6-17 and 6-18, relating CBR and R-values to the subgrade soil resilient modulus, do not apply to granular untreated aggregate base and subbase materials according to the Asphalt Institute's MS-1 (1991).

To obtain a design value for the subgrade's representative resilient modulus, the following procedure can be used in conjunctions with Figure 6.47.

(a) Obtain M_r values from various soil tests conducted to determine the subgrade's resilient modulus. These data can be obtained by converting CBR or R-values to M_r values.

(b) List the test results in descending numerical order.

(c) Calculate for each M_r (beginning with the lowest value) the percentage of total values that it is equal to or greater than.

(d) Plot the results and fit a smooth curve through the points.

(e) Obtain from the graph prepared in step 4 the design subgrade resilient modulus, using Figure 6.47 as a guide.

For example, five random soil tests obtained from the proposed roadbed indicate the following M_r values: 6000; 7500; 8800; 9100; and 14,700 psi. Determination of the design M_r is shown in Figure 6.48 and Figure 6.49, following steps a through e above.

The Asphalt Institute's pavement design procedures using MS-1 (1991) assume the following:

(a) That cohesive subgrades will be compacted to a minimum of 95 percent of maximum density determined by AASHTO T 180-90, Method D for the top 12 inches of fill, and a minimum of 90 percent in fill areas below the top 12 inches.

Test Value, psi.	No Values Equal or Greater	Percent Equal to or Greater
14700	1	1/5 x 100 or 20%
9100	2	2/5 x 100 or 40%
8800	3	3/5 x 100 or 60%
7500	4	4/5 x 100 or 80%
6000	5	5/5 x 100 or 100%

Figure 6.48 Determining the M_r value for pavement thickness design.

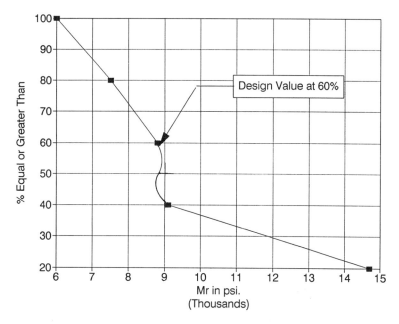

Figure 6.49 Plot of field M_r values for pavement thickness design.

(b) That cohesionless subgrade soils will be compacted to 100 percent of the maximum density determined by AASHTO T 180-90, Method D for the top 12 inches of fill, and a minimum of 95 percent in fill areas below the top 12 inches.

(c) That asphalt concrete mix designs for both surface and base courses will be for dense graded aggregates and made in accordance with the Asphalt Institute's *Mix Design Methods for Asphalt Concrete and Other Hot Mix Types (MS-2)*. Coarse aggregate for surface mixes should be comprised of at least 50 percent crushed material with two or more fractured faces.

Test	Test Requirements	
	Subbase	Base
CBR, minimum*	20	80
or		
R-value, minimum*	55	78
Liquid Limit, maximum	25	25
Plasticity Index, maximum, or	6	NP
Sand Equivalent, minimum	25	35
Passing No. 200 sieve, maximum	12	7

Equations 6-17 and 6-18 relating CBR and R-value to subgrade soil resilient modulus do not apply to untreated aggregate base and subbase.

Figure 6.50 Untreated aggregate base and subbase quality requirements for pavement thickness design. Source: The Asphalt Institute. *Thickness Design—Asphalt Pavements for Highways and Streets (MS-1)*, 1991. Reprinted by permission of the Asphalt Institute, Lexington, Ky.

Temperature Condition	Asphalt Grades *	
Cold, mean annual air temperature less than and including 45-deg F.	AC-5 AR-2000 120/150 pen.	AC-10 AR-4000 85/100 pen.
Warm, mean annual air temperature between 45-deg F and 75-deg F.	AC-10 AR-4000 85/100 pen.	AC-20 AR-8000 60/70 pen.
Hot, mean annual temperature greater than and including 75-deg F.	AC-20 AR-8000 60/70 pen.	AC-40 AR-16000 40/50 pen.

* Both medium setting (MS) and slow setting (SS) emulsified
asphalts are used in emulsified asphalt base mixes.
They can be either of two types: cationic (ASTM D 2397 or
AASHTO M 208) or anionic (ASTM D 977 or AASHTO M 140).

The grade of emulsified asphalt is selected primarily on the
basis of its ability to satisfactorily coat the aggregate. This
is determined by coating and stabilty tests (ASTM D 244, AASHTO T 59).
Other factors important in the selection are the water availability
at the jobsite, anticipated weather at the time of construction,
the mixing process to be used, and the curing rate.

Figure 6.51 Selecting asphalt grade for pavement thickness design. Source: The Asphalt Institute. *Thickness Design—Asphalt Pavements for Highways and Streets (MS-1),* 1991. Reprinted by permission of the Asphalt Institute, Lexington, Ky.

(d) That the prepared asphalt mix will be properly placed and compacted in the field to at least 96 percent of the average laboratory density (based on Marshall compacted specimens) or will achieve 92 percent of the maximum theoretical specific gravity.

(e) That emulsified asphalt base mixes will meet requirements set forth in Chapter VII of the Institute's *A Basic Asphalt Emulsion Manual (MS-19).* For dense graded emulsified aggregate base mixes, plant mixing is required to insure uniform aggregate blending.

(f) That field compaction of emulsified asphalt base mixes shall be accomplished in accordance with procedures set forth in the Asphalt Institute's *A Basic Asphalt Emulsion Manual (MS-19).*

(g) That untreated aggregate base and subbase materials will be compacted to 100 percent of their maximum density as determined by AASHTO T 180, Method D, and will conform to the requirements contained in Figure 6.50, otherwise ASTM D 2940.

(h) That appropriate grades of asphalt cement will be incorporated into the mix design using guidelines included in Figure 6.51. This will help reduce low temperature cracking and rutting in the installed pavement.

6.5.4 Pavement Structural Design Procedures

Recommended procedures for designing a pavement's thickness include initially determining the design traffic in terms of ESALs, the subgrade's design strength, M_r, and the types of surface and base courses potentially available for use.

The pavement thickness can be determined from this initial data. If incremental project phasing is proposed, stage construction can be analyzed and evaluated for economical results. Stage construction in some areas can take the form of initially installing the pavement's base or binder course, then waiting until near the end of the regulatory warranty period to install the remaining surface courses. This allows repairs to be made to distressed areas and at the same time provides a quality riding surface that will not be subjected to heavy construction traffic during the intervening time period.

Using procedures set forth in the Asphalt Institute's MS-1 (1991), the resulting pavement thickness satisfies two strain design criteria (the vertical compressive strain at the subgrade's surface, and the horizontal tensile strain on the underside of the lowest bituminous layer). When asphaltic concrete wearing courses are placed over emulsified asphalt base materials classified as Type II or III (see Section 6.5.1), the minimum thickness of asphaltic concrete to be used is shown in Figure 6.52.

Emulsified asphalt mix Type I can be substituted in accordance with Figure 6.52; however, when the Type I base course is used, a surface treatment must be provided.

Nomographs for asphalt pavement design using emulsified asphalt base courses (allowing six month curing periods) include the following for three sets of environmental conditions where the mean annual air temperature (MAAT) is less than or equal to 45°F, 60°F, or greater than or equal to 75°F:

(a) Type I—Emulsified asphalt mixes made with processed, dense graded aggregates.
(b) Type II—Emulsified asphalt mixes made with semiprocessed, crusher run, or bank run aggregates.
(c) Type III—Emulsified asphalt mixes made with sands or silty sands.

Figure 6.53 illustrates a typical design chart for emulsified pavements where Type I is shown. This and other included charts provide combined thicknesses of asphaltic concrete surface and base or emulsified asphalt surface with surface treatment and emulsified asphalt base.

For example, if the design ESAL is 10^5 and the design subgrade modulus, M_r, is 6000 psi, the required pavement thicknesses (MAAT = 60°F) are shown in Figure 6.54 for the three types of emulsified bases.

For full depth asphaltic concrete pavement, minimum thickness requirements can be

Traffic Level EAL	Type II and Type III * Inches
10,000	2
100,000	2
1,000,000	3
10,000,000	4

* Asphalt concrete, or Type I Emulsified asphalt mix with a surface treatment, may be used over Type II or Type III emulsified asphalt base courses.

Figure 6.52 Minimum thickness of asphalt concrete over emulsified bases. Source: The Asphalt Institute. *Thickness Design—Asphalt Pavements for Highways and Streets (MS-1),* 1991. Reprinted by permission of the Asphalt Institute, Lexington, Ky.

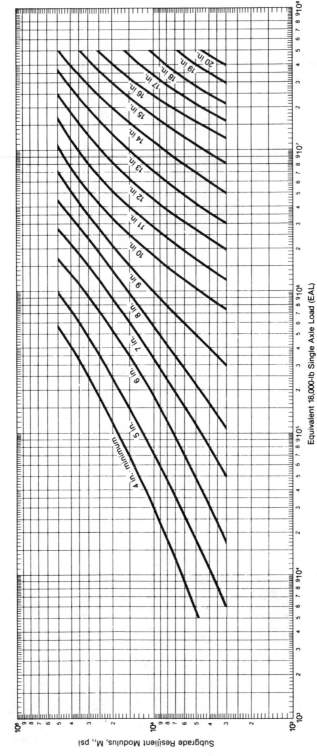

Figure 6.53 Emulsified asphalt mix Type I. Source: The Asphalt Institute. *Thickness Design—Asphalt Pavements for Highways and Streets (MS-1)*, 1991. Reprinted by permission of the Asphalt Institute, Lexington, Ky.

226

Emulsified Asphalt Pavement Thickness Design				
Type Base	Minimum Thickness, In.	Surface Course	Final Design in Inches	
			Emul. Base	Surface
I	6.5	Surface Treatment	6.5	Surface Treatment
II	7.0	2 Inches of Asphaltic Conc.	5.0	2 Inches A.C.
III	9.5	2 Inches of Asphaltic Conc.	7.5	2 Inches A.C.

Figure 6.54   Examples of emulsified asphalt pavement thickness design.

obtained using a similar procedure with other design charts contained in the Asphalt Institute's MS-1 (1991). Figure 6.55 illustrates one of these chart types.

Full depth pavement thicknesses can be obtained by entering the appropriate nomograph with the design ESAL and M_r values. For example, the minimum full depth asphaltic concrete pavement for a design ESAL of 10^5 and M_r equal to 6000 psi (MAAT = 60°F) is 6.5 inches of asphaltic concrete surface and base.

Thickness determination for asphaltic concrete surface and base courses placed over untreated aggregate bases can be obtained by using nomographs similar to the one for six inches of untreated base shown in Figure 6.56. Minimum allowed thickness of asphaltic concrete over untreated base material is shown in Figure 6.57. Charts for pavement thicknesses using other depths of untreated base are contained in the Asphalt Institute's MS-1 (1991). To illustrate the use of Figure 6.57 a design ESAL of 10^5 and an M_r value of 6000 psi (MAAT = 60° F) can be used to determine that the minimum thickness of asphaltic concrete pavement is 4.8 inches. The total pavement depth is the 6 inches of untreated aggregate base and the 4.8 inches of asphaltic concrete, or approximately 11 inches total. The engineer should develop several pavement structural designs and make an economic analysis of each. In many cases, alternate bids are invited to determine the least expensive design. Other factors, such as ease of construction and traffic conditions, should be considered in the final determination of pavement courses.

Design conditions not covered by the supplied nomographs in the Asphalt Institute's MS-1 (1991) require use of computer analysis to obtain the desired results. The reader is referred elsewhere for additional information from the Asphalt Institute (IS-91 (1981) and IS-96 (1981)) pertaining to pavement thickness design.

6.6   PEDESTRIAN FACILITY DESIGN

The most common pedestrian facilities are sidewalks. They are usually four to five feet wide, unless unusually high usage is anticipated, such as near schools and public transportation stops. The *Highway Capacity Manual,* (TRB, 1985) contains information on determining various levels of service for pedestrian travel.

In suburban areas more often than in urban ones sidewalks are usually offset from the curb several feet to allow a planting area. This offset of several feet also allows protection by placing the pedestrian away from vehicular movements. Considerations for handicap access must be included, as indicated in Section 2.5.7, in addition to other pedestrian considerations contained in Section 2.5.6.

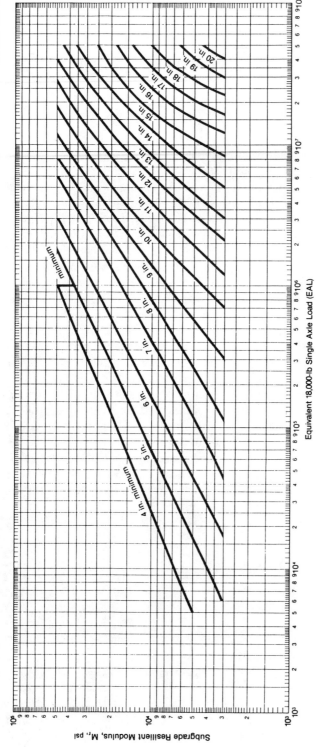

MAAT 60°F

Figure 6.55 Full-depth asphalt concrete. Source: The Asphalt Institute. *Thickness Design—Asphalt Pavements for Highways and Streets (MS-1)*, 1991. Reprinted by permission of the Asphalt Institute, Lexington, Ky.

228

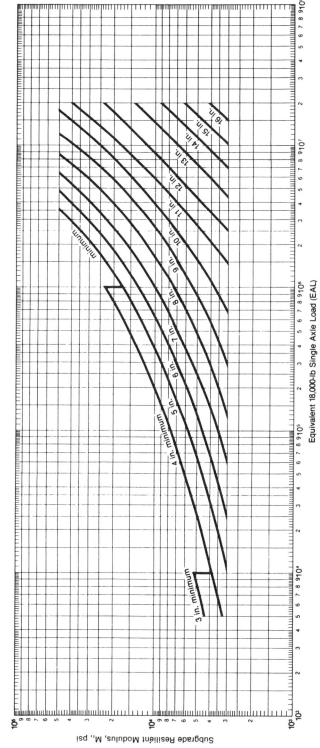

Figure 6.56 Untreated aggregate base—six inch thickness. Source: The Asphalt Institute. *Thickness Design—Asphalt Pavements for Highways and Streets* (MS-1), 1991. Reprinted by permission of the Asphalt Institute, Lexington, Ky.

229

Traffic EAL	Traffic Condition	Minimum Thickness of Asphalt Concrete
10,000 or less	Light traffic *	3 inches **
Between 10,000 and 1,000,000	Medium truck traffic	4 inches
1,000,000 or more	Heavy truck traffic	5 inches, or greater

* Includes parking lots, and driveways.

** For Full-Depth asphalt concrete or emulsified asphalt pavements
   a minimum thickness of 4-inches applies in this traffic region, as shown
   in Figures 6.53 and 6.55 .

Figure 6.57   Minimum thickness of asphalt concrete over untreated aggregate base. Source: The Asphalt Institute. *Thickness Design—Asphalt Pavements for Highways and Streets (MS-1)*, 1991. Reprinted by permission of the Asphalt Institute, Lexington, Ky.

6.7   BICYCLE FACILITY DESIGN

Depending on the project's scope, the design professional can include various site improvements to accommodate bicycling. Section 2.5.8 includes some of these considerations.

Design of bicycle facilities includes provisions for safety and access. Consideration for the separation of bicyclists from pedestrian and motor vehicle movements is warranted to help prevent accidents. Bicycle paths are one method to promote safety.

6.7.1   Bicycle Path Design

Two way bicycle paths require an eight foot paved area with two feet of shoulder along each edge. Pavements should be smooth and free from depressions, bumps, and irregular surface patterns. The minimum pavement radii that can be safely used for horizontal control are shown in Figure 6.58, where a two percent superelevation is assumed to exist.

Most bicycle paths are kept to less than a five percent gradient to minimize down-slope speeds and up-hill fatigue. Of concern to the designer is the distance required to allow a

Design Speed - V (MPH)	Friction Factor * - f	Design Radius - R (Feet)
20	0.27	95
25	0.25	155
30	0.22	250
35	0.19	390
40	0.17	565

* e = 2.0%

Figure 6.58   Design radii for paved bicycle paths. Taken from *Guide for the Development of New Bicycle Facilities*. Copyright © 1981 by the American Association of State Highway and Transportation Officials. Reproduced with permission.

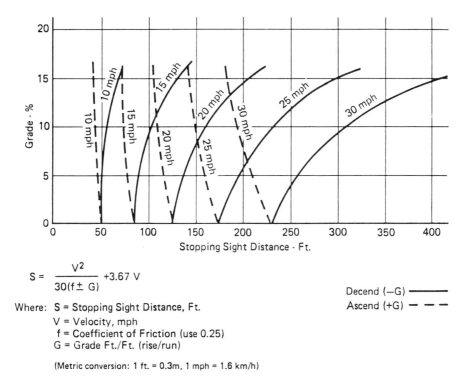

$$S = \frac{V^2}{30(f \pm G)} + 3.67\,V$$

Decend (−G) ——————
Ascend (+G) — — —

Where: S = Stopping Sight Distance, Ft.
   V = Velocity, mph
   f = Coefficient of Friction (use 0.25)
   G = Grade Ft./Ft. (rise/run)

(Metric conversion: 1 ft. = 0.3m, 1 mph = 1.6 km/h)

Figure 6.59 Bicycle stopping sight distances. Taken from *Guide for the Development of New Bicycle Facilities.* Copyright © 1981 by the American Association of State Highway and Transportation Officials. Reproduced with permission.

cyclist to stop. Figure 6.59 shows various stopping sight distances for various bicycle speeds and pavement gradients.

Stopping sight distances for bicycles can be limited by vertical curves and horizontal curves. Figures 6.60 and 6.61 depict examples of design constraints for each condition.

6.7.2 Miscellaneous Considerations

Bicycle paths require pavement markings to separate opposing bicycle traffic and to provide safety. Proper route markers also are necessary to warn bicyclists of potential hazards and general information.

Night lighting can reduce chances for accidents and provide user safety. This is essential if high volumes of bicycle traffic are expected during certain nighttime periods. Restraints and barriers are needed to prevent accidents from falls when crossing bridges, traveling close to a steep embankment, or near the edge of an area being supported by a retaining wall.

The bicycle path should also include drainage facilities to remove potential runoff while avoiding grates and manhole covers, which are not compatible with narrow bicycle tires.

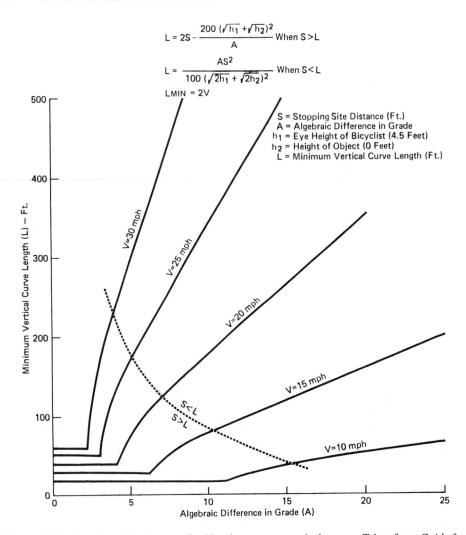

$$L = 2S - \frac{200 \left(\sqrt{h_1} + \sqrt{h_2} \right)^2}{A} \quad \text{When } S > L$$

$$L = \frac{AS^2}{100 \left(\sqrt{2h_1} + \sqrt{2h_2} \right)^2} \quad \text{When } S < L$$

$$L_{MIN} = 2V$$

S = Stopping Site Distance (Ft.)
A = Algebraic Difference in Grade
h_1 = Eye Height of Bicyclist (4.5 Feet)
h_2 = Height of Object (0 Feet)
L = Minimum Vertical Curve Length (Ft.)

Figure 6.60 Stopping sight distances for bicycles on crest vertical curves. Taken from *Guide for the Development of New Bicycle Facilities.* Copyright © 1981 by the American Association of State Highway and Transportation Officials. Reproduced with permission.

REFERENCES

American Association of State Highway and Transportation Officials (AASHTO). *Guide for the Development of New Bicycle Facilities.* Washington: AASHTO, 1981, pp. 19, 21–23.

American Association of State Highway and Transportation Officials (AASHTO). *Guide for Design of Pavement Structures—Appendix D.* Washington: AASHTO, 1986.

American Association of State Highway and Transportation Officials (AASHTO). *A Policy on Geometric Design of Highways and Streets.* Washington: AASHTO, 1990, pp. 13–15, 90–91, 119–124, 223, 276–295, 421–425, 433–448, 469, 470–472, 474. Copyright © 1990 by the American Association of State Highway and Transportation Officials. Referenced material reproduced with permission.

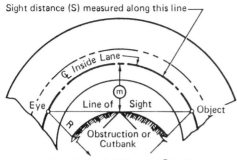

Sight distance (S) measured along this line

Line of sight is 2.0' above ℄ inside lane at point of obstruction.

S = Sight distance in feet.
R = Radius of ℄ inside lane in feet.
m = Distance from ℄ inside lane in feet.
V = Design speed for S in mph.

Angle is expressed in degrees

$$m = R \left[vers \left(\frac{28.65S}{R} \right) \right]$$

$$S = \frac{R}{28.65} \left[cos^{-1} \left(\frac{R-m}{R} \right) \right]$$

Formula applies only when S is equal to or less than length of curve.

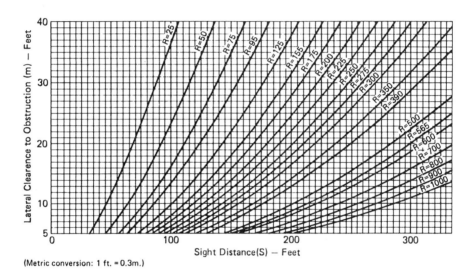

(Metric conversion: 1 ft. = 0.3m.)

Figure 6.61 Lateral clearances for bicycles on horizontal curves. Taken from *Guide for the Development of New Bicycle Facilities*. Copyright © 1981 by the American Association of State Highway and Transportation Officials. Reproduced with permission.

American Association of State Highway and Transportation Officials. *Standard Specifications for Transportation Materials and Methods of Sampling and Testing—AASHTO T 180-90 Standard Method of Test for Moisture-Density Relations of Soils Using a 10-Lb. [4.54 kg] Rammer and an 18-in [457 mm] Drop.*Washington: 1990, pp. 455–459.

American Association of State Highway and Transportation Officials. *Standard Specifications for Transportation Materials and Methods of Samples, and Testing—AASHTO T 190-90 Standard Method of Test for Resistance R-Value and Pressure Expansion of Compacted Soils.* Washington: AASHTO, 1990, pp. 472–476.

American Society of Civil Engineers (ASCE). *Bicycle Transportation—A Civil Engineer's Notebook for Bicycle Facilities.* New York: ASCE, 1980.

American Society for Testing and Materials (ASTM). Standard specification for Graduated Aggregate Material for Bases or Subbases for Highways on Airports (ASTM D2940-74). Philadelphia: ASTM, 1974.

American Society of Civil Engineers (ASCE), National Association of Home Builders, and Urban Land Institute. *Residential Streets,* 2nd ed. New York: ASCE, 1990, p. 73.

Asphalt Institute. *Basic Asphalt Emulsion Manual 2nd ed. (MS-19).* College Park, Md.: Asphalt Institute, 1987.

Asphalt Institute. *Full-Depth Asphalt Pavements for Parking Lots, Service Stations, and Driveways (IS-91).* College Park, Md.: Asphalt Institute, 1981.

Asphalt Institute. *How to Design Full-Depth Asphalt Pavements for Streets (IS-96).* College Park, Md.: Asphalt Institute, 1981.

Asphalt Institute. *Thickness Design—Asphalt Pavements for Highways and Streets (MS-1).* Lexington, Ky.: The Asphalt Institute, 1991.

Asphalt Institute. *Mix Design Methods for Asphalt Concrete and Other Hot-Mix Types, 4th (MS-2).* College Park, Md.: Asphalt Institute, 1988.

Asphalt Institute. *Soils Manual* 5th. ed. *(MS-10).* Lexington, Ky.: Asphalt Institute, 1991.

Barton-Aschman Associates, Inc. *Shared Parking.* Washington: The Urban Land Institute, 1983.

Box, Paul C. and Joseph C. Opperlander. *Manual of Traffic Engineering Studies,* 4th ed. Washington: Institute of Transportation Engineers, 1976.

Corwin, Margaret A. *Parking Lot Landscaping.* Chicago: American Society of Planning Officials, 1978.

Homburger, Wolfgang S., Ed., and Louis E. Keefer and William R. McGrath, Assoc. Eds. *Transportation and Traffic Engineering Handbook,* 2nd ed. Englewood Cliffs, N.J.: Prentice-Hall, 1982.

Homburger, Wolfgang S., Elizabeth A. Deakin, Peter C. Bosselmann, Daniel T. Smith, Jr., and Bert Beukers. *Institute of Transportation Engineers (ITE) Residential Street Design and Traffic Control.* Englewood Cliffs, N.J.: Prentice-Hall, 1989.

Institute of Transportation Engineers (ITE). *Guidelines for Planning and Designing Access Systems for Shopping Centers.* Washington: ITE, 1975.

Institute of Transportation Engineers (ITE). *Recommended Guidelines for Subdivision Streets.* Washington: ITE, 1984, p. 1. Referenced material is reprinted by permission of the Institute of Transportation Engineers.

Institute of Transportation Engineers (ITE). *Proper Location of Bus Stops.* Washington: ITE, 1986.

Institute of Transportation Engineers (ITE). *Parking Generation.* Washington: ITE, 1987.

Institute of Transportation Engineers (ITE). *Guidelines for Driveway Location and Design.* Washington: ITE, 1987, pp. 18–19. Referenced material is reprinted by permission of the Institute of Transportation Engineers.

Institute of Transportation Engineers (ITE). *Trip Generation,* 4th ed. Washington, ITE, 1987.

Institute of Transportation Engineers (ITE). *A Recommended Practice—Traffic Access and Impact Studies for Site Development.* Washington: ITE, 1989, pp. 1–50. Referenced material reprinted by permission of the Institute of Transportation Engineers.

Institute of Transportation Engineers (ITE). *Proposed Revisions to a Recommended Practice— Guidelines for Residential Subdivision Street Design.* Washington: ITE, 1990, pp. 16, 30, 32, 38, 40. Referenced material is reprinted by permission of the Institute of Transportation Engineers.

Institute of Transportation Engineers (ITE) (Paul C. Box, Ed.). *A Proposed Recommended Practice—ITE Guidelines for Parking Facility Location and Design.* Washington: ITE, 1990, pp. 7–10. Referenced material is reprinted by permission of the Institute of Transportation Engineers.

Kell, James H. and Iris J. Fullerton. *Manual of Traffic Signal Design,* 4th ed. Englewood Cliffs, N.J.: Prentice-Hall, 1991.

Leisch, Jack E. and Associates. *Planning and Design Guide At-Grade Intersections.* An unpublished report. Evanston, Ill.: Jack E. Leisch and Associates, 1890 Maple Street, 1990.

Mannerling, Fred L. and Walter P. Kilareski. *Principles of Highway Engineering and Traffic Analysis*. New York: Wiley, 1990.

McShane, William R. and Roger P. Roess. *Traffic Engineering*. Englewood Cliffs, N.J.: Prentice-Hall, 1990.

Paquette, Radnor J., Norman J. Ashford, and Paul H. Wright. *Transportation Engineering, Planning, and Design*, 2nd ed. New York: Wiley, 1982.

Smith, Thomas P. *The Aesthetics of Parking*. Chicago: The American Planning Association, 1988.

Transportation Research Board (TRB). *Highway Capacity Manual, Special Report 209*. Washington: TRB, 1985, pp. 9-4, 10-9.

Urban Land Institute (ULI) and International Council of Shopping Centers. *Parking Requirements for Shopping Centers*. Washington: ULI, 1982.

Urban Land Institute (ULI) and National Parking Association (NPA). *The Dimensions of Parking*, 2nd ed. Washington: ULI, 1990.

U.S. Department of Transportation—Federal Highway Administration. *Site Impact Traffic Evaluation (S.I.T.E.) Handbook*. Washington: Department of Transportation, 1985.

Wright, Paul H. and Radnor J. Paquette. *Highway Engineering*, 5th ed. New York: Wiley, 1987.

7 Drainage System Design

Once the road system is fairly well defined, design of the drainage system can begin. A civil engineering professor once said that:

> There are three cardinal rules to remember when dealing with a construction project: RULE 1. Get rid of the water; RULES 2 AND 3. Get rid of the water.

Several reasons for wanting to remove water, in addition to minimizing flooding, are:

(a) Excess moisture can weaken a soil's strength and even make it plastic if it is cohesive material.

(b) Excess moisture can elevate the soil's moisture content to a level that makes it impossible to attain proper field compaction.

(c) Excess moisture can result in seepage problems.

(d) High groundwater can complicate excavation operations by destabilizing the soil.

(e) High groundwater can cause pavement to fail as a result of hydrostatic pressure and loss of subgrade stability.

(f) Excess water resulting from inadequate permanent drainage facilities can result in standing water in yards or water in basements and crawl spaces, providing an environment that invites mildew, rot, and possibly insect damage.

Today, it is not a simple task to control and remove water associated with a construction project.

The water to which these cardinal rules are referring contributes to the hydrologic cycle. This cycle includes water evaporating from oceans and surface waters, along with that which is transpiring from plants to form clouds. Water vapor carried by the air masses can cool into droplets under certain meteorological conditions, resulting in precipitation. Some of the precipitation that falls on the earth can temporarily be stored by surface soil until evaporation or transpiration occurs. Other water can become surface runoff by flowing into streams and lakes, while the remainder can percolate into the soil and become groundwater. The hydrologic cycle plays an important role in determining the water supply demands attributed to lawn sprinkling. (See Chapter 11.)

Drainage design involves both surface and subsurface (ground) water. Surface water resulting from storm water runoff, snow melt, rainfall, hail, or sleet must be controlled both during and after construction. Erosion mitigation and storm drainage collection systems can be effectively used to remove excess water.

Each drainage system is designed for a particular storm intensity and duration based on a certain return period. During that time period, one storm having a particular magnitude can be expected to occur, on the average. This does not mean that another storm of the same magnitude cannot occur within the same period of time, but chances are small. Typically, land development design is based on a storm of a magnitude associated with a 10 year return period. This gives a 10 percent probability of a storm that size occurring in any given year. Since drainage systems are usually designed to flow at capacity under open channel conditions for return periods ranging from 5 to 10 years, it is advisable to analyze the designed system's performance using larger return periods to investigate potential flooding. A 25 year return period can be used to investigate the impact of a sizable storm on the development, while using a 50 or 100 year period allows insight into drainage conditions while the system is under duress.

Local regulations usually prescribe which return periods must be utilized for storm system design. The design professional would have to obtain this information while conducting preliminary engineering activities.

7.1 PRECIPITATION AND DURATION

To design a drainage system the expected amount of rainfall must first be known. Procedures for determining this are now given.

7.1.1 Precipitation Frequency Determination

The U.S. Weather Bureau, a part of the National Oceanic and Atmospheric Administration (NOAA), has gathered rainfall data pertaining to this country for many years. Statistical procedures developed by E. J. Gumbel were used to fit "annual series" values to the extreme value Type I distribution (Frederick et al., 1977). The one largest event occurring during each year of a particular duration was identified, and the group of these events constituted the annual series. The annual series values were converted into equivalent partial duration values in those cases where a single year in a series contained several significant storm events that were greater than the largest event during other years in the series. This information has been recorded in various publications associated with the probability distribution of precipitation in the United States. Information pertaining to geographical areas located generally west of 103° W longitude is included in the NOAA Atlas 2 (Miller et al., 1973). For areas generally east of 103° W longitude, NWS Hydro-35 (Frederick et al., 1977) contains precipitation data for durations of one hour or less, and USWB TP-40 (Hershfield, 1961) includes durations of up to 24 hours. The amount of precipitation is depicted using isopluvial lines that show lines of equal precipitation depths for specified return periods and durations. For example, Figure 7.1 depicts precipitation data adjusted to the partial duration series from USWB TP-40.

Space limitations prohibit the inclusion of the other rainfall charts; therefore, the reader is referred to the referenced sources for this information.

Precipitation data in inches for a sample site shown in Figure 7.2 are obtained from Figure 7.1 and the other precipitation charts referenced previously. The design professional, however, must use specific data corresponding to the actual geographic location of the development site.

Partial duration values for return periods not included in the published data and for areas generally east of 103° W longitude can be computed using procedures and charts contained in NWS Hydro 35 and USWB TP-40 (Frederick et al., 1977; Hershfield, 1961).

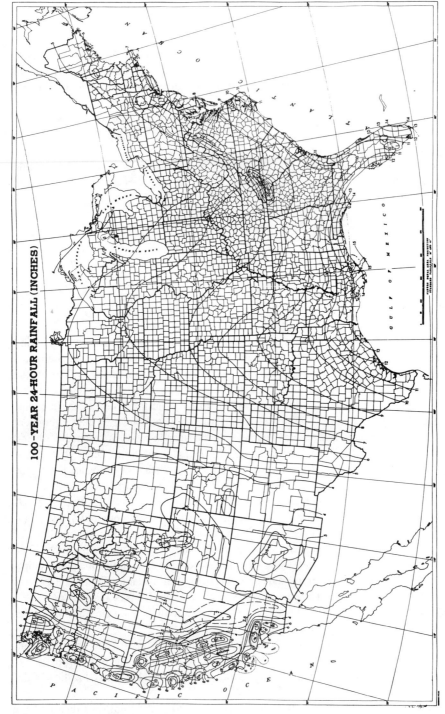

Figure 7.1 100 year, 24 hour rainfall. Source: P. M. Hershfield. U.S. Weather Bureau Technical Paper No. 40, 1961.

Return	Duration	Inches
2 Year	60 Minute Precipitation	2.20
100 Year	60 Minute Precipitation	4.25
2 Year	5 Minute Precipitation	0.51
100 Year	5 Minute Precipitation	0.80
2 Year	15 Minute Precipitation	1.13
100 Year	15 Minute Precipitation	1.85
2 Year	24 Hour Precipitation	4.50
100 Year	24 Hour Precipitation	9.90

Figure 7.2   Precipitation data for a sample project site.

To calculate these values, the following equations, contained in NWS Hydro 35, can be used:

$$5 \text{ year precipitation} = 0.278(100 \text{ years}) + 0.674(2 \text{ years}) \tag{7-1}$$

$$10 \text{ year precipitation} = 0.449(100 \text{ years}) + 0.496(2 \text{ years}) \tag{7-2}$$

$$25 \text{ year precipitation} = 0.669(100 \text{ years}) + 0.293(2 \text{ years}) \tag{7-3}$$

$$50 \text{ year precipitation} = 0.835(100 \text{ years}) + 0.146(2 \text{ years}) \tag{7-4}$$

To empirically compute values for the intervening durations shown in Figure 7.3, the following equations, from NWS Hydro 35, can be used:

$$10 \text{ minute value} = 0.59(15 \text{ minute value}) + 0.41(5 \text{ minute value}) \tag{7-5}$$

$$30 \text{ minute value} = 0.49(60 \text{ minute value}) + 0.51(15 \text{ minute value}) \tag{7-6}$$

The remaining durations can be empirically calculated, using the following equations adapted from Figure 2 of USWB TP-40 (Bradford et al., 1979):

$$2 \text{ hour value} = 0.11(24 \text{ hour value}) + 0.89(1 \text{ hour value}) \tag{7-7}$$

$$3 \text{ hour value} = 0.22(24 \text{ hour value}) + 0.78(1 \text{ hour value}) \tag{7-8}$$

$$6 \text{ hour value} = 0.50(24 \text{ hour value}) + 0.50(1 \text{ hour value}) \tag{7-9}$$

$$12 \text{ hour value} = 0.75(24 \text{ hour value}) + 0.25(1 \text{ hour value}) \tag{7-10}$$

The completed depth duration values for this example are included in Figure 7.3.

Period	Duration									
Years	5 min.	10 min.	15 min.	30 min.	1 hr.	2 hr.	3 hr.	6 hr.	12 hr.	24 hr.
2	0.51	0.88	1.13	1.65	2.20	2.45	2.71	3.35	3.93	4.50
5	0.57	0.98	1.28	1.96	2.66	3.00	3.35	4.43	5.01	5.79
10	0.61	1.07	1.39	2.18	3.00	3.40	3.81	4.84	5.76	6.68
25	0.68	1.21	1.57	2.51	3.49	3.98	4.47	5.72	6.83	7.94
50	0.74	1.31	1.71	2.77	3.87	4.43	4.98	6.40	7.66	8.92
100	0.80	1.42	1.85	3.03	4.25	4.87	5.49	7.08	8.49	9.90

Figure 7.3   Completed depth-duration frequency table in inches.

Period	Duration									
Years	5 min.	10 min.	15 min.	30 min.	1 hr.	2 hr.	3 hr.	6 hr.	12 hr.	24 hr.
2	6.12	5.28	4.52	3.30	2.20	1.23	0.90	0.56	0.33	0.19
5	6.84	5.88	5.12	3.92	2.66	1.51	1.12	0.70	0.42	0.24
10	7.32	6.42	5.56	4.36	3.00	1.70	1.27	0.81	0.48	0.28
25	8.16	7.26	6.28	5.02	3.49	1.99	1.49	0.95	0.57	0.33
50	8.88	7.86	6.84	5.54	3.87	2.22	1.66	1.07	0.64	0.37
100	9.60	8.52	7.40	6.06	4.25	2.44	1.83	1.18	0.71	0.41

Figure 7.4   An example of rainfall intensity values.

7.1.2   Defining a Storm

If actual data are not available a design storm can be defined by using either an intensity-duration curve or a depth duration relationship with synthetic rainfall distribution. The intensity-duration curve does not show rainfall sequentially according to time; it shows average intensities for a particular duration. A rainfall intensity curve can be constructed in the following manner:

(a) Divide each rainfall depth included in Figure 7.3 by its corresponding duration (in terms of hours) to obtain average rainfall intensity values. For example, if a 100 year return period and corresponding 5 minute duration are accompanied by a rainfall depth of 0.80 inch, the average intensity is calculated by (0.8 inch/5 minutes) × 60 minutes per hour, or 9.6 inches per hour. Figure 7.4 contains the calculation results.

(b) Plot each average intensity value based on its duration and return period to produce a series of curves similar to Figure 7.5. For clarity only the 10 and 100 year return periods are included.

Although a number of ways to distribute rainfall are available, the Soil Conservation Service has developed four synthetic rainfall distribution curves that can be used to distribute storm precipitation having periods of 24 hours. These curves have been constructed in a manner that allows use for determining not only peak runoff discharges but also runoff volumes for a variety of drainage area sizes (USDA—SCS TR-55, 1986). Figure 7.6 shows the four SCS rainfall distribution curves and their geographic application.

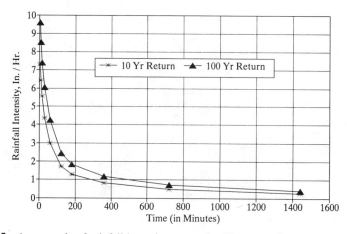

Figure 7.5   An example of rainfall intensity curves for 10 year and 100 year return periods.

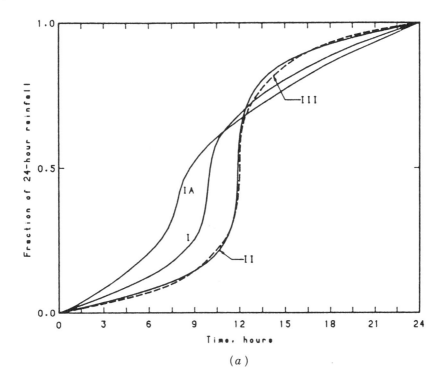

(a)

(b)

Figure 7.6 (a) SCS 24 hour rainfall distributions. (b) Approximate geographic boundaries for SCS rainfall distribution. Source: U.S. Department of Agriculture—Soil Conservation Service. *TR-55*, 1986.

A 10 year design storm using the SCS Type II curve (when the 24 hour rainfall is 6.68 inches) yields the rainfall distribution contained in Figure 7.7 for the 24 hour period.

7.2 RAINFALL ABSTRACTIONS

Once rainfall is defined, an estimate must be made as to what portion will be retained and what will become excess precipitation, or runoff. Some of the initial rainfall may land in channelized areas, contributing to the initial runoff flows, while the majority is likely to fall on soil, vegetation, and man-made surfaces. Rainfall occurring on dry porous soil will initially be absorbed to a point where soil voids near the surface become saturated, causing depressed areas to fill. It is not until the surface depressions reach capacity that runoff begins to occur. Rain falling on vegetation can be held by leaves and returned to the atmosphere by evaporation and transpiration rather than becoming runoff.

When rainfall is no longer capable of being abstracted, it becomes runoff. Of particular concern is the amount of moisture present in the soil at the storm's beginning. A storm having a particular return period can provide varying amounts of excess rainfall, depending on whether the exposed soil was initially dry, moist, or saturated. The saturated condition produces the most excess precipitation.

Time,hrs.	Px/P24	Depth, in.
0.00	0.0000	0.0000
1.00	0.0106	0.0708
2.00	0.0220	0.1470
3.00	0.0344	0.2298
4.00	0.0480	0.3206
5.00	0.0631	0.4215
6.00	0.0800	0.5344
7.00	0.0982	0.6560
8.00	0.1200	0.8016
9.00	0.1470	0.9820
10.00	0.1810	1.2091
10.50	0.2040	1.3627
11.00	0.2350	1.5698
11.50	0.2830	1.8904
12.00	0.6630	4.4288
12.50	0.7350	4.9098
13.00	0.7720	5.1570
13.50	0.7990	5.3373
14.00	0.8200	5.4776
15.00	0.8542	5.7061
16.00	0.8800	5.8784
17.00	0.9029	6.0314
18.00	0.9220	6.1590
19.00	0.9381	6.2665
20.00	0.9520	6.3594
21.00	0.9644	6.4422
22.00	0.9760	6.5197
23.00	0.9876	6.5972
24.00	1.0000	6.6800

Figure 7.7   An example of rainfall distribution for a 10 year Type II storm where P_x/P_{24} is the ratio of accumulated rainfall to the 24 hour total.

7.2.1 Coefficient of Runoff

One method of defining the estimated amount of runoff is by using a coefficient of runoff, which is utilized in the rational method. The coefficient of runoff is the ratio of the amount of excess precipitation expected to total amount of expected rainfall. Figure 7.8 shows typical coefficient values.

Should the contributory area be comprised of various land uses, a composite coefficient of runoff must be obtained based on the ratio of individual contributory subbasin characteristics to that of the total area. The composite coefficient of runoff can be calculated by:

$$C_{\text{comp}} = \frac{C_1 A_1 + C_2 A_2 + C_3 A_3 + \cdots + C_n A_n}{A_1 + A_2 + A_3 + \cdots + A_n} \qquad (7\text{-}11)$$

where:

C_n = The coefficient of runoff associated with subbasin area A_n.

A_n = The subbasin drainage area.

7.2.2 SCS Runoff Curve Number

The Soil Conservation Service estimates the amount of runoff by a runoff curve number (CN). The CN is a function of land use and prevailing soil conditions; the higher the CN, the higher the runoff potential if the soil is not frozen.

Soil properties affect the amount of runoff as a result of varying infiltration rates. SCS has defined four hydrologic soil groups:

Group A Soils   These have low runoff potential and high infiltration rates even when thoroughly wetted. They consist chiefly of deep, well to excessively drained sands or gravels and have a high rate of water transmission (greater than 0.30 inch per hour).

Group B Soils   These have moderate infiltration rates when thoroughly wetted and consist chiefly of moderately deep to deep, moderately well to well drained soils with moderately fine to moderately coarse textures. These soils have a moderate rate of water transmission (0.15–0.30 inch per hour).

Group C Soils   These have low infiltration rates when thoroughly wetted and consist chiefly of soils with a layer that impedes downward movement of water and soils with moderately fine to fine texture. These soils have a low rate of water transmission (0.05–0.15 inch per hour).

Group D Soils   These have high runoff potential. They have very low infiltration rates when thoroughly wetted and consist chiefly of clay soils with a high swelling potential, soils with a permanently high water table, soils with a claypan or clay layer at or near the surface, and shallow soils over nearly impervious material. These soils have a very low rate of water transmission (0.00–0.05 inch per hour).

Compaction and soil mixing attributed to construction can affect soil conditions in a land development and can alter the runoff curve number. Figure 7.9 contains various runoff curves for various land uses.

Soil types can be identified from SCS soils maps (referred to in Section 4.1.1) containing the project area. Figure 7.10*a* shows a typical soils map used to identify the predomi-

Slope	Land-Use	Sandy Soils **			Clay Soils **		
		Min.		Max.	Min.		Max.
	Woodlands	0.10		0.15	0.15		0.20
	*Pasture, grass & farmland	0.15		0.20	0.20		0.25
	Rooftops and pavement		0.95			0.95	
	Single family residential:						
	1/2 acre lots & larger	0.30		0.35	0.35		0.45
Flat	Smaller lots	0.35		0.45	0.40		0.50
	Multi-family residential:						
(0-2%)	Duplexes	0.35		0.45	0.40		0.50
	Apartments, townhouses,						
	and condominiums	0.45		0.60	0.50		0.70
	Commercial and Industrial	0.50		0.95	0.50		0.95
	Woodlands	0.15		0.20	0.20		0.25
	*Pasture, grass & farmland	0.20		0.25	0.25		0.30
	Rooftops and pavement		0.95			0.95	
	Single family residential:						
	1/2 acre lots & larger	0.35		0.50	0.40		0.55
Rolling	Smaller lots	0.40		0.55	0.45		0.60
	Multi-family residential:						
(2%-7%)	Duplexes	0.40		0.55	0.45		0.60
	Apartments, townhouses,						
	and condominiums	0.50		0.70	0.60		0.80
	Commercial and Industrial	0.50		0.95	0.60		0.95
	Woodlands	0.20		0.25	0.25		0.30
	*Pasture, grass & farmland	0.25		0.35	0.30		0.40
	Rooftops and pavement		0.95			0.95	
	Single family residential:						
	1/2 acre lots & larger	0.40		0.55	0.50		0.65
Steep	Smaller lots	0.45		0.60	0.55		0.70
	Multi-family residential:						
(7%+)	Duplexes	0.45		0.60	0.55		0.70
	Apartments, townhouses,						
	and condominiums	0.60		0.75	0.65		0.85
	Commercial and Industrial	0.60		0.95	0.65		0.95

* Coefficients assume good ground cover and conservation treatment.
** Weighted coefficient based on percentage of impervious surfaces and
green areas must be selected for each site.

Figure 7.8 Runoff coefficients for various land uses. Source: *De Kalb County Drainage Proce-dures Manual,* Dekalb County, Ga. Copyright © 1976 by W. L. Jorden & Company, Inc. Repro-duced with permission.

nant soil type, while Figure 7.10*b* provides the hydrologic grouping and other characteris-tics.

The factors affecting *CN* are the hydrologic soil group, cover type, treatment, hydro-logic condition, antecedent runoff condition (ARC), and land use. Should the contributory area contain different land uses, a weighted *CN* value must be obtained. For example, a drainage basin comprised of woods, residential area, and open space has a weighted *CN* value of:

Land Use	Percent	Soil Type	Curve	Product
Woods (good)	25	C	70	$0.25 \times 70 = 17.5$
Residential ($\frac{1}{3}$ acre)	60	A	57	$0.60 \times 57 = 34.2$
Open space (good condition)	15	D	80	$0.15 \times 80 = 12.0$
			Sum	63.7

Cover description		Curve numbers for hydrologic soil group[1]-			
Cover type and hydrologic condition	Average percent impervious area	A	B	C	D

Fully developed urban areas (vegetation established)

Open space (lawns, parks, golf courses, cemeteries, etc.)[3]:

Poor condition (grass cover < 50%)		68	79	86	89
Fair condition (grass cover 50% to 75%)..........		49	69	79	84
Good condition (grass cover > 75%)		39	61	74	80
Impervious areas:					
Paved parking lots, roofs, driveways, etc. (excluding right-of-way).		98	98	98	98
Streets and roads:					
Paved; curbs and storm sewers (excluding right-of-way)................................		98	98	98	98
Paved; open ditches (including right-of-way)		83	89	92	93
Gravel (including right-of-way)		76	85	89	91
Dirt (including right-of-way)		72	82	87	89
Western desert urban areas:					
Natural desert landscaping (pervious areas only) ...		63	77	85	88
Artificial desert landscaping (impervious weed barrier, desert shrub with 1- to 2-inch sand or gravel mulch and basin borders).		96	96	96	96
Urban districts:					
Commercial and business.........................	85	89	92	94	95
Industrial..	72	81	88	91	93
Residential districts by average lot size:					
1/8 acre or less (town houses).....................	65	77	85	90	92
1/4 acre	38	61	75	83	87
1/3 acre	30	57	72	81	86
1/2 acre	25	54	70	80	85
1 acre	20	51	68	79	84
2 acres	12	46	65	77	82

Developing urban areas

Newly graded areas (pervious areas only, no vegetation)		77	86	91	94

[1]Average runoff condition, and $I_a = 0.2S$.

Figure 7.9 Runoff curve numbers for urban areas. Source: U.S. Department of Agriculture—Soil Conservation Service, *TR-55*, 1986.

The weighted *CN* is taken as 64.

The *CN* value is altered by the amount of moisture present in the soil. The ARC reflects estimated variations in *CN* from storm to storm; however, for design purposes in urban areas, SCS recommends using the average *CN* value.

7.3 TIME OF CONCENTRATION

The time of concentration (T_c) is the time required for excess precipitation to travel hydraulically from the remotest point within the drainage basin to the point under investigation. This time is influenced by land use, slope, channel conditions, and ponding. T_c in hydrograph analysis is, however, defined as "the time from the end of excessive rainfall to the point on the falling limb of the hydrograph (point of inflection) where the recession curve begins" (USDA—SCS, NEH 4-102, 1972, Chapter 15).

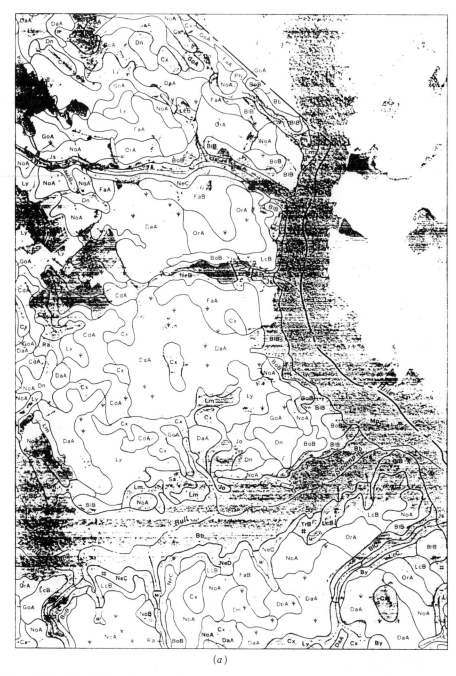

(a)

Figure 7.10 (a) SCS soils map, Orangeburg County, S.C. Source: U.S. Department of Agriculture—Soil Conservation Service. Soil Survey of Orangeburg County, South Carolina (A National Cooperative Survey) Washington: USDA, 1988. (b) An example of soil groupings found in typical SCS soil surveys. Source: U.S. Department of Agriculture—Soil Conservation Service and Forest Service, 1980. *Soil Survey of Berkeley County, South Carolina* (A National Cooperative Soil Survey) Washington: USDA, 1980.

Soil name and map symbol	Hydro-logic group	Flooding			High water table		
		Frequency	Duration	Months	Depth Ft	Kind	Months
Aquic Udifluvents: AU-----------	D	Common--------	Long----------	Dec-Mar	1.0-3.0	Apparent	Dec-Mar
Bayboro: Ba--------------	D	Common--------	Brief----------	Dec-Mar	0-1.0	Apparent	Nov-Apr
Bethera: Be--------------	D	Common--------	Brief----------	Dec-Apr	0-1.5	Apparent	Dec-Apr
Bohicket: [1]BH-----------	D	Frequent------	Very brief-----	Jan-Dec	(3)-0	Apparent	Jan-Dec
Bonneau: BoA, BoB--------	A	None----------	---	---	3.5-5.0	Apparent	Dec-Apr
Borrow pits: Bp.							
Byars: By--------------	D	Common--------	Brief----------	Dec-Mar	0-1.0	Apparent	Nov-Apr
Cainhoy: CaB-------------	A	None----------	---	---	>6.0	---	---
Capers: [2]CP-------------	D	Frequent------	Very long------	Jan-Dec	(1)-1.0	Apparent	Jan-Dec
Caroline: CoA, CoB--------	C	None----------	---	---	>6.0	---	---
Chastain: [2]CS-------------	D	Common--------	Very long------	Dec-Apr	0-1.0	Apparent	Nov-May
Chipley: [2]Ct: Chipley part----	C	None----------	---	---	2.5-5.0	Apparent	Nov-Apr
Echaw part------	B	None----------	---	---	2.5-5.0	Apparent	Dec-Apr
Coxville: Cu--------------	D	Rare----------	---	---	0-1.5	Apparent	Nov-Apr
Craven: CvA, CvB---------	C	None----------	---	---	2.0-3.5	Apparent	Dec-Mar
Duplin: DuA, DuB--------	C	None----------	---	---	2.0-3.5	Apparent	Dec-Mar
Goldsboro: GoA-----------	B	None----------	---	---	2.5-3.5	Apparent	Dec-Mar
Lenoir: Le--------------	D	None----------	---	---	1.0-2.5	Apparent	Dec-Mar
Leon: Lo--------------	A/D	None----------	---	---	0-1.0	Apparent	Nov-Mar
Lucy: LuB-------------	A	None----------	---	---	>6.0	---	---
Lynchburg: Ly--------------	B/D	None----------	---	---	0.5-1.5	Apparent	Nov-Apr
Meggett: Mg, Mp-----------	D	Common--------	Brief----------	Dec-Mar	0-1.0	Apparent	Jun-Apr
Norfolk: NoA, NoB--------	B	None----------	---	---	>6.0	---	---

(*b*)

Figure 7.10   (continued)

7.3.1 Travel Time

Most drainage basins are comprised of various elements that affect the time required for runoff to travel to a specified downstream point. This requires the contributory area to be analyzed by subbasins where elements affecting the time of concentration within each subbasin are fairly uniform. The individual subbasin travel times (T_n) are then summed for the contributory area's time of concentration. More specifically,

$$T_c = T_1 + T_2 + T_3 + \cdots + T_n. \tag{7-12}$$

Within each subbasin that is analyzed the surface roughness greatly affects the velocity of runoff. For example, areas comprised of grass retard runoff movement, while paved areas promote more rapid movement. The type of flow also affects travel times. Velocities associated with sheet flow are usually less than those associated with channelized flow, because hydraulic features associated with channels promote the higher velocities. The other factor that affects travel time is the slope of the surface carrying the runoff. The steeper the slope, the shorter the time of travel.

7.3.2 Overland Flow

Water flowing freely across plane surfaces is usually referred to as sheet flow. Depths may approach one inch, and the time of travel associated with sheet flow is dependent upon a roughness coefficient that includes the effect of erosion, sedimentation transport, drag, and rainfall impact. (See Chapter 8.) According to SCS *Technical Note—Hydrology No. 4* (USDA—SCS, 1986), the time of travel for overland flow can be approximated using the Manning-kinematic solution, where the maximum flow level can be up to 300 feet, but is more likely to be 100 feet, using the following equation:

$$T_t = \frac{0.93 \ (nL)^{0.6}}{i^{0.4} \ S^{0.3}} \tag{7-13}$$

where:

T_t = Travel time in minutes.
n = Friction factor, Manning's n for sheet flow.
L = Slope length in feet.
i = Average rainfall excess intensity for a storm duration equal to T_c (inches per hour).
S = Slope in feet per foot.

According to the Technical Note, the assumptions for using Equation 7-13 include:

Those for Manning's equation (steady uniform turbulent flow), a flat, wide plane surface, constant rainfall excess intensity, and duration of rainfall equaling T_t. Overton (1972) found that the Manning-kinematic solution for the rising overland flow hydrograph produced a 15 percent standard of error in fitting the observed data. Use of this equation requires an estimate of the rainfall excess intensity and trial and error calculations until T_t equals rainfall duration.*

According to SCS TR-55 (USDA—SCS, 1986) the overland time of travel can be estimated using Equation 7-14, where the rainfall excess intensity term in Equation 7-13 has been replaced by a 24 hour precipitation term:

$$T_t = \frac{0.007 \ (nL)^{0.8}}{(P_2)^{0.5} \ (S)^{0.4}} \tag{7-14}$$

*Taken from USDA—SCS, *Technical Note: Hydrology No. 4* (1986).

where:

T_t = Travel time in hours for sheet flow for distances up to 300 feet (although 50 feet is probably more representative).

n = Manning's coefficient of roughness as shown in Figure 7.11 for sheet flow conditions.

L = Flow length in feet.

P_2 = The 2 year, 24 hour rainfall in inches.

S = Slope of the drainage surface in feet per foot.

According to TR-55 (USDA—SCS, 1986), once sheet flows become shallow concentrated flows, the following equation, based on Manning's formula, can be used to estimate flow velocities:

$$\text{Unpaved} \qquad V = 16.1345 \, (S)^{0.5} \qquad (7\text{-}15)$$

$$\text{Paved} \qquad V = 20.3282 \, (S)^{0.5} \qquad (7\text{-}16)$$

where:

V = The average velocity in feet per second.

S = The surface slope in feet per foot.

The assumptions used for these two equations include Manning's roughness coefficient values of 0.05 for unpaved areas and 0.025 for paved areas. In addition, the hydraulic radius (included in the numerical coefficients in Equations 7-15 and 7-16) is assumed to be 0.4 for unpaved areas and 0.2 for paved areas.

Surface Description	Value of n
Smooth surfaces (concrete, asphalt, or bare soil)	0.011
Fallow (no residue)	0.050
Cultivated soils:	
Residue cover $\leq$ 20%	0.060
Residue cover $\geq$ 20%	0.170
Grass:	
Short grass prairie	0.150
Dense grasses	0.240
Bermuda grass	0.410
Range (natural)	0.130
Woods:	
Light underbrush	0.400
Dense underbrush	0.800[a]

[a]May be revised to 0.600.

Figure 7.11 Manning's coefficient of roughness, n, for sheet flow. Source: U.S. Department of Agriculture—Soil Conservation Service. *TR-55*, 1986.

Once concentrated shallow flows enter channelized areas, velocities of travel can be determined utilizing Manning's equation for open channel flow, which is:

$$V = \frac{1.49 r^{0.67} S^{0.5}}{n} \qquad (7\text{-}17)$$

where:

V = The average flow velocity in feet per second.
r = The hydraulic radius in feet defined by the ratio of the cross-sectional flow area to the corresponding wetted perimeter. The wetted perimeter is the exposed length of a channel cross section where contact is made with the flow.
S = The slope of the channel in feet per foot.
n = Manning's roughness coefficient for open channel flow. Typical values are shown in Figure 7.12.

Often Manning's equation is rearranged for use in design by incorporating the continuity equation relationship with:

$$Q = AV \qquad (7\text{-}18)$$

where:

Q = The rate of flow in cubic feet per second.
A = The cross-sectional flow area in square feet.
V = The velocity of flow in feet per second.

The rearranged equation is:

$$Q = \frac{1.49 A r^{0.67} S^{0.5}}{n} \qquad (7\text{-}19)$$

When travel velocities are obtained for shallow concentrated flows or open channel flows using Equations 7-15 through 7-17, the corresponding time of travel can be obtained using:

$$T_n = \frac{L}{3600 \times V} \qquad (7\text{-}20)$$

where:

T_n = The travel time in hours.
L = The length of flow in feet.
V = The average velocity in feet per second.
3600 = The factor to convert seconds into hours.

Another method to determine the time of concentration is the lag method. The lag equation was developed from limited agricultural watershed data. It was later adjusted and

Type of Channel or Structure	Value of n
Open Channels for Type of Lining Shown	
Smooth concrete	0.013
Rough concrete	0.022
Riprap	0.03–0.04
Asphalt, smooth texture	0.013
Good stand, any grass—depth of flow	
More than 6 inches	0.09–0.30
Less than 6 inches	0.07–0.20
Earth, uniform section, clean	0.016
Earth, fairly uniform section, no vegetation	0.022
Channels not maintained, dense weeds	0.08
Natural Stream Channels *(Surface Width at Flood Stage is 100 feet)*	
Fairly regular section	
Some grass and weeds, little or no brush	0.030–0.035
Dense growth of weeds, depth of flow materially greater than weed height	0.035–0.05
Some weeds, light brush on banks	0.035–0.05
Some weeds, heavy brush on banks	0.05–0.07
Some weeds, dense willows on banks	0.06–0.08
For trees within channel, with branches submerged at high stage, increase all values above by	0.01–0.02
Irregular sections with pools, slight channel meander: increase values given above by approximately	0.01–0.02
Culverts	
Concrete pipe and boxes	0.012
Corrugated metal	
Unpaved	0.024–0.027
25% paved	0.021–0.026
Fully paved	0.012

Figure 7.12   Typical values for Manning's roughness coefficient. Source: Paul H. Wright and Radnor J. Paquette. *Highway Engineering*, 5th ed. Copyright © 1987 by John Wiley & Sons, Inc. Reprinted by permission of John Wiley & Sons, Inc.

now has an upper drainage area limit of 2000 acres. Use of the lag method should be confined to agricultural watersheds, while the velocity method included here is better suited in urban areas.

7.4   RUNOFF

After the design storm is established, rainfall abstractions are addressed, and the time of concentration is determined, the expected runoff can be calculated. If the drainage area is

100 acres or less, the rational method is satisfactory for computing peak watershed runoff if care is exercised. The SCS method may be used for small and medium watersheds up to 25,000 acres. SCS TR-55 (USDA—SCS, 1986) however, is limited to a T_c that does not exceed a maximum value of 10 hours, rather than being limited by a maximum drainage area.

7.4.1   The Rational Method

According to ASCE Manual No. 37, the rational method is based on the following three assumptions (ASCE and WPCF, Manual 37, 1969):

(a) The peak rate of runoff at any point is a direct function of the average rainfall intensity during the time of concentration to that point.

(b) The frequency of the peak discharge is the same as the frequency of the average rainfall intensity.

(c) The time of concentration is the time required for the runoff to become established and flow from the most remote part of the drainage area to the point under design.

The rational formula takes the form of:

$$Q = CIA \tag{7-21}$$

where:

Q = Maximum rate of runoff, in cubic feet per second, from a given drainage basin.
C = Coefficient of runoff, which relates runoff to rainfall.
I = Intensity in inches per hour, for a duration equal to the time of concentration.
A = Contributory drainage basin, in acres.

This formula is not dimensionally correct; however, a one inch depth of rainfall applied at a uniform rate for one hour on an area of one acre yields 1.008 cubic feet per second of runoff if no loss occurs.

To illustrate the use of the rational method, the peak runoff from a 10 acre drainage basin will be determined. For this example, a coefficient of runoff of 0.45 is used, along with a time of concentration of 30 minutes. If a 10 year return period is specified, and Figure 7.5 is assumed to contain rainfall data applicable to this site's geographical area, the rainfall intensity of 4.36 inches per hour can be obtained. The peak discharge using Equation 7-21 is: $Q = CIA = 0.45 \times 4.36 \times 10 = 19.62$, or 20 cubic feet per second.

7.4.2   Peak Discharge Using the SCS Graphical Method

The U.S. Soil Conservation Service has developed a method, contained in TR-55 (USDA—SCS, 1986), to estimate peak runoff using graphical means; this method takes into account precipitation and initial abstractions. The generalized SCS runoff equation is:

$$Q = \frac{(P - I_a)^2}{(P - I_a) + S} \tag{7-22}$$

where:

Q = Runoff in inches.
P = Rainfall in inches.
S = Potential maximum retention after runoff begins in inches.
I_a = Initial abstractions in inches. This term has been further defined empirically from field observations by:

$$I_a = 0.2S \qquad (7\text{-}23)$$

Therefore the combined equation is:

$$Q = \frac{(P - 0.2S)^2}{(P + 0.8S)} \qquad (7\text{-}24)$$

where:

P = Total precipitation in inches.
S = Watershed retention factor.
Q = Excess rainfall in inches.

$$S = \frac{1000}{CN} - 10 \qquad (7\text{-}25)$$

where:

CN = SCS curve number defined in Section 7.2.2.

Figure 7.13 or Figure 7.14 can be used, depending on the type of rainfall distribution, to graphically obtain the unit peak discharge, q_u, knowing the time of concentration and the I_a/P ratio. Once the unit peak discharge is obtained, the peak discharge of the drainage basin can be calculated using:

$$q_p = q_u A_m Q F_p \qquad (7\text{-}26)$$

where:

q_p = The peak discharge in cubic feet per second.
q_u = The unit peak discharge (CSM per inch).
    [Note: CSM = cubic feet per second/square mile.]
A_m = The drainage area in square miles.
Q = Runoff in inches.
F_p = Pond or swamp adjustment factor.

In summary, peak discharges using the SCS graphical method are obtained by:

(a) Establishing a storm return period and obtaining the corresponding 24-hour precipitation (see Figure 7.1).
(b) Calculating the initial abstractions, I_a, using Equations 7-23 and 7-25.

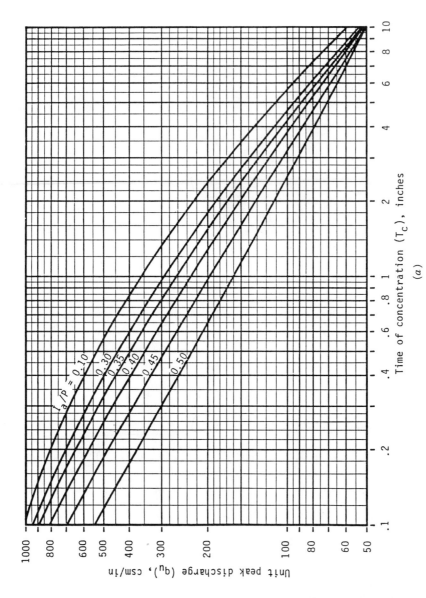

Figure 7.13 Unit peak discharges. (*a*) For SCS Type II rainfall distribution. (*b*) For SCS Type III rainfall distribution. Source: U.S. Department of Agriculture—Soil Conservation Service. *TR-55*, 1986.

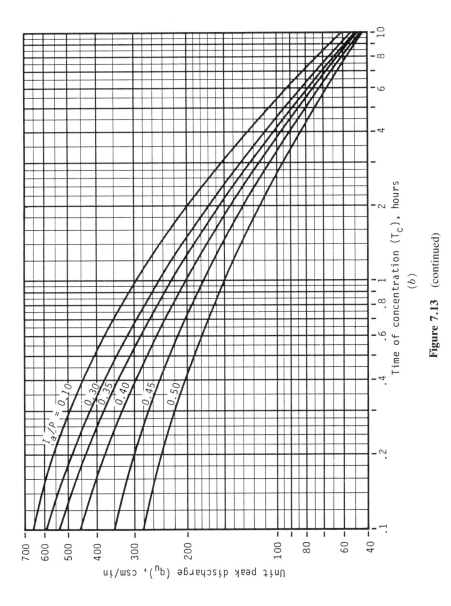

Figure 7.13 (continued)

(b)

255

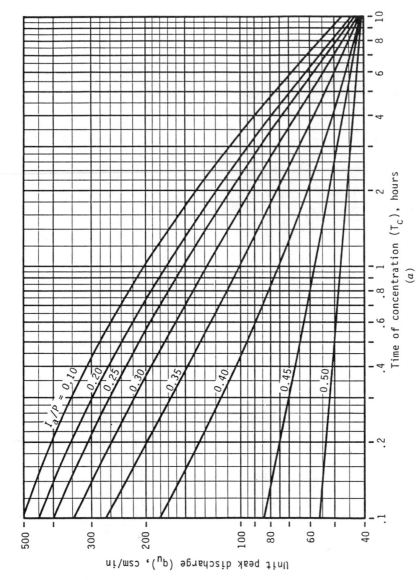

Figure 7.14 Unit peak discharges. (*a*) For SCS Type I rainfall distribution. (*b*) For SCS Type IA rainfall distribution. Source: U.S. Department of Agriculture—Soil Conservation Service. *TR-55*, 1986.

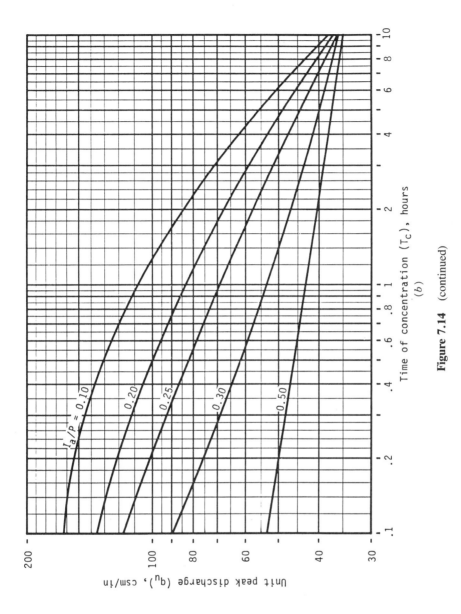

Figure 7.14 (continued)

257

Percentage of Pond and Swamp Areas	F_p
0	1.00
0.2	0.97
1.0	0.87
3.0	0.75
5.0	0.72

Figure 7.15 Adjustment factor, F_p, for ponds and swamp areas that are spread throughout the watershed. Source: U.S. Department of Agriculture—Soil Conservation Service. *TR-55*, 1986.

(c) Calculating I_a/P to obtain the unit peak discharge from either Figure 7.13 or 7.14.

(d) Calculating the runoff, Q, using Equation 7-24.

(e) Obtaining the pond or swamp factor, if applicable, using Figure 7.15.

(f) Calculating the peak discharge, q_p, using Equation 7-26.

According to TR-55 (USDA—SCS, 1986), the graphical method has the following limitations:

(a) Watershed areas must be hydrologically homogeneous with uniform soils, cover, and land use.

(b) Although the watershed can have more than one stream or branch, all must have approximately the same T_c.

(c) F_p applies only to ponds or swamps not in the T_c flow path.

7.4.3 The SCS Unit Hydrograph

A unit hydrograph is a hydrograph that represents one inch of runoff produced over a specified duration. The assumption is made that the rate of excess precipitation is constant. The duration allowing the one inch of excess precipitation is important because a different duration for the same watershed will produce a different unit hydrograph. Figure 7.16 illustrates the SCS dimensionless unit hydrograph.

This dimensionless hydrograph has the following characteristics*:

$$L = 0.6T_c \tag{7-27}$$

$$(D/2) + L = T_p \tag{7-28}$$

$$D + T_c = 1.7T_p \tag{7-29}$$

and

$$T = 5T_p \tag{7-30}$$

where:

L = The time of lag from the centroid of the inch of excess precipitation to the peak of the unit hydrograph.

*Only if the peak flow factor (PFF) is 484.

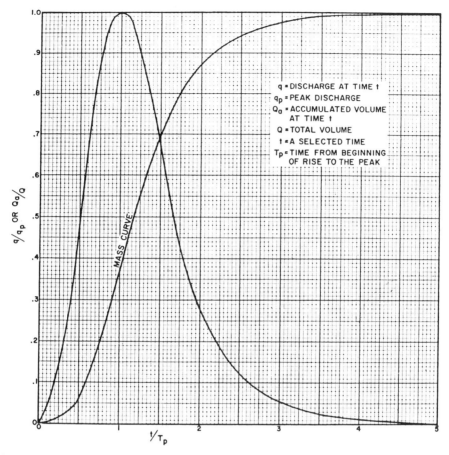

Figure 7.16 SCS dimensionless unit hydrograph and mass curve. Source: U.S. Department of Agriculture—Soil Conservation Service. *National Engineering Handbook*, 1972, Chapter 16.

D = The duration of excess precipitation.

T_p = The time from the beginning of excess to the peak of the unit hydrograph in hours.

T = Total time of direct runoff in hours.

T_c = The watershed time of concentration in hours.

The peak flow q_{up} of the unit hydrograph can be determined by:

$$q_{up} = \frac{484AQ}{T_p} \qquad (7\text{-}31)$$

where:

484 = A constant for a hydrograph with three eighths of its included area under its leading edge. This is called the peak flow factor (PFF). If the PFF is modified, then the unit hydrograph shape will change.

A = The drainage basin area in square miles.

Q = Excess precipitation in inches.

T_p = Time for the hydrograph to peak in hours, $(D/2 + L)$, or $(D/2 + 0.6T_c)$.

If the duration, D, is $0.133 \times T_c$, the unit hydrograph can be constructed. The following example illustrates the process using an assumed $T_c = 9$ minutes and a drainage basin area of 25 acres:

(a) L = $0.6T_c$, or 0.6 $\times (9/60) = 0.09$ hour.

(b) D = $0.133T_c$, or $0.133 \times (9/60) = 0.02$ hour.

(c) T_p = $(0.02/2 + 0.09)$ = 0.10 hour.

(d) q_{up} = $\dfrac{(484)(25 \text{ acres}/640 \text{ acres/square mile})(1 \text{ inch})}{0.10}$

or 189.1 cubic feet per second.

(e) The completed unit hydrograph is calculated by multiplying the q/q_P factors obtained from Figure 7.16 with the peak flow of 189.1 cubic feet per second. The results are included in Figure 7.17.

t (hours)	t/t_p	q/q_p	q (cfs)
0.00	0.0	0.000	0.00
0.02	0.2	0.100	18.91
0.04	0.4	0.310	58.62
0.06	0.6	0.660	124.81
0.08	0.8	0.930	175.86
0.10	1.0	1.000	189.10
0.12	1.2	0.930	175.86
0.14	1.4	0.780	147.50
0.16	1.6	0.560	105.90
0.18	1.8	0.390	73.75
0.20	2.0	0.280	52.95
0.22	2.2	0.207	39.14
0.24	2.4	0.147	27.80
0.26	2.6	0.107	22.23
0.28	2.8	0.077	14.56
0.30	3.0	0.055	10.40
0.32	3.2	0.040	7.56
0.34	3.4	0.029	5.48
0.36	3.6	0.021	3.97
0.38	3.8	0.015	2.84
0.40	4.0	0.011	2.08
0.42	4.2	0.009	1.70
0.44	4.4	0.006	1.13
0.46	4.6	0.004	0.76
0.48	4.8	0.002	0.38
0.50	5.0	0.000	0.00

Figure 7.17 Unit hydrograph ordinate calculations.

Figure 7.18*a* illustrates a unit hydrograph with an assumed unit duration of one hour. If one inch of excess precipitation falls on the watershed during the period of one hour, the unit hydrograph also represents the direct runoff hydrograph for that duration. If, however, two inches of excess precipitation fall within the hour, Figure 7.18*b* would represent the direct runoff hydrograph where each unit hydrograph is added linearly. Figure 7.18*c* shows a different situation, where one inch of direct runoff during the period of one hour is followed by another one inch of runoff during the subsequent hour. In this case the direct runoff hydrograph for the two durations is the sum of the two unit hydrographs with the second being lagged by one hour. Each direct runoff hydrograph for each time increment is therefore a ratio of the excess precipitation for a particular duration to that of one inch.

7.4.4 Direct Runoff Calculations

Before the direct runoff hydrograph can be calculated, the duration of the unit hydrograph and the time step used to determine excess precipitation must be the same. This would require the data contained in Figure 7.7 to be recomputed using a suitable time increment. A recalculated portion is shown in Figure 7.19 using a 0.02 hour increment. For this example *CN* is assumed to be 72; therefore, *S* equals 3.888 inches when Equation 7-25 is applied. Runoff, *Q*, is obtained with Equation 7-24 for each time period and corresponding rainfall depth (in inches). Incremental excess precipitation is shown between each time step as Delta *Q*.

A direct runoff hydrograph for each time step of the storm can then be calculated by multiplying each incremental excess precipitation (Delta *Q*) by every ordinate of the unit hydrograph, as shown in Figure 7.20. The 11.98 to 12 hour direct runoff hydrograph where Delta *Q* equals 0.138437 in Figure 7.19 is used in this example.

7.4.5 Composite Direct Runoff Hydrographs

Once each direct runoff hydrograph for each time step has been calculated, all of them can be added and lagged by time increment *D*, and any base flows can be included, to produce the total runoff hydrograph for the storm. Figure 7.18*c* illustrates the process graphically. This total runoff hydrograph can now be routed through downstream channels or reservoirs.

7.4.6 Hydrograph Routing

Hydrograph routing is associated with either stream or reservoir routing. Streamflow routing involves the process of moving a hydrograph from one upstream point to a downstream location. The stream's channel storage affects the amount of hydrograph attenuation.

Although there are a number of ways to route hydrographs through channels, the convex routing method developed by SCS is presented here. Figure 7.21 illustrates the fundamental relationship between the inflow and outflow hydrographs using this routing method. The basic equation for this relationship is:

$$O_{t+dt} = (1 - C)O_t + CI_t \qquad (7\text{-}32)$$

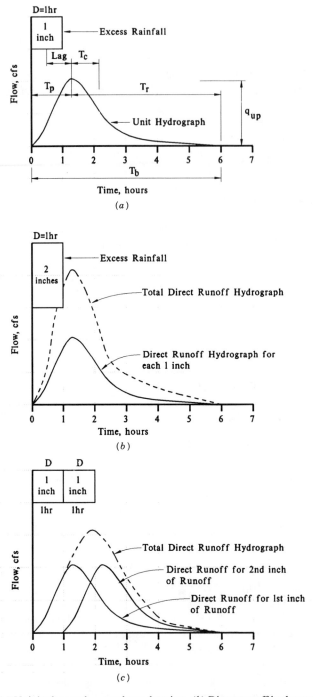

Figure 7.18 (*a*) Unit hydrograph—one hour duration. (*b*) Direct runoff hydrograph for two inches of excess precipitation. (*c*) Direct runoff hydrograph for subsequent one inch of excess precipitation.

Time (hours)	P_x/P_{24}	Depth (inches)	Q (inches)	Delta Q (inches)
11.96	0.6060	4.0481	1.494177	——
11.98	0.6345	4.2385	1.629852	> 0.135675
12.00	0.6630	4.4288	1.768289	> 0.138437
12.02	0.6678	4.4606	1.791616	> 0.023327
12.04	0.6725	4.4923	1.815013	> 0.023397

Figure 7.19 Calculations showing a partial development of the excess precipitation.

t (hours)	q-unit (cfs)	Direct Runoff (cfs)
0.00	0.00	0.0
0.02	18.91	2.6
0.04	58.62	8.1
0.06	124.81	17.3
0.08	175.86	24.3
0.10	189.10	26.2
0.12	175.86	24.3
0.14	147.50	20.4
0.16	105.90	14.7
0.18	73.75	10.2
0.20	52.95	7.3
0.22	39.14	5.4
0.24	27.80	3.8
0.26	22.23	3.1
0.28	14.56	2.0
0.30	10.40	1.4
0.32	7.56	1.0
0.34	5.48	0.8
0.36	3.97	0.5
0.38	2.84	0.4
0.40	2.08	0.3
0.42	1.70	0.2
0.44	1.13	0.2
0.46	0.76	0.1
0.48	0.38	0.1
0.50	0.00	0.0

Figure 7.20 Direct runoff hydrograph calculations for time increment 11.98–12.00 and Delta Q = 0.138437 inch.

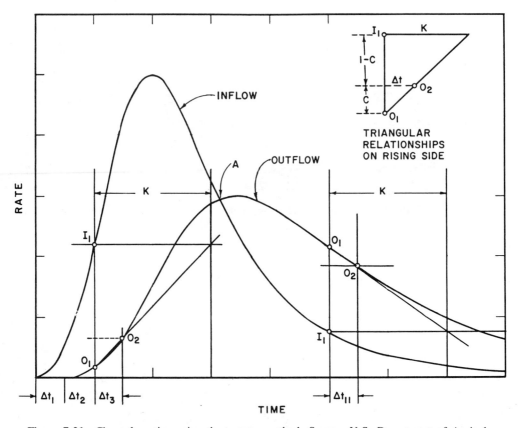

Figure 7.21   Channel routing using the convex method. Source: U.S. Department of Agriculture—Soil Conservation Service. *National Engineering Handbook*, 1972, Chapter 17.

where:

O_{t+dt} = The outflow at time $t + dt$.
O_t = The outflow at time t.
I_t = The inflow at time t.
C = The routing coefficient where $0 \le C \le 1$.

The term dt should be no more than about 20 percent of the time required for the inflow hydrograph to peak. The routing coefficient, C, can be estimated if the steady state velocity of drainage is known using:

$$C = \frac{V}{V + 1.7} \qquad (7\text{-}33)$$

where:

V = The velocity of travel in feet per second.
$V + 1.7$ = The approximate speed of a kinematic wave traveling through the reach.

To illustrate the routing of a hydrograph where the inflow hydrograph is assumed to be a direct runoff hydrograph with ordinates the same as the unit hydrograph shown in Figure 7.17. The steady state channel flow velocity is estimated at 1.5 feet per second. Using Equation 7-33, the routing coefficient C is equal to 0.46875. The routing calculations are shown in Figure 7.22.

Reservoir routing, however, considers the following general relationship:

$$\frac{\text{storage volume change}}{dt} = \text{inflow} - \text{outflow} \qquad (7\text{-}34)$$

In an expanded form this equation appears as:

$$\frac{I_1 + I_2}{2} - \frac{O_1 + O_2}{2} = \frac{S_2 - S_1}{dt} \qquad (7\text{-}35)$$

where:

I_1 = The inflow in cubic feet per second at time t.
I_2 = The inflow in cubic feet per second at time $t + dt$.

t (hour)	Inflow (cfs)	Routing $(1 - C)Ot + C It$		Outflow (cfs)
0.00	0.0	$(1 - 0.46875)$ 0.0 + (0.46875) 0.0		0.0
0.02	18.9	$(1 - 0.46875)$ 0.0 + (0.46875) 0.0		0.0
0.04	58.6	$(1 - 0.46875)$ 0.0 + (0.46875) 18.9		8.9
0.06	124.8	$(1 - 0.46875)$ 8.9 + (0.46875) 58.6		32.2
0.08	175.9	$(1 - 0.46875)$ 32.2 + (0.46875)124.8		75.6
0.10	189.1	$(1 - 0.46875)$ 75.6 + (0.46875)175.9		122.6
0.12	175.9	$(1 - 0.46875)$122.6 + (0.46875)189.1		153.8
0.14	147.5	$(1 - 0.46875)$153.8 + (0.46875)175.9		164.2
0.16	105.9	$(1 - 0.46875)$164.2 + (0.46875)147.5		156.4
0.18	73.8	$(1 - 0.46875)$156.4 + (0.46875)105.9		132.7
0.20	53.0	$(1 - 0.46875)$132.7 + (0.46875) 73.8		105.1
0.22	39.1	$(1 - 0.46875)$105.1 + (0.46875) 53.0		80.7
0.24	27.8	$(1 - 0.46875)$ 80.7 + (0.46875) 39.1		61.2
0.26	22.2	$(1 - 0.46875)$ 61.2 + (0.46875) 27.8		45.5
0.28	14.6	$(1 - 0.46875)$ 45.5 + (0.46875) 22.2		34.6
0.30	10.4	$(1 - 0.46875)$ 34.6 + (0.46875) 14.6		25.2
0.32	7.6	$(1 - 0.46875)$ 25.2 + (0.46875) 10.4		18.3
0.34	5.5	$(1 - 0.46875)$ 18.3 + (0.46875) 7.6		13.3
0.36	4.0	$(1 - 0.46875)$ 13.3 + (0.46875) 5.5		9.6
0.38	2.8	$(1 - 0.46875)$ 9.6 + (0.46875) 4.0		7.0
0.40	2.1	$(1 - 0.46875)$ 7.0 + (0.46875) 2.8		5.0
0.42	1.7	$(1 - 0.46875)$ 5.0 + (0.46875) 2.1		3.6
0.44	1.1	$(1 - 0.46875)$ 3.6 + (0.46875) 1.7		2.7
0.46	0.8	$(1 - 0.46875)$ 2.7 + (0.46875) 1.1		2.0
0.48	0.4	$(1 - 0.46875)$ 2.0 + (0.46875) 0.8		1.4
0.50	0.0	$(1 - 0.46875)$ 1.4 + (0.46875) 0.4		0.9

Figure 7.22 An example of stream flow routing using the convex method with $C = 0.46875$.

O_1 = The outflow in cubic feet per second at time t.
O_2 = The outflow in cubic feet per second at time $t + dt$.
S_1 = The storage volume in cubic feet at time t.
S_2 = The storage volume in cubic feet at time $t + dt$.
dt = The time interval.

The outflow and storage volumes are dependent upon the surface level of the water in the reservoir. Two curves shown in Figure 7.23a and Figure 7.23b represent these relationships based on the reservoir's discharge structure and geometry. These two curves can be combined into a third curve, Figure 7.23c, which represents the relationship between storage and discharge.

One method that can be used for reservoir routing is called the "storage-indication" method. This method is often used in land development projects because discharge rates are used for input and output instead of using mass curves.

According to NEH 4, Chapter 17 (USDA—SCS, 1972), the storage-indication method utilizes the following equation:

$$I_{ave+} + \frac{S_1}{dt} - \frac{O_1}{2} = \frac{S_2}{dt} + \frac{O_2}{2} \tag{7-36}$$

where:

$I_{ave} = (I_1 + I_2)/2$; the values for I can be taken from regular intervals or at the midpoints of routing intervals.

The working relationship between O_2 and $[(S_2/dt) + (O_2/2)]$ must be available in either numerical or graphical form to solve Equation 7-36. Care must be exercised to insure that an acceptable routing interval is used. According to NEH 4 (USDA—SCS, 1972):

Trial routings show that negative outflows will occur during recession periods of the outflow whenever dt is greater than $2S_2/O_2$ (or whenever $O_2/2$ is greater than S_2/dt). This also means that rising portions of hydrographs are being distorted.*

These problems can be avoided by plotting the working O_2 versus $[(S_2/dt) + (O_2/2)]$ values on a log-log graph, along with a second curve where each O_2 value equals each $[(S_2/dt) + (O_2/2)]$ value. If any portion of the working O_2 versus $[(S_2/dt) + (O_2/2)]$ curve plots above the second curve (representing a plot of equal values), then a sufficiently small dt increment must be utilized to produce a plot that is completely below the curve of equal values.

To illustrate the storage-indication reservoir routing procedure, a reservoir that has a square surface will be used. Each side is assumed to be 100 feet long and the reservoir sides are assumed to be vertical (only for simplicity in this example). The outlet structure provides discharge rates that are shown in Figure 7.24. Once the inflow hydrograph and initial reservoir discharges are known, hydrograph routing can be performed.

The inflow hydrograph for this example is the same as the outflow hydrograph contained in Figure 7.22, thus requiring a time step, dt, of 0.02 hour. The working relationship between O_2 and $[(S_2/dt) + (O_2/2)]$ is determined using data contained in Figure 7.24.

*Taken from USDA—SCS, NEH 4, 1972.

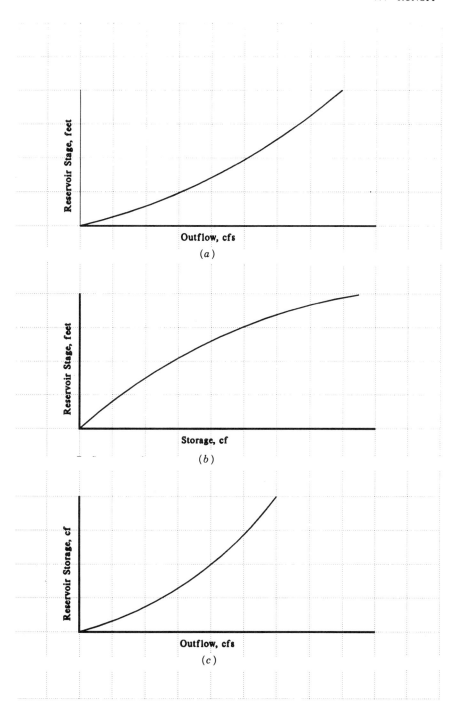

Figure 7.23 Reservoir outflow and storage relationships. (*a*) Stage versus discharge. (*b*) Stage versus storage. (*c*) Storage versus outflow.

Stage (feet)	Disharge (cfs)	Storage (cft)
0	0	0
2	12	20,000
4	26	40,000
6	34	60,000
8	42	80,000

Figure 7.24 An example of reservoir stage, discharge, and storage relationships.

The discharge values in cubic feet per second are taken directly from the figure, while the values for S_2 must be in terms of "cfs-hours", therefore requiring the storage values shown in cubic feet to first be converted into acre-feet, and then into "cfs-hours" by multiplying each storage value in equivalent acre-feet by a factor of "12.10." Figure 7.25 shows the completed working relationships.

The O_2 and $[(S_2/dt) + (O_2/2)]$ values shown in Figure 7.25 can be used to numerically solve the storage-indication equation, or they can be plotted as a curve for graphical use. Equation 7-36 is solved numerically in Figure 7.26. Initially when solving the routing equation for this example, the beginning time, inflow, average inflow, $[(S_2/dt) + (O_2/2)]$, and outflow are known. Also known are other times, inflows, and average inflow values.

To solve for the $[(S_2/dt) + (O_2/2)]$ value at time equal to 0.02 hour, subtract the value of column (5) from the value of column (4) for time 0.00, and add to this difference the value of column (3) when the time is 0.02 hour. This yields $(0 - 0 + 0) = 0$ for $[(S_2/dt) + (O_2/2)]$ when time equals 0.02 hour. According to Figure 7.25, the corresponding value of O_2 is 0.0 cubic foot per second.

To solve for the $[(S_2/dt) + (O_2/2)]$ value when time equals 0.04 hour, subtract the value of column (5) from the value of column (4) in Figure 7.26 for time 0.02, and add to this difference the value of column (3) when the time is 0.04 hour. This yields $(0 - 0 + 4.45) = 4.45$ cubic feet per second for $[(S_2/dt) + (O_2/2)]$ when time equals 0.04 hour. The corresponding value of O_2 can be obtained by interpolating (or graphically) using data found in Figure 7.25. In this case $O_2 = (4.45/283.75)(12) = 0.188$ cubic foot per second when time equals 0.04 hour.

By repeating these steps, $[(S_2/dt) + (O_2/2)]$ for time 0.06 hour is found to be $(4.45 - 0.188 + 20.55)$, or 24.812 cubic feet per second. The corresponding outflow can be found using Figure 7.25 as $(24.812/283.75)(12)$, or 1.049 cubic feet per second. The remainder of the outflow hydrograph is obtained using these same procedures.

O_2 (cfs)	$O_2/2$ (cfs)	S_2 (cfs-hours)	S_2/dt (cfs)	$[(S_2/dt) + (O_2/2)]$ (cfs)
0	0	0	0	0
12	6	12.1(20,000/43560) = 5.555	277.75	283.75
26	19	12.1(40,000/43560) = 11.111	555.55	574.55
34	30	12.1(60,000/43560) = 16.667	833.35	863.35
42	38	12.1(80,000/43560) = 22.222	1111.10	1149.10

Figure 7.25 An example of working relationships between O_2 and $(S_2/dt) + (O_2/2)$.

T (hours) (1)	I (cfs) (2)	I_{ave} (cfs) (3)	$[(S_2/dt) + (O_2/2)]$ (cfs) (4)	O (cfs) (5)
0.00	0.0	0.00	0.000	0.000
0.02	0.0	0.00	0.000	0.000
0.04	8.9	4.45	4.450	0.188
0.06	32.2	20.55	24.812	1.049
0.08	75.6	53.90	77.663	3.284
0.10	122.6	99.10	173.479	7.337

Figure 7.26   The storage-indication method of hydrograph routing through a reservoir.

The reservoir stage and storage values can be obtained for each time step using Equation 7-34 along with the routed hydrograph from the reservoir and the reservoir's geometry. An example is given later in this chapter where computer analysis provides hydrograph routing through a reservoir.

7.5   FACTORS AFFECTING SITE DRAINAGE

The engineer must establish various criteria to follow during the drainage design process including:

(a) Projects located on ridge lines that receive no off-site storm water allow the design engineer to concentrate strictly on drainage design within the site.

(b) Sites located away from ridge lines require the engineer to determine whether off-site storm runoff will contribute to on-site design flows. If so, provisions must be made to allow the off-site runoff to pass through the development and, if necessary, to intercept these flows for flood prevention.

(c) Generalized lot grading plans (see Chapter 9) must be established to provide positive site drainage.

(d) Regulatory discharge limits for storm water must be obtained to establish allowable post-construction site discharge rates. Requirements for storm water retention or detention facilities are also necessary.

(e) The maximum distance that runoff can be carried in a roadside ditch or curb and gutter before it must be removed must also be determined. Some development regulations specify maximum placement distances for standard inlet devices. These distances can be used when making the preliminary development plan; however, inlet performance should be verified during the final design. If local regulations do not specify a maximum interval, an analysis must be made to determine maximum inlet spacing based on performance.

7.5.1   Inlet Spacing Interval

A maximum spacing interval for curb inlets on a project can be estimated by analyzing a typical road cross section (similar to Figure 7.27a) on minimum grade using a typical inlet device. The simplified road section shown in Figure 7.27b is used in this example for inlet

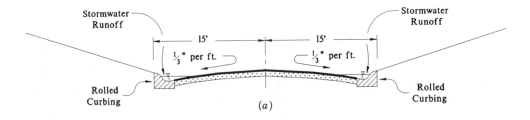

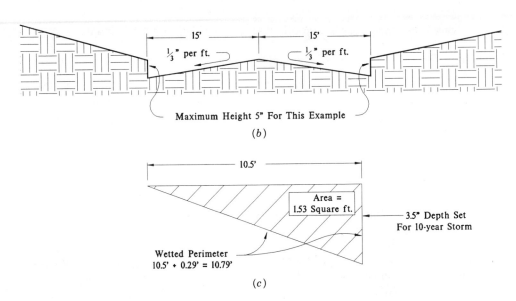

Figure 7.27 Roadway drainage analysis using a rolled curb section. (*a*) Road cross section for storm drainage analysis. (*b*) Simplified road cross section for storm drainage analysis. (*c*) Flow area when 3.5 inches depth of flow occurs in the gutter. This will allow a single 9-foot wide traffic lane for use during a 10 year storm.

spacing analysis. Manning's equation, Equation 7-19, is used to determine the maximum flow capacity when a predetermined depth of flow is established in the gutter.

If the maximum flow depth in the curb and gutter half section is arbitrarily set for this example at approximately 3.5 inches for a 10 year return period, the corresponding flow cross section would appear as shown in Figure 7.27*c* for half of the street. If the minimum centerline slope of the road is 0.5 percent, the amount of runoff that can be carried by half of the street section can be found where:

(a) The area of the flowing triangle equals (1/2) (3.5/12 feet depth of flow)(10.5 feet wide), or 1.53 square feet.

(b) The hydraulic radius, A/P, equals (1.53 square feet)/{(3.5/12) + [10.5^2 + (3.5/12)^2]$^{0.5}$}, or 0.142 foot.

(c) The minimum road slope is given as 0.5 percent, or 0.005 feet per foot of roadway.

(d) Manning's roughness coefficient, n, is taken as 0.013 from Figure 7.12.

Using Equation 7-19, the flow is determined as:

$$Q = \frac{1.49(1.53)(0.142)^{0.67}(0.005)^{0.5}}{0.013} = 3.35 \text{ cubic feet per second}$$

This is the capacity of curb flow for the 10 year return period that will be used to estimate inlet spacing.

If a weighted runoff coefficient of 0.54 is used with the rational formula, Equation 7-21, the relationship of rainfall intensity to drainage area is established by:

$$Q = CIA$$
$$3.35 = (0.54)(I)(A)$$
$$(I)(A) = 6.2 \text{ acre-inches per hour}$$

The corresponding velocity of flow for a 10.5 foot wide by 3.5 inch deep triangular shaped cross section is computed using Equation 7-18 by:

$$Q = AV.$$
$$V = 3.35 \text{ cubic feet per second}/(1.53 \text{ square feet})$$
$$V = 2.19 \text{ feet per second}$$

If a typical lot is to be 100 feet wide and the typical lot area is one-third acre, the corresponding lot depth is approximately 145 feet. The contributory drainage area encompassed by one lot and half the fronting road (assuming a 50 foot right of way) is [100 × (145 + 25)]/43,560, or 0.39 acre. This assumes that the entire lot drains toward the street. The time of travel from the remotest point in the lot (at the rear lot line) to the street can be estimated using the sheet flow equation, Equation 7-13, and Figures 7.5 (for the 10 year return period curve) and 7.11, where:

$$T_t = \frac{(0.93)(n \times L)^{0.6}}{(i)^{0.4}(S)^{0.3}} = \frac{(0.93)(0.15 \times 145)^{0.6}}{(i)^{0.4}(0.005)^{0.3}}$$

By trial and error, it is determined that the above equation is satisfied when $i_{10} = 5.73$ inches per hour, and the duration is 14.38 minutes. The amount of time required for the runoff to travel 100 feet in the street gutter corresponding to one lot width is (100 feet per lot width)/(2.19 feet per second), or 45.66 seconds. This value is rounded to 0.75 minute for these preliminary calculations. Figure 7.28 is constructed to aid in determining the maximum number of contributory lots. This maximum number provides the largest drainage area that can be handled by this road section using a 10 year return period design.

The $(I)(A)$ based on minimum slopes for this roadway is 6.2 acre-inches per hour; therefore, three lots approximate the maximum contributory drainage area to produce an approximate flood depth of 3.5 inches. This would indicate a maximum spacing between inlet devices of 3 × 100 feet (typical lot width for this development) or 300 feet, unless a sag vertical curve low point is present, as illustrated in Figure 7.29.

The performance of all inlet structures should be verified to insure their adequacy. To

Number of Lots	T_c (minutes)	I (inches per hour)	A (acres)	$(I)(A)$ (acre-inches per hour)
1	(14.4 + 0.75 = 15.15); use 15.2	5.54	0.39	2.16
2	15.15 + 0.75 = 15.9	5.48	0.78	4.27
3	(15.9 + 0.75 = 16.65); use 16.7	5.42	1.17	6.34
4	16.65 + 0.75 = 17.4	5.37	1.56	8.38
5	(17.4 + 0.75 = 18.15); use 18.2	5.32	1.95	10.37
6	18.15 + 0.75 = 18.9	5.25	2.34	12.29

Figure 7.28   Inlet spacing capacities for various contributory areas.

determine the required length of throat of a curb inlet that depends on weir flow, the generalized weir formula can be utilized (Walesh, 1989):

$$Q = C_w L h^{1.5} \tag{7-37}$$

where:

Q = The flowrate in cubic feet per second.
C_w = The weir coefficient.
L = The throat length in feet.
h = The depth of flow in the gutter in feet.

If an estimated weir coefficient of 3 is assumed, based on free discharge conditions, the corresponding length of inlet throat to accommodate a 10 year return period would be:

$$3.35 \text{ cubic feet per second} = 3 \times L \times (3.5/12)^{1.5},$$

or

$$L = 7.09 \text{ feet}$$

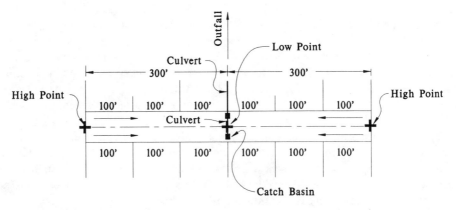

Figure 7.29   An example of positioning of curb inlets at a vertical sag curve low point.

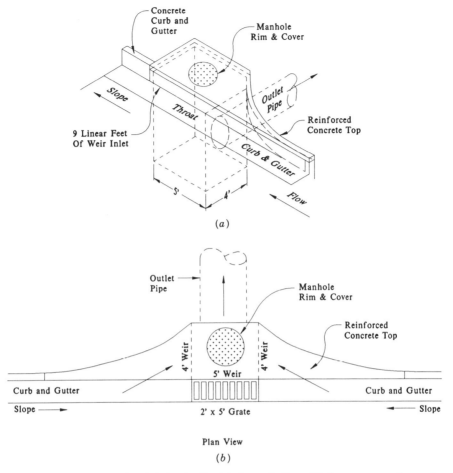

Figure 7.30 Examples of curb inlet details. (*a*) On an incline. (*b*) At a sag curve low point.

A curb inlet similar to Figure 7.30*a* would provide adequate inlet capacity with a five foot inlet parallel to the street and another four foot inlet perpendicular to the street positioned on the uphill side. If the inlet, however, is located at the low point of a sag curve where the contributory flow totals 6.70 cubic feet per second (from both directions), a different inlet configuration must be utilized. By including an additional four foot inlet perpendicular to the street, a total throat length of 13 feet can be obtained. The required throat length for the 6.70 cubic foot per second inflow, however, is 14.18 feet. A grate casting that is two feet by five feet can be installed as shown in Figure 7.30*b* to increase the inlet's capacity using orifice flow through the grate. The generalized orifice equation is (Walesh, 1989):

$$Q = C_d A(2gh)^{0.5} \tag{7-38}$$

where:

Q = The orifice flow in cubic feet per second.

C_d = The coefficient of discharge, taken as 0.6 for the grate entrance.

A = The cross-sectional area of the opening. In the case of a grate, the sum of all openings is used.

g = 32.2 feet per second per second.

h = The depth of flow in feet.

If the proposed grate contains 16 slots, each 1.5 inches × 16 inches, the total grate opening area is 16 slots × (1.5 inches × 16 inches)/144 square inches per square foot, or 2.66 square feet. Using the approximate flood depth of 3.5 inches for the 10 year return period, the grate inflow may be computed through application of Equation 7-38 as:

$$Q = 0.6 \times 2.66 \times [2 \times 32.2 \times (3.5/12)]^{0.5} = 6.92 \text{ cubic feet per second}$$

The curb inlet and grate should be adequate for sag locations. As an alternative to using the weir and orifice equations, the design professional can use inlet capacity charts or nomographs that are available to determine inflow rates. If charts or nomographs are used, the configuration of the inlet used in the design aid must be the same as that proposed for installation in the field.

7.5.2 Retention/Detention Area Requirements

The engineer should determine the approximate size of any required retention/detention facilities based on local ordinances since adequate land area must be set aside for that use. The difference between retention and detention is that a retention facility has zero discharge while a detention facility has a limited discharge. Some regulatory agencies may require that:

(a) Maximum discharges from storage facilities are to be limited to the peak storm water runoff rates that prevailed prior to site development.

(b) Storm water runoff is to be stored and its release controlled when a site's modification results in the peak runoff rate being increased by more than one cubic foot per second for a five year return period.

(c) Areas having runoff that is directly channeled into receiving streams are to be required to provide on-site retention volumes equivalent to at least 1.5 inches of runoff depth. This volume can include a portion of a detention basin where the basin's outlet structure is elevated to provide the necessary storage. To eliminate standing water (if the basin does not have a permanent pool as in a lake), a subsurface drain can be installed to allow the water to slowly drain and at the same time be filtered. Natural wetland areas also can be used to provide the required on-site retention volume.

Other ways to provide on-site retention can include raising inlet elevations to allow runoff to pond and infiltrate if the soil is porous; otherwise this method is not used. Dutch drains (trenches filled with granular material encased in filter fabric) also can be used to temporarily store runoff underground until it can percolate into the soil, if the groundwater table is low enough.

Retention of parking lot runoff might be achieved by installing a sump into each catch basin and including a baffle inlet to keep any oil in the runoff from leaving the sump. Sumps such as these require regular cleaning to perform adequately.

An approximate detention facility size can be determined once the maximum allowable discharge from the developed site is known. Assuming that the allowable rate will be the predevelopment rate, the allowable discharge is computed using the rational formula, Equation 7-21. The watershed area, the coefficient of runoff, and the time of concentration must be determined. The rainfall intensity (I) can be obtained from the rainfall intensity curve for the area. Figure 7.5 is assumed to be applicable for the following example.

To illustrate the process, if the contributory area of a development consists of 5.13 acres (Figure 4.31), the predevelopment rate of runoff can be calculated. The site is moderately flat and has an estimated coefficient of runoff of 0.20.

The time of concentration is the maximum time required for a drop of water to travel from the remotest point within the drainage basin to the point under investigation, which in this case is the discharge point for the site. Figure 4.40 shows the existing topography, which can be used as a basis for time of concentration determination.

For this example, the time of concentration is divided into three separate travel times to include sheet flow, shallow concentrated flow, and channel flow. Figure 4.40 indicates where estimated sheet flow will occur between points A and B. Shallow concentrated flows are expected between points B and C, while channelized flows will occur between points C and D.

The time of travel, T_1, for sheet flow is calculated using Equation 7-13. Manning's roughness coefficient, n, is estimated from Figure 7.11 as 0.15. The scaled length of travel is approximately 300 feet, and the ground slope is estimated as 0.0033 feet per foot. T_1 must be calculated for each return period using the rainfall intensity curve for that particular return period. T_1 for the two year return period is estimated, with Equation 7-13 being solved by trial and error, as:

$$T_1 = \frac{(0.93)(n \times L)^{0.6}}{(i_2)^{0.4}(S)^{0.3}} = \frac{(0.93)(0.15 \times 300)^{0.6}}{(i_2)^{0.4}(0.0033)^{0.3}}$$

This equation is satisfied when the intensity is 3.227 inches per hour for the two year storm, and the corresponding duration is 31.72 minutes. Other T_1 values are shown in Figure 7.31.

The velocity of travel, V_2, for shallow concentrated flow is calculated using Equation 7-15. The travel distance is estimated as 120 feet, and the corresponding ground slope at 0.0125 feet per foot.

$$V_2 = 16.1345 \,(0.0125)^{0.5} = 1.8 \text{ feet per second}$$

The time of travel, T_2, can be obtained by utilizing Equation 7-20:

$$T_2 = \frac{120}{3600 \times 1.8} = 0.02 \text{ hour}$$

The velocity of travel, V_3, for channelized flow for line C–D is calculated using Equation 7-17. The cross-sectional area is estimated at 5 square feet, and the wetted perimeter is estimated at 10 feet. Manning's roughness coefficient, n, is taken as 0.05 from Figure 7.12 for a ditch bank with weeds and brush, while the slope is estimated as 0.0125 feet per foot.

Return Period (Years)	2	5	10	25	50	100
T_1 (minutes)	31.7	29.1	27.7	25.9	24.8	23.8

Figure 7.31 Travel time, T_1, for six return periods.

$$V_3 = \frac{1.49(0.5)^{0.67}(0.0125)^{0.5}}{0.05} = 2.1 \text{ feet per second}$$

The time of travel, T_3, can again be obtained by utilizing Equation 7-20 and the scaled distance between points C and D as 280 feet.

$$T_3 = \frac{280}{3600 \times 2.1} = 0.04 \text{ hour}$$

The time of concentration for the two year return period, T_c, is equal to $T_1 + T_2 + T_3$, or $0.53 + 0.02 + 0.04 = 0.59$ hour (or approximately 35 minutes). Other times of concentration are shown in Figure 7.32.

The site's predevelopment discharge rates, shown in Figure 7.33 for various return periods, are found using Equation 7-21, where the coefficient of runoff is estimated to be 0.2, a contributory area to be 5.13 acres, and an intensity for each return period to correspond to the time of concentration shown in Figure 7.32.

The FAA method of ponding can be used to approximate the volume of required detention storage (U.S. Department of Transportation—FAA, 1970). Should the discharge be regulated at zero (for a retention pond), the procedure is identical with zero discharge. The volume of inflow is computed using:

$$V_{\text{storage}} = CIAT \tag{7-39}$$

where:

V_{storage} = The required storage volume in cubic feet.
$\phantom{V_{\text{storage}}}C$ = The coefficient of runoff.
$\phantom{V_{\text{storage}}}I$ = The rainfall intensity in inches per hour for time T.
$\phantom{V_{\text{storage}}}T$ = The time of runoff volume accumulation in seconds.
$\phantom{V_{\text{storage}}}A$ = The drainage area in acres.

A coefficient of runoff is assumed for this example to be 0.54 for post-construction conditions. Using Figure 7.4 the rainfall intensities for durations of 5, 10, 15, 30, 60, 120, 180, and 360 minutes are obtained for the 100 year return period. The corresponding volumes are calculated using Equation 7-39. Volumes of runoff are plotted as ordinates

Return Period (years)	2	5	10	25	50	100
T_c (minutes)	35	33	31	30	28	27

Figure 7.32 Times of concentration of a sample undeveloped site for six return periods.

Return Period (years)	Predevelopment Q (cfs)
2	3.20
5	3.89
10	4.43
25	5.15
50	5.86
100	6.49

Figure 7.33 Examples of predevelopment discharge rates for six return periods.

and corresponding times as abscissas. The allowable discharge for the 100 year return period is then plotted as a straight line (starting at time zero with zero discharge). The accumulated discharge volume when time is equal to 360 minutes (360 minutes × 60 seconds per minute × 6.49 cubic feet per second, or 140,184 cubic feet) is also plotted, where this point and the initial point are connected. When the curves are completed, the time that provides the maximum vertical separation between the two curves yields the required storage volume. Examples of calculations and curves are shown in Figure 7.34.

100 Year Storm- Allowable Discharge: 6.49 cfs.			
Time Min.	Intensity In/hr.	Inflow Volume CFT	Discharge CFT
0	0.00	0.00	0.00
5	9.60	7978.18	1947.00
10	8.52	14161.26	3894.00
15	7.40	18449.53	5841.00
30	6.06	30217.34	11682.00
60	4.25	42384.06	23364.00
120	2.44	48666.87	46728.00
180	1.83	54750.23	70092.00
360	1.18	70606.86	140184.00

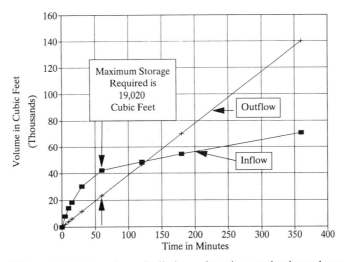

Figure 7.34 FAA method of ponding—Preliminary detention pond volume determination for a sample 100 year return period.

An approximate basin surface area can be obtained by scaling the proposed detention area on the preliminary plan (or using a planimeter), as shown in Figure 7.35. This surface area is necessary to compute an approximate basin depth to check outfall elevations.

The maximum required storage for the 100 year storm is 19,020 cubic feet; however, a value of twice this should initially be used to compensate for approximations and to allow flexibility while designing. The estimated basin size (assuming a 90 foot by 90 foot surface area) is 10 feet deep with 2 : 1 side slopes. The resulting bottom would be 50 feet by 50 feet. The approximate water depth to provide 38,040 cubic feet of storage is 8.15 feet. Based on the topography, this preliminary depth appears to be satisfactory, based on contour lines shown in Figure 7.35, which indicate an elevation difference of approximately 13 feet between the pond site and the downstream invert of the outfall ditch. This elevation difference allows proposed culverts to be placed with sufficient cover and slope. If the estimated storage depth is large where gravity flow into the site's storm water outfall cannot be utilized, then the basin's surface area could be increased to allow a corresponding decrease in detention pond depth. This may accommodate the terrain requirements; otherwise additional basins may be required.

7.6   DRAINAGE SYSTEM DESIGN

After adequate surface area has been reserved for basin sites, design of the storm drainage collection system can begin. If the site is subject to storm water runoff from adjacent property because of the surrounding terrain, then the amount of runoff expected from off-site must be determined. Two situations are possible alone or in combination:

(a) The site is subject to sheet or overland flow from the adjoining property. If this occurs, a swale (shallow vee bottom ditch) or trapezoidal ditch can be provided parallel and contiguous to the property line where the sheet flows cross to the site for interception and diversion. This ditch system can provide additional benefits by lowering the groundwater table if high groundwater is present.

(b) The site is subject to channelized flow from the adjoining property. In this case adequate provisions must be made to allow this off-site water to flow downstream without creating on-site problems.

7.6.1   Contributory Drainage Areas

Interior drainage areas can be identified by using different colored pencils on a transparent paper overlay placed on top of the preliminary plan or by using computer aided drafting (CAD) techniques similar to this. Either system provides an excellent method to delineate the different contributory areas while determining how each lot will be individually graded to provide positive drainage. The centerline of the road (pavement crown) is usually a limiting boundary for drainage areas where curb and gutter road sections are used. Figure 7.36 shows three contributory areas that generate runoff on the 5.13 acre site. Areas A_1 and A_2 drain into the frontage road, while area A_3 drains into the swale ditch along the rear property line. Pipe 1's inflow comes from contributory area A_1. Pipe 2 has a contributory flow from areas A_1 and A_2, while pipe 3 has a contributory flow from A_1, A_2, and A_3.

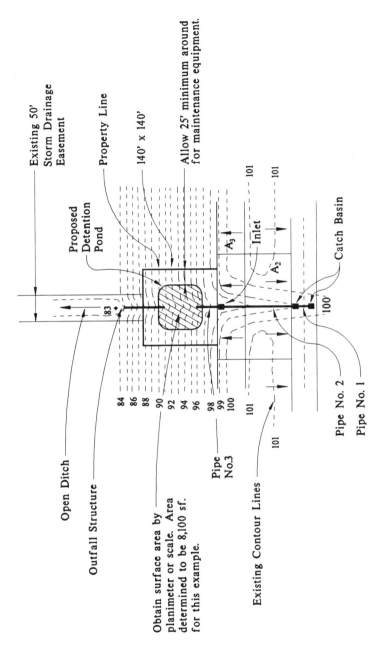

Figure 7.35 Initial detention basin plan for the sample drainage system.

Existing 50'
Storm Drainage
Easement

Property Line

140' x 140'

Allow 25' minimum around
for maintenance equipment.

Proposed
Detention
Pond

Open Ditch

Outfall Structure

Obtain surface area by
planimeter or scale. Area
determined to be 8,100 sf.
for this example.

Existing Contour Lines

Pipe
No.3

84
86
88
90
92
94
96
98
99
100

101

101

101

101

101

A_3

Inlet

A_2

Catch Basin

Pipe No. 2

Pipe No. 1

100'

83

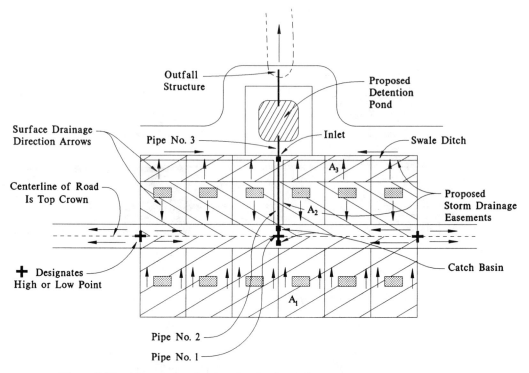

Figure 7.36 An example of a site plan for the drainage system design and analysis.

7.6.2 The Design Process

The availability of computer software, such as HYDROS (Debo, 1987), which utilizes hydrograph generation and routing, allows the design professional to first design the storm drainage system using the rational method and then analyze the designed system's performance using computer generated hydrographs. Should imperfections surface while routing various hydrographs through the various components with the computer software, adjustments can be made and the hydrographs rerun to refine the design.

Initially several rules should be followed that, with experience, can be modified by the designer. The rules are:

(a) Determine the minimum allowable slopes that produce a self-scouring velocity for all culverts. Some regulations stipulate minimum slopes that should be used to initially establish the culvert alignment throughout the system.

(b) Use a uniform material of construction for all culverts to ease calculations where Manning's roughness coefficient is constant. Alterations can be made later, if required.

(c) Size culverts beginning at the furthest point upstream and progress downstream in logical sequence. The time of concentration must be recalculated for each culvert using the longest time of travel within the basin to the point under consideration. In addition, the composite coefficient of runoff must be determined. If a junction is encountered where other tributary flows enter the system, do not proceed downstream until all other upstream channels have been sized. The process continues until the outfall structure is reached.

Pipes are usually sized to flow full while carrying the peak design runoff under gravity flow conditions. If ponding is allowed at culvert entrances, then a capacity chart similar to Figure 7.37 can be utilized to determine flows.

(d) Establish all invert elevations (the elevation of the bottom inside of the channel or pipe where flow first begins) commencing with the outfall, or discharge, and proceed upstream in logical sequence to the most elevated contributory area, using the same pipe or channel slopes that were included for capacity determination.

7.6.3 An Example of Drainage System Design

To illustrate the design process, an example using the rational method follows, using a 10 year return period for design. Figure 7.5 is assumed to represent local rainfall conditions, while Figure 7.36 represents the site conditions.

(a) Compute the composite coefficient of runoff using Equation 7-11 for A_1 and using a typical lot for calculations.

Typical lot size	= 0.33 acre
$\frac{1}{2}$ road R/W	= 0.05 acre
Total area	= 0.388 acre

APPROXIMATE IMPERVIOUS AREAS

House roof	= 1600 square feet
Paved driveway	= 600 square feet
Paved patio	= 400 square feet
Roadway pavement	= 1500 square feet
Total impervious	= 4100 square feet = 0.094 acre

A coefficient of runoff of 0.4 is used for nonimpervious areas and a value of 0.95 is used for impervious areas.

$$C_{w1} = \frac{C_1 A_1 + C_2 A_2}{A_1 + A_2} = \frac{0.094 \times 0.95 + 0.4 \times 0.294}{0.388} = 0.54$$

(b) Compute the composite coefficient of runoff using Equation 7-11 for A_1 and A_2 using a typical lot. Since A_2 has less contributory area, a typical lot in A_2 needs to be evaluated.

Typical lot portion contributing to A_2 = 90 × 100	= 9000 square feet
$\frac{1}{2}$ road R/W (25 feet × 100)	= 2500 square feet
Total	= 11,500 square feet = 0.264 acre

$$C_{w12} = \frac{0.094 \times 0.95 + 0.094 \times 0.95 + 0.4 \times 0.294 + 0.4 \times 0.17}{0.094 + 0.094 + 0.294 + 0.17} = 0.56$$

(c) Calculate the composite coefficient of runoff for A_1, A_2, and A_3, using Equation 7-11, by:

$$C_{w123} = \frac{0.094 \times 0.95 \times 2 + 0.4 \times 0.294 + 0.4 \times 0.17 + 0.4 \times 0.126}{0.778} = 0.54$$

Figure 7.38 depicts the contributory area features required for design calculations.

(d) Compute T_c for A_1. T_c is the time for a drop of water to travel 145 linear feet perpendicular to the road and then 300 linear feet in the road gutter to the curb inlet. With a ground and road slope of 0.5 percent, T_c is obtained from previous calculations included in Section 7.5.1, where the lot flow time for sheet flow is 14.4 minutes and the street travel time is 300 feet/ 2.19 feet per second, or 2.3 minutes. The time of concentration is estimated as 16.7 minutes.

(e) Use Figure 7.5 to determine the intensity corresponding to a time of concentration of 16.7 minutes for a 10 year return period; this intensity is 5.42 inches per hour.

(f) Compute the peak discharge for A_1 using Equation 7-21:

$$Q_1 = 0.54 \times 5.42 \times 2.34, \text{ or } 6.85 \text{ cubic feet per second}$$

(g) Determine the size of pipe 1 using $S_{min} = 0.3$ percent and Manning's roughness coefficient equal to 0.012 from Figure 7.12. Manning's equation, Equation 7-17, can be rearranged to allow a direct solution for the minimum required pipe diameter, in inches, assuming full flow by using:

$$d_{inches} = 2 \left[\frac{258.68 \times Q \times n}{S^{0.5}} \right]^{0.375} \tag{7-40}$$

where:

d_{inches} = The required pipe diameter in inches to carry the required flow for full flow under open channel conditions.

Q = The required flow rate in cubic feet per second.

n = Manning's roughness coefficient.

S = The pipe slope in feet per foot.

Using Equation 7-40, the required minimum diameter is:

$$d_{inches} = 2 \left[\frac{258.68 \times 6.85 \times 0.012}{(0.003)^{0.5}} \right]^{0.375}$$

$$= 18.70 \text{ inches}$$

The closest pipe size locally available is assumed to have a 24 inch diameter. This size will initially be used for pipe 1. The maximum capacity for a full 24 inch pipe using Manning's equation, Equation 7-19, is:

$$Q = \frac{1.49 A r^{0.67} S^{0.5}}{n}$$

where:

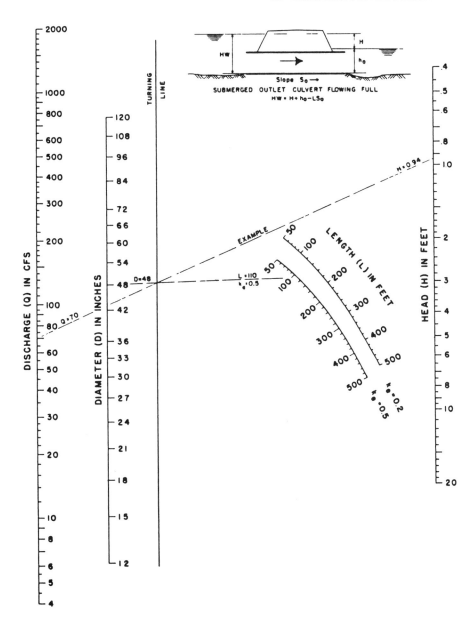

**HEAD FOR
CONCRETE PIPE CULVERTS
FLOWING FULL
n = 0.012**

Figure 7.37 Culvert capacity chart. Source: Bureau of Public Roads (1965), "Hydraulic Charts for the Selection of Highway Culverts," Hydraulic Engineering Circular No. 5, U.S. Dept. of Commerce, Washington, D.C.

$$A = (3.14)(2/2)^2 = 3.14 \text{ square feet}$$
$$P = (3.14)(2) \quad = 6.28 \text{ feet}$$
$$r = A/P = 0.5 \text{ foot}$$

$$Q = \frac{1.49(3.14)(0.5)^{0.67}(0.003)^{0.5}}{0.012}$$

$$= 13.42 \text{ cubic feet per second}$$

The corresponding velocity at full flow, using Equation 7-18, is:

$$V_{\text{full}} = \frac{13.42 \text{ cubic feet per second}}{3.14 \text{ square feet}}$$

$$= 4.27 \text{ feet per second}$$

Figure 7.39 can be used to graphically determine partial flow relationships when pipe diameters are used that are larger than those that allow full flow to occur. Since a 24 inch pipe is being used, the corresponding velocity for 6.85 cubic feet per second is needed; however, Q/Q_{full} must first be computed assuming that Manning's n varies with depth. Q/Q_{full} = 6.85 cubic feet per second/13.42 cubic feet per second, or 51 percent. Figure 7.39 indicates a percent depth of flow of approximately 56 percent (using the variable n curve for discharge). The corresponding velocity ratio using the same 56 percent depth of flow is approximately 86 percent of full flow velocity. Velocity for the 6.85 cubic feet per second flow is therefore 0.86 × 4.27, or 3.67 feet per second.

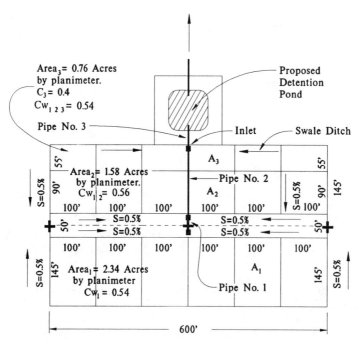

Figure 7.38 Delineation of various contributory drainage areas in a sample land development.

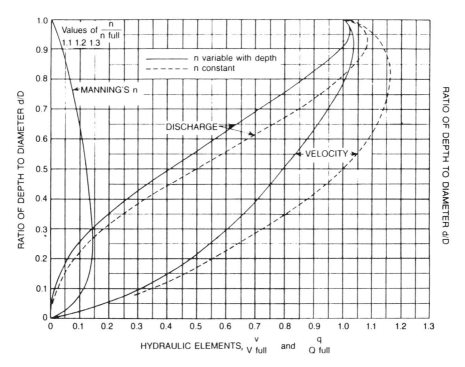

Figure 7.39 Hydraulic elements for a circular section. Source: The American Society of Civil Engineers (ASCE) and the Water Pollution Control Federation (WPCF). *Gravity Sanitary Sewer Design and Construction—ASCE Manual No. 60, WPCF No. FD-5,* 1982. Reproduced by permission of the American Society of Civil Engineers.

(h) Compute the contributory area for pipe 2:

$$A_1 + A_2 = 2.34 + 1.58 = 3.92 \text{ acres}$$

(i) Determine the longest time of travel considering the combined areas for T_c. Two possibilities occur: First, the T_c may equal the previous T_c of 16.7 minutes plus the time of travel through pipe 1. The approximate pipe 1 time of travel = 30 feet/3.67 feet per second = 8.2 seconds. Therefore, T_c is approximately 16.8 minutes. The second possibility is the time for a drop of water to travel perpendicular to the road for 90 feet in A_2, and then for 300 feet within the road gutter to the inlet. Based on these conditions this time would be less than the previously calculated time of 16.8 minutes; therefore, use 16.8 minutes for T_c. The rainfall intensity is estimated as 5.42 inches per hour.

(j) Compute the peak discharge from A_1 and A_2 using Equation 7.21 and a weighted coefficient of 0.56:

$$Q_2 = 0.56 \times 5.42 \times 3.92, \text{ or } 11.9 \text{ cubic feet per second}$$

(k) Size pipe 2 using $S_{min} = 0.3$ percent and Manning's coefficient of roughness of 0.012 in conjunction with Equation 7-40:

$$d_{\text{inches}} = 2 \left[\frac{258.68 \times 11.9 \times 0.012}{(0.003)^{0.5}} \right]^{0.375}$$

$$= 23.00 \text{ inches.}$$

Due to local supply availability, the closest diameter that can be used for pipe 2 is assumed to be a 24 inch culvert. The corresponding full flow of a 24 inch culvert is the previously calculated 13.42 cubic feet per second and the velocity is 4.27 feet per second. Figure 7.39 can again be used to determine the partial flow relationships; however, Q/Q_{full} must first be computed. $Q/Q_{\text{full}} = 11.9$ cubic feet per second/13.42 cubic feet per second, or 89 percent. Figure 7.39 indicates a percent depth of flow of approximately 82 percent. The corresponding velocity ratio using the same 82 percent depth of flow is approximately 102 percent of full flow velocity. Velocity for the 11.9 cubic foot per second flow is therefore 1.02×4.27 or 4.35 feet per second.

(l) Compute the contributory area for pipe 3:

$$A_1 + A_2 + A_3 = 4.68 \text{ acres}$$

(m) Determine the longest time of travel for T_c for pipe 3. First, T_c may be equal to the previous 16.8 minutes plus the time of travel through pipe 2. This time of travel is approximately 145 feet/4.35 feet per second = 0.56 minute.

$$T_c = 16.8 + 0.56 = 17.36, \text{ or } 17.4 \text{ minutes}$$

The second possibility is the time for a drop of water to travel perpendicular to the road for 90 feet in A_2, then in the road gutter for 300 feet to the curb inlet, and finally the time required to travel through pipe 2. This time can be seen as less than the previously computed 17.4 minutes.

The third possibility would be for the time of travel required for a drop of water to go 300 feet in A_3 to the inlet at Pipe 3. This is also less than the 17.4 minutes. The time of concentration is therefore 17.4 minutes, and the corresponding rainfall intensity is 5.37 inches per hour.

(n) Compute the peak discharge for A_1, A_2, and A_3 using Equation 7-21 and a weighted coefficient of runoff of 0.54:

$$Q_{123} = 0.54 \times 5.37 \times 4.68 = 13.57 \text{ cubic feet per second}$$

(o) Size pipe 3 using $S_{\text{min}} = 0.3$ percent and Manning's coefficient of roughness of 0.012 using Equation 7-40 by:

$$d_{\text{inches}} = 2 \left[\frac{258.68 \times 13.57 \times 0.012}{(0.003)^{0.5}} \right]^{0.375}$$

$$= 24.17 \text{ inches.}$$

Therefore, it is assumed that the next largest size culvert locally available is 30 inches in diameter.

(p) Compute the preliminary invert elevations of each culvert beginning with the discharge of the detention basin. From the topographic map (Figure 7.35), the outlet invert elevation could be preliminarily set at 83.00. If an elevation of 83.5 is selected for the pond bottom, the estimated maximum height of water based on twice the FAA method of ponding results (8.15 feet) would be 91.65 feet, which can be used for the invert of pipe 3's discharge elevation.

Since pipe 3 is 30 feet long and on a slope of 0.3 percent, the inlet invert elevation of pipe 3 would be 91.74 feet. To account for the diameter differences in pipes 2 and 3, the outlet of pipe 2 can be set at elevation (91.74 + 0.5), or 92.24 feet.

Pipe 2's inlet can now be calculated as 0.003 × 145 + 92.24 = 92.68. Since pipe 2 is proposed to be 24 inches in diameter, the topography indicates that approximately 5.3 feet of cover over the top of the pipe will be available, which appears acceptable.

Pipe 1's invert elevation at the pipe inlet is computed by 92.68 + 0.003 × 30 = 92.77. The elevation at the centerline of the road where the pipe crosses needs to be checked to insure that proper cover is available over the crown of the pipe.

To aid the designer in determining pipe cover, drainage information must be plotted on the road plan and profile where the road centerline has been drawn from PVI to PVI as described in Section 6.4.2. By plotting the elevation view of any culvert that crosses a road (by plotting the station, invert elevation, and pipe diameter), one can view the relationship of the top of the culvert to the road grade. Culverts need a minimum amount of cover for protection from traffic loads (usually a minimum of 12 inches), as indicated in Section 7.7.

If sufficient cover is not available, then either the road grade must be adjusted upward, or the drainage system must be lowered, which means the outfall ditch may need to be lowered and resloped for some distance downstream. Sometimes neither of these options is available, and arched pipe must be utilized in lieu of the circular culvert to obtain the proper clearance. Corrugated metal pipes have a different Manning's coefficient of roughness than reinforced concrete pipe. This must be considered when sizing for a substitute pipe. (One other option would be to increase the surface area of the detention pond, thus reducing the amount of working depth required in the pond so that the pond's inlet pipe elevation can be lowered along with upstream contributory pipes.

Should the storm drainage system indicate excessive installment depths, the invert elevations can be increased appropriately and the slopes adjusted accordingly. Sometimes drop structures, similar to the one shown in Figure 7.40, can be utilized to minimize construction costs.

Once the initial refinements have been made and the storm drainage system appears to fit adequately within the road system and terrain, the designer can undertake preliminary design of the sanitary sewer collection system as delineated in Chapter 10. Should future design problems arise from elevation conflicts between the sanitary sewer and storm drainage systems, revisions may need to be made to both systems. This, in turn, will require the engineer to recheck the road alignment, which may also require further adjustment.

The storm drainage hydraulic grade line should be plotted to determine whether local flooding might occur and whether the system will perform as expected. This is presented in the next section.

It is recommended that once the road, storm drainage, and sanitary sewer systems are fairly well established where the most economical costs of construction are obtained, the storm drainage system should then be analyzed using hydrograph generation and routing.

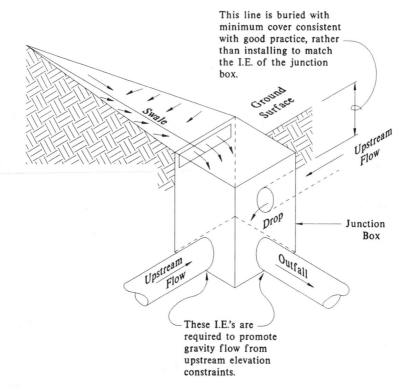

This line is buried with minimum cover consistent with good practice, rather than installing to match the I.E. of the junction box.

Ground Surface

Swale

Upstream Flow

Drop

Junction Box

Upstream Flow

Outfall

These I.E.'s are required to promote gravity flow from upstream elevation constraints.

Figure 7.40 Typical stormwater junction box details with an inlet drop. I.E. stands for invert elevation.

7.6.4 Example of a Drainage System Analysis

An analysis of a designed system using computer software illustrates hydrograph generation and routing. The software utilized for this analysis is HYDROS (Debo, 1987) where SCS methods are used for hydrograph generation while Muskingum procedures (alternative to the convex method) are used for routing.

The system being analyzed is the same 5.13 acre tract's previously designed storm drainage system; however A_1, A_2, and A_3 must be further subdivided into A_{1a} and A_{1b}, A_{2a} and A_{2b}, and A_{3a} and A_{3b} because a runoff hydrograph is generated from each of these subareas.

To fully investigate the performance of the storm drainage system under several storm return periods, areas contiguous to inlet structures can be treated as small reservoirs. This gives the designer a chance to investigate potential localized flooding under different storm conditions. The drainage subbasins and flow network are shown in Figure 7.41.

Channel velocities for hydrographs are estimated as those occurring during the design storm. Stage, storage, discharge relationships for the various reservoirs are used as input data. The program interpolates between data points during calculations to obtain approximate intervening values. The input data are determined by:

A. Calculation of Stage/Storage for Reservoirs 1 and 6. These reservoirs consist of the swale sections in drainage areas A_{3a} and A_{3b}. Temporary storage is provided there while runoff enters the inlet, as illustrated in Figure 7.42.

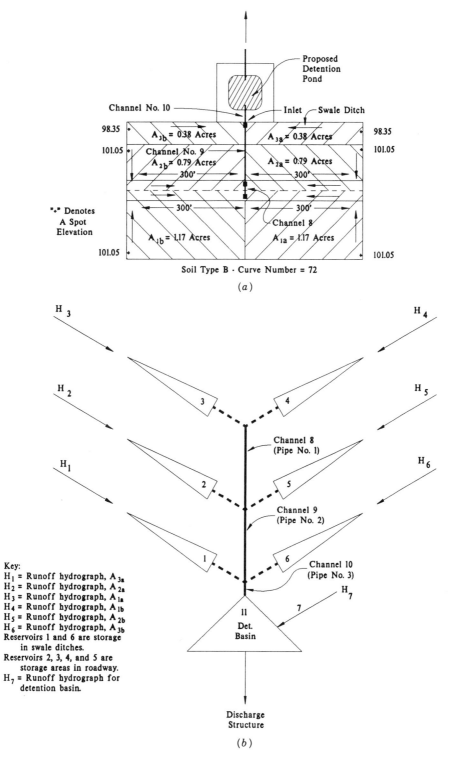

Figure 7.41 An example of drainage subbasin flow network. (*a*) Drainage basins for hydrograph generation. (*b*) Network for hydrograph analysis.

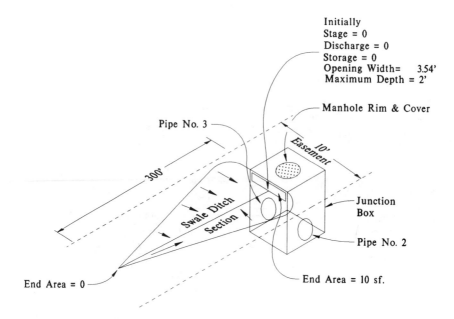

Figure 7.42 The geometry for sample reservoirs 1 and 6—swale ditch sections.

Using the average end area method (Equation 9-1) to estimate the storage volume for the tapered sections (see Section 9.3), the volume is calculated as:

$$\text{Vol} = [(a_1 + a_2)/2](L)$$
$$= [(0 + 0.5 \times 2 \times 5 \times 2)/2](300) = 1500 \text{ cubic feet}$$

The weir inflow of the inlet is determined using Equation 7-37 with an assumed broad crested weir coefficient of 3, a length of 3.54 feet, and a maximum flow depth of 2 feet at the junction box entrance because the swale's maximum depth is expected not to be exceeded.

$$Q = 3.0 \times 3.54 \times 2^{1.5} = 30 \text{ cubic feet per second}$$

Therefore two stage, discharge, and storage relationships for reservoirs 1 and 6 are:

Stage (feet)	Discharge (cubic feet per second)	Storage (cubic feet)
0	0	0
2	30	1500

B. Calculation of the Stage/Storage Relationships for Reservoirs 2, 3, 4, and 5. Figure 7.43 illustrates the stage and storage relationship in the roadway where maximum flooding 0.41 foot deep is estimated based on the road's cross section and anticipated topography. These reservoirs will drain into curb inlets, then through storm drainage culverts to the detention basin. Additional storage volumes attributed to the space behind the curb where the soil slopes to the road is neglected for this example.

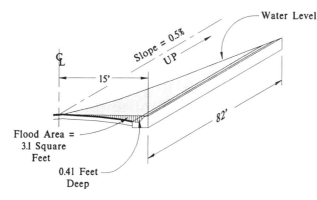

Approximate Storage Volume With
Flood Depth At 0.41 Feet is:

$$\left[\frac{0 + 3.1}{2} \right] 82 = 127.1 \text{ Cubic Feet}$$

Figure 7.43 Street detention volumes for a five-inch gutter flood depth.

Using a sag type curb inlet similar to Figure 7.30*b* with an estimated maximum flood elevation of 0.41 feet, the total inflow capacity of the inlet is calculated using the weir and orifice equations by:

(a) Weir flow using Equation 7-37 and a constant of 3 for throat flow:

$$Q = 3 \times 13 \text{ feet of weir length} \times (0.41)^{1.5}$$
$$= 10.24 \text{ cubic feet per second}$$

(b) Orifice flow using Equation 7-38 with a constant equal to 0.6 for grate flow and a cross-sectional area of 2.66 square feet:

$$Q = 0.6 \times 2.66 \times (2 \times 32.2 \times 0.41 \text{ feet})^{0.5}$$
$$= 8.2 \text{ cubic feet per second}$$

The total inlet flow is therefore 18.44 cubic feet per second. One half of this inflow is allocated to each contributory area with 9.22 cubic feet per second being allocated to reservoir 2 and 9.22 cubic feet per second to reservoir 5. (The same is true for reservoirs 3 and 4.) Two stage, storage, and discharge relationships are shown below that are applicable to each of the four reservoirs:

Stage (feet)	Discharge (cubic feet per second)	Storage (cubic feet)
0	0	0
0.41	9.22	127.1

C. Calculation of the Stage, Discharge, and Storage Relationships for the Detention Pond. This calculation uses procedures contained in Section 7.5.2. For this example, the design objective is to size the detention pond to accommodate the 100 year flood. This may be achieved by placing the pond in the location shown in Figure 7.35, selecting side slopes of 2 : 1, and using 10 feet as the maximum depth. Using bottom dimensions of 50 feet by 50 feet allows the top to extend to 90 feet by 90 feet. The limiting discharge out of the pond for the 100 year storm is 6.49 cubic feet per second; therefore, using 10 feet as the conservative maximum stage value (a 8.15 foot depth is the preliminary value obtained in Section 7.5.2 with the corresponding water surface area being 82.6 feet by 82.6 feet), an orifice is sized for use as the discharge structure with Equation 7-38. (Note: A more representative value of the reservoir's maximum storage height can be obtained after hydrograph generation and routing is accomplished, as indicated in Figure 7.46.) The orifice coefficient is taken as 0.58 for a sharp edged entrance.

$$6.49 = 0.58 \times A \times (2 \times 32.2 \times 10 \text{ feet headwater})^{0.5}$$
$$A = 0.44 \text{ square foot of opening}$$

The corresponding diameter of the orifice is approximately nine inches. This can be manufactured from a plate and installed at the discharge entrance. Eleven detention basin stage, discharge, and storage values are determined beginning with 0, 0, 0 at the basin bottom with discharge values corresponding to a nine inch orifice opening.

Stage (feet)	Discharge (cubic feet per second)	Storage (cubic feet)
0	0.00	0
1	2.06	2,708
2	2.91	5,864
3	3.56	9,516
4	4.11	13,712
5	4.60	18,500
6	5.04	23,928
7	5.44	30,044
8	5.82	36,896
9	6.17	44,532
10	6.50	53,000

Computer input data required for each drainage subbasin to perform and be analyzed include the drainage area, high and low elevations, distance of runoff travel including the amount channelized, the amount of impervious area, and the SCS runoff curve number. Data required for each reservoir include stage, discharge, and storage nomenclature. Channels are identified by contributory areas and are defined by invert slopes, length, and reach (channel) velocities. Lack of space prohibits the inclusion of this input data and intermediate calculations; however, Figure 7.44 shows hydrograph data for the six reservoirs (site symmetry allows dual use of the data) using a 10 year design storm.

These reservoir routed hydrographs are then routed through the system's channels as indicated in Figure 7.45, where detention basin (reservoir 11) routing is also shown.

Using Figure 7.41 as a guide, channel 8 (pipe 1) includes inflow hydrographs from both reservoirs 3 and 4, resulting in a combined inflow hydrograph twice that shown in Figure 7.44. Due to number rounding, however, the inflow into channel 8 varies slightly.

Time	Reservoir 1 and Reservoir 6				Reservoir 2 and Reservoir 5				Reservoir 3 and Reservoir 4			
	I,cfs.	O,cfs.	V,cft.	H,ft.	I,cfs.	O,cfs.	V,ctf.	H,ft.	I,cfs.	O,cfs.	V,ctf.	H,ft.
0.00	0	0	0.00	0.00	0	0	0.00	0.00	0	0	0.00	0.00
0.04	0	0	13.40	0.02	0	0	4.11	0.01	0	0	1.23	0.00
0.08	1	1	42.18	0.06	1	1	10.51	0.03	0	0	3.46	0.01
0.12	2	1	68.82	0.09	1	1	17.73	0.06	0	0	6.28	0.02
0.16	2	2	87.45	0.12	2	2	26.44	0.09	1	1	10.67	0.03
0.21	2	2	85.84	0.11	3	3	34.81	0.11	1	1	16.43	0.05
0.25	1	1	58.56	0.08	3	3	42.78	0.14	2	2	22.96	0.07
0.29	0	1	27.22	0.04	4	4	48.77	0.16	2	2	31.20	0.10
0.33	0	0	7.56	0.01	4	4	49.62	0.16	3	3	40.13	0.13
0.37	0	0	0.39	0.00	3	3	43.92	0.14	4	4	49.24	0.16
0.41	0	0	0.07	0.00	2	2	33.19	0.11	4	4	56.54	0.18
0.45	0	0	0.01	0.00	2	2	22.18	0.07	5	4	61.91	0.20
0.49	0	0	0.00	0.00	1	1	13.59	0.04	4	4	61.98	0.20
0.53	0	0	0.00	0.00	0	1	6.70	0.02	4	4	58.27	0.19
0.57	0	0	0.00	0.00	0	0	2.30	0.01	4	4	51.61	0.17
0.62	0	0	0.00	0.00	0	0	0.06	0.00	3	3	42.15	0.14
0.66	0	0	0.00	0.00	0	0	0.04	0.00	2	2	32.94	0.11
0.70	0	0	0.00	0.00	0	0	0.03	0.00	2	2	24.18	0.08
0.74	0	0	0.00	0.00	0	0	0.02	0.00	1	1	16.87	0.05
0.78	0	0	0.00	0.00	0	0	0.01	0.00	1	1	10.32	0.03
0.82	0	0	0.00	0.00	0	0	0.01	0.00	0	0	5.81	0.02
0.86	0	0	0.00	0.00	0	0	0.01	0.00	0	0	2.02	0.01
0.90	0	0	0.00	0.00	0	0	0.00	0.00	0	0	0.07	0.00
0.94	0	0	0.00	0.00	0	0	0.00	0.00	0	0	0.05	0.00
0.98	0	0	0.00	0.00	0	0	0.00	0.00	0	0	0.03	0.00
1.03	0	0	0.00	0.00	0	0	0.00	0.00	0	0	0.02	0.00
1.07	0	0	0.00	0.00	0	0	0.00	0.00	0	0	0.02	0.00

Figure 7.44   An example of reservoir routing—10 year storm. Source: Thomas N. Debo, HYDROS, 1987.

Channel 9 (pipe 2) includes inflow hydrographs from channel 8 and reservoirs 2 and 5. Channel 10 (pipe 3) includes inflow from channel 9 and reservoirs 1 and 6.

Reservoir 11, the detention basin, includes inflow from channel 10 and additional inflow resulting from precipitation falling on the basin's surface designated as hydrograph 7 (feeding into a fictitious channel 7 required by the software), as indicated in Figure 7.41. The hydrograph values for fictitious channel 7 are not included here.

The maximum detention basin depth necessary to accommodate a 10 year return period storage requirement is estimated at 4.30 feet.

Figure 7.46 is a summary sheet depicting peak hydrograph flows, maximum storage volumes, and maximum storage depths. For example, the 100 year return period predicts an estimated flood depth in the street at the inlet of pipe 1 to be approximately 0.36 foot, while a flood depth of approximately 0.28 foot is expected in the street at the inlet of pipe 2. Since hydrograph routing indicates that the approximate working depth in member 11 (the main detention pond) is 7.12 feet (rather than the initial estimate of 8.15 feet for the 100 year storm) and pipe 2 has more than the minimum cover, the detention pond's outlet pipe can be raised from its preliminary elevation, which was estimated at 83.00 feet, to an invert elevation of 85.00 feet for the discharge side and 85.60 feet for the inlet where the nine inch orifice is located. This will allow the discharge invert of pipe 3 to be increased from its preliminary elevation, which was estimated at 91.65 feet, to an elevation of 92.72 feet (85.60 + 7.12).

Time	Channel 8 Routing I,cfs.	Channel 8 Routing O,cfs.	Channel 9 Routing I,cfs.	Channel 9 Routing O,cfs.	Channel 10 Routing I,cfs.	Channel 10 Routing O,cfs.	Reservoir 11 Routing I,cfs.	Reservoir 11 Routing O,cfs	Reservoir 11 Routing V,cft.	Reservoir 11 Routing H,ft.
0.00	0	0	0	0	0	0	0	0	0.00	0.00
0.04	0	0	1	1	1	1	4	0	283.30	0.10
0.08	1	0	2	2	3	3	4	1	788.82	0.29
0.12	1	1	3	3	6	6	6	1	1360.24	0.50
0.16	2	2	5	5	8	8	8	2	2195.56	0.81
0.21	2	2	7	7	10	10	10	2	3277.94	1.18
0.25	3	3	9	9	11	11	11	3	4521.45	1.57
0.29	5	4	12	11	12	12	12	3	5848.55	2.00
0.33	6	6	13	13	13	13	13	3	7251.28	2.38
0.37	7	7	13	13	13	13	13	3	8710.96	2.78
0.41	8	8	13	13	13	13	13	4	10150.60	3.15
0.45	9	9	12	12	12	12	12	4	11486.42	3.47
0.49	9	9	11	11	11	11	11	4	12663.38	3.75
0.53	8	8	9	10	10	10	10	4	13635.78	3.98
0.57	7	8	8	8	8	8	8	4	14369.29	4.14
0.62	6	6	6	7	7	7	7	4	14855.53	4.24
0.66	5	5	5	5	5	5	5	4	15104.82	4.29
0.70	4	4	4	4	4	4	4	4	15148.81	4.30
0.74	2	2	2	3	3	3	3	4	15016.19	4.27
0.78	1	2	2	2	2	2	2	4	14729.59	4.21
0.82	1	1	1	1	1	1	1	4	14319.54	4.13
0.86	0	0	0	0	0	0	0	4	13817.85	4.02
0.90	0	0	0	0	0	0	0	4	13255.01	3.89
0.94	0	0	0	0	0	0	0	4	12668.65	3.75
0.98	0	0	0	0	0	0	0	4	12087.66	3.61
1.03	0	0	0	0	0	0	0	4	11517.87	3.48
1.07	0	0	0	0	0	0	0	4	10959.05	3.34

Figure 7.45 An example of channel routing—10 year storm. Source: Thomas N. Debo, HYDROS, 1987.

From Figure 7.46, the computer generated detention basin discharges are found to be less than the predevelopment release rates for all return periods previously computed in Figure 7.33. Figure 7.47 illustrates a plot of the 100 year return period inflow and outflow hydrographs in the detention basin. This graph shows the value of detaining runoff to reduce peak discharges.

To evaluate the performance of pipes 1, 2, and 3 over a range of storm return periods, a culvert analysis sheet (Figure 7.48) is constructed. Maximum open channel flows available for each culvert are recorded. Peak runoff flows generated by various return periods (Figure 7.46) that are less than these maximum available open channel flows allow open channel flows to prevail. Analysis of the expected performance of the 24 inch pipe proposed for pipes 1 and 2 indicates that open channel conditions will prevail for the chosen 10 year design storm. Pipe 3, the 30 inch pipe, will also perform under open channel flow conditions for the same return period (including the 100 year storm). When return periods greater than 10 years are expected, the analysis indicates that pipe 1 will flow under pressurized conditions during the 50 and 100 year storms. Pipe 2 is also expected to flow under pressurized conditions for storms beginning with the 25 year return period and larger. System adequacy for return periods greater than the 10 year design

Peak Discharges (CFS)	2-Yr	5-Yr	10-Yr	25-Yr	50-Yr	100-Yr
Sub-Watershed No. 1	1	1	2	2	3	3
Sub-Watershed No.2	2	3	4	5	6	6
Sub-Watershed No.3	2	4	5	6	7	8
Sub-Watershed No. 4	2	4	5	6	7	8
Sub-Watershed No.5	2	3	4	5	6	6
Sub-Watershed No.6	1	1	2	2	3	3
Sub-Watershed No.7	2	3	3	4	4	5
Detention Basin No.1						
Inflow Peak, cfs.	1	1	2	2	3	3
Outflow Peak, cfs.	1	1	2	2	3	3
Maximum Stage,ft.	0.06	0.09	0.12	0.15	0.18	0.21
Maximum Storage, cft.	43	69	87	115	136	157
Detention Basin No.2						
Inflow Peak, cfs.	2	3	4	5	6	6
Outflow Peak, cfs.	2	3	4	5	6	6
Maximum Stage,ft.	0.08	0.13	0.16	0.21	0.25	0.28
Maximum Storage, cft.	25	39	50	65	76	88
Detention Basin No.3						
Inflow Peak, cfs.	2	4	5	6	7	8
Outflow Peak, cfs.	2	4	4	6	7	8
Maximum Stage,ft.	0.1	0.16	0.2	0.26	0.31	0.36
Maximum Storage, cft.	31	49	62	81	96	111
Detention Basin No.4						
Inflow Peak, cfs.	2	4	5	6	7	8
Outflow Peak, cfs.	2	4	4	6	7	8
Maximum Stage,ft.	0.1	0.16	0.2	0.26	0.31	0.36
Maximum Storage, cft.	31	49	62	81	96	111
Detention Basin No.5						
Inflow Peak, cfs.	2	3	4	5	6	6
Outflow Peak, cfs.	2	3	4	5	6	6
Maximum Stage,ft.	0.08	0.13	0.16	0.21	0.25	0.28
Maximum Storage, cft.	25	39	50	65	76	88
Detention Basin No.6						
Inflow Peak, cfs.	1	1	2	2	3	3
Outflow Peak, cfs.	1	1	2	2	3	3
Maximum Stage,ft.	0.06	0.09	0.12	0.15	0.18	0.21
Maximum Storage, cft.	43	69	87	115	136	157
Channel No.7						
Inflow Peak, cfs.	2	3	3	4	4	5
Outflow Peak, cfs.	2	3	3	4	4	5
Channel No.8						
Inflow Peak, cfs.	4	7	9	12	14	16
Outflow Peak, cfs.	4	7	9	12	14	16
Channel No.9						
Inflow Peak, cfs.	7	11	13	18	21	24
Outflow Peak, cfs.	7	11	13	18	21	24
Channel No.10						
Inflow Peak, cfs.	7	11	13	18	21	24
Outflow Peak, cfs.	7	11	13	18	21	24
Detention Basin No. 11						
Inflow Peak, cfs.	7	11	13	18	21	24
Outflow Peak, cfs.	3	3	4	5	5	5
Maximum Stage,ft.	2.11	3.41	4.3	5.48	6.33	7.12
Maximum Storage, cft.	6261	11246	15149	21118	25948	30898

Figure 7.46 Summary of hydrologic data for the sample 5.13 acre site. Source: Thomas N. Debo, HYDROS, 1987.

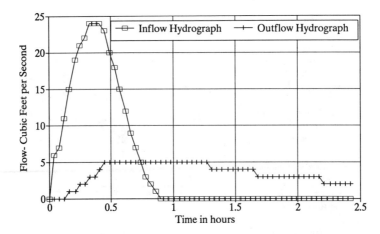

Figure 7.47   Examples of detention basin hydrographs for 100 year return period. Source: Thomas N. Debo, HYDROS, 1987.

storm is demonstrated in Figure 7.49, where the hydraulic grade line (HGL) resulting from a 100 year storm is plotted.

In the final analysis, the hydraulic grade line indicates that all three culverts can be raised approximately one foot without adversely affecting the drainage system's performance or providing inadequate cover over the pipes. This assumes that the designer wishes the hydraulic grade line to be approximately six inches below the ground surface when a 100 year storm occurs. Raising the culverts means less excavation will be required during construction. This can result in reduced installation costs. If the invert of pipe 3's outlet is raised (along with that of the emergency spillway shown in Figure 7.50) one foot from the shown 92.72 foot elevation to 93.72, and if the maximum headwater for the 100 year storm is estimated at 7.12 feet in the detention basin, the elevation of the basin's bottom can be raised from the shown 85.60 feet to (93.72 − 7.12), or 86.60 feet. This will also provide additional economy by reducing the amount of earthwork associated with building the detention basin. The basin's discharge pipe outlet could then be kept at 85.0 feet to provide adequate outfall, as shown in Figure 7.50.

The stability of the detention basin's slopes should be checked (Section 9.6) to insure that they will remain safe when subjected to inflows expected for 100 year return periods or greater.

If local regulatory agencies require that a retention volume (the volume of water not allowed to leave the site) equal to 1.5 inches of runoff be provided for this development, a retained volume of 27,932 cubic feet must be provided. This would require:

(a) Increasing the elevation of the discharge structure to allow 27,932 cubic feet of retained storage below the outlet.

******Pipe Design*******Peak Flow In CFS For Various Storms***										
Pipe	Dia.	n	Slope	Qfull o.c.	2	5	10	25	50	100
1	24"	0.012	0.3%	13.4 cfs.	4	7	9	12	14	16
2	24"	0.012	0.3%	13.4 cfs.	7	11	13	18	21	24
3	30"	0.012	0.3%	24.4 cfs.	7	11	13	18	21	24
[Note: Shaded area signifies pressure flow conditions exist.]										

Figure 7.48   An example of culvert performance analysis.

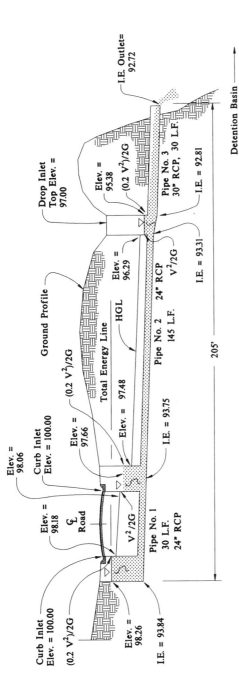

Pipe No.	1	2	3
Flow, CFS	16	24	24
Length, ft.	30	145	30
Hazen-Williams			
Roughness- C	120	120	Open Channel
Diameter, in.	24	24	30
Friction Loss, ft.	0.12	1.19	----
Velocity, ft/sec	5.09	7.64	4.88
Velocity Head, ($V^2/2g$)	0.40	0.91	0.37

Figure 7.49 An example of a hydraulic grade line for a 100 year storm.

297

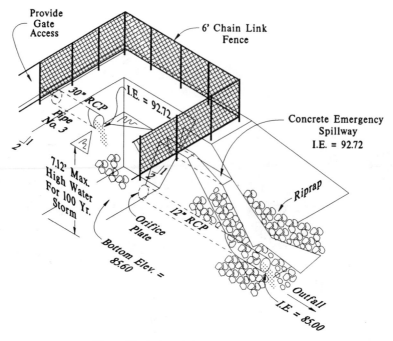

Note: Vertical scale is exaggerated for clarity.

Figure 7.50 Tentative detention basin configuration before final bottom elevation of basin is established.

(b) Redesigning the outlet structure to provide the desired discharge flows.

(c) Installing a granular bottom to cover an included subsurface drainage system that will allow the detained water to be filtered and seep into the outfall ditch over a period of as much as 72 hours.

7.6.5 Control Discharge Structures

The types of discharge structures that can be utilized on detention basins vary based on need and site conditions. Some of the most common discharge structures in addition to rectangular weirs and orifices are trapezoidal weirs, V notch weirs, and culverts.

Normally an auxiliary emergency discharge structure is set approximately at the 100 year maximum pond storage elevation to allow free discharge of runoff accumulations in excess of the 100 year storm. Generally this type of structure is in the form of a broad crested weir.

The exit velocity of the outfall discharge must be examined to insure that downstream erosion will not occur. Recommended maximum velocities are shown in Figure 7.51. Normally culvert outfalls have riprap (stones) placed as a blanket between the erodible soil and the exiting water. The riprap also acts as an energy dissipator to slow down the discharge. Baffle devices can also be utilized.

7.6.6 Trapezoidal Ditches for Discharge Outfalls

Often outfall ditches have trapezoidal shapes to minimize bank erosion. Manning's equation can be used to determine the flow properties of this type of ditch; however, uses of

Material	Mean Velocity, Clear Water (feet per second)	Mean Velocity, Silty Water (feet per second)
Fine sand, colloidal	1.50	2.50
Sandy loam, noncolloidal	1.75	2.50
Silt loam, noncolloidal	2.00	3.00
Alluvial silts, noncolloidal	2.00	3.50
Ordinary firm loam	2.50	3.50
Volcanic ash	2.50	3.50
Stiff clay, very colloidal	3.75	5.00
Alluvial silts, colloidal	3.75	5.00
Shales and hardpans	6.00	6.00
Fine gravel	2.50	5.00
Graded loam to cobbles, noncolloidal	3.75	5.00
Graded silts to cobbles, colloidal	4.00	5.50
Coarse gravel, noncolloidal	4.00	6.00
Cobbles and shingles	5.00	5.50

(a)

Material	Loose (feet per second)	Fairly Compact (feet per second)	Compact (feet per second)	Very Compact (feet per second)
Lean clayey soils	1.2	2.5	3.4	4.4
Clays	1.3	2.8	3.9	5.4
Heavy clayey soils	1.5	3.0	4.1	5.8
Sandy clays (sand < 50%)	1.7	3.2	4.3	6.0

(b)

Average Depth (feet)	Correction Factor
1	0.81
2	0.91
3	0.99
4	1.04
5	1.11
6	1.14
7	1.18
8	1.21
9	1.24
10	1.26

(c)

Figure 7.51 Maximum permissible velocities for various channels. (a) Maximum permissible mean velocities recommended by Fortier and Scobey for straight, aged channels on small slope. Taken from Fortier and Scobey, 1926 American Society of Civil Engineers Special Committee on Irrigation. Reproduced by permission of the American Society of Civil Engineers. (b) Permissible velocities in cohesive soils. (c) Correction factors to be applied to Figure 7.51b for cohesive and noncohesive soils. Source: Portland Cement Association, 1964. Reproduced with permission; however the Portland Cement Association does not verify the accuracy.

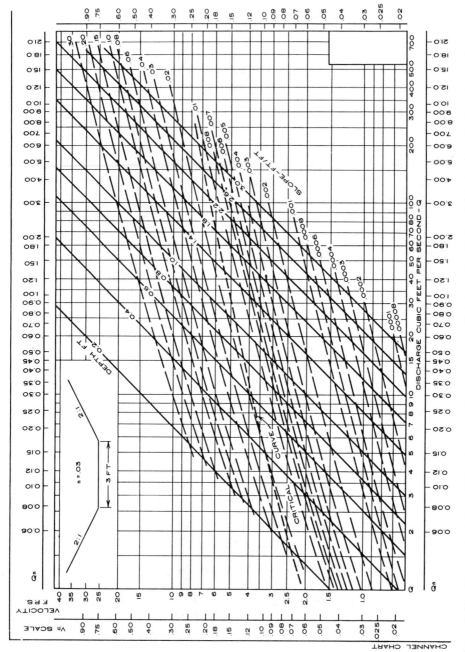

Figure 7.52 Graphical solution of Manning's equation for a trapezoidal channel, with 2 : 1 side slopes and a three foot wide bottom. Source: U.S. Department of Transportation, Federal Highway Administration. Design charts for open channel flow. Hydraulic Design Series No. 3, 1973.

nomographs such as the one shown in Figure 7.52 are common. This particular chart is suitable if the trapezoidal ditch has a three foot bottom and 2 : 1 side slopes. Direct chart use is available if the channel's coefficient of roughness is 0.03. Other coefficients of roughness require the calculation of Q_n or V_n values prior to entering the chart.

7.7 PIPE BEDDING AND PROTECTION

Figure 7.53 shows four types of bedding conditions used to install culverts. Each provides varying support conditions that allow the pipe to withstand different loads. Earth and

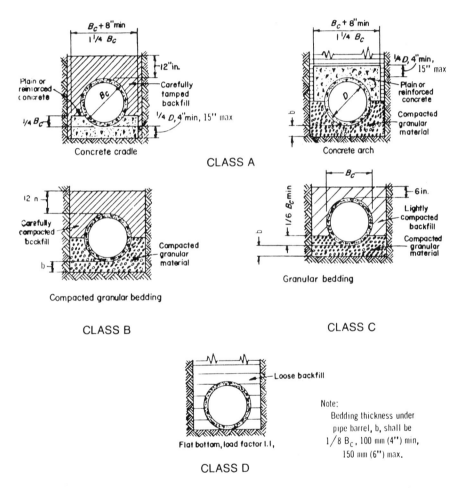

NOTE: In rock trench, excavate at least 15 cm (6 in.) below bell of pipe except where concrete cradle is used (in. × 25.4 = mm).

Figure 7.53 Pipe bedding classifications. Source: American Society of Civil Engineers (ASCE) and the Water Pollution Control Federation (WPCF). *Gravity Sanitary Sewer Design and Construction—ASCE Manual No. 60, WPCF No. FD-5*, 1982. Reproduced by permission of the American Society of Civil Engineers.

surcharge loads are usually small in land developments, whereas the live loads imposed by vehicular traffic are relatively large. The pipe must support both the live and dead loads along with a factor of safety. According to Fair et al. (1966), Anson Marston and others developed various relationships pertaining to external loads imposed upon rigid conduits laid in trenches or otherwise surrounded by earth. Some of these relationships are presented in the following sample problem.

To illustrate the analysis of external loads on a buried culvert, a 24 inch, class II reinforced concrete pipe, having the characteristics shown in Figure 7.54, will be used along with class B bedding. If the design professional wishes to establish as a design standard the minimum cover over the crown of pipes crossing roads as one foot, the pipe should be checked to insure it will maintain its integrity when subjected to traffic wheel loads. An H-20 wheel load, similar to that shown in Figure 7.55, is utilized in this analysis.

The strength test requirements in pounds-force per linear foot of pipe under the three-edge-bearing method shall be either the D-load (test load expressed in pounds-force per linear foot per foot of diameter) to produce a 0.01-in. crack, or the D-loads to produce the 0.01-in. crack and the ultimate load as specified below, multiplied by the internal diameter of the pipe in feet.

D-load to produce a 0.01-in. crack	1,000
D-load to produce the ultimate load	1,500

	Reinforcement, in.²/linear ft of pipe wall														
	Wall A					Wall B					Wall C				
	Concrete Strength, 4,000 psi					Concrete Strength, 4,000 psi					Concrete Strength, 4,000 psi				
		Circular Reinforcement[C]					Circular Reinforcement[C]					Circular Reinforcement[C]			
Internal Designated Diameter, in.	Wall Thickness, in.	Inner Cage	Outer Cage	Elliptical Reinforcement[D]		Wall Thickness, in.	Inner Cage	Outer Cage	Elliptical Reinforcement[D]		Wall Thickness, in.	Inner Cage	Outer Cage	Elliptical Reinforcement[D]	
12	1¾	0.07[B]	—	—		2	0.07[B]	—	—		2¾	0.07[B]	—	—	
15	1⅞	0.07[B]	—	—		2¼	0.07[B]	—	—		3	0.07[B]	—	—	
18	2	0.07[B]	—	0.07[B]		2½	0.07[B]	—	0.07[B]		3¼	0.07[B]	—	0.07[B]	
21	2¼	0.12	—	0.10		2¾	0.07[B]	—	0.07[B]		3½	0.07[B]	—	0.07[B]	
24	2½	0.13	—	0.11		3	0.07[B]	—	0.07[B]		3¾	0.07[B]	—	0.07[B]	
27	2⅝	0.15	—	0.13		3¼	0.13	—	0.11		4	0.07[B]	—	0.07[B]	
30	2¾	0.15	—	0.14		3½	0.14	—	0.12		4¼	0.07[B]	—	0.07[B]	
33	2⅞	0.16	—	0.15		3¾	0.15	—	0.13		4½	0.07[B]	—	0.07[B]	
36	3	0.14	0.08	0.15		4[E]	0.12	0.07	0.13		4¾[E]	0.07	0.07	0.08	
42	3½	0.16	0.10	0.18		4½	0.15	0.09	0.17		5¼	0.10	0.07	0.11	
48	4	0.21	0.13	0.23		5	0.18	0.11	0.20		5¾	0.14	0.08	0.15	
54	4½	0.25	0.15	0.28		5½	0.22	0.13	0.24		6¼	0.17	0.10	0.19	
60	5	0.30	0.18	0.33		6	0.25	0.15	0.28		6¾	0.22	0.13	0.24	
66	5½	0.35	0.21	0.39		6½	0.31	0.19	0.34		7¼	0.25	0.15	0.28	
72	6	0.41	0.25	0.45		7	0.35	0.21	0.39		7¾	0.30	0.18	0.33	
78	6½	0.46	0.28	0.51		7½	0.40	0.24	0.44		8¼	0.35	0.21	0.39	
84	7	0.51	0.31	0.57		8	0.46	0.28	0.51		8¾	0.41	0.25	0.46	
90	7½	0.57	0.34	0.63		8½	0.51	0.31	0.57		9¼	0.48	0.29	0.53	
96	8	0.62	0.37	0.69		9	0.57	0.34	0.63		9¾	0.55	0.33	0.61	
	Concrete Strength, 5,000 psi														
102	8½	0.76	0.46	Inner Circular	0.30	9½	0.68	0.41	Inner Circular	0.27	10¼	0.62	0.37	Inner Circular	0.25
				Plus Elliptical	0.46				Plus Elliptical	0.41				Plus Elliptical	0.37
108	9	0.85	0.51	Inner Circular	0.34	10	0.76	0.46	Inner Circular	0.30	10¾	0.70	0.42	Inner Circular	0.28
				Plus Elliptical	0.51				Plus Elliptical	0.46				Plus Elliptical	0.42
114	A	—	—	—	—	A	—	—	—	—	A	—	—	—	—
120	A	—	—	—	—	A	—	—	—	—	A	—	—	—	—
126	A	—	—	—	—	A	—	—	—	—	A	—	—	—	—
132	A	—	—	—	—	A	—	—	—	—	A	—	—	—	—
138	A	—	—	—	—	A	—	—	—	—	A	—	—	—	—
144	A	—	—	—	—	A	—	—	—	—	A	—	—	—	—

Figure 7.54 Design requirements for class II reinforced concrete pipe. Taken from *Standard Specifications for Transportation Materials and Methods of Sampling and Testing*. Copyright © 1990 by the American Association of State Highway and Transportation Officials. Reproduced with permission.

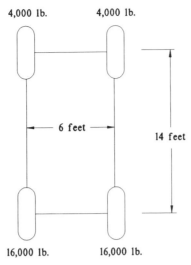

Figure 7.55 AASHTO H-20 wheel loads. Source: American Concrete Pipe Association. *Concrete Pipe Handbook*, 1980. Reproduced with permission.

First, the maximum load per linear foot of pipe can be computed using:

$$D\text{-load} = \frac{W_{tot} \times \text{F.S.}}{SB_f} \tag{7-41}$$

where:

D-load = D-load of pipe (three edge bearing test load expressed in pounds per linear foot per foot of diameter to produce a 0.01 inch crack according to AASHTO M 170-89, 1990).

S = Internal pipe diameter in feet.

W_{tot} = Dead load plus live load.

B_f = Bedding factor (1.9 for class B).

F.S. = Factor of Safety.

Therefore, using Equation 7-41 and Figure 7.54 to obtain the D-load that produces a 0.01 inch crack equal to 1000 pounds, the maximum imposed load (W_{tot}) equals:

$$D\text{-load} = \frac{W_{tot} \times \text{F.S.}}{2 \times 1.9} = 1000 \text{ pounds}$$

From the above relationship, W_{tot} is defined as:

$$W_{tot} = \frac{3800 \text{ pounds per linear foot of pipe}}{\text{F.S.}}$$

This expression for W_{tot} will be utilized to calculate the estimated factor of safety of the pipe while being subjected to the H-20 load. The actual load imposed on the pipe in the

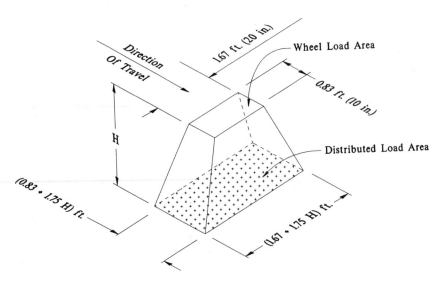

Figure 7.56   Distribution load area—single dual wheel. Source: American Concrete Pipe Association. *Concrete Pipe Handbook*, 1980. Reproduced with permission.

field must be less than 3800 pounds where a factor of safety greater than one is produced. The approximate live load pressure distribution can be seen in Figure 7.56.

According to the *Concrete Pipe Handbook* (American Concrete Pipe Association, 1980), the live load pressure on a buried concrete pipe can be determined by using the average pressure intensity at the pipe's outside crown elevation. The total live load acting on the pipe can be calculated using the average pressure intensity. The live load pressure per linear foot of pipe can then be obtained by dividing the total live load on the pipe by its effective supporting length.

The average pressure intensity at the pipe's outside crown is calculated using:

$$w_1 = \frac{P(1 + I_f)}{A_{11}} \tag{7-42}$$

where:

w_1 = Average pressure intensity in pounds per square foot.
P = Total applied surface wheel load in pounds.
A_{11} = Distributed live load area on the subsoil plane at the outside top of the pipe in square feet.
I_f = Impact factor (use 30 percent in this case).

For this analysis using Equation 7-42 the average pressure is:

$$w_1 = \frac{(16,000)(1 + 0.30)}{(0.83 + 1.75)(1.67 + 1.75)} = 2357 \text{ pounds per square foot at one foot depth}$$

The total load transferred to the pipe can now be computed based on the affected pipe area. For this example the length, L, parallel to the longitudinal axis of the pipe is 1.67 +

1.75, or 3.42 feet. The effective width is the projected outside diameter of the pipe, or the base area of the fustrum pyramid at the one foot depth, whichever is smaller. In this case, the 24 inch diameter pipe has a total wall thickness of five inches; therefore the outside projected distance, as shown in Figure 7.57, would be 29 inches, or 2.42 feet, which will be used in the calculations.

The total live load applied to the pipe is:

$$2357 \times 3.42 \times 2.42 = 19,480 \text{ pounds}$$

The live load per foot of pipe is equal to the total live load in pounds divided by the effective supporting length of pipe in feet, L_e, where:

$$w_{11} = \frac{19,480}{3.42 + (1.75 \times 0.75 \times 2.42)}$$

$$= 2953 \text{ pounds per linear foot}$$

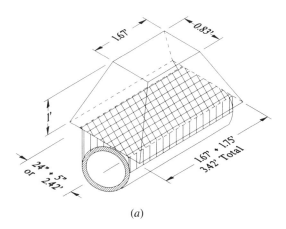

(a)

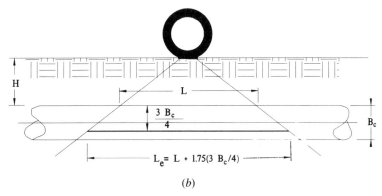

(b)

Figure 7.57 (a) Effective pipe pressure. (b) Effective supporting length of pipe. Source: American Concrete Pipe Association. *Concrete Pipe Handbook*, 1980. Reproduced with permission.

Marston's formula provides a means to estimate the dead load applied to the culvert by utilizing the following:

$$W_c = C_d w B_d^2 \tag{7-43}$$

where:

W_c = The load that the culvert is subjected to in pounds per linear foot.
C_d = Soil structure factor obtained from Figure 7.58 while using H as the depth of cover over the pipe's crown in feet.
w = The unit weight of backfill soil in pounds per cubic foot.
B_d = The width of the trench at the top of the culvert in feet.

Assuming that a typical three foot wide excavation bucket is used when installing the culvert, and that the soil is assumed to have a unit weight of 120 pounds per cubic foot, C_d can be estimated using Figure 7.58 (where $H/B_d = 1/3$ and the soil is assumed to be granular without cohesion) as 0.30. The load on the culvert is then estimated, using Equation 7-43, as:

$$W_c = 0.30 \times 120 \times 3^2 = 324 \text{ pounds per linear foot of pipe}$$

The total pipe load is therefore:

$$W_{\text{tot-actual}} = w_{ll} + w_c = 2953 + 324, \text{ or } 3277 \text{ pounds}$$

The allowable load was previously defined, using Equation 7-41, by the expression:

$$W_{\text{tot}} = \frac{3800 \text{ pounds per linear foot of pipe}}{\text{F.S.}}$$

The factor of safety is determined by dividing the maximum load by the actual load (3800/3277), or 1.16. If a factor of safety that is greater than one is acceptable for buried culverts in a residential area, then the H-20 loading will not damage the pipe used in this example. For a more complete treatment of this important topic, the reader is referred elsewhere (American Concrete Pipe Association, 1980; ASCE and WPCF, 1982).

7.8 SUBSURFACE WATER

In addition to surface runoff, drainage systems can handle subsurface water removal. Some precipitation becomes subsurface water by infiltration. The amount of infiltration is a function of the permeability of the soil, the slope, and the surface cover characteristics. Subsurface water flows in much the same manner as surface water does—toward lower elevations. The most common method of determining laminar flow through porous media is by using Darcy's equation (Equation 11-6), contained in Section 11.3.2.

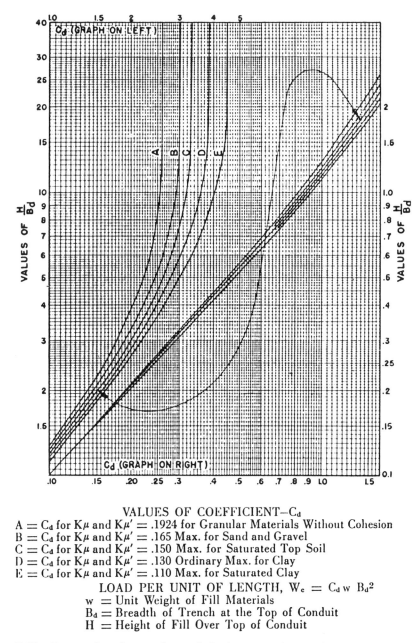

VALUES OF COEFFICIENT–C_d

A = C_d for Kμ and Kμ' = .1924 for Granular Materials Without Cohesion
B = C_d for Kμ and Kμ' = .165 Max. for Sand and Gravel
C = C_d for Kμ and Kμ' = .150 Max. for Saturated Top Soil
D = C_d for Kμ and Kμ' = .130 Ordinary Max. for Clay
E = C_d for Kμ and Kμ' = .110 Max. for Saturated Clay

LOAD PER UNIT OF LENGTH, $W_c = C_d w B_d^2$
w = Unit Weight of Fill Materials
B_d = Breadth of Trench at the Top of Conduit
H = Height of Fill Over Top of Conduit

Figure 7.58 Computation diagram for earth loads on trench conduits (completely buried in trenches). Source: American Concrete Pipe Association. Reproduced with permission.

Accurate calculations are difficult because of the nonhomogeneity of soils. Permeability coefficients can be estimated by field pumping, laboratory permeability tests, or from gradation analysis as indicated in Chapter 11. Typical permeability values for various soils are included in Figure 7.59.

AASHO Class.	Liquid Limit Percent	Plasticity Index Percent	Amount Passing # 200 Sieve Percent	Consoli- dating Load Kips/sq.ft.	Compacted Wet Permeability Coefficient Ft/day	Compacted Dry Permeability Coefficient Ft/day
A - 1	23	8	26	1	0.00045	0.00480
				2	0.00036	0.00200
				4	0.00024	0.00170
A - 2	28	11	40	1	0.00048	0.00470
				2	0.00028	0.00160
				4	0.00018	0.00090
A - 3	NP	NP	Patted with spatula		200.00000	226.00000
A - 4	33	12	99	1	0.00030	0.00075
				2		0.00060
				4	0.00020	0.00024
A - 5	35	6	35	1	0.02100	0.37000
				2	0.01500	0.25000
				4	0.01200	0.17000
A - 6	72	45	86	1	0.00004	0.00011
				2	0.00002	0.00011
				4	0.00001	0.00010
A - 7	67	34	71	1	0.00051	0.26000
				2	0.00028	0.00840
				4	0.00020	0.00120
A - 8	78	8	38	1	0.00390	0.04100
				2	0.00100	0.01800
				4	0.00046	0.01000

Figure 7.59 Permeability of soils. Source: E. S. Barber and C. L. Sawyer. *Proceedings—U.S. Highway Research Board*, Vol. 31, 1952. Reproduced by permission of Transportation Research Board, National Research Council, Washington, D.C.

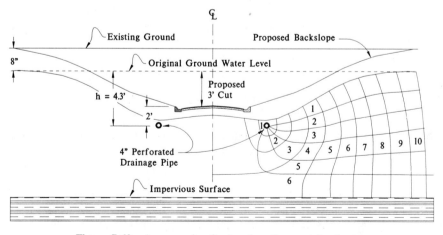

Figure 7.60 An example of subsurface flow net development.

Normally high groundwater tables in a development can be lowered by open ditching or by installing subsurface drains, commonly called french drains. For example, if the proposed roadbed shown in Figure 7.60 is to be located in an area using a curb and gutter section, and if it is desired to keep the stabilized groundwater table two feet below the road surface, a french drain must be properly designed.

To compute the flow per foot of roadbed:

$$Q = \text{coefficient of permeability} \times \text{head} \times \left[\frac{\text{number of flow lines}}{\text{number of equipotential lines}}\right] \qquad (7\text{-}44)$$

$$Q = K_s h(N_f/N_d) = K_s h(6/10)$$

Assuming an A-2 type soil (AASHTO classification), with a permeability of 0.0047 foot per day, the estimated flow for 100 linear feet of perforated pipe is:

$$Q = \frac{0.0047 \text{ foot per day} \times 4.3 \text{ feet of head} \times 0.6 \times 100 \text{ feet of pipe}}{24 \times 60 \times 60 \text{ seconds per day}}$$

$$= 0.000014 \text{ cubic feet per second}$$

A four inch diameter perforated pipe on a slope of 0.5 percent can carry 0.12 cubic foot per second, which would be more than adequate to provide the necessary drainage. Normal installation of the perforated pipe used as a french drain is shown in Figure 7.61.

Other subsurface flow conditions may require installation of a layer of graded granular material as part of the road subgrade. It is covered with filter fabric to provide a porous blanket for larger subsurface water removal requirements. For more in-depth coverage on subsurface water and seepage, the reader is referred to Cedergren (1989).

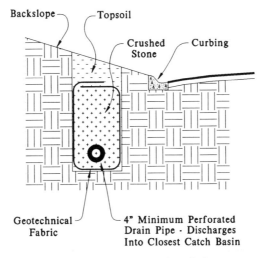

Figure 7.61 French drain installation.

REFERENCES

American Association of State Highway and Transportation Officials (AASHTO). *Standard Specifications for Transportation Materials and Methods of Sampling and Testing—AASHTO M 170-89, Standard Specifications for Reinforced Concrete Culvert, Storm Drain, and Sewer Pipe.* Washington: AASHTO, 1990, p. 263.

American Concrete Pipe Association. *Concrete Pipe Handbook.* Vienna, Va.: 1980, pp. 4-39–4-43.

American Society of Civil Engineers (ASCE) and Water Pollution Control Federation (WPCF). *Gravity Sanitary Sewer Design and Construction—ASCE Manual No. 60, WPCF No. FD-5.* New York: ASCE, 1982, pp. 109, 201–204. Referenced material is reproduced by permission of the American Society of Civil Engineers.

American Society of Civil Engineers (ASCE) and Water Pollution Control Federation (WPCF). *Design and Construction of Sanitary and Storm Sewers—ASCE Manual No. 37, MOP-9.* New York: ASCE, 1969, pp. 41–62. Referenced material is reproduced by permission of the American Society of Civil Engineers.

Asphalt Institute. *Drainage of Asphalt Pavement Structures—MS-15.* College Park, Md.: Asphalt Institute, 1966, pp. 68, 70.

Barber, E. S. and C. L. Sawyer. "Highway Subdrainage" in *Proceedings, U.S. Highway Research Board,* Vol. 31, 1952.

Bradford, Bruce H., Neil S. Grigg, and L. Scott Tucker. *Storm Water Management: Vol. I Urban Hydrology.* Atlanta: Water Management Science, Inc. 1979, pp. III–5.

Brown, L. A., Director. *Dekalb County—Drainage Procedures Manual.* An unpublished manual. Atlanta, Ga.: W. L. Jorden & Co., Inc. 1908 Cliff Valley Way, 1976, p. 4.

Cedergren, Harry R. *Seepage, Drainage, and Flow Nets,* 3rd ed. New York: Wiley, 1989.

Debo, Thomas N. *HYDROS* (computer software). Atlanta: Georgia Institute of Technology, 1987.

Fair, Gordon M., John C. Geyer, and Daniel A. Okun. *Water and Wastewater Engineering, Vol. I.* New York: Wiley, 1966, pp. 12–17. Copyright © 1966 by John Wiley & Sons, Inc. Reprinted by permission of John Wiley & Sons, Inc.

Fortier and Scobey. 1926 American Society of Civil Engineers (ASCE) Special Committee on Irrigation. New York: ASCE, 1926.

Frederick, R. H., V. A. Myers, and E. P. Anciello. *Five to 60 Minute Precipitation Frequency for the Eastern and Central United States, NWS Hydro-35.* Silver Springs, Md.: National Weather Service, NOAA, U.S. Department of Commerce, 1977.

Hershfield, P. M. *Rainfall Frequency Atlas of the United States for Durations from 30 Minutes to 24 Hours and Return Periods from 1 to 100 Years, Technical Paper No. 40 (USWB TP-40).* Washington: U.S. Weather Bureau, U.S. Department of Commerce, 1961.

Miller, J. F., R. H. Frederick, and R. J. Traley. *Precipitation-Frequency Atlas of the Western United States, Vol. I–XI, NOAA Atlas 2.* Silver Springs, Md.: National Weather Service, NOAA, U.S. Department of Commerce, 1973.

Overton, Donald E. and Michael E. Meadows. *Stormwater Modeling.* New York: Academic Press, 1976, p. 89.

Peckworth, H. F. *Concrete Pipe Field Manual.* Arlington, Va.: American Concrete Pipe Association, 1962, p. 189.

Portland Cement Association (PCA). *Handbook of Concrete Pipe Hydraulics.* Skokie, Ill.: PCA, 1964, p. 142.

U.S. Department of Agriculture (USDA)—Soil Conservation Service (SCS). *National Engineering Handbook Section 4, Notice NEH 4-102.* Washington: SCS, 1972, Chapters 15, 16, 17.

U.S. Department of Agriculture (USDA)—Soil Conservation Service (SCS). *A Method for Estimating Volume and Rate of Runoff in Small Watersheds—SCS TP 149.* Washington: SCS, 1973.

U.S. Department of Agriculture (USDA)—Soil Conservation Service (SCS). *Urban Hydrology for Small Watersheds—SCS TR-55*. Washington: SCS, 1986.

U.S. Department of Agriculture—Soil Conservation Service and Forest Service. *Soil Survey of Berkeley County, S.C.* (A National Cooperative Soil Survey). Washington: USDA, 1980.

U.S. Department of Agriculture (USDA)—Soil Conservation Service (SCS). *Technical Note: Hydrology No. N4—Time of Concentration*. Chester, Pa.: NENTC, 1986.

U.S. Department of Agriculture (USDA)—Soil Conservation Service (SCS). *Soil Survey—Orangeburg County, South Carolina*. (A National Cooperative Soil Survey). Washington: USDA, 1988, p. 30.

U.S. Department of Transportation—Federal Aviation Administration (FAA). *Airport Drainage*. FAA Advisory Circular 150/5320-5B, Washington: FAA, 1970.

U.S. Department of Transportation—Federal Highway Administration. Design charts for open channel flow. *Hydraulic Design Series No. 3*. Washington: Federal Highway Administration, 1973.

U.S. Department of Commerce—Bureau of Public Roads. "Hydraulic Charts for Selection of Highway Culverts," *Hydraulic Engineering Circular No. 5*, December 1965.

Walesh, Stuart G. *Urban Surface Water Management*. New York: Wiley, 1989, pp. 195–199. Copyright © 1989 by John Wiley & Sons, Inc. Reprinted by permission of John Wiley & Sons, Inc.

Wright, Paul H. and Radnor J. Paquette. *Highway Engineering*, 5th ed. New York: Wiley, 1987, p. 289.

8 Sedimentation and Erosion Control

Corollary to the storm drainage system design, considerations for the control of sedimentation and erosion must be addressed. During construction natural cover is removed, leaving underlying soils exposed to erosive elements. Rain and wind can mobilize this material and transport it great distances. Control of this unwanted activity is necessary to maintain improvements as they are constructed and to eliminate adverse environmental impacts on surrounding areas. The EPA, through its NPDES General Permit for storm water discharges associated with construction activities where five or more acres of land are disturbed, requires that sedimentation and erosion control provisions be utilized on-site. Determining the rate of sediment accumulation in detention or retention facilities is also important for scheduling post-construction maintenance.

Sedimentation involves three basic processes: erosion, transportation, and deposition. Erosion increases when the protecting vegetation is disturbed, the mantle of the surface soil is destroyed, or the stability of the surface water runoff system is altered.

8.1 GENERAL CONSIDERATIONS

Several topics that should be considered for minimizing or eliminating sedimentation and erosion on construction sites include:

(a) Initially selecting the development site where drainage patterns, topography, and soils are suitable for development activities.

(b) Incorporating into the construction specifications (Chapter 13) provisions for:

(1) Exposing the smallest possible area for the least amount of time during construction.

(2) Retaining topsoil for covering graded areas and protecting natural vegetation.

(3) Including provisions for temporary planting, mulching, chemical bonding, stone surfacing, and/or structures to control runoff and protect areas subject to erosion during construction.

(4) Accommodating increased runoff resulting from changed soil and surface conditions during construction.

(5) Using sediment basins and other forms of silt traps to remove sediment from runoff.

(6) Installing permanent vegetative covering and long-term erosion protection structures as soon as practical.

Before commencing work, the contractor, in conjunction with the engineer, should locate storage areas and shops where the potential for erosion is slight. If satisfactory areas are not available, the plans and specifications should include provisions for paving and the use of erosion control practices to minimize potential erosion. In other areas the contractor may be required to saturate the ground or apply suppressors to minimize dust. In addition, if fording of streams is objectionable, the contractor may be required to install and use temporary bridges or culverts.

The contractor and project engineer should give attention to protecting potential borrow areas from erosion after the borrow has been removed. The contractor must also protect streams and the ground from chemicals, fuel, lubricants, sewage, or other pollutants and avoid disposing waste materials in drainage ways or floodplains. Other improvements that can be utilized during construction to minimize erosion follow (USDA—SCS, 1977).

8.1.1 Temporary Construction Entrance

A temporary construction entrance is a stone filled area located where construction vehicles enter public rights of way, streets, or parking lots. This entrance is used to reduce and eliminate mud migration from the construction site. Aggregates up to three inches in size can be used to construct the stone filled mat. Entrances can exceed 50 feet in length to provide adequate space for washing the underside of vehicles prior to their departure. It may be necessary to add new stone periodically to maintain a relatively porous surface. Provisions to divert rinse water into an appropriate trap or basin are necessary to prevent sediment migration.

8.1.2 Temporary Perimeter Dike

Temporary perimeter dikes are usually installed for one year or less to divert sediment laden runoff to trapping facilities. Normally dikes such as these are placed around the perimeter of an area to control runoff from small watersheds of five acres or less. The dike can have a 2 foot wide top with 2 : 1 side slopes and be 18 inches tall.

8.1.3 Temporary Diversionary Dike

Temporary diversionary dikes are usually in place for one year or less to divert sediment laden runoff from running over unstabilized slopes. Typically these dikes are placed parallel to the top of a cut or fill slope and are constructed in a manner similar to perimeter dikes.

8.1.4 Temporary Interceptor Dike

Temporary interceptor dikes are similar to perimeter and temporary diversionary dikes except that they are placed at intervals across disturbed areas (such as cleared rights of way) to shorten the distance sediment laden runoff must travel before reaching a stabilized area.

8.1.5 Temporary Level Spreaders

Temporary level spreaders are storm water outlets constructed with no slope to spread concentrated discharge runoff where a nonerosive velocity can be obtained. This dispersed water should be prevented from reconcentrating until it reaches a suitable outfall. Level spreader lengths should be based on a 10 year return period for determining the peak runoff generated by the contributory area.

8.1.6   Temporary Gravel Outlet Structure

A temporary gravel outlet structure is an auxiliary facility installed as part of a diversionary, interceptor, or perimeter dike to provide for the removal of storm runoff that collects upstream while retaining the sediment. This outlet structure is usually constructed of gravel that has a maximum size of 3 inches and is well graded. Precautions must be taken to insure that the gravel does not become clogged with trapped silt, in which case the gravel would have to be replaced. An emergency spillway should be provided to accommodate the runoff from at least a 10 year storm.

8.1.7   Temporary Pipe Drop

Temporary pipe drops are used to convey storm water from the top of a bank to the bottom without causing erosion. The pipe should be at least 8 inches in diameter and have the capability of accommodating the runoff from at least a 5 year 24 hour storm. Standard flared pipe sections can be installed at each end to decrease entrance and exit velocities.

8.1.8   Straw (or Hay) Bale Barriers

Bale barriers are used to intercept and retain sediment in silt-laden runoff from small drainage areas that are usually less than half an acre in size. These barriers produce excellent results in reducing sheet erosion where large quantities of sediment are not produced. The bales are usually placed in single rows, and anchored by wooden stakes or reinforcing steel rods that are driven into the soil. The bales must be replaced if they become laden with silt.

8.1.9   Silt Fencing

Silt fencing provides a system of sediment control similar to bale barriers. The fencing material can be made of burlap or geotechnical fabric that is supported by wire stays fastened to posts. Care must be exercised to protect the fence from tearing and ripping during construction.

8.1.10   Temporary Sediment Trap

Temporary sediment traps are small impoundments for trapping sediment being transported by storm runoff. These traps prevent sediment from entering natural drainage systems while the construction area is unstabilized. Usually these temporary traps are constructed at the inlet of catch basins and drop inlets and require periodic cleaning during construction.

8.1.11   Sediment Basin

Sediment basins are used to trap suspended solids in storm runoff to prevent siltation in channels and reservoirs downstream. This improves the quality of storm water runoff by removing possible pollutants contained in the sediment. The basin can be either temporary or permanent, depending on local regulations and site conditions, as well as for single or multiple purposes. A sediment basin can be designed as an integral part of a detention facility so that runoff control is simultaneous with sediment reduction. The basins can even be placed ahead of retention facilities to reduce sediment pool loadings, or after a detention facility to take advantage of fairly controlled flows that will help promote quiescence in the basin. One drawback (or benefit) to placing the basin after the detention facility is that some settling may occur in the detention facility, resulting in a possible reduction in working detention volume. If this is of concern, the detention facility may be overexcavated to provide adequate storage for any sediment and cleaned along with the sediment basin.

The design of sedimentation basins can be based on empirical relationships such as the one set forth in the EPA's NPDES General Permit for storm water discharges. This permit, for example, requires a minimum of 3600 cubic feet of sediment storage be provided per acre drained on construction sites having 10 or more disturbed acres. Sediment basin design, however, usually addresses the trap efficiency of the pool, the volume/weight relationship of the sediment, and the sediment grain size distribution and storage allocation. The trap efficiency of a basin is its ability to allow sediment to settle and remain within the basin, expressed as a percentage of the delivered sediment. Most flood water retarding structures with permanent retarding pools have an efficiency of about 85 to 95 percent. For smaller basins or basins where water is not permanently impounded, a lower efficiency may exist.

When the efficiency of the basin is impaired trapped sediment must be removed and disposed of properly, as it possibly could contain organic materials high in biochemical oxygen demands (BOD_5) or other pollutants. If the sediment basin is to be incorporated as part of a permanent lake after construction, then a minimum of six feet of headwater should be provided at the deep end in addition to any required volume for storage of accumulated sediment. This combined volume can be considered dead storage if the lake is to be used as a detention pond. The working volume for detention plus any flood volume requirement would be added to this dead storage allocation to determine the total lake volume. Figure 8.1 illustrates these relationships.

Most temporary sediment basins used on construction sites have a perforated vertical pipe spillway riser for conveying normal storm water runoff discharges and a separate emergency discharge spillway similar to that used in detention basins.

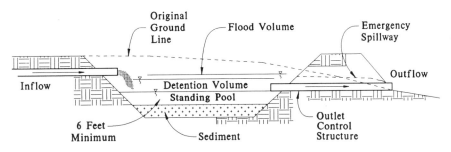

Figure 8.1 Volume relationships for combined flood, detention, permanent pool, and sediment storage.

8.2   PREDICTING RAINFALL EROSION LOSS

The erosional relationship between soils and water has been studied since about 1940 and has yielded one generally acceptable method for predicting soil losses with the universal soil loss equation (USLE). According to Wischmeier and Smith (1978), the USLE is an erosion model designed to predict long term average soil losses in runoff from specific field areas using specific crop and management systems for sheet and rill erosion under certain conditions. (It does not include the effects of concentrated runoff.) It can be applied with due caution to construction sites and other nonagricultural situations, but it does not predict deposition or sediment yields from gully, streambank, and streambed erosion. The USLE is:

$$A = RKLSCP \qquad\qquad (8\text{-}1)$$

where:

A = The computed soil loss per unit area, expressed in the units selected for K and for the period selected for R. In practice, these are usually selected to compute A in tons per acre per year, but other units can be selected.

R = The rainfall and runoff factor and is the number of rainfall erosion index units, plus a factor for runoff from snowmelt or applied water where such runoff is significant. Figure 8.2 shows these values throughout the United States.

K = The soil erodibility factor, the soil loss rate per erosion index unit for a specified soil as measured on a unit plot, which is defined as a 72.6 foot length of uniform 9 percent slope continuously in clean-tilled fallow. Figure 8.3 contains typical values available in SCS soil surveys.

L = The slope-length factor, the ratio of soil loss from the field slope length to that from a 72.6 foot length under identical conditions.

S = The slope-steepness factor, the ratio of soil loss from the field slope gradient to that from a nine percent slope under otherwise identical conditions.

C = The cover and management factor, the ratio of soil loss from an area with specified cover and management to that from an identical area in tilled continuous fallow.

P = The support practice factor, the ratio of soil loss with a support practice such as contouring, strip-cropping, or terracing, to that with straight-row farming up and down slope.

Each factor is now discussed in detail.

8.2.1   Rainfall and Runoff Factor

R in the soil loss equation represents not only the impact effect of the raindrops, but also amounts and rate of runoff. The EI parameter for any rainstorm is the total energy in the storm (E) times the maximum 30 minute intensity (I_{30}), where E is in hundreds of foot tons per acre and I_{30} is in inches per hour. EI is a function of the amount of rain and of all the storm's component intensities by:

$$E = 916 + 331 \log_{10} I_{30} \qquad\qquad (8\text{-}2)$$

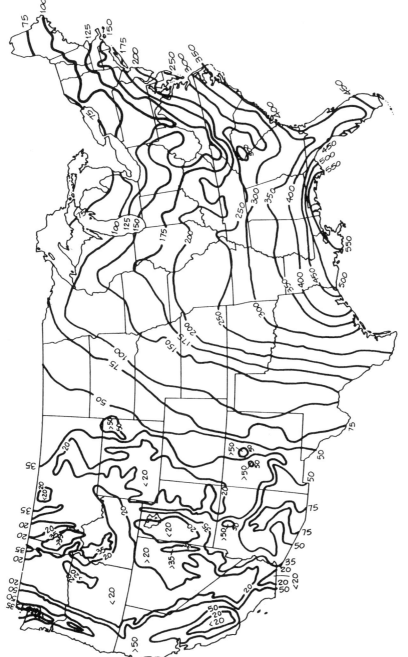

Figure 8.2 Average annual values of the rainfall erosion index. Source: W. H. Wischmeier and D. D. Smith. *Predicting Rainfall Erosion Losses*, 1978.

Series	Texture	K	T
Blanton	Sandy, Fine Sand	0.10	5
	Loamy Sand, Loamy Fine Sand	0.15	5
Bohicket	Silty Clay Loam, Silty Clay, Clay	0.28	5
Bonneau	Loamy Sand, Sandy, Fine Sand	0.15	5
Brevard	Loamy, Silty Loam	0.24	5
	Sandy Loam	0.15	4
	Sandy Clay Loam	0.24	4
Brogdon	Loamy Sand, Loamy Fine Sand	0.15	5
	Sandy, Fine Sand	0.10	5
Brookman	Clay Loam, Loam	0.24	4
Buncombe	Loamy Sand, Sandy	0.10	5
Byars	Silty Clay Loam, Clay Loam, Loamy	0.28	5
Cahaba	Silty Loam, Fine Sandy Loam	0.24	5
	Loamy Sand, Loamy Fine Sand	0.15	5
	Loamy	0.28	5
Cainhoy	Sandy Loam, Fine Sandy Loam	0.24	5
Candor	Sandy, Loamy Sand	0.10	5
Cantey	Sandy Loam, Fine Sandy Loam	0.24	5
	Loamy	0.28	5
Cape Fear	Loamy, Silty Loam	0.15	5
Capers	Silty Clay Loam, Clay, Silty Clay	0.28	5

Note: The soil erodibility factor K is the rate of soil loss per erosion unit index. "T" is the soil loss tolerance value which indicates the rate of soil loss in tons per acre per year.

Figure 8.3   Typical erosion K and T values for various soil types. Source: U.S. Department of Agriculture—Soil Conservation Service. *Predicting Soil Losses Using the Universal Soil Loss Equation—South Carolina*, 1974.

Local values can be obtained from **isoerodent maps,** which show points of equal rainfall erosivity as illustrated by Figure 8.2.

These parameters account for erosive forces associated with rainfall and runoff but not surface thaws or snowmelt. In areas where these factors are of consequence, an additional subfactor (R_s) must be added to R, where $R_s = 1.5 \times$ local December through March precipitation, measured in inches of water.

R does not indicate seasonal rainfall patterns affecting soil erosion; therefore the distribution of rain throughout the year is required. This is referred to as the erosion index distribution (EI). Records for each rainfall on a specific area have been recorded over various observation periods from which accumulated percentages of rainfall are determined. An EI curve is a plot of the monthly accumulated rainfall percentages throughout the year on an average basis. EI distributions can vary based on information contained in Figure 8.4. To determine the amount of erosion that is expected to occur between various months of the year, take the difference between the EI of the final month and the initial month, and account for the end of the year if it is included.

8.2.2 Soil Erodibility Factor (K)

K is an experimentally obtained factor that represents the rate of soil loss per erosion index unit as measured on a "unit plot" described in Section 8.2. This differs from the erosion rate factor A because the A variable is a function of rainfall, slope, cover, and manage-

(a)

Area No.	Jan. 1	Jan. 15	Feb. 1	Feb. 15	Mar. 1	Mar. 15	Apr. 1	Apr. 15	May 1	May 15	June 1	June 15	July 1	July 15	Aug. 1	Aug. 15	Sept. 1	Sept. 15	Oct. 1	Oct. 15	Nov. 1	Nov. 15	Dec. 1	Dec. 15
1	0	0	0	0	0	0	1	2	3	6	11	23	36	49	63	77	90	95	98	99	100	100	100	100
2	0	0	0	0	1	1	2	3	6	10	17	29	43	55	67	77	85	91	96	98	99	100	100	100
3	0	0	0	0	1	1	2	3	6	13	23	37	51	61	69	78	85	91	94	96	98	99	99	100
4	0	0	1	1	2	3	4	7	12	18	27	38	48	55	62	69	76	83	90	94	97	98	99	100
5	0	1	2	3	4	6	8	13	21	29	37	46	54	60	65	69	74	81	87	92	95	97	98	99
6	0	0	0	0	1	1	1	2	6	16	29	39	46	53	60	67	74	81	88	95	99	99	100	100
7	0	1	1	2	3	4	6	8	13	25	40	49	56	62	67	72	76	80	85	91	97	98	99	99
8	0	1	3	5	7	10	14	20	28	37	48	56	61	64	68	72	77	81	86	89	92	95	98	99
9	0	2	4	6	9	12	17	23	30	37	43	49	54	58	62	66	70	74	78	82	86	90	94	97
10	0	1	2	4	6	8	10	15	21	29	38	47	53	57	61	65	70	76	83	88	91	94	96	98
11	0	1	3	5	7	9	11	14	18	27	35	41	46	51	57	62	68	73	79	84	89	93	96	98
12	0	0	0	0	1	1	2	3	5	9	15	27	38	50	62	74	84	91	95	97	98	99	99	100
13	0	0	0	1	1	2	3	5	7	12	19	33	48	57	65	74	82	88	93	96	98	99	100	100
14	0	0	0	0	2	3	4	6	9	14	20	28	39	52	63	72	80	87	91	94	97	98	99	100
15	0	0	1	2	3	4	6	8	11	15	22	31	40	49	59	69	78	85	91	94	96	98	99	100
16	0	1	2	3	4	6	8	10	14	18	25	34	45	56	64	72	79	84	89	92	95	97	98	99
17	0	1	2	3	4	5	6	8	11	15	20	28	41	54	65	74	82	87	92	94	96	97	98	99
18	0	1	2	4	6	8	10	13	19	26	34	42	50	58	63	68	74	79	84	89	93	95	97	99
19	0	1	3	6	9	12	16	21	26	31	37	43	50	57	64	71	77	81	85	88	91	93	95	97
20	0	2	3	5	7	10	13	16	19	23	27	34	44	54	63	72	80	85	89	91	93	95	96	98
21	0	3	6	10	13	16	19	23	26	29	33	39	47	58	68	75	80	83	86	88	90	92	95	97
22	0	3	6	9	13	17	21	27	33	38	44	49	55	61	67	71	75	78	81	84	86	90	94	97
23	0	3	5	7	10	14	18	23	27	31	35	39	45	53	60	67	74	80	84	86	88	90	93	95
24	0	3	6	9	12	16	20	24	28	33	38	43	50	59	69	75	80	84	87	90	92	94	96	98
25	0	1	3	5	7	10	13	17	21	24	27	33	40	46	53	61	69	78	89	92	94	95	97	98
26	0	2	4	6	8	12	16	20	25	30	35	41	47	56	67	75	81	85	87	89	91	93	95	97
27	0	1	2	3	5	7	10	14	18	22	27	32	37	46	58	69	80	89	93	94	95	96	97	99
28	0	1	3	5	7	9	12	15	18	21	25	29	36	45	56	68	77	83	88	91	93	95	97	99
29	0	1	2	3	4	5	7	9	11	14	17	22	31	42	54	65	74	83	89	92	95	97	98	99
30	0	1	2	3	4	5	6	8	10	14	19	26	34	45	56	66	76	82	86	90	93	95	97	99
31	0	0	0	1	2	3	4	5	7	12	17	24	33	42	55	67	76	83	89	92	94	96	98	99
32	0	1	2	3	4	5	6	8	10	13	17	22	31	42	52	60	68	75	80	85	89	92	96	98
33	0	1	2	4	6	8	11	13	15	18	21	26	32	38	46	55	64	71	77	81	85	89	93	97

[1] For dates not listed in the table, interpolate between adjacent values.

(b)

Figure 8.4 *EI* distribution for various geographical areas. (*a*) Map for use with Figure 8.4*b*. (*b*) Percentage of average annual *EI* which normally occurs between January 1 and the indicated date. Source: W. H. Wischmeier and D. D. Smith. *Predicting Rainfall Erosion Losses*, 1978.

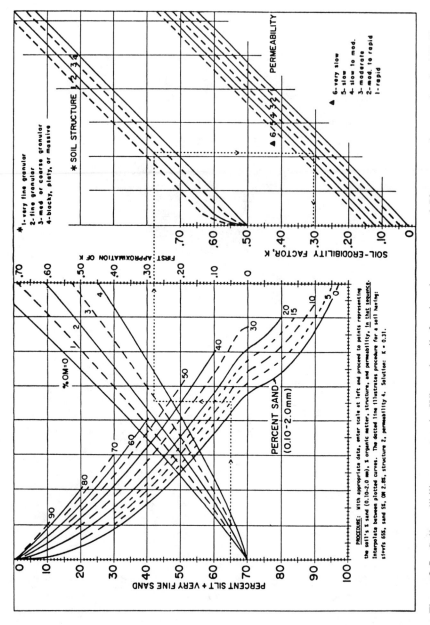

Figure 8.5 Soil-erodibility nomograph. Where the silt fraction does not exceed 70 percent, the equation is: $100 \text{ K} = 2.1 \text{ M}^{1.14} (10^{-4}) (12 - a) + 3.25 (b - 2) + 2.5 (c - 3)$ where

$\text{M} = (\text{percent silt} + \text{very fine sand}) \times (100 - \text{percent c})$

a = percent organic matter

b = soil structure code used in soil classification

c = profile permeability class

Source: W. H. Wischmeier and D. D. Smith. *Predicting Rainfall Erosion Losses*, 1978.

320

ment. *K* values for most soil types can be obtained from tables or by use of various computational aids similar to the chart shown in Figure 8.5. (Note: *T* is the soil loss tolerance value, which indicates the rate of soil loss in tons per acre per year.)

8.2.3 Topographic Factor (*LS*)

The rate of soil erosion is greatly affected by the length and steepness of the runoff slope. Both factors usually are considered together, and as a result the *LS* value is the ratio of soil loss from a given field slope to that of a unit plot. The slope length is the distance from the point where overland flows originate to the point where the slope gradient decreases resulting in the deposition of materials or where the runoff water enters a well defined channel. *LS* values predict average soil loss over the entire slope; however, this loss may not be evenly distributed over its total length. Figure 8.6 can be used to estimate relative soil losses from equal length segments having uniform slopes.

The second component of the *LS* term affects the amount of runoff from a given runoff surface. Soil losses increase much more rapidly than runoff as the slope increases. The slope-steepness factor (*S*) can be computed by:

$$S = 65.41\sin^2\Theta + 4.56\sin\Theta + 0.065 \tag{8-3}$$

where Θ is the angle of slope.

Many construction sites exhibit nonuniform gradients. Some gradients steepen toward the lower end (convex slope) while others flatten toward the lower end (concave slope). An underestimate of soil movement would be attained for the convex slope, and an overestimate of movement would be attained for the concave slope by using the average slope. These slopes can be divided into uniform segments that contain uniform gradients; however, when the runoff flows from one segment to the next the slopes must be evaluated together.

A complex slope can be divided into equal length segments to determine a composite *LS* value. First the segments must be listed in their order of progression downstream beginning with the most elevated. Next the *LS* factor must be obtained from Figure 8.6 using the total slope length and then adjusted by multiplying it by factors from Figure 8.7. The sum gives a composite *LS* value.

For example, if a 1500 foot convex slope has a gradient of 4 percent for the upper one-third of the slope, an 8 percent gradient for the middle one-third, and a 12 percent gradient for the lower one-third, the computed composite *LS* value would be as shown in Figure 8.8.

If various soil types are encountered, they too can be included in the calculations. If segment 1 has a *K* value of 0.25, segment 2 a *K* of 0.34, and segment 3 a *K* of 0.39, the composite *KLS* value would be as indicated in Figure 8.9.

8.2.4 Cover and Management Factor (*C*)

The cover and management factor (*C*) is the ratio of soil loss from land cropped under specified conditions to a corresponding loss from clean-tilled, continuous fallow. Because growing seasons change and ground cover varies based on management practices, the *C* factor is usually evaluated along with the seasonal rainfall patterns (*EI* distribution data).

For a construction site the *C* factor is usually ignored, since the ground cover has been

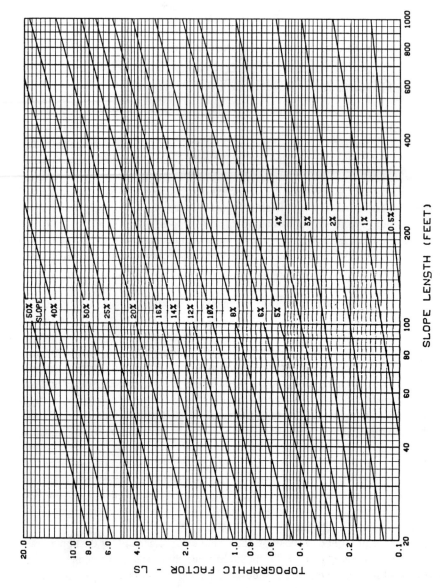

Figure 8.6 *LS* graph for soil erosion determination. Source: W. H. Wischmeier and D. D. Smith. *Predicting Rainfall Erosion Losses,* 1978.

322

Estimated Relative Soil Losses From Uneven Slopes				
Number of Segments	Sequence Number of Segment	Fraction of Soil Loss		
		m=0.5	m=0.4	m=0.3
2	1	0.35	0.38	0.41
	2	0.65	0.62	0.59
3	1	0.19	0.22	0.24
	2	0.35	0.35	0.35
	3	0.46	0.43	0.41
4	1	0.12	0.14	0.17
	2	0.23	0.24	0.24
	3	0.3	0.29	0.28
	4	0.35	0.33	0.31
5	1	0.09	0.11	0.12
	2	0.16	0.17	0.18
	3	0.21	0.21	0.21
	4	0.25	0.24	0.23
	5	0.28	0.27	0.25

Derived by the formula:

$$\text{soil loss fraction} = \frac{i^{m+1} - (i-1)^{m+1}}{N^{m+1}}$$

where i = segment sequence number; m = slope-length exponent (0.5 for slopes $\geq$ 5 percent, 0.4 for 4 percent slopes, and 0.3 for 3 percent or less); and N = number of equal-length segments into which the slope is divided.

Figure 8.7 Estimated relative soil losses from successive equal-length segments of a uniform slope. Source: W. H. Wischmeier and D. D. Smith. *Predicting Rainfall Erosion Losses*, 1978.

Segment	Percent Slope	Figure 8.6	Figure 8.7	Product
1	4	0.762	0.19	0.145
2	8	2.220	0.35	0.777
3	12	4.040	0.46	1.858
			LS =	2.780

Figure 8.8 An example of *LS* factor determination on a nonuniform slope.

Segment	L.S. Factor Figure 8.6	Rel. Loss Figure 8.7	K	Product
1	0.762	0.19	0.25	0.036
2	2.220	0.35	0.34	0.265
3	4.040	0.46	0.39	0.725
			KLS =	1.026

Figure 8.9 An example of *KLS* factor determination on a nonuniform slope with varying soil types.

cleared and grubbed and no appreciable cover management is practiced. In some areas regulatory agencies are requiring some type of cover management to be used. Construction specifications should provide for mulching and temporary planting to protect denuded areas. Corresponding C factors can be obtained utilizing Figure 8.10.

Should analysis be required to estimate long term sedimentation volumes that are expected to accumulate in a permanent pond or lake, appropriate C factors must be used as indicated in Figure 8.11.

8.2.5 Support Practice Factor (P)

The support practice factor (P) is the ratio of soil loss using a specific type of planting scheme as compared to up and down hill cultivation. Contouring, terracing, and strip-cropping are methods of providing this activity. Normally in a construction situation the P factor is considered unity.

8.2.6 Revisions to USLE

The universal soil loss equation is currently being revised and called "the revised universal soil loss equation" (RUSLE). Basic elements of the USLE will be retained; however it will be in computerized format. The RUSLE reflects new technology for evaluating factor values and new data to evaluate equation terms under specified conditions. Although the RUSLE calculations will be more detailed than those presented here for the USLE, computer software should facilitate its use (Renard et al., 1991).

Mulching Factors for Construction Slopes

Type of Mulch	Mulch Rate Tons/Acre	Land Slope Percent	Factor C	Length Feet
None	0	all	1.00	----
Straw or Hay, Anchored	1	1-5	0.20	200
	1	6-10	0.20	100
	1.5	1-5	0.12	300
	1.5	6-10	0.12	150
	2	1-5	0.06	400
	2	6-10	0.06	200
	2	11-15	0.07	150
	2	16-20	0.11	100
Crushed Stone 1/4 to 1.5 in.	135	<16	0.05	200
	135	16-20	0.05	150
	135	21-33	0.05	100
	240	<21	0.02	300
	240	21-33	0.02	200
Wood Chips	7	<16	0.08	75
	7	16-20	0.08	50
	12	<16	0.05	150
	12	16-20	0.05	100
	12	21-33	0.05	75
	25	<16	0.02	200
	25	16-20	0.02	150
	25	21-33	0.02	100

Figure 8.10 Mulch factors and length limits for construction slopes. Source: W. H. Wischmeier and D. D. Smith. *Predicting Rainfall Erosion Losses*, 1978.

C Values For Various Cover Treatments

Type	Annual Yield	C-Value
Grasses	3-Tons/Ac	0.004
Clover	2-Tons/Ac	0.006
Clover	1-Ton/Ac	0.010
Lawns-Good Condition		0.008
Lawns-Fair Condition		0.010
Lawns-Poor Condition		0.020

Type and Height of Raised Canopy	% Canopy Cover	Type*	Percent of Perennial Ground Cover					
			0	20	40	60	80	95-100
No appreciable canopy of weeds		G	0.450	0.200	0.100	0.042	0.013	0.003
		W	0.450	0.240	0.150	0.090	0.043	0.011
Canopy of tall weeds or short brush- 20 inches fall height	25	G	0.360	0.170	0.090	0.038	0.012	0.003
	25	W	0.360	0.200	0.130	0.082	0.041	0.011
	50	G	0.260	0.130	0.070	0.035	0.012	0.003
	50	W	0.260	0.160	0.110	0.075	0.039	0.011
	75	G	0.170	0.100	0.060	0.031	0.011	0.003
	75	W	0.170	0.120	0.090	0.067	0.038	0.011
Appreciable brush or bushes 6.5 feet fall height	25	G	0.400	0.180	0.090	0.040	0.013	0.003
	25	W	0.400	0.220	0.140	0.085	0.042	0.011
	50	G	0.340	0.160	0.085	0.038	0.012	0.003
	50	W	0.340	0.190	0.130	0.081	0.041	0.011
	75	G	0.280	0.140	0.080	0.036	0.012	0.003
	75	W	0.280	0.170	0.120	0.077	0.040	0.011
Trees but no appreciable low brush- 13 feet fall height	25	G	0.420	0.190	0.100	0.041	0.013	0.003
	25	W	0.420	0.230	0.140	0.087	0.042	0.011
	50	G	0.390	0.180	0.090	0.040	0.013	0.003
	50	W	0.390	0.210	0.140	0.085	0.042	0.011
	75	G	0.360	0.170	0.090	0.039	0.012	0.003
	75	W	0.360	0.200	0.130	0.083	0.041	0.011

* G- grass type cover, W- broadleaf type cover

Figure 8.11 Cover factor values for various land uses including woodlands. Source: U.S. Department of Agriculture—Soil Conservation Service. *Predicting Soil Losses Using the Universal Soil Loss Equation—South Carolina,* 1974.

8.2.7 Erosion from Concentrated Flows

Erosion on construction sites that is not attributable to the sheet and rill erosion accounted for in the USLE can be the result of ephemeral gully erosion, permanent gully erosion, and stream channel erosion.

Ephemeral gully erosion forms where runoff concentrates and produces features that are temporary in nature and are obscured during construction operations such as fine grading. The amount of erosion attributed to this, if left uncorrected, can be significant. This type of erosion produces cross sections that are usually wide in relation to the depth, while head cuts are not readily visible.

Ephemeral erosion can be effectively reduced by controlling runoff. The rate of runoff must be kept within the critical shear value of the soil; otherwise erosion will occur. Ephemeral gully erosion usually can be prevented by establishing temporary grassed waterways in areas susceptible to this type of activity (Laflen et al., 1986). Should ephemeral gully erosion continue unchecked, permanent gullies will form; therefore, effective management of this type of erosion is paramount.

Should permanent gully erosion occur, well defined drainage walls with narrow cross sections are formed. Permanent gullies usually have steep side slopes that progress to the head of the gully, resulting in a prominent head cut. Permanent gully channels advance upstream if uncorrected.

Although there are methods available for predicting erosion caused by concentrated flows (including stream channel erosion), contract specifications should provide for preventive measures to inhibit the formation of ephemeral gullies (Watson et al., 1986). If they form, construction specifications should provide for rectification activities. With stringent field control, erosion from concentrated flows should be minimal.

8.3   CONSTRUCTION SITES

Since the USLE yields anticipated soil losses for agricultural applications on an annual basis, *EI* values may be required to obtain *R* for construction durations of less than a year. Erosion calculations should also address the various subsoil horizon *K* factors that will be encountered during construction excavation. If chemical additives are to be applied to the soil to make it less erodible, the *K* values can be adjusted to reflect the improved performance. By analyzing various applications of the USLE to a construction site, the production of erosion can be studied by the design engineer for determining alternatives that minimize erosion potential and include preventive measures given in Section 8.1.

8.3.1   Upslope Contributions to Watershed Sediment Yield

The gross erosion within a watershed is the sum of all erosion that has taken place, including sheet, rill, gully, streambed, and streambank erosion. Not all eroded material calculated by the USLE leaves the site, as eroded soils often are redeposited after only a short distance of travel. To ascertain the amount of sediment delivered to any point within the system, the sediment delivery ratio must be determined using curves similar to those contained in Figure 8.12. This information is necessary when calculating the amount of sediment being delivered to a pond, lake, or basin.

8.3.2   Determination of Sediment Production for Ponds, Lakes, and Basins

The USLE can be used with caution to estimate the production of sediment from a denuded construction site. To illustrate an example using the site depicted in Figure 4.31, the following parameters apply:

Construction site area denuded (assuming for this example that the
   site was completely cleared and grubbed: 5.13 acres minus the 140
   foot × 140 foot detention pond site)   = 4.68 acres
Exposed soil horizon   = Cahaba
Slope length   = 340 feet
Duration of exposure   = 12 months
Slope   = 0.5 percent
Assumed *R* value   = 400

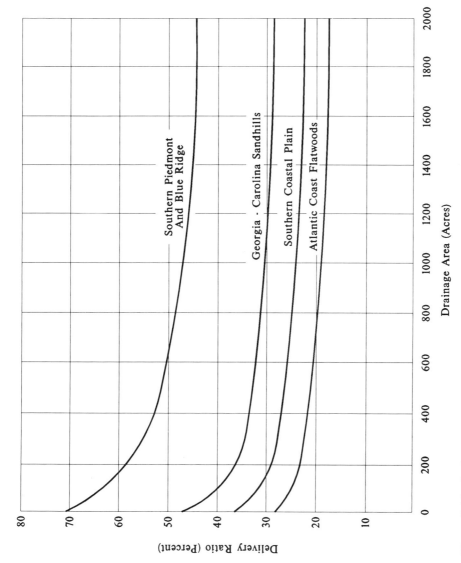

Figure 8.12 Sediment delivery ratio curves. Source: W. G. Groce. *USDA—SCS—SC Technical Note, Engineering 3 (Sedimentation)*, 1981.

327

First the following factors must be determined:

$$K = 0.28 \text{ for the Cahaba series from Figure 8.3}$$
$$LS = 0.12 \text{ from Figure 8.6}$$
$$C \text{ and } P = 1$$

Using Equation 8-1 the following calculation is made:

$$A = 400 \times 0.28 \times 0.12 = 13.44 \text{ tons per acre per year}$$

This gives the total annual runoff-related erosion estimated for this site as 62.9 tons per year. To convert tons per acre to cubic yards per acre, use the factors contained in Figure 8.13.

Since the Cahaba series falls into the loam category, an estimated conversion factor of 0.87 is used to determine the number of cubic yards lost. The total annual volume equates to 0.87×62.9, or 54.7 cubic yards.

If the area is only to be exposed between 15 October and 1 May of the following year, then the proportional amount of erosion is determined as a percentage of the annual rainfall expected between these dates by using a graph of the cumulative erosion index. For this example geographic area 29 is chosen in Figure 8.4, and the resulting EI graph is Figure 8.14.

$$EI \text{ October } 15 = 92$$
$$EI \text{ May } 1 = 11$$
$$\text{total } EI = (100 - 92) + 11 = 19$$
$$\text{erosion}_{15\text{Oct-1May}} = 0.19 \times 54.7 = 10.4 \text{ cubic yards}$$

If soil losses other than average are desired, then data similar to those contained in Figure 8.15 must be utilized.

If the maximum 5 year frequency (20 percent probability or 1 year out of 5) erosion quantity is required while referring to the above example, the corresponding (the closest recorded) R value for Charleston, S.C. is 559. An adjustment factor for the site is $559/400 = 1.398$. The adjusted erosion quantity at the site for 20 percent of the years would be $1.398 \times 54.7 = 76.5$ cubic yards.

To determine the amount of soil loss attributed to an individual storm (exceeded only once in five years for example), an adjusted R value again must be obtained from Figure

| Factors To Convert Eroded Materials From Weight To Volume ||
Texture	Factor
Sands, Loamy Sands, Sand Loam	0.70
Sand Clay Loam, Silt Loams, Loams, and Silty Clay	0.87
Clay Loams, Sandy Clays, and Silty Clays	1.02

Figure 8.13   Factors to convert tons per acre to cubic yards per acre of eroded materials. Source: U.S. Department of Agriculture—Soil Conservation Service. *Predicting Soil Losses Using the Universal Soil Loss Equation—South Carolina*, 1974.

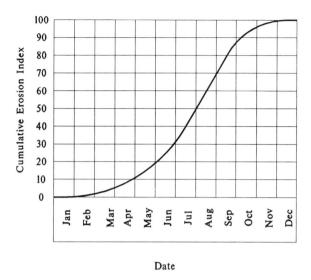

Figure 8.14 *EI* graph for geographic area 29 of Figure 8.4.

8.15, which in this case is 154. If erosion is being determined where the C value is other than 1, then an adjusted C must be utilized based on the season when the anticipated storm is to occur; however, C for this construction site is assumed to be unity.

$$V = RKLSCP \times \text{area} \times (\text{factor from Figure 8.13})$$
$$V = 154 \times 0.28 \times 0.12 \times 1 \times 1 \times 4.68 \times 0.87 \tag{8-4}$$

or, 21.1 cubic yards of erosion for a single storm exceeded only once in 5 years.

The amount of sediment estimated to be delivered to the low elevation of the site is determined by utilizing sediment delivery ratio curves similar to those shown in Figure 8.12. This example's contributory area is 4.68 acres; therefore, the delivery ratio is 0.29 if the site is assumed to be located in coastal flatland. The corresponding amount of sediment delivered to the low area is $0.29 \times 54.7 = 15.9$ cubic yards per year.

Since the buoyant unit weight of the saturated sediment is less than the moist unit weight, the sediment basin must have adequate volume to accommodate this changed condition. Sediment exposed to alternating drying and wetting action within the basin will have a unit weight that approximates its moist condition. Figure 8.16 illustrates this relationship.

8.3.3 Sediment Basin Trap Efficiency

According to Ward et al. (1977), in 1904 Hazen developed what has become a classical approach for determining trap efficiencies of reservoirs when he studied the settling of soil particles under various hydraulic conditions. He determined that trap efficiencies should be based on detention times, the velocities that particles fall, and their size and other conditions. Since then various other methods for estimating trap efficiencies of reservoirs have evolved, such as the empirical method proposed by Brune (1953). Brune's method, which appears to provide little correlation to semidry reservoirs, utilizes a capacity-inflow

| Locations | Probability Values Of Erosion Index | | | Expected Magnitudes Of Single-Storm Erosion Index Values Index Values Normally Exceeded Once In | | | | |
	50% Probability (1 Year) (out of 2)	20% Probability (1 Year) (out of 5)	5% Probability (1 Year) (out of 20)	1 Year	2 Years	5 Years	10 Years	20 Years
South Carolina								
Charleston	387	559	795	74	106	154	196	240
Clemson	280	384	519	51	73	106	133	163
Columbia	213	298	410	41	59	85	106	132
Greenville	249	350	487	44	65	96	124	153
North Carolina								
Asheville	135	175	223	28	40	58	72	87
Charlotte	229	322	443	41	63	100	131	164
Georgia								
Augusta	229	308	408	34	50	74	94	118
Savannah	412	571	780	82	128	203	272	358

Figure 8.15 Probabilities and magnitudes of erosion indexes. Source: U.S. Department of Agriculture—Soil Conservation Service. *Erosion and Sediment Control in Developing Areas*, 1977.

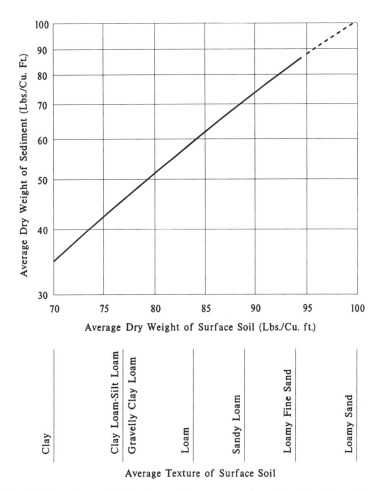

Figure 8.16 Relation of watershed surface soil texture to surface soil and sediment. Source: W. G. Groce. *USDA—SCS—SC Technical Note, Engineering 3 (Sedimentation), 1981.*

relationship based on annual inflow, which may render the use of Brune's method questionable for small sediment basins as shown in Figure 8.17.

Brune's procedure, based on data collected from 44 reservoirs, has nevertheless been used to generally predict trap efficiencies of pooled basins. To illustrate an application of the use of these curves the capacity-inflow ratio must first be determined.

If, for example, the previously designed detention basin illustrated in Chapter 7 is to be used as a temporary sediment basin during construction, a perforated standpipe must be installed to act as the outlet. This outlet could be positioned close to the side of the detention pond, away from pipe 3. The 100 year storm analyzed in Chapter 7 for detention volume determination was found to require a working volume of 30,898 cubic feet (or 0.709 acre-foot).

The total annual rainfall inflow can be estimated by utilizing an average amount of precipitation for the development area, in this case assumed to be 48 inches. Applying a coefficient of runoff of 0.56, the inflow contributed from the 4.68 acre portion of the site is: 4.68 acres × 48 inches × 0.56 coefficient of runoff/12 inches per foot, or 10.48 acre-

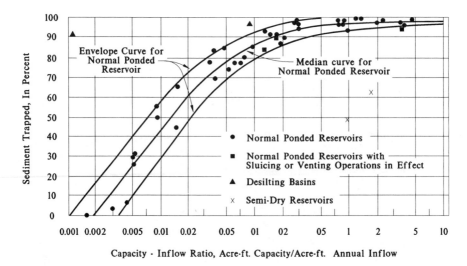

Figure 8.17 Trap efficiency as related to capacity-inflow ratio, type of reservoir, and method of operation. Source: Gunnar M. Brune, *Transactions of the American Geophysical Union.* Volume 34(3), 1953. Copyright © 1953 by the American Geophysical Union.

feet of estimated annual inflow. The capacity inflow ratio would be 0.709 acre-foot of storage/10.48 acre-feet of annual inflow, or 0.068. Figure 8.17 graphically indicates a corresponding trap efficiency of approximately 82 percent.

If the pond must, by local regulations, maintain 75 percent trap efficiency according to Brune's curves, the minimum volume required to produce that efficiency can be determined by obtaining the corresponding capacity-inflow required for 75 percent. By referring to Figure 8.17, one can see that a capacity-inflow rate of 0.047 corresponds to approximately 75 percent trap efficiency. The minimum storage volume would be 0.047 × 10.48 acre-feet of annual inflow, or 0.49 acre-feet. The volume of sediment that can be allowed to accumulate is 0.709 acre-foot − 0.49 acre-foot, or 0.219 acre-foot (9540 cubic feet). Based on the geometry of the sediment basin, the elevation that allows 9540 cubic feet of storage in its bottom can be calculated, and this value can be used to set a field stake with its top at this elevation. Using the top of the stake as a visual indicator, the basin would require cleaning when the stake is just covered.

Another empirical method to estimate trap efficiency is the use of curves produced by Churchill (1948). This method is event oriented and relates the percentage of sediment passing through a reservoir to the sediment index of the facility as defined by the detention time/mean velocity.

Figure 8.18 can be used to estimate the trap efficiency of the previous example while considering a two year storm return period suggested by Walesh (1989). The mean outflow rate through the riser pipe's elevated entrance can be arbitrarily established at 1.5 cubic feet per second using the orifice equation (7-38). The corresponding detention time would be 0.709 acre-foot × 43,560 square feet per acre/1.5 cubic feet per second, or approximately 20,590 seconds. The average cross-sectional area of the basin can be estimated using the volume of the basin divided by its length (using the water surface)— 0.709 acre-foot × 43,560 square feet per acre/78.5 feet of assumed length when the pond is at capacity, or 393 square feet of cross-sectional area. In this particular situation, a

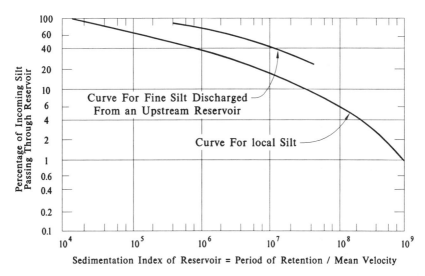

Figure 8.18 Churchill's trap efficiency curves for reservoirs. Source: M. A. Churchill, Discussion of "Analysis and Use of Reservoir Sedimentation Data" by L. C. Gottschalk. *Proceedings of the Federal Inter-Agency Sedimentation Conference,* Springfield, Va: U.S. Department of Commerce—National Technical Information Service (NTIS) 1948.

baffle would probably have to be installed to prevent short circuit flow patterns. The mean velocity of travel would be the flow rate/the average cross-sectional area. In this example the mean velocity would be 1.5 cubic feet per second/393 square feet, or 0.004 foot per second. The corresponding sediment index (detention time/mean velocity) is 20,590 seconds/0.004 foot per second, or 5.1 × 106 seconds^2 per foot. The corresponding estimated percent of sediment passing through the basin is obtained graphically from Figure 8.18 as 20 percent, resulting in an estimated trap efficiency of 80 percent. As can be seen, neither the Brune nor the Churchill method considers sediment characteristics.

Settling is usually categorized into four different types. Type I settling involves discrete particles that settle in low concentration solutions where particles tend to fall independently as defined by Stokes law. Type II settling, or flocculation settling, occurs when dilute solutions of particles flocculate to form larger settlable particles. Type III, or zone, settling occurs where particles are so concentrated that forces between the particles hinder settlement activity, resulting in uniform settling within the zone. Type IV settling, or compression settling, refers to settling in which the particles are so close together that inter-particle bridging has formed a stable structure, requiring compression for further settling activities.

Settling basins installed in land developments and on construction sites generally utilize Type I settling; therefore, suspended particles requiring flocculation to promote settling will pass through the basin where high turbidity can be expected. A Type I settling basin that conveys continuous flows is divided into four zones: the inlet zone, the settling zone, the bottom zone, and the outlet zone, as shown in Figure 8.19. Settlement paths of discrete particles in a horizontal flow basin are the vector sum of their settling velocities V_{settling} and the displacement velocity V of the basin. If particle settling velocities are less than V_{crit}, they will not all be removed because V_{crit} is the minimum velocity that allows a

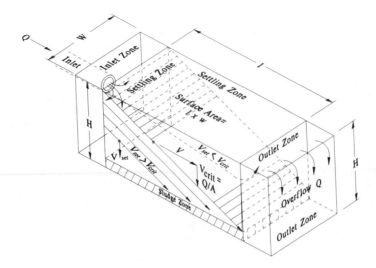

Figure 8.19 Typical Type I settling zones in a rectangular basin.

minimum size particle to fall a vertical distance H equal to the basin depth during the time of detention.

The following example will serve to illustrate the design of a basin utilizing this basic concept. A rectangular settling basin is proposed to be placed where it would receive the relatively controlled discharge of the normally configured detention basin, as shown in Figure 8.20b. The maximum two year return period flow rate obtained from Figure 7.46 of 3 cubic feet per second will be utilized, along with a proposed sediment basin surface area of 25 feet by 75 feet. The overflow rate is Q/A or 3 cubic feet per minute/(25×75). This gives a velocity, V_{crit}, of 0.0016 foot per second. If a temperature of 68°F, and a Reynolds number of less than 0.5 are assumed, the corresponding sphere diameter that is capable of being trapped is (Barfield et al., 1987):

$$d = \left[\frac{V_{settling}}{2.81} \right]^{0.5}, \text{ or } 0.024 \text{ mm} \tag{8-5}$$

where:

$V_{settling}$ Is in feet per second and in this calculation is assumed to be V_{crit}.
    d Is in mm.

The fractional size of particles smaller than 0.024 mm (found using Equation 8-5) must be obtained from soil analysis representative of the site. Figure 4.49 will be used to illustrate the procedure. Only that portion of the semilog curve from two mm size and smaller will be considered. If 100 percent of the material is finer than the two mm size, the percent smaller than 0.024 mm is found to be ten. Figure 8.21 shows only the two mm and smaller portion of the grain size distribution curve. The corresponding trap efficiency is the percentage of material greater than 0.024 mm plus proportional amounts of soils under 0.024 mm, based on the ratio of their settling velocity as compared to the critical. The

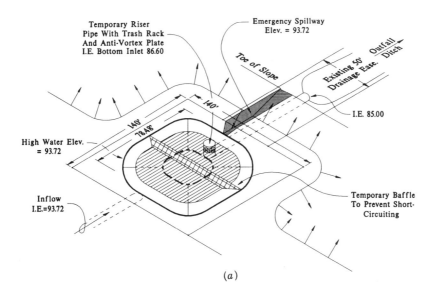

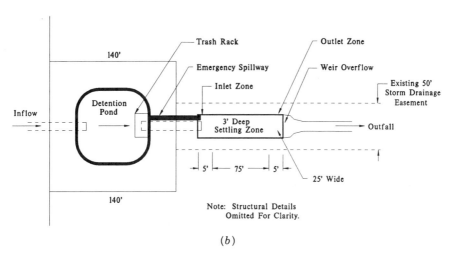

Figure 8.20 Examples of sedimentation facilities. (*a*) Detention basin (showing final design elevations) being used temporarily as a sediment basin during the construction period. (*b*) Alternative approach using a three foot deep sediment basin after the detention facility.

incremental diameters shown in Figure 8.21 correspond to the average soil particle diameter of each two percent increment used in this example. Trap efficiency calculations are included as part of Figure 8.21.

If a three foot deep basin is used, the detention time would be the basin's volume divided by the flow rate, which is (3 × 25 × 75)/3 cubic feet per second, or 1875 seconds. The corresponding velocity of plug flow would be the flow rate divided by the cross-sectional area, which is 3 cubic feet per second/(3 feet × 25 feet), or 0.04 foot per

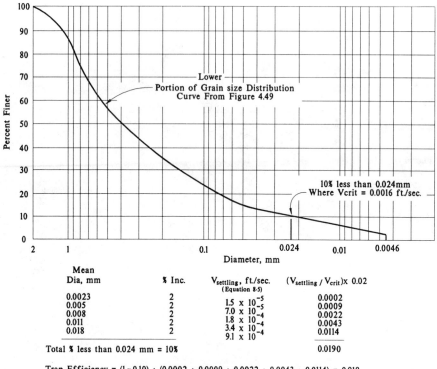

Mean Dia, mm	% Inc.	$V_{settling}$, ft./sec. (Equation 8-5)	$(V_{settling} / V_{crit})$x 0.02
0.0023	2	1.5×10^{-5}	0.0002
0.005	2	7.0×10^{-5}	0.0009
0.008	2	1.8×10^{-4}	0.0022
0.011	2	3.4×10^{-4}	0.0043
0.018	2	9.1×10^{-4}	0.0114

Total % less than 0.024 mm = 10%   0.0190

Trap Efficiency = (1 − 0.10) + (0.0002 + 0.0009 + 0.0022 + 0.0043 + 0.0114) = 0.919
or 91.9 % Efficient

Figure 8.21   An example of trap efficiency determination for Type I settling basin.

second. To verify plug flow conditions, Reynolds number can be calculated using the flow through velocity of 0.04 foot per second and Equation 8-6:

$$R_e = (4R_h V)/\text{kinematic viscosity of water} \qquad (8\text{-}6)$$

where:

R_h = The hydraulic radius, A/P, in feet.
V = The mean (flow through) velocity in feet per second.

Kinematic viscosity for water is assumed for this example as 1.059 square feet per second.

$$R_e = (4)[(3 \times 25)/(3 + 3 + 25)](0.04)/1.059,$$

or 0.37, which indicates plug flow conditions. When a 25 (or even 100) year storm's controlled discharge (5 cubic feet per second) passes through this basin (Figure 7.46), the corresponding Reynolds number would be:

Volume and Weight Relationships For Sediments In Selected Areas				
Land Resources Area	Aerated Sediment		Submerged Sediment	
	Volume (1)	Weight (2)	Volume (3)	Weight (4)
Blue Ridge	82	1786	55	1198
Southern Piedmont	82	1786	55	1198
Carolina-Georgia Sandhills	92	2004	80	1742
Southern Coastal Plain and Atlantic Coast Flatlands	88	1917	69	1503
Note: Columns (1) and (3) are given in pounds per cubic foot. Columns (2) and (4) are given as tons per acre foot.				

Figure 8.22 Volume/weight relationships of sediment. Source: W. G. Groce. *USDA—SCS—SC Technical Note, Engineering 3 (Sedimentation)*, 1981.

$$R_e = (4)[(3 \times 25)/(3 + 3 + 25)](0.066)/1.059,$$

or 0.60, which still approximates the Reynolds number of 0.5 that was assumed in Equation 8-5.

If this basin were to be used during the course of a year while construction was under way, the annual estimated sediment delivery rate based on USLE (Section 8.3.2) would be 18.24 tons (62.9 tons per year × 0.29 delivery ratio). This equates to approximately 530 submerged cubic feet of sediment using a factor of 69 pounds per cubic foot obtained from Figure 8.22 (18.24 tons × 2000 pounds per ton/69 pounds per cubic foot). This would amount to 530 cubic feet/(25 feet × 75 feet), or 0.28 foot of sediment sludge accumulation per year in the sediment basin. Since the sediment basin is to be positioned after the detention facility in this case, the bottom of the preceding detention facility should be undercut by approximately 0.2 foot [(530 cubic feet/(50 feet × 50 feet)], to allow for some additional sediment accrual during the year.

8.3.4 Summary

The placement and design of sediment basins are varied and must be tailored to individual site requirements. What has been included here are only examples of possible applications. More complete presentations on this important topic are included elsewhere (Barfield et al., 1987; Malcom et al., 1975; Walesh, 1989).

8.4 CONSIDERATIONS FOR PERMANENT PONDS OR LAKES

If the detention pond is to be incorporated with a permanent standing pool of water (lake), then the detention basin should include dead storage for long term sedimentation accumulation, possibly for as long as 20 years.

Total Drainage Area   4.68 Acres
Assumed R Valued   400
Soil Type   Cahaba

Watershed Erosion Data Calculations									
Land Use	Treat. Practice	K	% Slope	Length Slope	L.S.	KRLSP	C	Area* Acres	Annual (Tons/year)
Yards A-1	Up and Down	0.28	0.5	145	0.10	11.20	0.01	2.34	0.26
Yards A-2	Up and Down	0.28	0.5	90	0.10	11.20	0.01	1.58	0.18
Drain A-3	Up and Down	0.28	0.5	300	0.12	13.44	0.02	0.76	0.20
						Total Annual Tons			0.64

*Areas can be further refined to account for impervious surfaces such as roads, roofs, driveways, etc.

Gully, Bank, and Bed Erosion Data Calculations								
Erosion Source	Width (ft)	Length (ft)	%Length Eroding In Feet	Ave. Ht. In Feet	Annual Rate in ft/yr	Volume Weight lbs/cft	Annual Erosion lbs/yr*	Annual Erosion Tons per year
Roadbed	N/A							
Ditches	N/A							
Road Banks	N/A							
Streams	N/A							

*Note: The annual erosion rate in pounds per year is the product of all previous columns.

Reservoir Sedimentation Design		
Sheet Erosion Sediment	0.64	Tons per Year
Gully Erosion	----	Tons per Year
Stream Bed Erosion	----	Tons per Year
Road Ditches and Banks	----	Tons per Year
Total	0.64	Tons per Year

Deposition of Sediment		
Sediment Delivered to Site (Delivery Ratio = 29%)	0.19	Tons per Year
Annual Deposition- Using 90% Trap Efficiency	0.17	Tons per Year
20 Year Deposition	3.4	Tons

Sediment Storage Requirements
Assuming 100% submergence with the unit weight = 69 pounds per cubic foot.
Storage Required = 3.4 x 2,000 / 69 = 98.6 cubic feet.
The dead storage volume is now estimated by taking into account the required sediment storage along with a minimum headwater depth of six feet.

Figure 8.23 An example of watershed sedimentation calculations.

By referring to Figure 7.38, the information shown in Figure 8.23 is obtained. The cover factor for lawns is obtained from Figure 8.11.

A permanent pond or lake has multiple benefits that can be an asset to a development. Most facilities of this nature improve the landscape quality of the site, and fire protection services can be augmented by installing dry hydrants into the pond for pumpers to have a source of replenishment water similar to that shown in Figure 8.24. In addition, recreational benefits can be obtained in the form of fishing, boating, and swimming. Wildlife usage and irrigation utilize permanent ponds to the benefit of all.

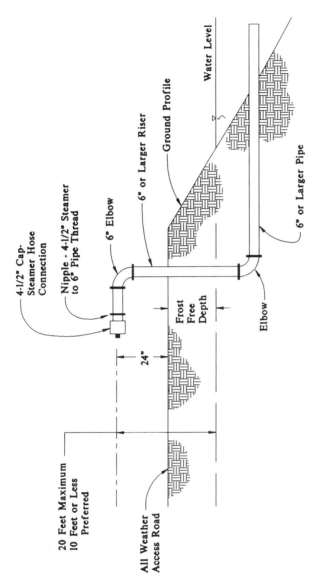

Figure 8.24 Dry hydrant construction. Reprinted with permission from *NFPA 1231, Water Supplies for Suburban and Rural Fire Fighting*, Copyright © 1989 by the National Fire Protection Association, Quincy, Mass. 02269. This reprinted material is not the complete and official position of the National Fire Protection Association on the referenced subject, which is represented only by the standard in its entirety.

REFERENCES

Barfield, B. J., R. C. Warner, and C. T. Haan. *Applied Hydrology and Sedimentology for Disturbed Areas*. Stillwater, Okla.: Oklahoma Technical Press, 1987, pp. 400–407.

Brune, G. M. "Trap Efficiency of Reservoirs," *Transactions of the American Geophysical Union*. Volume 34(3). 1953, pp. 407–418.

Churchill, M. A. Discussion of "Analysis and Use of Reservoir Sedimentation Data" by L. C. Gottschalk, *Proceedings of the Federal Inter-Agency Sedimentation Conference*. Springfield, Va.: U.S. Department of Commerce—National Technical Information Service (NTIS), 1948, pp. 131–140.

Groce, W. G. *USDA—SCS—SC Technical Note, Engineering 3 (Sedimentation)*. Columbia, S.C.: SCS, 1981.

Laflen, J. M., D. A. Watson, and T. G. Franti. "Ephemeral Gully Erosion," *Proceedings of the Fourth Federal Inter-Agency Sedimentation Conference*. Volume I, Section 3. Springfield, Va.: U.S. Dept. of Commerce—National Technical Information Service (NTIS), 1986, pp. 29–37.

Malcom, H. R. and V. New. *Approaches for Stormwater Management in Urban Areas, Part 4*. Raleigh, N.C.: North Carolina State University, 1975.

National Fire Protection Association (NFPA). *Suburban and Rural Fire Fighting, NFPA 1231*. Quincy, Mass.: NFPA, 1989, p. 1231-23.

Renard, K. G., G. R. Foster, G. A. Weesies, and J. P. Porter. "RUSLE—Revised Universal Soil Loss Equation," *Journal of Soil and Water Conservation*. Volume 46. Ankeny, Iowa. January–February 1991, pp. 30–33.

U.S. Department of Agriculture (USDA)—Soil Conservation Service (SCS). *Predicting Soil Losses Using the Universal Soil Loss Equation—South Carolina*. Columbia, S.C.: SCS, 1974.

U.S. Department of Agriculture (USDA)—Soil Conservation Service (SCS). *Erosion and Sediment Control in Developing Areas*. An unpublished guide. Columbia, S.C.: SCS, 1977.

U.S. Department of Agriculture (USDA)—Soil Conservation Service (SCS). *Ponds—Planning, Design, and Construction*. Washington: U.S. Government Printing Office, 1982.

Walesh, S. G. *Urban Surface Water Management*. New York: Wiley, 1989, pp. 297–313.

Ward, A. J., C. T. Haan, and B. J. Barfield. *Simulation of the Sedimentology of Sediment Detention Basins—Research Report No. 103*. Lexington, Ky.: University of Kentucky, 1977, pp. 8–15.

Watson, D. A., J. M. Laflen, and T. G. Franti. *Estimating Ephemeral Gully Erosion*. St. Joseph, Mich.: American Society of Agricultural Engineers, 1986 summer meeting.

Wischmeier, W. H., and D. D. Smith. *Predicting Rainfall Erosion Losses*. Washington: U.S. Government Printing Office, 1978.

9 Site Grading and Appurtenance Design

Soil is the material most used to support and connect improvements made in a land development. It gives buildings and roads elevation to minimize the effects of flooding and provides continuity between points of unequal elevation. Soil is used to support imposed loads and its surface can be shaped to form channels that control water movement. Naturally occurring granular soil usually seeks a surface slope that approaches its natural angle of repose, while erosion and weathering activities tend to smooth and round all exposed soil surfaces. Design professionals include rigorous requirements in their construction drawings to alter the shape of the soil's surface from its natural state into a reshaped (or graded) surface. This allows the various improvements made within a development to function properly and be pleasing to the eye. The developer's goals and constraints (Chapter 2) are considered by the design professional while evaluating site grading options. A site grading plan is used for this purpose. The plan is dependent upon the transportation, storm drainage, and the sanitary sewer system designs.

9.1 IMPROVED LAND DEVELOPMENTS

Improved land developments consisting of detached single-family dwellings usually provide for the developer to arrange and pay for the grading of rights of way and drainage facilities, while requiring the builders of the dwelling units to undertake the grading of their respective lots. Because independent parties may provide lot grading services later, an overall plan is necessary to indicate how each lot will be graded. This plan should define general methods to grade lots with the storm drainage flowing to the street, to the lot rear, or in both directions. A lot grading detail sheet can be included in the construction drawings (similar to the one included in Chapter 16) keyed to a plan view drawing showing each lot.

9.1.1 Lot Drainage Types

The U.S. Department of Housing and Urban Development has categorized three general types of lot surface runoff drainage plans (U.S. HUD, 1973). In instances where the rear lot elevations are higher than the front, the lot should be graded to allow drainage toward the front of the lot and should incorporate provisions for diverting water around the dwelling unit. This general plan could be labeled type "A" on the detail sheet.

   If the lot is subject to high elevations at its center with low elevations at the street and rear lot line, another type of grading plan must be used. A rear lot line drainage easement can be incorporated to receive accumulated storm water that is directed toward the lot rear. This general grading plan could be labeled type "B" on the detail sheet.

Should a lot be oriented where the topography is falling toward the rear of the lot, drainage must then generally follow in this direction. By including a rear lot line drainage easement, storm water can be intercepted and redirected before it has an opportunity to travel on abutting parcels. This grading scheme could be labeled type "C".

9.1.2   Lot Drainage Plans

To effectively identify and provide general guidance for the grading of each lot, a development plan is prepared that shows each lot along with the proposed type of drainage plan, signified by labeling with "A", "B", or "C". Including this information in the construction plans (and later sharing it with individual lot builders) ensures the coordination of building lot improvements that complement and function in unison with the entire road and storm drainage systems.

9.1.3   Initial Grading Improvements

In a single-family development, grading improvements provided by the developer are a function of the road and storm drainage systems. The design professional would attempt to balance earthwork cut and fill volumes, as indicated in Section 9.3, when excavated material is suitable for construction use. Exceptions would be cases where proposed gravity ties to existing sanitary sewers or storm drains would necessitate the use of borrow material to provide adequate elevation for fall. The amount of borrow would be a function of cost. If the cost of borrow material is significant, pumping facilities to compensate for nongravity conditions would have to be considered. A specific site grading plan for this type of development normally is prepared only to show the erosion control plan, as pertinent grading information is already included on the road and storm drainage plans. The disposition of possible excess cut materials would be determined by the engineer in accordance with provisions included in the construction specifications. Any borrow material requirements would be similarly handled.

9.2   TOTAL DEVELOPMENT ACTIVITIES

A formal site grading plan is usually prepared when total development activities are undertaken to convert unimproved land into a finished product that includes all site improvements and structures.

9.2.1   Site Grading Plan Preparation

To prepare a site grading plan, the design professional must have the road and storm drainage plan well defined. In addition, the siting of proposed structures must have been decided to properly allocate space for final contour lines. These are necessary to define topographically the final surface configuration.

The site grading plan should provide a minimum of soil cutting and filling operations consistent with good engineering practice. Advantage should be taken of natural site features when considering the plan and gravity flow systems. Most large sites cannot be clean cut and stripped because of tree and landscape ordinances, erosion control plans, and preservation of natural site features.

Many developments include concrete curb and gutter type road cross sections that require placement in a "cut" environment to allow proper drainage. The design professional can estimate the volume of any excess soil expected when constructing the road and storm drainage systems. If this excess soil is suitable for construction use, the designer can determine where to place it on site. This may require adjusting the final grades and contours on the grading plan to accommodate earthwork balancing.

9.2.2 Small Sites Associated with Total Development Activities

Some small sites associated with total development activities call for clearing the entire site. Landscaping provisions normally provide a means to reestablish natural growth in designated areas to enhance the finished product. In such cases the amount of earthwork associated with cut and fill operations is usually balanced. However, requirements for tying into an existing gravity sanitary sewer main or a storm drainage outfall may necessitate the inclusion of borrow material to elevate the site to provide for gravity flow to occur. Otherwise pumping facilities, which are expensive to construct and operate, may be required. Small development sites associated with total development activities, such as one containing several units of multifamily attached row or townhouses, may not require separate plan and profile sheets for the road and storm drainage systems. Information normally contained on those drawings could be incorporated on the site grading plan.

9.3 EARTHWORK CALCULATIONS

The amount of soil that a contractor moves and shapes directly affects the cost of construction. The goal is to minimize the amount of earthwork consistent with good engineering practice. To achieve this goal, an estimate of the amount of soil excavation (cutting) and embankment (filling) must be made. If the volume of cut material equals the volume of fill material (after soil expansion or shrinkage is considered), a balanced condition is achieved. Hence, there will be no soil surplus and no requirement for possibly expensive off-site borrow.

9.3.1 Earthwork Volume Determination Using Contour Lines

Since a contour line represents points of equal elevation, the volume of soil encompassed between sequential contour lines can be determined using the average end area method as expressed by the following formula:

$$\text{Volume} = (D_{1\text{-}2})(A_1 + A_2)/2 \tag{9-1}$$

where:

D_{1-2} = The vertical distance between end areas 1 and 2, in feet.
A = The appropriate contributory end area, in square feet, that establishes the volume that is computed.

Figure 9.1 illustrates an application of Equation 9-1 to determine the volume of soil between elevations 10 and 15 feet.

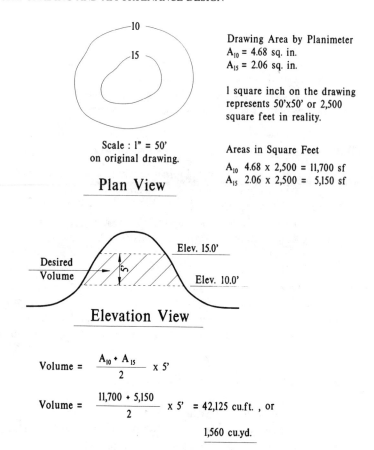

Figure 9.1   An example of earthwork volume calculations using contour lines.

9.3.2 Earthwork Volume Determination Using Cross Sections

Instead of calculating volumes based on horizontal reference surfaces, which is the case when using contour lines, vertical reference planes showing cross sections can be utilized. The procedure, however, is the same except that the distance, D, is the horizontal distance between end sections (cross sections), and A is the area of each cross section.

The cross-sectional area can be determined by plotting each section to scale and graphically determining the enclosed area using a planimeter. Usually the resulting area determined by the planimeter is in square inches and must be converted to square feet using a factor based on the scale with which the cross sections are plotted. For example, if the horizontal scale is 1 inch = 50 feet, and the vertical scale is 1 inch = 10 feet, each square inch in the cross section represents 50 feet × 10 feet, or 500 square feet. The equivalent end area in square feet is determined by multiplying the planimeter reading in square inches by the 500 equivalent square feet in this case.

Another method to determine the area of a cross section is to use section elevations as northings and horizontal reference distances as eastings while calculating the enclosed area using the coordinate method found in Figure 4.27. This procedure is illustrated in Figure 9.2.

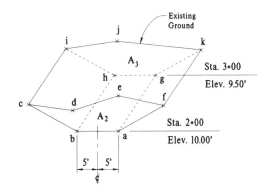

End Area for Sta. 2+00				
Point	N	E	+	−
a	10	+5		
b	10	-5	-50	+50
c	14	-13	-130	-70
d	12	-2	-28	-156
e	13	+6	+72	-26
f	12.5	+10	+130	+75
a	10	+5	+62.5	+100
Summation			+56.5	-27

End Area for Sta. 3+00				
Point	N	E	+	−
g	9.5	+5		
h	9.5	-5	-47.5	+47.5
i	15	-15	-142.5	-75
j	16	-4	-60	-240
k	14	+13	+208	-56
g	9.5	+5	+70	+123.5
Summation			+28	-200

Double Area = 83.5   Double Area = 228

A_2 = 41.75 square feet A_3 = 114 square feet

$$\text{Volume} = \left[\frac{A_2 + A_3}{2}\right](100') = \left[\frac{41.75 + 114}{2}\right](100') = 7787.5 \text{ cubic feet}$$

Figure 9.2 An example of earthwork calculations using cross sections.

9.3.3 Transition and Taper End Areas

The designer must only include in the calculations end areas that contribute to the desired volume. Often there are cross sections that require the use of different end areas depending on whether the volume is being calculated behind or ahead of the section. For example Figure 9.3 illustrates both tapered and transition sections where an access road joins the end of a borrow pit. The amount of soil removed from the access road (a tapered section) is calculated using a beginning end area of zero cut at Station 0 + 00. At Station 2 + 00, the end area is 15 feet × 10 feet, or 150 square feet. Using Equation 9-1, the corresponding volume of excavation is:

$$\text{Volume}_{0-200} = (200 \text{ feet})(0 + 150)/2$$

or 15,000 cubic feet (555.55 cubic yards). If the volume between Stations 2 + 00 and 3 + 00 were determined, the corresponding end area for the transition section at Station 2 + 00 would be 10 feet × 300 feet, or 3000 square feet instead of the previously used 150

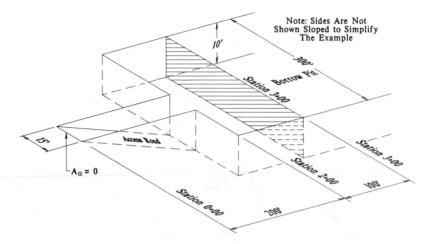

Figure 9.3   Transition and taper end sections associated with earthwork.

square feet. Transition sections require special attention to insure that the appropriate end area is used.

9.3.4   Earthwork Summary Sheet

The earthwork summary sheet shown in Figure 9.4 depicts calculations necessary to determine earthwork volumes in cubic yards and their balances. These calculations are prerequisites for constructing a mass diagram. End areas are shown at stations, while volumes are shown between stations. In these calculations, cut is designated as a positive volume, while fill is denoted as a negative value. Shrinkage is the volume reduction that soil will undergo when compacted in the field. In this example shrinkage is assumed to be 10 percent. The mass ordinates shown are the cumulative amount of all earthwork from the beginning of the project up to the point under investigation.

9.3.5   Earthwork Mass Diagram

Earthwork mass diagrams are often used on large earth moving projects such as highways, railroads, and airports. Occasionally they are used in conjunction with land developments to provide a means to identify the amount of excavation and embankment, the location of balance points, economical haul directions, and borrow placement.

There are numerous ways to estimate overhaul distances and volumes; however, the method of moments is illustrated in the mass diagram shown in Figure 9.5. Most land development contracts call for earthwork to be performed on a lump-sum basis where costs associated with transporting soil long distances (overhaul costs—costs for hauling soil distances in excess of the freehaul distance) are included in the project's base cost. To know what the estimated amount of earthwork overhaul can be, however, is useful to the design professional when estimating the cost of the project. A reasonable distance for an earthwork contractor to haul soil without incurring undue costs associated with hauling is called the freehaul distance. Often one-half mile is used for the freehaul distance to estimate the best distribution of soil placement, although the freehaul distance can be defined by stations or miles.

| | End Areas | | Between | | Volumes, Cubic Yards | | | | | Volume | Mass Ordinates | |
| | Cut,S.F. | Fill,S.F. | Stations | Cut | Fill | Shrinkage | Unusable | Embankment | Balance | Station | Ordinant |
Station											
0	0	0	0-5	231.48	0.00	0.00	NONE	0.00	231.48	0	0.00
5	25	0	5-10	1064.81	138.89	13.89	NONE	152.78	912.04	5	231.48
10	90	15	10-15	2175.93	648.15	64.81	NONE	712.96	1462.96	10	1143.52
15	145	55	15-20	3148.15	1666.67	166.67	NONE	1833.33	1314.81	15	2606.48
20	195	125	20-25	3750.00	2129.63	212.96	NONE	2342.59	1407.41	20	3921.30
25	210	105	25-30	3453.70	2824.07	282.41	NONE	3106.48	347.22	25	5328.70
30	163	200	30-35	2324.07	4166.67	416.67	NONE	4583.33	-2259.26	30	5675.93
35	88	250	35-40	953.70	3175.93	317.59	NONE	3493.52	-2539.81	35	3416.67
40	15	93	40-45	138.89	1092.59	109.26	NONE	1201.85	-1062.96	40	876.85
45	0	25	45-50	0.00	231.48	23.15	NONE	254.63	-254.63	45	-186.11
50	0	0								50	-440.74
Summation				17240.74	16074.07	1607.41		17681.48			

Figure 9.4 An example of an earthwork summary sheet.

347

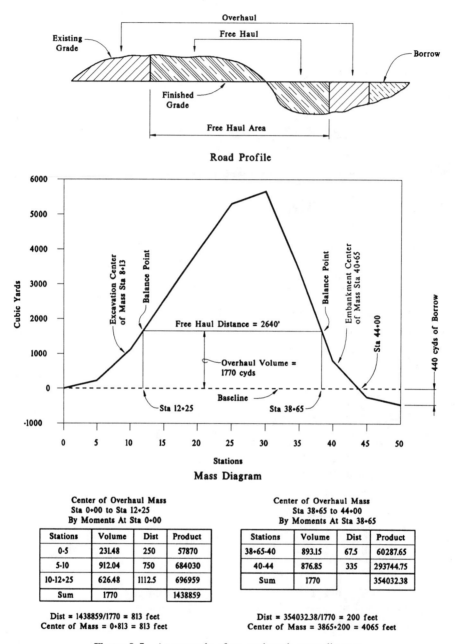

Figure 9.5   An example of an earthwork mass diagram.

In a mass diagram, a balance point is a station where the summation of volumes of excavation and embankment (including shrinkage) are equal. Any line that is constructed parallel to the mass diagram's base indicates balance points wherever the construction line intersects the diagram. For example, all earthwork conducted between Stations 0 + 00 and 44 + 00 balances in Figure 9.5.

If balance lines are constructed parallel to the diagram's base using the freehaul distance to scale, all earthwork located between the balance points is to be accomplished with no overhaul being involved. In Figure 9.5 this occurs between Stations $12 + 25$ and $38 + 65$. Overhaul, therefore, can be expressed as the number of cubic yards of soil moved outside of the freehaul distance and is expressed in this case as cubic yard half-miles. The required overhaul distance is found by using the mass diagram to obtain the distance between mass centers for excavation and embankment.

To determine the center of mass by moments, each volume within the overhaul area (on each side of the freehaul zone) is multiplied by the distance from a reference station (zero for one, 3865 feet for the other) to the midpoint between sequential volumes. Figure 9.5 shows the necessary calculations to determine the centers of mass by moments. The overhaul is calculated by multiplying the overhaul volume by the distance between mass centers while excluding the freehaul distance. The overhaul is therefore:

$$OH = \frac{(OH \text{ volume})(\text{center of mass right} - \text{center of mass left} - \text{freehaul})}{\text{freehaul}}$$

$$OH = \frac{1770 \, (4065 - 813 - 2640)}{2640},$$

or 410 cubic yard half-miles. This value can aid the design engineer in estimating earthwork costs. Also indicated on the corresponding profile in Figure 9.5 are directions of haul and borrow placement.

9.4 ELEMENTS OF THE GRADING PLAN

The site grading plan is an excellent place to show sedimentation and erosion control improvements along with temporary drainage improvements. This plan also includes existing contour lines and their proposed realignment, spot elevations, finished floor elevations for proposed buildings, elevations for recreational playing courts, parking lots, drainage channels, lakes, and ponds. Often a site's topography requires additional improvements to minimize design slope gradients and permit user access. Appurtenances such as stairs, ramps, and retaining walls can be used for such purposes.

The plan should also include considerations for laterally supporting adjoining property, especially if permanent excavations could affect this absolute right of the abutting landowner. Lateral support required for construction, however, is the responsibility of the contractor. Other elements of the grading plan could include delineation of areas requiring various amounts of compaction effort that would augment compaction requirements set forth in the construction specifications. The plan also could include provisions for stabilizing slopes and easement areas to protect slopes that would have to be shown on the final recorded plat.

Locations where poor soil conditions are expected to be found (as a result of geotechnical explorations and testing) can be delineated on the grading plan, along with provisions to improve or remove this undesirable soil. If the area contains expansive soils, then the grading plan could possibly:

(a) Indicate replacement of the upper soil layer along with provisions for backfilling with nonexpansive low permeability soil. Provisions for the backfill to be compacted to a specified amount would be included in the construction specifications, as would requirements for shaping the compacted soil's surface to conform with the grading plan.

(b) Indicate the placement and mixing of nonexpansive soils with the existing expansive soils to reduce the effects of expansion. Provisions for the soil mixture to be compacted to a specified amount would be included in the construction specifications, along with requirements for shaping the compacted soil's surface to conform with the grading plan.

9.4.1   Finished Floor Elevations

The establishment of finished floor elevations is important to insure that the proposed structure will not flood, is elevated enough to allow gravity flow sanitary sewer lines to drain into the installed main, and is high enough to keep soil from being transported into the structure.

Where the terrain is relatively flat, it is good practice to establish the finished floor elevation of a ground slab at least a foot above the elevation of the frontage road unless the expected flood elevation for the site requires the elevation to be set higher. If the site slopes appreciably, the finished floor elevation is a function of the driveway gradient. This driveway gradient should not usually exceed 10 percent. However, a slightly higher gradient can be used if the drive is upgrade from the street.

Slab on grade structures also require a minimum of three feet from the finished floor elevation to the invert of the sanitary sewer service line for the fittings and cleanout to have adequate space for installation. The invert elevation at the cleanout also must be high enough to allow the gravity service to slope to the main at a gradient equal to or greater than the minimum allowable slope for service lines according to local authority.

If the grading plan requires substantial earthwork operations to shape the site, extra care must be used if expansive soils are present to eliminate the possibility of differential heave under building foundations. This is especially true where a hilly site is regraded. Proposed building foundations located partly on fill and cut areas could experience this problem, according to Jones and Jones (1987). The grading plan should call for excavating problem areas to a depth greater than the one desired (called undercutting) so that nonexpansive backfill can be placed and compacted. Other methods of controlling the detrimental effects of expansive soils can include provisions for controlling the soil moisture or using chemicals. The reader is referred elsewhere for a more complete presentation on expansive soils (Nelson and Miller, 1992).

9.4.2   Spot Elevations

Spot elevations can represent either existing or proposed elevations on a grading plan. A legend is required to distinguish one elevation type from the other. Spot elevations are useful to indicate elevations where other means are not suitable. For example, if the design professional wishes to signify that the finished floor slab of a structure is eight inches above the surrounding ground, spot elevations are used, as shown in Figure 9.6.

9.4.3   Slope Gradients

Maximum grades for finished surfaces within a development are based on functional use. Maximum gradients for curb cuts and ramps to accommodate handicap access are contained in Section 2.5. The gradients for streets are included in Chapter 6.

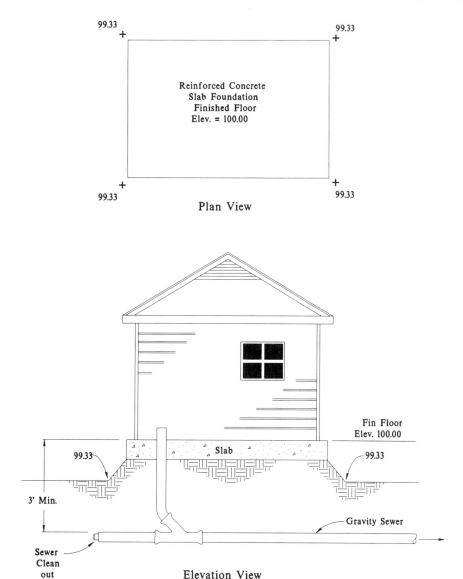

Figure 9.6 Spot elevations signifying finished grades.

Driveways should be designed to prevent vehicular exhaust pipes and trailer hitches from dragging when accessing a steep grade. Should the drive be oriented on a negative gradient from the street, care must be taken to prevent street runoff from flowing down the driveway. Also, runoff that accumulates on the driveway must be properly diverted before it enters the structure or an unwanted place.

Slopes in other areas are a function of stability and erosion potential. The use of retaining walls can reduce the slope gradient.

9.4.4 An Example of a Grading Plan

To illustrate an example of a site grading plan, the project shown in Figure 4.40 will be used. Existing contours are shown as fine lines, while contours requiring alterations (new

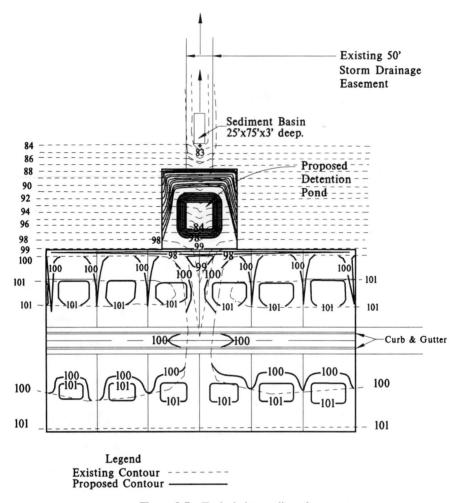

Figure 9.7   Typical site grading plan.

contours) are shown thicker. By applying the principles in Section 4.5 concerning contour lines, the grading plan shown in Figure 9.7 is constructed.

Development sites can include topography that has great differences in elevation. To provide flexibility in accommodating these elevation changes within a development, appurtenances are often required, including stairs, ramps, handrails and retaining walls (see section 9.7 for retaining walls).

9.5   STAIRS, RAMPS, AND HANDRAILS

Steps and ramps provide pedestrians access between levels of unequal elevation. Steps provide incremental vertical movement while ramps provide uniform movements. Both steps and ramps must be designed to provide user comfort and safety.

9.5.1 Steps

Steps can be constructed of wood, reinforced concrete, brick, metal, or stone. The proportions are usually based on local regulatory code requirements. Steps must be placed so that their support is obtained below the frost line. Handrails are included to aid pedestrians; they provide security and balance if the gradient is not relatively flat.

9.5.2 Ramps

Ramps are usually made of materials similar to those for steps but allow access to either pedestrians or wheeled vehicles. As in the case of steps, ramps usually include handrails for user protection and restraint.

9.5.3 Handrails

Handrails provide positive restraint to pedestrians to help prevent accidental falls. They also give stability and guidance to the user. Handrails are usually constructed of metal; however, brick, stone, reinforced concrete, and wooden rails (walls) are also utilized. As with stairs, handrails must comply with local regulatory codes. In addition to being placed along steps and ramps, handrails are also used along the top edge of retaining walls if pedestrian exposure is contemplated.

9.6 SLOPE STABILITY

A soil slope may be stable under certain conditions, but highly unstable if conditions change. Slope failure is usually expressed by sliding and occurs when the internal resistance of the soil to movement is less than the forces attempting to induce movement.

9.6.1 Slope Stability Analysis

Typically if a one foot thick section of embankment is analyzed as shown in Figure 9.8a, the length of failure zone, L, is the corresponding length of arc. The moment attempting to induce sliding along the arc is the product of the weight of soil (W_t) and its lever arm (a). If the strength of the soil is expressed by its shear strength, S_s, in tons per square foot, the total shearing force is $S_s \times L \times 1$ foot. The total resisting moment is $S_s \times L \times 1 \times R$, as shown in Figure 9.8a, where R is the moment arm. The factor of safety against sliding is then the resisting moment divided by the overturning moment.

An increase in soil moisture from rain or the addition of overburden pressures at the top of the slope can increase the moment attempting to induce soil movement. This will result in a decrease in the factor of safety. Likewise, a decrease in the soil's shear strength can result in possible slope failure. More specific reasons for slope failure can include:

(a) An increase in W_t without a corresponding increase in S_s.
(b) Additional fill or structures being placed at the top of the slope.
(c) Increasing moisture in the soil.

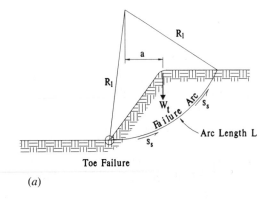

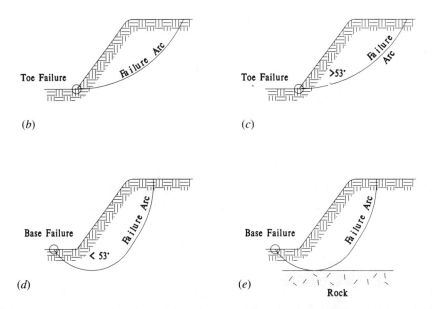

Figure 9.8   Slope stability. (*a*) General relationships for slope stability analysis. (*b*) Typical slope failure for soil containing both cohesive and noncohesive materials. (*c*) Typical slope failure for cohesive soil with a steep slope. (*d*) Typical slope failure for cohesive soil with a mild slope. (*e*) Typical slope failure with underlying rock. Source: Karl Terzaghi and Ralph B. Peck. *Soil Mechanics in Engineering Practice*. Copyright © 1966 by John Wiley & Sons, Inc. Reprinted by permission of John Wiley & Sons, Inc.

   (d) Hydrodynamic forces such as piping attributed to running water or sudden drawdown (especially in earthen dams).

   (e) Slope failure attributed to frost action, chemical changes, and foundation failure.

9.6.2   Typical Slope Failures

A stable slope comprised of cohesionless materials such as sand and gravel would take the form of the soil's natural angle of repose (or less) based on field conditions. If the soil is moist, it will stand on a slope greater than the natural angle of repose under dry conditions. However, this may not always be safe.

If the soil is cohesive, slope stability is difficult to determine and the analysis may be undertaken on a trial and error basis. Generally when the soil obtains its strength from both cohesion and shear (both clay and sand or gravel are present), slope failures usually occur at the toe of slope, as shown in Figure 9.8*b*. If the soil is entirely comprised of cohesive material, where it is essentially in a quick condition, two failures normally occur, as shown in Figure 9.8*c* and *d*:

(a) If the angle of slope is greater than 53°, the slope face is considered steep, and the failure usually occurs at the toe of slope.

(b) If the angle of slope is less than 53°, the failure usually occurs at the base of the toe.

Underlying soil conditions such as a rock layers can also cause failure at the base of the toe, as shown in Figure 9.8*e*.

Various methods are available to analyze slope stability. Interested readers are referred to Lambe and Whitman (1969) and Terzaghi and Peck (1966).

9.7 RETAINING WALLS

Structures to hold soil or other materials into position are called retaining walls. They can be constructed of wood, concrete, stone, masonry block, or steel. Figure 9.9 depicts various types of retaining structures.

9.7.1 Wall Pressures

The earth pressure exerted by a soil on a wall is not a constant value, but depends on the yield of the wall. There are an infinite number of values, as depicted in Figure 9.10; however, one of three general conditions will occur:

(a) Earth pressure at rest occurs when there is no wall movement.

(b) Active earth pressure occurs when the wall moves outward.

(c) Passive earth pressure occurs when the wall moves inward.

In the case of no wall movement either toward or away from the soil, the coefficient of at rest earth pressure is expressed as:

K_0 = Coefficient of earth pressure at rest or horizontal pressure/vertical pressure

This value can range from 0.4 for loose sands to 0.5 for dense sands (Terzaghi and Peck, 1966). The horizontal pressure, as illustrated in Figure 9.10, is calculated at any given depth by:

$$p_0 = K_0 wh \qquad (9\text{-}2)$$

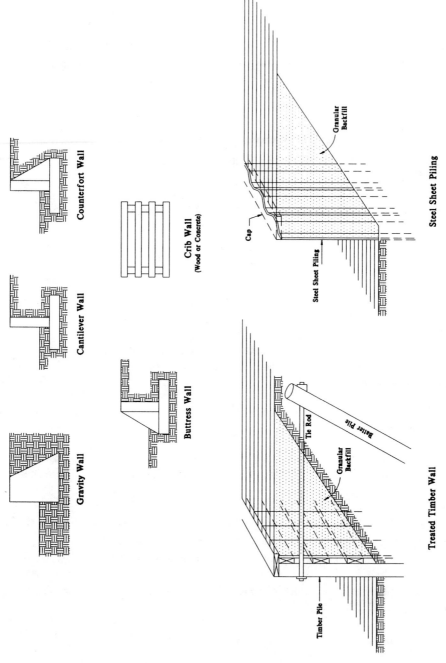

Figure 9.9 Examples of retaining structures. Source: Whitney Clark Huntington. *Earth Pressures and Retaining Walls.* John Wiley & Sons, Inc., 1957.

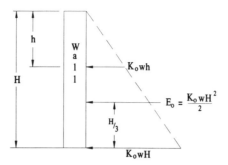

Figure 9.10   Earth pressure relationships.

where:

K_0 = The coefficient of earth pressure at rest.
w = The unit weight of soil in pounds per cubic foot.
h = Any depth specified in feet.

If the pressure at the bottom of the wall is K_0wH, where H is the total depth, the total horizontal force exerted on the wall's face is:

$$E_0 = (K_0wH)\ (H/2) \text{ per foot of wall length} \qquad (9\text{-}3)$$

E_0 acts $H/3$ from the wall's bottom.

Should the wall be allowed to yield slightly outward, the soil will follow the wall until on the verge of failure. At this point the soil's shear is maximum, and the pressure exerted on the wall has diminished since the soil is better able to hold itself. The resulting wall pressure is active earth pressure, which is commonly used in retaining wall analysis. In this case:

K_a = coefficient of active earth pressure or horizontal pressure/vertical
         pressure

The horizontal pressure at any depth is calculated by:

$$p_a = K_awh \qquad (9\text{-}4)$$

where:

K_a = The coefficient of active earth pressure.
w = The unit weight of soil in pounds per cubic foot.
h = Any depth specified in feet.

The total horizontal force exerted on the wall's face using H as the total depth is:

$$E_a = (K_awH)(H/2) \text{ per foot of wall} \qquad (9\text{-}5)$$

Should the wall be forced into the soil so that the soil is on the verge of failure, the soil resists, resulting in passive earth pressure. In the passive case:

K_p = coefficient of passive earth pressure where, horizontal pressure/vertical pressure

The horizontal pressure at any depth

$$p_p = K_p wh \tag{9-6}$$

where:

K_p = The coefficient of passive earth pressure.
w = The unit weight of soil in pounds per cubic foot.
h = Any depth specified in feet.

The total horizontal force on the wall using H as the total depth is:

$$E_p = (K_p wH)(H/2) \text{ per foot of wall} \tag{9-7}$$

Two methods are commonly used to predict earth pressures: Coulomb's and Rankine's theories. In both theories assumptions are made that the wall is free to move and that water contained in the voids of the soil does not exert any appreciable seepage pressure (Terzaghi and Peck, 1966).

9.7.2   Coulomb's Theory

According to Huntington (1957), C. A. Coulomb published in France during the year 1776 a theory for predicting earth pressures against a wall where an assumption is made that the wall's face is rough and that friction exists between the wall and backfill (see Figure 9.11).

The active earth pressure exerted against the wall's surface results from the tendency of the soil wedge to slide along A–B. The soil weight, W, is held in equilibrium by a force on the plane of the retaining wall (P) and a force on the rupture surface of the soil (R). The

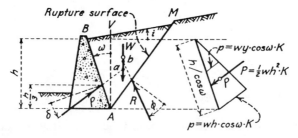

Figure 9.11   Coulomb's active earth pressure relationships. Source: Whitney Clark Huntington. *Earth Pressures and Retaining Walls.* John Wiley & Sons, Inc., 1957.

angle of inclination of the rupture surface is assumed to be the plane that makes P a maximum value. To exert this pressure the wedge must slide as the wall deflects away from the wedge, which further allows the soil to slide along the wall A–B, as illustrated in Figure 9.11. As a result Coulomb assumed that the wedge has a failure surface A–M and a pressure surface A–B. In practice, the failure surface is curved rather than straight; however, the effect is negligible for active earth pressure calculations. To estimate the active earth pressure using Coulomb's theory, the following relationships apply:

$$P_a = \frac{1}{2} wh^2 \frac{\cos^2(\phi - \omega)}{\cos^2 \omega \cos(\delta + \omega)\left[1 + \sqrt{\dfrac{\sin(\delta + \phi)\sin(\phi - i)}{\cos(\delta + \omega)\cos(\omega - i)}}\right]^2} \qquad (9\text{-}8)$$

where:

ω = The wall's angle referenced from the vertical.
ϕ = The soil's internal angle of friction.
δ = The angle of friction between the soil and pressure surface.
i = The backfill slope reference to the horizon.
w = The unit weight of soil.
h = The vertical height equivalent of the pressure surface.

Although Coulomb did not include the analysis of passive earth pressures, his theory can be applied, with the resulting frictional forces acting in opposite directions. The equation for passive earth pressure would then become:

$$P_p = \frac{1}{2} wh^2 \frac{\cos^2(\phi + \omega)}{\cos^2 \omega \cos(\omega - \delta)\left[1 - \sqrt{\dfrac{\sin(\phi + \delta)\sin(\phi + i)}{\cos(\omega - \delta)\cos(\omega - i)}}\right]^2} \qquad (9\text{-}9)$$

9.7.3 Rankine's Theory

According to Huntington (1957) in 1857 W. J. M. Rankine developed a theory to estimate earth pressures pertaining to cohesionless soil deposits with a plane ground surface unlimited in lateral extent and depth. Rankine's theory is based upon the soil backfill having a horizontal surface and being supported by a frictionless vertical wall. Active earth pressure occurs as thin layers of soil parallel to the backfill's surface expand laterally. This expansion results in shear planes developing similar to those shown in Figure 9.12.

The coefficient of active earth pressure for Rankine's dry granular soil is defined by:

$$K_a = \tan^2 (45° - \phi/2) \qquad (9\text{-}10)$$

and the passive coefficient by:

$$K_p = \tan^2 (45° + \phi/2) \qquad (9\text{-}11)$$

where ϕ is the soil's internal angle of friction. For a more complete treatment of active and

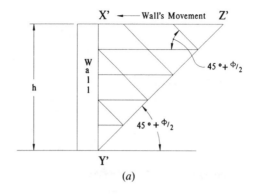

(a)

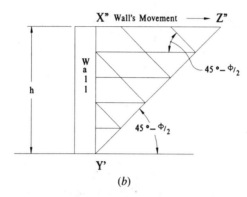

(b)

Figure 9.12   Rankine's failure planes behind a frictionless wall. (*a*) Active condition. (*b*) Passive condition. Source: Karl Terzaghi and Ralph B. Peck. *Soil Mechanics in Engineering Practice.* Copyright © 1966 by John Wiley & Sons, Inc. Reprinted by permission of John Wiley & Sons, Inc.

passive earth pressures, the reader is referred elsewhere (Lambe and Whitman, 1969; Perloff and Baron, 1976; Terzagi and Peck, 1966).

9.7.4   Uncertainty of Cohesive Soil Backfills

Should cohesive soils be placed behind a retaining wall, active earth pressures would decrease while passive pressures would increase over those generated by cohesionless materials of similar unit weights. The presence of moisture could cause these soils to swell, creating very large pressures. Also, if adequate drainage is not maintained behind the wall, accumulated hydrostatic pressures can be significant in the analysis. It is therefore necessary that porous backfill be provided along with proper drainage design to eliminate these adverse field conditions.

9.7.5   Other Considerations

Most walls tilt outward when backfill is being installed. When walls are isolated, a small amount of wall tilting is usually acceptable. When the wall is long or consists of various height segments, wall joists may be necessary to allow differential movements. Extra care should be exercised to anticipate tilting if other rigid structures are to be connected to the retaining wall.

9.7.6 Stability Analysis Procedures

In analyzing a retaining wall several conditions must be satisfied, including safety against:

(a) Overturning.

(b) Foundation failure.

(c) Sliding.

The normal procedure for checking the stability of a retaining wall begins by choosing tentative wall dimensions and then estimating the earth pressures acting on the wall section developed.

The actual factor of safety against overturning is determined by comparing the total moment of forces resisting overturning about the toe of the wall with the moment of forces tending to cause overturning about that point.

The determination of safety against foundation failure involves three steps: (1) determining if the resultant of the footing contact pressure falls within the middle third of the footing width (but preferably within the middle sixth segment); (2) determining the actual factor of safety against bearing capacity failure by comparing the ultimate bearing capacity of the soil with the actual maximum footing contact pressure (or insuring that the actual maximum footing contact pressure is less than the allowable soil bearing capacity); and (3) determining if anticipated total or differential settlements obtained using the principles of settlement analysis exceed criteria limits for settlement or tilt.

The actual factor of safety against sliding is determined by comparing the total horizontal forces resisting sliding with those tending to cause sliding.

Comparison of actual factors of safety determined above with values that are appropriate to the existing conditions are then made. Typical, acceptable factors are published in a number of current design texts. Should these calculations indicate an inadequate design, modifications to the initial wall's dimensions must be made and the process repeated until the design is found to be satisfactory. The structural design of the wall can then be undertaken using other guidance (Ferguson, 1979). Once the design is completed, it should be checked for slope stability as indicated in Section 9.6.

9.7.7 An Example of a Retaining Wall Stability Analysis

Using Figure 9.13 as a guide, the footing thickness, t, can usually range from $H/8$ to $H/12$, with 12 inches customarily being the minimum. If $H/12$ is used for estimation purposes, the corresponding thickness is 1.42 feet. For ease of initial calculations 18 inches will be used. The corresponding stem height would be 15.5 feet (17.0 feet − 1.5 feet). If the preliminary stem cap is assumed to be 12 inches, the stem base dimension can be estimated as 15.88 inches (12 inches + 15.5 feet × 0.25 inch per foot). For convenience 16 inches will be preliminarily used. The footing width, W, should be as small as possible for economic reasons; however, it must be large enough to provide stability and can range from $0.4H$ to $0.65H$, depending on the soil and surcharge conditions. To determine W using $0.65H$, calculate 0.65×17 feet, or 11.05 feet; therefore use 11 feet. The preliminary wall dimensions are shown in Figure 9.14.

The effective soil height ($A–B$) can now be determined by $H = 17$ feet + 5.67 × tan 18°, or 18.84 feet. The active earth pressure can now be calculated using Equation

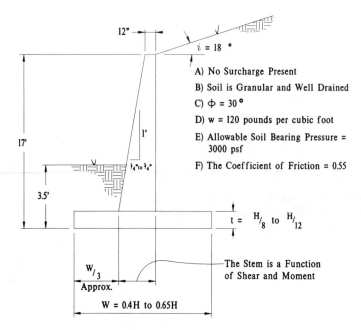

12"

$i = 18°$

A) No Surcharge Present

B) Soil is Granular and Well Drained

C) $\phi = 30°$

D) w = 120 pounds per cubic foot

E) Allowable Soil Bearing Pressure = 3000 psf

F) The Coefficient of Friction = 0.55

17'

1'

3.5'

¼" to ¾"

$t = H/_8$ to $H/_{12}$

The Stem is a Function of Shear and Moment

$W/_3$
Approx.

$W = 0.4H$ to $0.65H$

Figure 9.13   An example of cantilever retaining wall stability analysis.

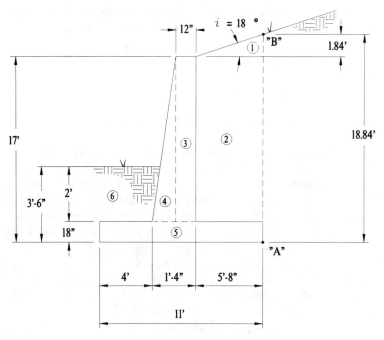

12" $i = 18°$

① "B"   1.84'

17'

18.84'

③   ②

3'-6"   2'

⑥   ④

18"   ⑤

4'   1'-4"   5'-8"

"A"

11'

Figure 9.14   Preliminary wall dimensions.

9-8, assuming that the wall friction (δ) has the same angle as the backfill surface slope (i):

$$P_a = \left[\frac{120 \times 18.84^2}{2} \right] \left[\frac{\cos^2(30 - 0)}{\cos^2(0)\cos(18 + 0)\left[1 + \sqrt{\dfrac{\sin(18 + 30)\sin(30 - 18)}{\cos(18 + 0)\cos(0 - 18)}} \right]^2} \right]$$

or 8408.2 pounds per foot of wall.

$$P_v = P_a \sin 18° = 2598.3 \text{ pounds}$$
$$P_h = P_a \cos 18° = 7996.6 \text{ pounds}$$

These forces act $H/3$ above the base (18.84 feet/3), or 6.28 feet.

Additional forces such as surcharge loads on the backfill surface, if present, would now be analyzed. Since none is included with this example, the active earth pressure forces would appear as shown in Figure 9.15.

The stem's base dimension now needs to be checked for preliminary structural adequacy. Interested readers are referred elsewhere for this analysis (Ferguson, 1979).

The wall's stability is determined taking moments about point A where all moments are counterclockwise:

Area	Vertical force (pounds)	Lever arm (feet)	Moment (foot-pounds)
1	$0.5 \times 5.67 \times 1.84 \times 120$	1.87	1,171
2	$5.67 \times 15.5 \times 120$	2.84	29,951
3	$1 \times 15.5 \times 150$	6.17	14,345
4	$(4/12) \times 0.5 \times 15.5 \times 150$	6.78	2,627
5	$11 \times 1.5 \times 150$	5.5	13,613
6	$2 \times 4 \times 120$	9.0	8,640
P_v	2598.3	0	0
P_h	7996.6	6.28	50,219
		Sum	120,566

To determine the location of the resultant using Figure 9.15, the following calculations are made:

summation of all vertical forces = 19,918 pounds
summation of moments at A = 120,566 foot-pounds
x = 120,566 foot-pounds/19,918 pounds

or 6.05 feet from A is where the resultant acts.

The resultant's eccentricity is calculated by 6.05 feet − 5.5 feet, or 0.55 feet, and is within the middle third of the base (11 feet/3 = 3.67 feet situated with 1.83 feet on each side of the centerline). Therefore, the resultant is located in the middle third, but more importantly within the middle sixth. The distribution of soil pressure under the footing can be determined using:

$$q = \left[\frac{\text{summation vertical forces}}{BW} \right] \left[1 \pm \frac{6e}{W} \right] \qquad (9\text{-}12)$$

where B is equal to one foot of wall (Huntington, 1957).

$$q_{max} = \left[\frac{19,918}{1 \times 11} \right] \left[1 + \frac{6(0.55)}{11} \right]$$

or 2354 pounds per square foot.

$$q_{min} = \left[\frac{19,918}{1 \times 11} \right] \left[1 - \frac{6(0.55)}{11} \right]$$

or 1268 pounds per square foot. Since $q_{allowable}$ was given as 3000 pounds per square foot, the supporting soil appears that it will not be overstressed.

The corresponding soil pressures at the base are shown in Figure 9.16a.

The wall must be checked for sliding, as shown in Figure 9.16b. Since the amount of embedment ($h = 3.5$ feet) is nominal and vegetation or erosion could affect the soil's strength, sliding should first be checked using the soil's friction with the sum of vertical forces constituting the normal load.

$$\text{Friction} = 19,918 \times 0.55 \text{ coefficient of friction} = 10,955 \text{ pounds per foot}$$
$$\text{of wall}$$

The horizontal component of the active earth pressure is 7996.6 pounds; therefore the factor of safety against sliding is:

$$\text{F.S.} = \frac{19,918}{7,996.6} = 1.37$$

Because the safety factor is less than the minimum recommendation of 1.5, a keyway needs to be incorporated to increase lateral sliding resistance, as illustrated in Figure 9.16c.

If it is desired to have a safety factor of 1.5 with soil removed to the bottom of the footing's toe, the keyway depth can be computed by:

$$\text{F.S.} = \frac{\text{resisting force}}{\text{sliding force}}$$

$$1.5 = \frac{\text{resisting force}}{7996.6} \text{ or } 11,995 \text{ pounds}$$

The keyway must produce 1040 pounds ($11,995 - 10,955$).

Although there are a number of ways to design the key's depth, a conservative approach would be to use Rankine's formula with K_p being defined by Equation 9-11:

$$E_p = \frac{wH^2}{2} \left[\tan^2 (45 + \phi/2) \right]$$

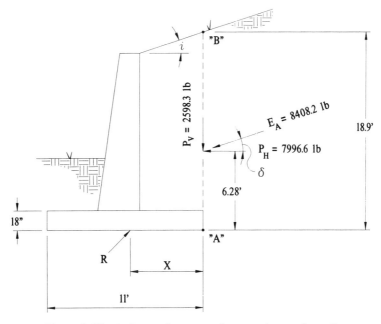

Figure 9.15 Active earth pressure forces acting on the wall.

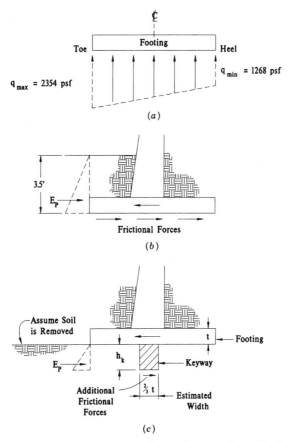

Figure 9.16 Footing analysis for an example of a cantilever wall. (*a*) Footing soil pressures. (*b*) Passive earth pressure for sliding analysis. (*c*) Keyway design.

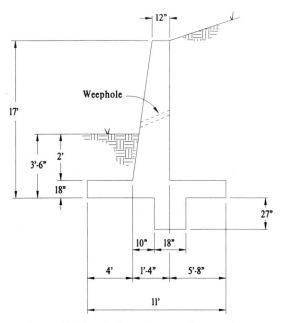

Figure 9.17 Final cantilever wall dimensions.

and the additional frictional force produced by the reinforced concrete keyway's weight (150 pounds per cubic foot) using two-thirds of the footing thickness for its width is:

$$1040 = \frac{120 \times H^2}{2} \ \tan^2 60° + (0.66)(1.5)(H)(150)(0.55)$$

$$1040 = 60 \, H^2 \tan^2(60°) + 81.7H$$

$$180H^2 + 81.7H - 1040 = 0$$

$$H = \frac{-81.7 + 869}{360} = h_k$$

or $h_k = 2.2$ feet deep. The width is rounded to 1.5 feet. The keyway is usually located where the stem's reinforcement can extend into it. That requires the key's front face to be about six inches in front of the interior wall's face, as shown in Figure 9.17.

If the passive earth pressure produced by the full 3.5 feet of soil is considered, the factor of safety against sliding would be:

$$E_p = \frac{wH^2}{2} \ [\tan^2(45 + \phi/2)]$$

or 5951 pounds, where H is comprised of the 3.5 feet of soil above the footing toe and 2.25 feet of soil in front of the keyway (5.75 feet total).

The resisting force is $(0.55)(19,918 + 150 \times 1.5 \times 2.25) + 5951$, or 17,185 pounds. The factor of safety is:

$$\text{F.S.} = \frac{17,185}{7997}, \text{ or } 2.14$$

which is adequate. The final resultant force is determined by using:

The summation of moments previously calculated   120,566 foot-pounds
$1.5 \times 150 \times 2.25 \times 5.42$ keyway   2,744 foot-pounds
Total of all moments at A 123,310 foot-pounds

summation vertical forces $= 19,918 + 1.5 \times 150 \times 2.25 = 20,424$ pounds

The resultant acts at distance x, or

$$123,310/20,424 = 6.04 \text{ feet} - 5.5 \text{ feet or } 0.54 \text{ feet}$$

$$q_{max} = \left[\frac{20,424}{1 \times 11} \right] \left[1 + \frac{6(0.54)}{11} \right]$$

or 2404 pounds per square foot.

$$q_{min} = \left[\frac{20,424}{1 \times 11} \right] \left[1 - \frac{6(0.54)}{11} \right]$$

or 1310 pounds per square foot.

Information about the size and placement of structural reinforcing steel following code requirements is available elsewhere (Ferguson, 1979).

The designed wall should now be checked for slope stability as outlined in Section 9.4.4.

REFERENCES

Das, Braja M. *Principles of Geotechnical Engineering,* 2nd ed. Boston: PWS-Kent, 1990.

Ferguson, P. M. *Reinforced Concrete Fundamentals,* 4th ed. New York: Wiley, 1979, pp. 242–278.

Huntington, Whitney C. *Earth Pressures and Retaining Walls.* New York: Wiley, 1957, pp. 5–8, 51, 59–60, 70, 132, 408.

Jones, D. Earl, Jr. and Karen A. Jones. "Treating Expansive Soils," *Civil Engineering.* New York: ASCE, August 1987, pp. 62–65. Referenced material is reproduced by permission of the American Society of Civil Engineers.

Lambe, T. William and Robert V. Whitman. *Soil Mechanics.* New York: Wiley, 1969, pp. 122–185, 352–373.

Liu, Cheng and Jack B. Evett. *Soils and Foundations,* 3rd ed. Englewood Cliffs, N.J.: Prentice-Hall, 1992.

Nelson, John D. and Debora J. Miller. *Expansive Soils—Problems and Practice in Foundation and Pavement Engineering.* New York: Wiley, 1992.

Perloff, William H. and William Baron. *Soil Mechanics—Principles and Applications.* New York: Wiley, 1976, pp. 100–149.

Sowers, George F. *Soil Mechanics and Foundations: Geotechnical Engineering*, 4th ed. New York: Macmillan, 1979.

Terzaghi, Karl and Ralph B. Peck. *Soil Mechanics in Engineering Practice*. New York: Wiley, 1966, pp. 140–141, 185, 188. Copyright © 1966 John Wiley & Sons, Inc. Reprinted by permission of John Wiley & Sons, Inc.

U.S. Department of Housing and Urban Development (HUD). *Land Planning Principles for Home Mortgage Insurance*. Washington: HUD, 1973, pp. 6-1–6-38.

10 Sanitary Sewer System Design

Design of the sanitary sewer system usually takes place after the transportation and storm drainage systems are initially formulated, unless some unusual circumstances exist. For example, if an existing gravity sanitary sewage system's elevation at a point of proposed connection is high in relation to the ground elevation of the development site, the design of the sanitary system might be undertaken first. The cost of obtaining and placing borrow material to elevate the site (Section 9.2.2) must be evaluated in order to achieve the use of gravity flow when a situation such as this occurs. The designer must also determine how the fill will affect proposed street alignment, storm drainage configuration, and building siting.

The availability of new technology, however, allows various forms of alternative wastewater systems to be considered when dealing with difficult sites where gravity flow is not easily achieved. The U.S. Environmental Protection Agency (EPA)'s Small Community Outreach and Education (SCORE) Program promotes lower utility rates, conservation (of energy, water, materials, and land), and a reduction in urban sprawl. Some of these technologies are included in Section 10.3 and are further defined in a U.S. EPA Bulletin (1992).

The design of a sanitary sewer system can begin with the preparation of an engineer's report and the determination of wastewater flows, followed by the design of either individual disposal systems or a centralized wastewater collection and treatment system.

10.1 ENGINEER'S REPORT

Most regulatory review agencies require the design engineer to prepare a written report for a new sanitary sewer system prior to granting permits for the project's construction and operation. This report usually contains general project information (including phasing plans if an incremental approach is to be utilized), information pertaining to the collection system, a description of the anticipated waste along with its treatability and possible treatment processes, and a description of the treated wastewater outfall and impact on the receiving stream (or soil if land application is contemplated) along with emergency provisions. The report can also address how the proposed system would integrate with the Area Wide 208 Water Quality Management Plan (Section 2.5) as provided in the Clean Water Act. In addition, Section 401 of the Clean Water Act requires that each new point source discharge obtain a National Pollution Discharge Elimination System (NPDES) permit.

10.1.1 General Information

The project's owner is identified in the report to notify the regulatory review agency of the party responsible for the development. A detailed description of the proposed facility, including collection, transport, and treatment, is included; all proposed unit operations and processes intended to be employed for treatment are described. Usually a flow diagram is provided to clarify the overall proposed sequencing of treatment units.

10.1.2   Description of the Waste

The report can include the type of anticipated waste and all potential sources of waste generation. The amount of wastewater can be quantified in terms of average daily flow, minimum flow, and peak flow. Where proposed flows are anticipated, representative characteristics are usually based on comparable wastewater measurements. If some existing flows will be diverted into the proposed system, measurements of these flows should be obtained. Land developments usually include only domestic wastes; however, commercial and industrial wastes could be present. Combined wastes are described in terms of their proportions based on volumes and strength (biochemical oxygen demand—BOD— and chemical oxygen demand—COD—or total organic carbon—TOC). If raw or pretreated industrial wastewater is expected, a four digit Standard Industrial code number must be provided, along with information on the process that generates the waste, its relative volume, and discharge frequency.

10.1.3   Characteristics of the Waste

Certain characteristics of the proposed untreated wastewater should be evaluated. (Section 10.10.1 includes typical domestic wastewater characteristics.) The wastewater's strength is described including the biochemical oxygen demand (BOD, five day, 20°C), and chemical oxygen demand (COD). Color can be included along with anticipated variations and treatability. The raw wastewater's pH is also of importance along with any anticipated fluctuations, as well as a description of the total and phenolphthalein alkalinity.

Should any heavy metals, noxious or toxic/hazardous compounds be anticipated, they must be identified and their concentrations quantified. Pretreatment of wastewater containing these types of materials is usually required.

Other materials resistant to biological degradation normally are identified along with their anticipated concentrations. Should the wastewater contain surfactants, nitrogen, or phosphorous, these should be identified and quantified.

10.1.4   Treatability of Waste

A description of the treatability of the anticipated wastewater is included where proposed pilot (field) or bench (laboratory) studies are described. Any applicable manufacturer's literature can be included along with a description of the anticipated quality and characteristics of the treated effluent.

10.1.5   Treatment Facility Location and Effluent Discharge

A description of the location of the proposed wastewater treatment plant is included along with the latitude and longitude of the proposed treated effluent discharge (including references to nearby landmarks). A location map may be provided for clarification. The discharge point could be some distance away from the treatment plant's site, in which case the effluent must be transported from the treatment plant in a long gravity outfall or pumped in a forcemain. Alternative effluent discharges can include irrigation or other land application practices.

10.1.6 Physical Characteristics of the Proposed Site

Should the project's scope include ground storage basins or provisions for spray irrigation, seepage, or composting, various prevailing site conditions must be investigated. These conditions can include identification of soils, percolation test results, drainage characteristics of the surrounding area, distance to the nearest existing wastewater treatment plant, distance to the nearest inhabited structure, distance to the nearest property line, and the location of all wells close to the water supply. Determination of the seasonal groundwater table depths also is important.

10.1.7 Development Configuration

The engineer's report usually includes a preliminary plan showing the layout of the proposed development. This plan can be a copy of the preliminary development plan that is described in Chapter 5.

10.1.8 Receiving Waters

A description of the receiving waters subjected to discharge of treated effluent usually is included, along with the minimum seven day low flow that occurs on the average of once every 10 years ($7 Q 10$). Chemical and biological characteristics of the receiving waters should be included along with the results of any field sample surveys.

The engineer must identify all downstream areas that can be influenced by the treated effluent, including downstream water intakes, swimming areas, impoundments, and shellfish areas. It is important to identify any other discharges, both up- and downstream, that might affect the receiving water's quality in combination with the proposed treated effluent.

10.1.9 Impact of Discharge on Receiving Waters

A description can be included of the quantitative and qualitative effects of the proposed discharge on the receiving stream based on local regulatory requirements as well as how the proposed discharge conforms to the Area Wide 208 Water Quality Management Plan. This assessment can consist of the amount of dissolved oxygen, dissolved solids, total suspended solids, and other parameters.

10.1.10 Emergency Provisions

The design engineer can describe provisions for standby power and routine maintenance when major system components are taken off-line. Should system failure or shutdown occur, a description of the anticipated consequences must include the expected quality of the effluent discharged during downtime and its effect on the receiving waters. The potential for health hazards, nuisance, and hazardous conditions to operators or the public also must be addressed.

10.1.11 Facility Consolidation

Other wastewater treatment facilities might exist close to the proposed new facility. In cases such as this, the Area Wide 208 Water Quality Management Plan may encourage consolidation of facilities; for example, the capacity of the existing plant might be upgraded to accommodate the additional flows. This could provide better service to end users and better environmental conditions. Alternatives such as this should be evaluated.

An analysis of the existing and ultimate plant capacities must be made to determine if an existing treatment facility is capable of accommodating the anticipated wastewater from the proposed development. Use of an existing facility can entail the installation of a wastewater pumping station to transport the waste either directly to the plant or to a sewer line that leads to the plant and that has adequate carrying capacity. Agreement with the existing facility's owner to accept and treat the proposed waste must be obtained by the developer. Evidence of this acceptance normally is included in the report.

10.2 WASTEWATER FLOWS

Wastewater flows can be comprised of domestic sewage (along with industrial wastes and processing water if applicable), inflow, infiltration, and possibly cooling water. In a new system, the engineer usually provides means to minimize the effects of inflow, infiltration, and cooling water. According to a Phase I Residential Water Use Study conducted in the Baltimore area (Geyer et al., 1963), approximately 89 percent of the water supplied for nonsprinkling domestic water use becomes contributory flows into sanitary sewers. Design of the wastewater system must account for anticipated peak flow conditions.

10.2.1 Variations in Domestic Wastewater Flows

In the same Baltimore study, wastewater flows were found to be at a minimum between the hours of 2 A.M. and 6 A.M., while a morning peak occurred just before noon, and an evening peak between 6 P.M. and 10 P.M., as illustrated in Figure 10.1.

According to Imhoff (Novotny et al., 1989), typical per capita sewage flows can range from 60 to 120 gallons per capita per day, with design peak flows between 1.5 and 10 times the average daily flow rate. Fair et al. (1966) suggest the ratios contained in Figure 10.2 can be used to estimate maximum and minimum flows, while other references suggest alternative values (Metcalf and Eddy and Tchobanoglous, 1981).

10.2.2 Flow Determination

Domestic wastewater flows can be estimated using fixture units when 1 fixture unit is equal to 7.5 gallons per minute. Figure 10.3 contains values for various fixtures.

Another method for estimating wastewater flows is the use of established unit loads similar to those contained in Figure 10.4.

The number of units of each establishment category gives the estimated average daily flow, which can be adjusted by factors similar to those contained in Figure 10.2 to obtain maximum or minimum design flows.

If land developments similar to the one being designed exist near the proposed site, water usage records and/or field gaging surveys can be conducted to obtain data that can

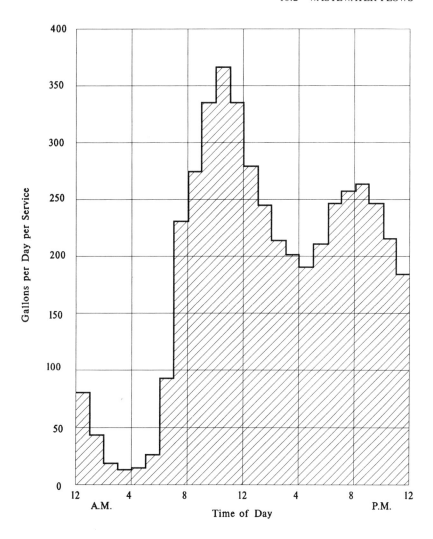

Figure 10.1   Mean hourly domestic sewage flows. Source: John C. Geyer, Jerome B. Wolff, F. P. Linaweaver, Jr., and Acheson J. Duncan. Report on Phase One, Residential Water Use Project, *Final and Summary Report on Phase I of the Residential Water Use Research Project, January 1, 1959 to January 1, 1963.* October 1963. Reproduced with permission.

Description Of Flow	Relationship To Average Daily Flow
Maximum Daily Flow	2 x Average Daily Flow
Maximum Hourly Flow	3 x Average Daily Flow
Minimum Daily Flow	0.67 x Average Daily Flow
Minimum Hourly Flow	0.33 x Average Daily Flow

Figure 10.2   Residential sewage flows as ratios to the average. Source: Gordon M. Fair, J. C. Geyer, and David A. Okun. *Water and Wastewater Engineering,* Vol. I. Copyright © 1966 by John Wiley & Sons, Inc. Reprinted by permission of John Wiley & Sons, Inc.

Fixture or Group	Fixture Unit Value
Bathroom group consisting of a lavatory, bathtub or shower stall, and a water closet (direct flush, valve)	8
Bathroom group consisting of a lavatory, bathtub or shower stall, and a water closet (flush tank).............................	6
Bathtub with $1\frac{1}{2}$ trap..	2
Bathtub with 2" trap	3
Bidet with $1\frac{1}{2}$ trap..	3
Combination sink and wash tray with $1\frac{1}{2}$ trap......................	3
Combination sink and wash tray with food waste grinder unit (separate $1\frac{1}{2}$ trap for each unit)	4
Dental unit or cuspidor...	1
Dental lavatory	1
Drinking fountain	$\frac{1}{2}$
Dishwasher, domestic type	2
Floor drain	1
Kitchen sink, domestic type	2
Kitchen sink, domestic type with food waste grinder unit.........	3
Lavatory with $1\frac{1}{2}$ waste plug outlet....................................	2
Lavatory with $1\frac{1}{4}$ or $1\frac{3}{8}$ waste plug outlet	1
Lavatory (barber shop, beauty parlor, or surgeon's)	2
Lavatory, multiple type (wash fountain or wash sink), per each equivalent lavatory unit	2
Laundry tray (1 or 2 compartments)	2
Shower stall	2
Showers (group) per head..	3
Sink (surgeon's)	3
Sink (flushing rim type, direct flush valve)	8
Sink (service type with floor outlet trap standard)	3
Sink (service type with P trap)..	2
Sink (pot, scullery, or similar type)	4
Urinal (1 flush valve)...	8
Urinal ($\frac{3}{4}$ flush valve)	4
Urinal (flush tank)..	4
Water closet (direct flush valve)	8
Water closet (flush tank)	4
Swimming pools, per each 1000-gal capacity	1
Unlisted fixture, $1\frac{1}{4}$ or less fixture drain or trap size...............	1
Unlisted fixture, $1\frac{1}{2}$ fixture drain or trap size........................	2
Unlisted fixture, 2 fixture drain or trap size..........................	3
Unlisted fixture, $2\frac{1}{2}$ fixture drain or trap size.......................	4
Unlisted fixture, 3 fixture drain or trap size..........................	5
Unlisted fixture, 4 fixture drain or trap size..........................	6

Note: Values for continuous or intermittent flow. For a continuous or intermittent flow into a drain, as from a pump, ejector, air-conditioning equipment or similar equipment, a fixture value of 2 shall be assigned for each gpm of flow at rated capacity. The total discharge flow in gpm for any single fixture, divided by 7.5, provides the fixture unit value for that particular fixture.

Figure 10.3 Sanitary drainage fixture units per fixture. Source: New York State Uniform Fire Prevention and Building Code, Table I-903, Division of Housing and Community Renewal, New York, N.Y. January 1984, pp. 276, 277.

Activity	Gallons Per Day Per Person	Lbs. 5-Day BOD Per Day Per Person
Apartments - 3 Bedroom 4 Persons Each	100	0.17
- 2 Bedroom 3 Persons Each	100	0.17
- 1 Bedroom 2 Persons Each	100	0.17
- With Garbage Disposal Units	100	0.23
Churches - Per Seat	3	0.02
Clinics - Per Staff	15	0.03
- Per Patient	5	0.02
Country Club - Each Member	50	0.10
Food Service Operations		
- Ordinary Restaurant		
(Not 24 Hours)(Per Seat)	70	0.20
- 24-Hour (Per Seat)	100	0.30
- Curb Service (Drive In)		
(Per Car Space)	100	0.20
- Vending Machine Restaurant	70	0.12
Hotels - Per Bedroom (No Restaurant)	100	0.17
Laundries - Self Service (Per Machine)	400	0.68
Mobile Homes - 3 Persons Each	100	0.17
Offices - Per Person (no Restaurant)	25	0.05
Picnic Parks - Average Attendance	10	0.06
Residences - 4 Persons Each	100	0.17
- With Garbage Disposal Units	100	0.23
Rest Homes - Per Bed (No Laundry)	100	0.17
- Per Bed (With Laundry)	150	0.20
Schools - Per Person (No Showers, Gym, Cafe.)	10	0.04
- Per Person With Cafe. Only	15	0.05
- Per Person (With Showers, Gym, Cafe.)	20	0.06
Service Stations- Each Car Served	10	0.06
- Each Car Washed	75	0.03
- First Bay (per Day)	1000	2.00
- Each Additional Bay (Per Day)	500	1.00
Shopping Centers - Per 1,000 sq. ft. Floor Space		
(No Restaurant)	200	0.40
Swimming Pools - Per Person (With Sanitary		
Facilities and Showers)	10	0.04

Figure 10.4 Typical unit contributory wastewater loadings. Source: South Carolina Department of Health and Environmental Control, 1972.

be used as a basis for design. This method is especially valuable if local jurisdictional requirements include the use of flow reducing and water saving devices. Other references contain fuller descriptions of this procedure (ASCE and WPCF, 1982; Metcalf and Eddy and Tchobanoglous, 1981).

10.2.3   Flows from Other Sources

Wastewater flows can originate from several sources including commercial, industrial and process water, inflow, infiltration, and cooling water. Quantities of industrial, process, and cooling water are dependent upon source activities, while some typical commercial flows are included in Figure 10.4. Inflow and infiltration often are difficult to determine.

Inflow can occur from storm water entering manhole covers or from downspouts and sump pumps, while infiltration can result from high groundwater entering the system through unsound joints in manholes or pipes. Older methods and materials used in the construction of sanitary sewer mains could produce pipe joints that are less tight than those customarily expected today. According to ASCE and WPCF (1982), infiltration rates as high as 60,000 gallons per day per mile have been recorded. ASCE and WPCF (1982) also states that it is common to allow 30,000 gallons per day per mile infiltration for sanitary sewers, laterals, and house connections without regard to size on sewers not exceeding 24 inches in diameter. Improved construction materials and joint fittings have greatly reduced this potential for infiltration.

10.3   INDIVIDUAL ON-SITE WASTEWATER DISPOSAL SYSTEMS

In areas where public wastewater treatment or collection facilities are not available, individual waste disposal systems can restore the wastewater prior to its return to groundwater. Adequate land area and soil depth are needed for conventional leaching (or absorption) systems. The site's groundwater must be deep enough to allow several feet of unsaturated flow below the absorption trenches, and must have adequate separation from bedrock. Granular type soils are most suitable for leaching system applications.

Safe disposal of domestic wastewater is necessary to protect the health of the public. An individual system such as this should not:

(a) Contaminate any drinking water supply.

(b) Allow insects, rodents, or other carriers to create any health hazard.

(c) Be accessible to children.

(d) Pollute or contaminate any surface waters including bathing, shellfish, or water supplies.

(e) Become a nuisance as a result of foul odor or unsightly appearances.

All components of an individual sewage disposal system, including septic tanks, absorption fields, and distribution boxes, should be at least 50 feet away from all ditches, wells, and bodies of water. No part of the individual disposal system should be closer than 10 feet to any building or parking area.

Suitable soil underlying leach fields can treat organic and inorganic materials and pathogens in the wastewater by filtration. The soil also acts as an exchanger, absorber, and a medium where chemical and biochemical processes can take place. These processes produce treated water that is acceptable for discharge into the groundwater under normal conditions of proper system operation. A thorough soils investigation must be conducted to study the soil characteristics in the proposed leaching area. This study can include the identification of the soil's texture, structure, and color as well as determining seasonal changes in the level of soil saturation. An observation well can also be installed for measuring seasonal groundwater fluctuations.

Individual wastewater disposal systems usually fail if the septic tank's sludge and septage are not pumped out periodically. Failure can also occur if the absorption field is subjected to more liquid than it is capable of assimilating or if the soil voids become biologically clogged. Systems subject to these failure conditions can exhibit undesirable attributes. The soil above the absorption field can remain soft and saturated and even hold standing black colored water that appears oily. Offensive odors usually accompany conditions such as these. A well designed system that is properly maintained will not have these symptoms. Other references are available for a more complete presentation on this important topic (Kaplan, 1987; Metcalf and Eddy et al., 1991; Salvato, 1992).

10.3.1 Septic Tanks

Individual wastewater disposal systems have usually been comprised of traditional septic tanks and absorption fields, similar to that illustrated in Figure 10.5.

The septic tank is a watertight container that receives wastewater from a dwelling unit. This wastewater contains (in addition to liquids and grease) solids such as grit, lint, food, and other materials, and is further defined in Section 10.10. Solids are separated from liquids in the tank. Floating material, including grease, is trapped along with accumulated scum. Organic materials are digested and stored while clarified liquid is discharged to the absorption field for final treatment in the soil. Septic tank volumes can be 900 gallons for a 1 or 2 bedroom dwelling unit, or 1000 gallons for a 3 bedroom dwelling. This provides enough volume to accumulate sludge and scum. The tank can be inspected after the first year of operation to insure that it is operating properly, and then every five years when it can be pumped out if needed. Most septic tanks are constructed of precast concrete, fiberglass, or polyethylene. They must be watertight and must have sufficient weight or be anchored to prevent floating as a result of any buoyancy forces attributed to high groundwater.

The user must limit grease, disposable diapers, garments, and other sanitary products from being introduced into the waste system. Solvents, paints, chemicals, and sand can promote operational problems, as can garbage disposal wastes. When septage removal is required, the septage must be properly handled and disposed of, using approved regulatory procedures.

10.3.2 Conventional Gravel Absorption Field

Although requirements can vary from locale to locale, a conventional absorption field consists of a trench between 12 and 36 inches wide filled with at least 14 inches of clean coarse stone aggregate, and a perforated 4 inch distribution pipe covered with a geotextile fabric. The trench, which is at least 24 inches deep, is covered by at least 9 inches of soil. The absorption field length is a function of the soil's ability to absorb and transport the wastewater away from the site. This can be determined by conducting a series of percolation tests on site and by analyzing the soil to determine the seasonal groundwater elevations. The amount of wastewater anticipated from the dwelling also affects the length of the absorption field. Local regulations usually provide guidance on the minimum leach field length based on percolation test results, the soil type, and the anticipated quantity of waste. Seasonal groundwater table elevation fluctuations also must be considered in the elevation and geometry of the absorption field. The maximum height of groundwater from the ground's surface should be at least three feet for proper treatment to take place.

The slope of the trench bottom is made level. Parallel trenches should be separated by at least five feet to allow proper soil absorption. Usually six inches of course aggregate are placed under the distribution pipe and about five inches are allowed to surround the distribution pipe. Three inches of coarse aggregate can be placed on top of the distribution

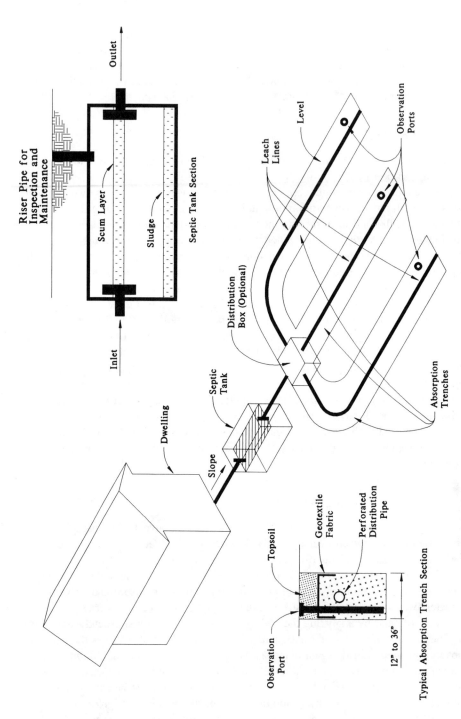

Figure 10.5 Individual wastewater disposal system—conventional method.

Outlet

Inlet

Riser Pipe for
Inspection and
Maintenance

Scum Layer

Sludge

Septic Tank Section

Dwelling

Septic
Tank

Slope

Distribution
Box (Optional)

Leach
Lines

Level

Observation
Ports

Absorption
Trenches

Topsoil

Geotextile
Fabric

Perforated
Distribution
Pipe

Observation
Port

12" to 36"

Typical Absorption Trench Section

pipe and covered with geotextile fabric to prevent fine soil particles from clogging the coarse aggregate voids.

A conventional gravel absorption field is placed where gravity flow conditions can be obtained from the effluent of the septic tank throughout the perforated distribution lines in the absorption field. This method of effluent delivery to the absorption field will result in localized ponding in the absorption trench because of nonuniform liquid applications to the field. As a result the absorption field develops a clogging layer on the soil interface. Clogging resulting from this phenomena usually begins close to the absorption field inflow and moves progressively throughout the field. The clogging layer insures unsaturated flow in the soil and helps reduce pathogens. One method used to reduce the clogging layer is to include a dosing tank (or pump) where the perforated distribution pipe is charged several times a day. This allows the distributed liquid to percolate into the soil and air to enter the absorption surface, thus reducing soil clogging.

A modification to the conventional system can be made where various absorption field trenches can be alternately used. This will prolong the useful life of the system and provide flexibility should a segment of the absorption field fail.

10.3.3 Serial Distribution Absorption Trenches

This system, which can be used on sloping sites, is depicted in Figure 10.6. A serial distribution system is comprised of a series of absorption fields that are filled with septic tank effluent beginning with the highest one. Once the trench completely fills, the effluent is allowed to overflow through one drop box to the next, which allows the effluent to flood all soil surfaces. This method allows the upper trenches to be used while not overusing the lower trenches. Upper trenches can be bypassed by capping to allow periods of rest.

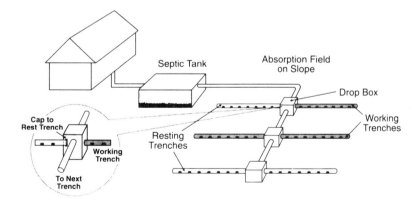

Figure 10.6 Septic tank with serial distribution. Source: U.S. EPA. *Small Wastewater Systems*, 1992.

10.3.4   Leaching Chambers

Septic tank effluent is conveyed to bottomless concrete or arched plastic chambers similar to those shown in Figure 10.7. While stored in the chambers, the effluent floods the soil surface and seeps vertically through the chamber bottom.

10.3.5   Shallow Trench Low-Pressure Pipe Distribution

This method utilizes small diameter pipe that is placed closer to the ground surface than a conventional system. The septic tank effluent is pumped under pressure using a septic tank effluent pump (STEP) through small orifices to soak the entire trench network area. This method can be used on sites where elevated groundwater or shallow soils exist. Figure 10.8 illustrates a typical installation.

The STEP is located in a tank that has sufficient volume to allow a pumping duration of enough time for the pump to attain proper operating temperatures and allow lubrication to mobilize throughout the pump's moving parts. The number of times the pump is energized can range from three to five times per day, depending on the soil type, thus providing several absorption field doses that allow proper percolation and unsaturated flow in the soil to take place. Typically, the pump delivers at least 15 gallons per minute at the design head.

STEP installations require pump controls for "pump on," "pump off," and a high water alarm. The high water alarm is set just above the "pump on" level to alert the dwelling occupant if problems occur. Storage above the alarm provides time to correct a problem with the STEP. Because this system is installed underground, a waterproof electrical system is required.

STEP discharge piping is designed to discharge into a pressurized distribution device that splits the flow onto the absorption field or a pressurized distribution network. The pressured distribution device discharges to the gravity flow absorption field. The pres-

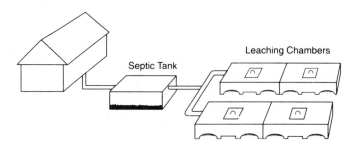

Figure 10.7   Septic tank and leaching chambers. Source: U.S. EPA. *Small Wastewater Systems,* 1992.

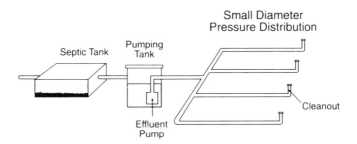

Figure 10.8 Shallow trench low-pressure pipe distribution. Source: U.S. EPA. *Small Wastewater Systems,* 1992.

surized system's distribution piping contains pipe that has openings placed at intervals to uniformly distribute the liquid through the absorption field. This is accomplished by using small diameter distribution piping and orifices that allow the distribution network to be charged with liquid prior to discharging into the absorption field. This provides uniform distribution, dosing features, and controlled absorption field loading, which promote aeration and absorption of the effluent.

When soil limitations prohibit the installation of a conventional gravel absorption field due to low percolation capability, high groundwater, or high bedrock, an artificially elevated absorption field can be considered. Aggregate in the raised bed must meet texture and uniformity limits while construction of the system requires specially trained and often regulatory certified contractors. This approach requires pumping the septic tank's effluent using a STEP from a low elevation to the elevated absorption field's higher elevation, as shown in Figure 10.9.

10.3.6 Evaporation and Absorption Bed

On difficult sites located in areas where the climate consistently promotes water evaporation at a rate that exceeds the rainfall rate, the system shown in Figure 10.10 can be considered. In this system, septic tank effluent flows into gravel trenches or chambers that are placed on a mound of sandy soil. Some of the effluent may seep into the soil, while trees and vegetation that grow close to the absorption field pull liquid from the sand and transpire the water into the air.

10.3.7 Constructed Wetlands

This method can be used on difficult sites where septic tank effluent passes through a bed of rocks. Reeds can be planted on top of the rocks, as shown in Figure 10.11. The liquid

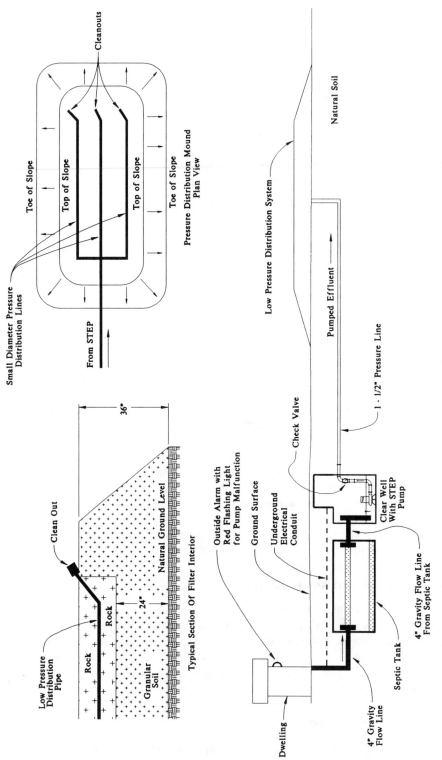

Cleanouts

Toe of Slope

Top of Slope

Top of Slope

Toe of Slope

Pressure Distribution Mound
Plan View

Small Diameter Pressure
Distribution Lines

From STEP

Low Pressure Distribution System

Natural Soil

Pumped Effluent

1 - 1/2" Pressure Line

Check Valve

Clear Well
With STEP
Pump

4" Gravity Flow Line
From Septic Tank

Septic Tank

4" Gravity
Flow Line

Dwelling

Outside Alarm with
Red Flashing Light
for Pump Malfunction

Ground Surface

Underground
Electrical
Conduit

36"

Clean Out

Rock

Rock

Rock

Low Pressure
Distribution
Pipe

Natural Ground Level

24"

Granular
Soil

Typical Section Of Filter Interior

Figure 10.9 Individual wastewater disposal system for high groundwater conditions.

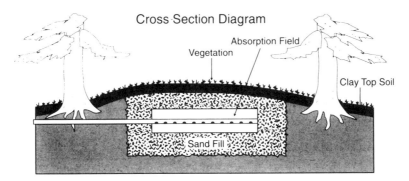

Figure 10.10 Evaporation and absorption bed. Source: U.S. EPA. *Small Wastewater Systems*, 1992.

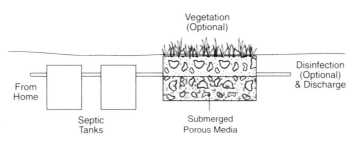

Figure 10.11 Constructed wetlands wastewater disposal. Source: U.S. EPA. *Small Wastewater Systems*, 1992.

evaporates and drains into a soil absorption system or is discharged after disinfection. This system can be used where additional treatment is required or where soils will not support absorption.

10.4   COLLECTION METHODS FOR OFF-SITE WASTEWATER TREATMENT

A centralized wastewater collection system can be considered in areas where individual wastewater disposal systems cannot be used due to limitations in topography, soil, groundwater table levels, space and setback requirements for drain field locations, or large numbers of dwelling units. If an existing centralized collection and treatment system is close to the site being developed, or if the project under design is of sufficient size to support a centralized wastewater treatment facility, a centralized pressurized flow collection system or a vacuum collection system can be considered where a gravity system is not feasible. A pressurized collection system is comprised of individual septic tanks and STEP units or vacuum interfaces that discharge into a common pressure sanitary sewer main for treatment off-site. The pressure main terminates at a wastewater treatment plant or another collection system.

An alternate system is comprised of a submersible grinder pump that pulverizes the fresh sewage and pumps the slurry into a common pressure main. Figure 10.12 illustrates a pressure flow collection system.

The design procedures for both STEP and grinder pumps are similar. The pressure main is the same in each case where small diameter plastic pipe is buried below the frost line at least 36 inches deep. In either case, the design engineer usually selects one size pump for the project. This helps minimize confusion and provides compatibility between units for maintenance and repair work.

The pressure main should be designed where it is not a looped system, and cleanouts should be placed within at least every 1000 feet of pressure main. Where the pressure line is placed at a high point, an air release valve should be installed to vent trapped air and gas to the atmosphere. A vacuum breaker valve placed at the discharge side of the pump may also be required to prevent siphoning. Siphoning is undesirable because the liquid in various pump sumps can be removed when it occurs. This may result in solids being forced into the pump's intake when it is not in operation, resulting in clogging. Also, the removal of the liquid could induce a vapor lock within the pump's impeller, resulting in a loss of prime when the pump is later energized, causing pump failure. Size of the pressure main can be based on several criteria. One criterion could be to assume that each pump is capable of discharging at least 15 gallons per minute into the common pressure collection line on a random basis that is governed by individual dwelling demands. The common pressure collection line is then designed to maintain self-scouring velocities using contributory flow rates that equal approximately 90 percent (see Section 10.2) of the design peak hourly potable water demands, independent of lawn sprinkling and fire flow requirements (Chapter 11). This pressure collection system would therefore be sized to receive approximately 90 percent of the design peak hourly flows (minus sprinkling), whereas the potable water distribution system is used to deliver design peak hourly flows.

10.4.1   STEP System

Effluent pumped by the STEP system is extremely corrosive above the liquid level where hydrogen sulfide can exist to form sulfuric acid. This requires all valves, piping, and

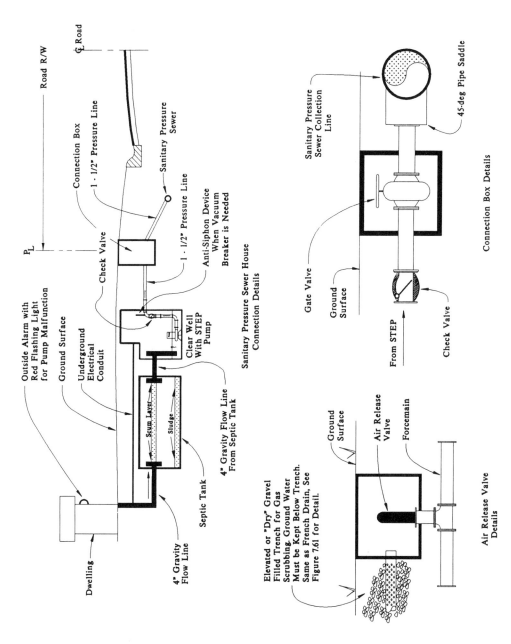

Figure 10.12 Centralized pressure flow wastewater collection system.

385

electrical controls to be made of materials not susceptible to degradation under these adverse conditions. Plastic is normally used.

Since the STEP is immersed in the liquid, standard cast iron parts can be used satisfactorily in the pump; however, the impeller should be made of bronze or plastic to prevent seizure. STEP system effluent has offensive odors that must be addressed where the delivered liquid is discharged, perhaps by the installation of a sewer gas scrubber. STEP effluent also gives off hydrogen sulfide and methane gas—both deadly in small quantities. This requires any work performed on a STEP system to incorporate forced clean air ventilation for worker safety.

Grease does not usually affect the operation of a STEP system because it is normally removed in the septic tank. A problem could result, however, if the septic tank is not periodically pumped of septage, resulting in the tank "burping" solids and/or grease into the STEP and pressure sewer. The discharge of solids may also affect the treatment systems, especially if a large soil absorption system is used.

10.4.2   Grinder Pump System

Grinder pumps are placed in a watertight chamber where the fresh wastewater is received from the dwelling. This wastewater can include grease that could possibly clog the pump and pressure lines. Velocities of flow greater than two feet per second must be maintained in the pressurized discharge lines during peak flows to help cleanse the pressure line.

One advantage of the grinder pump is that no sludge is formed because all wastes are evacuated; therefore, there is no need to periodically pump out the grinder pump's basin. Repair of grinder pumps can, however, be more costly than that for STEP pumps and can be required more often, especially if a high content of grit is present in the wastewater.

10.4.3   Vacuum Sewer Collection Lines

This system utilizes a storage well where wastewater accumulates until enough is present to activate a valve that allows air and sewage to enter the collection line and be transported toward the vacuum station. Normally one valve box serves a number of homes. (This is not the case with STEP or grinder pumps, which require individual power supplies). A central vacuum pump in the vacuum station (shown in Figure 10.13) maintains a reduced pressure in collection lines. Sewage pumps are used to transport wastewater from the vacuum station to the treatment facility. Because both the vacuum pumps and sewage pumps exist in the vacuum station, electrical power requirements are centrally located. This system can offer certain economic advantages over traditional gravity collection systems.

10.5   CENTRALIZED GRAVITY FLOW WASTEWATER COLLECTION SYSTEMS

For reasons similar to those cited in Section 10.4 to establish a need for centralized wastewater collection, a traditional approach can be to use a gravity flow collection system, if the site's topography is adequate and it is economically feasible.

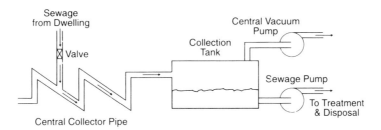

Figure 10.13 Vacuum sewer collection system. Source: U.S. EPA. *Small Wastewater Systems,* 1992.

10.5.1 General Considerations for Gravity Flow Collection Systems

Sanitary sewer collection systems can be designed as gravity flow systems that operate under open channel flow conditions. Factors affecting the design of a gravity flow collection system include horizontal control, vertical control, and hydraulic control.

Most regulations require the gravity sewage system to be placed some distance away from the potable water distribution system to protect the distributed water from contamination. Often local regulations provide for the sanitary sewer main to be located along one shoulder of the road while the potable water is placed along the other. For example, if a local road's geometric cross section provides for a 10 foot wide strip between the back of curb to the right of way line, the sanitary sewer main could be placed along the center of this area, as shown in Figure 10.14.

10.5.2 Horizontal Control for Gravity Sewer Collection Systems

The horizontal alignment of the proposed system is established by placing manholes wherever the sewer main changes horizontal direction (and/or slope or pipe size). The road's centerline alignment can be used to aid calculations. Local regulations usually limit the maximum spacing between manholes to 300 to 400 feet on small sewers; therefore, on long straight runs of sewer main, manholes must be placed at prescribed intervals along the line.

Where sanitary sewer mains are located along street curves, chord bearings and distances between manhole centers can be determined that are referenced to the street's centerline alignment. Particular care must be exercised to insure that the proposed gravity line segments do not cross over the edge of the right of way onto proposed building sites.

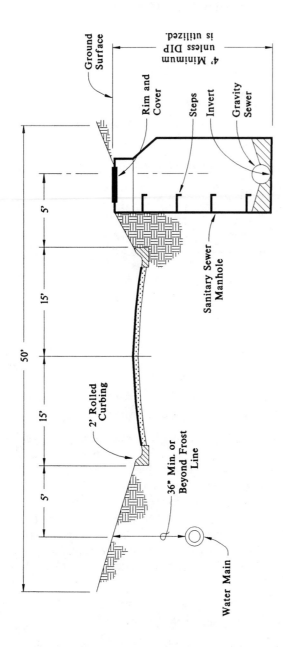

(Note differences between this gravity system and a pressure system shown in Figure 10.12)

Figure 10.14   An example of the placement of a gravity sanitary sewer main.

388

Figure 10.15 illustrates the establishment of horizontal control data for a sanitary sewer collection system located along a road. Key road stations such as PC's and PT's are excellent places to locate manholes because calculations and field staking during construction are simplified.

Gravity flow sanitary sewer mains need not extend the full length of all roads if smaller diameter lines can be used to provide service to each building site. The designer should keep in mind the slope of the topography in order to best locate the sewer mains for gravity flow. The site topography may require the engineer to place a portion of the proposed sewer main along various interior property lines of proposed building lots to take advantage of minimizing installation cuts. Mains placed in areas similar to this (off the road right of way) require a permanent easement for line protection and maintenance. Figure 10.16 illustrates several such conditions.

When completed, the horizontal alignment for the proposed sanitary sewer's centerline should be checked to insure that the bearings and distances from centerline of manhole to centerline of manhole produce a closed traverse when tied to appropriate street horizontal control data. Additional horizontal control requirements for the wastewater system may exist if the point of discharge (connection) is into an existing sanitary sewer manhole several hundred feet away from the site along an existing road or through an adjoining landowner's property. In either case permission must be obtained for installation of the proposed new main prior to construction. Most highway departments require an encroachment permit to be submitted and approved that describes and shows the proposed system's features and its relationship to the road. Where the connection must be made through another landowner's property, an easement must be obtained from that property owner. This easement usually entails a temporary construction easement where blanket ingress-egress is allowed during construction to provide working space for the contractor's workers and equipment, while a permanent easement of established width (possibly 20 feet with 10 feet along each side of the installed line) is provided. Final procurement of this easement is usually done by the developer with the technical assistance of the engineer.

10.5.3 Vertical Control for Gravity Sewer Collection Systems

Once horizontal control is established, the engineer can start design of the system's vertical control. According to Metcalf and Eddy and Tchobanoglous* (1981) sanitary sewer mains are usually designed for a minimum self-cleansing (scouring) velocity of at least two feet per second when flow is one-half or full depth, to prevent solids from settling. Some regulatory agencies recommend an eight inch diameter gravity sewer as a minimum size between manholes on all main lines. The minimum allowed slope for an eight inch line may be regulated at 0.4 percent to provide this minimum self-scouring velocity. The maximum slope may be regulated to a slope that would minimize pipe erosion attributed to solids transport. Usually a maximum velocity of 8 to 10 feet per second will provide safety, according to Metcalf and Eddy and Tchobanoglous (1981). Another regulatory requirement might include provisions for at least 36 inches of cover to be placed on top of the main. If the soil cover is less than 36 inches, ductile-iron must be utilized to protect against pipe collapse attributable to live traffic loads.

Vertical control is established by beginning with the invert elevation of the proposed

---

*Source: Metcalf and Eddy, Inc. and George Tchobanoglous. *Wastewater Engineering: Collection and Pumping of Wastewater.* Copyright © 1981 McGraw-Hill, Inc. Reprinted by Permission of McGraw-Hill, Inc.

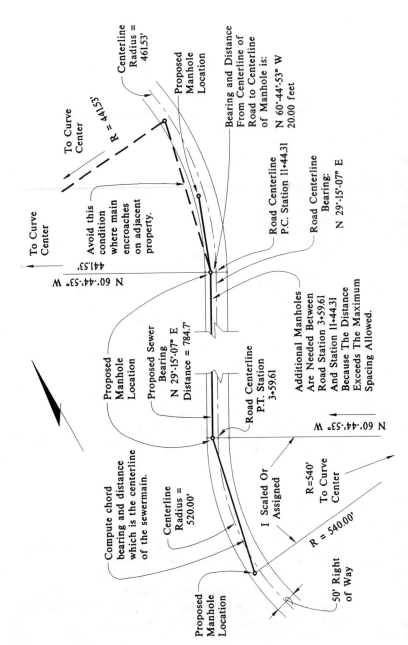

Figure 10.15 An example of horizontal control relationships for gravity sewer main layout.

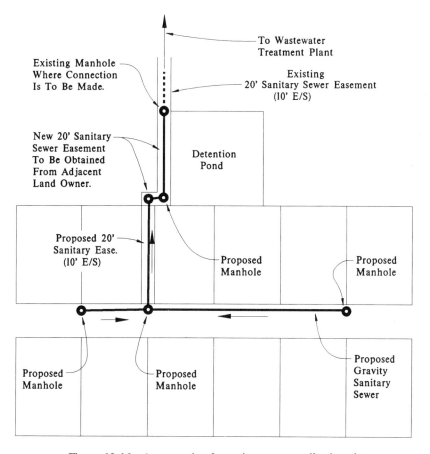

Figure 10.16   An example of a sanitary sewer collection plan.

point of connection. Each sewer manhole's invert elevation is calculated in sequence using an acceptable pipe slope and the horizontal distance between successive manholes. Figure 10.17*a* illustrates this process.

Should some economy of construction be available, the engineer can utilize a drop manhole to reduce the depth of cut on some lines, as illustrated in Figure 10.17*b*.

When a smaller diameter upstream sewer main joins a larger diameter downstream main at a manhole, the top inside pipe crowns are usually set at the same elevation to promote efficient hydraulic flows. In some instances where the site's topography is flat and the design engineer is attempting to conserve available fall, the pipe invert elevations can be matched if the partial flow characteristics of the larger downstream main can provide adequate capacity. The full flow discharge of the smaller upstream pipe must be conveyed by the larger line while adverse hydraulic flow conditions are prevented.

10.5.4   Hydraulic Considerations for Gravity Sewer Collection Systems

The hydraulic flow requirements must be examined once the system's horizontal and vertical control are established. Normally in a small land development the anticipated

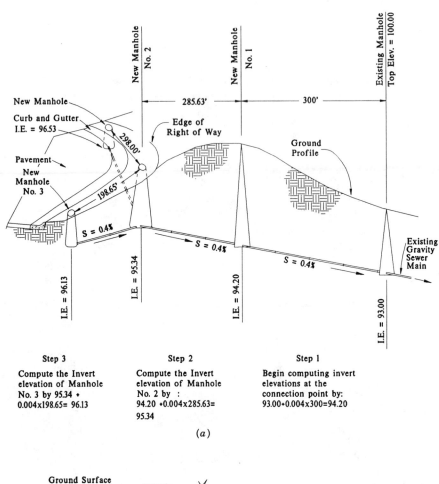

<table>
<tr><td>Step 3</td><td>Step 2</td><td>Step 1</td></tr>
</table>

Step 3	Step 2	Step 1
Compute the Invert elevation of Manhole No. 3 by 95.34 + 0.004x198.65= 96.13	Compute the Invert elevation of Manhole No. 2 by : 94.20 +0.004x285.63= 95.34	Begin computing invert elevations at the connection point by: 93.00+0.004x300=94.20

(a)

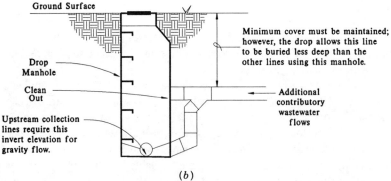

(b)

Figure 10.17 Sanitary sewer gravity collection system—vertical control. (a) A sample layout of a gravity sanitary sewer showing vertical control calculations. (b) Drop manhole application.

wastewater flows are much less than the full flow capacity of a minimum size eight inch sewer main. In large developments, however, flows can increase dramatically, and line sizing is critical. Each main must be sized for peak flows, and some regulatory agencies require a minimum peaking factor of 250 percent of the average daily flow to be used.

A certain quantity of water should be included along with domestic flows to account for groundwater infiltration. According to ASCE and WPCF (1982), it is common practice to design for the peak anticipated rate of wastewater flow plus 30,000 gallons per day infiltration per mile of small to medium sewers and house connections. This allowance represents average conditions and can be revised by the engineer based on the physical characteristics of the area and the type of pipe joints used.

Figure 10.18 is used to illustrate the determination of peak flows in a proposed gravity system. For this example, it is assumed that inflow will not be a factor because proposed manhole covers will be constructed where storm water cannot enter, and system users will not be allowed to place storm water into the sanitary sewer system. The volume per capita per day for this example is based on Figure 10.4 where an amount of 100 gallons is obtained. Infiltration is assumed to be 200 gallons per inch of pipe diameter per mile of pipe per day based on pipe manufacturer's guidelines.

Collection line flows are calculated beginning at the remotest upstream point in the system. Contributory flows are accumulated while progressing downstream in the system to the point of connection. For example, the contributory flows for each line segment from manhole 1 to manhole 4 are determined as follows:

Flow from Manhole 1 to Manhole 2

> 50 single family dwellings $\times$ 100 gallons per person per day $\times$ 4 persons per dwelling = 20,000 gallons per day
> Q_{ave} = 13.88 gallons per minute
> Q_{peak} = 2.5 $\times$ 13.88 = 34.7 gallons per minute

The length between manhole 1 and manhole 2 is 295.14 linear feet. Infiltration for an eight inch diameter line equals (200 $\times$ 8 $\times$ 295.14)/(24 $\times$ 60 $\times$ 5280), or 0.062 gallons per minute. The total peak flow from manhole 1 to manhole 2 equals:

> Q_{rpeak} = 34.76 gallons per minute

Flow from Manhole 2 to Manhole 3

> 75 single family dwellings $\times$ 100 gallons per person per day $\times$ 4 persons per dwelling = 30,000 gallons per day
> Q_{ave} = 20.83 gallons per minute
> Q_{peak} = 2.5 $\times$ 20.83 = 52.08 gallons per minute

The length between manhole 2 and manhole 3 is 300 linear feet. Infiltration for this segment of eight inch line equals (200 $\times$ 8 $\times$ 300)/(24 $\times$ 60 $\times$ 5280) = 0.063 gallons per minute. The total peak flow from manhole 2 to manhole 3 equals:

> Q_{rpeak} = 34.76 + 0.06 + 52.08 = 86.9 gallons per minute

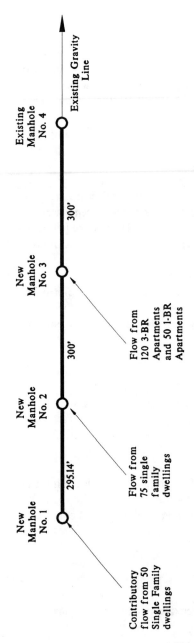

Figure 10.18 An example of contributory flows into a sanitary sewer gravity collection system.

Flow from Manhole 3 to Manhole 4

120 units apartments—3 BR × 100 × 4 = 48,000 gallons per day
50 units apartments—1 BR × 100 × 2 = 10,000 gallons per day
Q_{ave} = 40.3 gallons per minute
Q_{peak} = 2.5 × 40.3 = 100.75 gallons per minute

The length from manhole 3 to manhole 4 is 300 linear feet. Infiltration for this segment of eight inch line equals (200 × 8 × 300)/(24 × 60 × 5280), or 0.063 gallons per minute. The total peak flow from manhole 3 to manhole 4 equals:

$$Q_{tpeak} = 0.063 + 100.75 + 86.9 = 187.71 \text{ gallons per minute}$$

The full flow rate under open channel conditions for an eight inch diameter sewer main, using Equation 7-18 (Manning's n = 0.011) and a 0.4 percent slope, is:

$$Q = \frac{1.486AR^{0.67}S^{0.5}}{n}$$

$$= \frac{1.486(0.349)(0.167)^{0.67}(0.004)^{0.5}}{0.011}$$

$$= 0.89 \text{ cubic feet per second}$$
(or 403 gallons per minute)

It is evident that the flow capacity of an eight inch diameter main on a 0.4 percent slope is more than adequate to accommodate each of the above Q_{tpeak} flows. Should additional contributory flows be encountered further downstream, the analysis process would continue until the capacity of the eight inch main was exceeded. Then the next available size main would need to be analyzed. If the regulatory agency having jurisdiction over the development requires gravity sewers less than 15 inches in diameter to be designed to carry peak flows when the pipe is half full, the designer must select a pipe size that accommodates this requirement.

10.5.5 Evaluation of the Preliminary Wastewater Collection System Design

After the required pipe sizes have been determined, the system must be studied relative to other previously designed improvements. This includes the road and storm drainage systems. First, the elevation of each storm drainage pipe (top outside of pipe and bottom outside of pipe) must be checked against the sanitary sewer main's top and bottom elevations where they cross on the plans. If there is an elevation conflict, either the storm drainage system must be raised or lowered, and/or the sanitary sewer system must be raised or lowered. In either case, the designer must repeat all design steps from the point of alteration to the point of solving the conflict in elevations. Usually if there is less than one foot of separation between the bottom of one pipe and the crown of a second, crossing pipe, a 20 foot section of cast iron or ductile-iron pipe is installed for the sewer main to protect the sewer main from being crushed.

The sanitary sewer should also be checked to insure that at least 36 inches of cover are included over the top of all pipes. Otherwise ductile-iron pipe must be used, unless

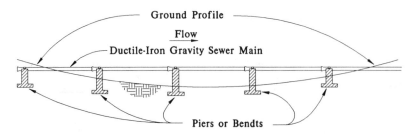

Figure 10.19   An example of a ductile-iron pipe elevated gravity line.

conditions warrant additional structural analysis of the pipe. On occasion ductile-iron pipe can be used to cross low topographical areas by placing the pipe on piers or pile bents for support at each pipe joint. Figure 10.19 illustrates this procedure.

An inverted siphon can be used in other areas such as ditch or river crossings (ASCE and WPCF, 1969).

A study of the relationships of the sanitary sewer gravity main elevations and those of proposed dwelling units should be undertaken to determine whether there is a potential for wastewater to back up into the dwellings if a stoppage occurs in the main line. Should the potential exist, check valves should be included on service lines leading to the vulnerable dwellings.

The design engineer should insure that all building sites can be serviced by gravity flow. When a development site is comprised of sloping topography, buildings on the low side of a street (and the sanitary sewer main) may not have sufficient elevation to allow gravity flow to the sewer main located along the fronting street. This situation requires special provisions where the design engineer should investigate providing gravity access to the rear of the lots by using an easement through abutting property. If this cannot be accomplished, a STEP system can be considered for these problem lots.

This gravity collection system design assumes there is an existing gravity tie-in close to the site that has adequate hydraulic capacity. Many times this is not the case, and a sanitary sewer pumping station must be installed.

10.6   WASTEWATER PRESSURE TRANSPORT SYSTEM

A wastewater pressure transport system includes a centralized pumping facility and pressurized forcemain. This type of system relies on a collection system, which gathers the wastewater at a central location, and a destination point, where the wastewater can be discharged. This destination can be a wastewater treatment plant, another collection system, or another pumping facility. Most pumping facilities are nonclog centrifugal pumps of the submersible, flooded suction, or self-priming types, although small stations can utilize ejectors. A pumping station has a wetwell that is used for temporary wastewater storage to supply the pump during periods of pumping. The pumps can be located in a drywell next to the wetwell, above the wetwell, or in the wetwell if they are submersible.

10.6.1   Centrifugal Pump Characteristics

A centrifugal pump is used to add energy to a fluid flow system and is usually rated at a certain flow rate (capacity) at a particular pressure head. The pump's rotating impeller (or propeller) forces water to the pump's casing by centrifugal action. A centrifugal pump provides smooth transitional flows when system pressure levels change. When system pressures increase, the pump's capacity and power decrease. Likewise, when the system pressure decreases, the pump's capacity and power increase. At start-up, the required shaft torque initially is small.

A pump's rating can be governed by the pump's configuration or the motor providing energy to the pump. The water horsepower (HP_{water}) of a pump is represented by:

$$HP_{water} = \frac{GPM \times TDH}{3960} \tag{10-1}$$

where:

GPM = The flow rate in gallons per minute.
TDH = The total dynamic head (when pumping at GPM) in feet.
3960 = A constant.

A pump's efficiency is defined as the ratio of energy output to input. The energy output is the water horsepower while the energy input is the brake horsepower (HP_{brake}) applied to the pump. A pump's efficiency is therefore defined as:

$$Eff = \frac{HP_{water} \times 100}{HP_{brake}} \tag{10-2}$$

Pump capacity curves provided by the manufacturer show the relationships between the rate of pump discharge and head, efficiency, and brake horsepower. An additional curve showing the net positive suction head (NPSH) required indicates the head necessary to cause water to flow into the impeller's eye. The NPSH required varies based on the pump's speed and capacity.

A particular pump at a given speed has a unique set of performance values. When a pump is operated at maximum efficiency, the corresponding head and discharge provide the best efficiency point (BEP).

Some pumps operate at constant rotational speeds when in service while others can be operated at variable speeds. When shaft rotational speeds vary, the relationship between the pump's capacity, head, and brake horsepower also varies according to the affinity laws and specific speed relationships contained in ASCE's *Pipeline Design for Water and Wastewater* (1975).* The affinity laws are:

$$\frac{Q_1}{Q_2} = \frac{N_1}{N_2} \tag{10-3}$$

$$\frac{H_1}{H_2} = \left[\frac{N_1}{N_2}\right]^2 \tag{10-4}$$

$$\frac{HP_{brake1}}{HP_{brake2}} = \left[\frac{N_1}{N_2}\right]^3 \tag{10-5}$$

*Reproduced by permission of the American Society of Civil Engineers.

where:

Q = The pump's discharge capacity in gallons per minute.
N = The pump's speed in revolutions per minute.
H = The pump's head in feet.
HP_{brake} = The pump's brake horsepower when pumping Q gallons per minute.

The specific speed of a centrifugal pump is the relationship of a pump's discharge, head, and rotational speed when operating at its BEP. Specific speed can be used to predict the performance of one pump based on other single stage pumps that are similar but of a different size:

$$\text{specific speed} = \frac{\text{RPM}(Q)^{0.5}}{(H)^{0.75}} \tag{10-6}$$

where:

Q = The pump's discharge at BEP in gallons per minute.
H = The pump's head in feet at the BEP.
RPM = The pump's rotational speed at the BEP.

A centrifugal pump can experience cavitation if the pressure at the edge of the impeller is reduced to the water's vapor pressure when vapor bubbles form. This reduces the pump's capacity and at the same time causes a pressure rise. The bubbles collapse and the cycle repeats rapidly with noise and vibrations resulting. Cavitation should be avoided because it can cause excessive impeller wear and premature pump failure.

10.6.2   General Considerations for Pumping Station Design

To design a sanitary sewer pumping station, several things must be known:

(a) The elevation of the ground where the pump station is to be placed.
(b) The invert elevation of the lowest influent line going into the wetwell.
(c) The peak inflow going into the wetwell.
(d) The distance the wastewater is to be pumped.
(e) The high and low points in the discharge line (forcemain), which are usually shown in the form of a profile.
(f) Criteria for pump control settings.

An example of the design of a suction lift (or self-priming) pumping station (pumps are located above the wetwell) follows, using as a general guide information contained in the DAVCO Equipment (1974) book:

A. Determine the Required Rate of Pumping.   If the average daily wastewater flow from a development is estimated at 200 gallons per minute, and a peaking factor of 250 percent is to be used to estimate peak flows, the peak flow can be calculated as:

$$Q_{\max} = 2.5 \times Q_{\text{ave}} = 2.5 \times 200$$
$$= 500 \text{ gallons per minute}$$

Most regulatory agencies require small pumping stations containing constant speed centrifugal pumps to maintain 100 percent standby pumping capability. To attain this, two pumps must be installed with each pump being capable of independently delivering 500 gallons per minute to the discharge point.

B. Compute the Storage Requirements for the Wetwell.   The wetwell should be sized for ultimate peak capacity (if economy dictates) even though the initial peak flows may be small. This can be accomplished by setting the pump control levels to accommodate the smaller flows initially. Normally, the wetwell should be pumped out every 10 to 15 minutes to prevent septic conditions; however, motors are not normally started more than three times per hour to minimize maintenance problems. Since this pumping station contains 2 pumps (providing 100% standby capacity in case of failure), a total of 6 cycles per hour are available to produce the maximum cycle rates. The time for one pump cycle is therefore:

$$T = \frac{V}{Q - S} + \frac{V}{S} \tag{10-7}$$

where:

T = The time for one pump cycle in minutes.
V = The effective volume of the wetwell in gallons.
Q = The pumping rate in gallons per minute.
S = The flow into the wetwell in gallons per minute.

If Q is assigned a value of 250 percent of the average daily flow, or $2.5S$, and T equals 12 minutes to account for the time to cycle, the effective wetwell volume (volume of wetwell storage available for pumping) must be $7.2 \times S$. Therefore, the working volume to meet the criteria would be $7.2 \times Q_{\text{ave}}$. For this example, the volume would be 200 gallons per minute $\times$ 7.2, or 1440 gallons of working storage in the wetwell.

It is common to use reinforced concrete pipe culverts or precast manhole sections to construct circular wetwells similar to the one illustrated in this example. Figure 10.20 shows various volumetric relationships for several possible wetwell diameters.

For this example a six foot diameter wetwell is selected. Using Figure 10.20 as a guide, each foot of vertical storage gives 212 gallons; therefore, the total required working depth is 1440 gallons of working volume / 212 gallons per foot of depth, or 6.8 feet.

C. Determine the Forcemain Diameter.   The forcemain should be at least four inches in diameter and maintain a self-cleansing velocity greater than two feet per second. The maximum velocity should be less than 10 feet per second to minimize waterhammer problems. Using the continuity equation (Equation 7-17) with a flow rate of 500 gallons per minute, a maximum cross-sectional area of 0.557 square foot is required to provide this minimum velocity. If an eight inch forcemain is chosen, the corresponding flow velocity at 500 gallons per minute is 3.19 feet per second, which is acceptable. This is determined by:

(a) Calculation of an eight inch pipe's cross-sectional area:

$$A = 3.1415 \times (8/(2 \times 12))^2 = 0.349 \text{ square feet}$$

(b) Calculation of the flow velocity using a rearranged form of Equation 7-17:

$$V = Q/A$$
$$= (0.002228 \times 500 \text{ gallons per minute})/0.349 \text{ square feet}$$
$$= 3.19 \text{ feet per second}$$

(The factor 0.002228 converts gallons per minute into equivalent cubic feet per second.)

Often the Hazen–Williams formula is used for pressure flow analysis where:

$$V = 1.318CR^{0.63}S^{0.54} \qquad\qquad (10\text{-}8)$$

where:

V = The velocity of flow in feet per second.
C = The Hazen–Williams coefficient of roughness.

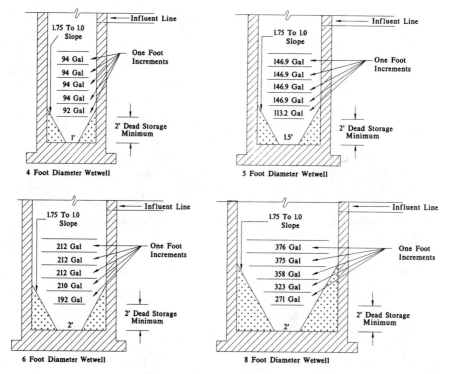

Figure 10.20 Volumetric relationships for various wetwell diameters. Source: DAVCO—Division of Davis Water and Waste Industries, Inc., 1828 Metcalf Ave., Thomasville, Ga. 31792. Reproduced with permission.

R = The hydraulic radius in feet equal to the cross-sectional area divided by the wetted perimeter.
S = The slope in feet per foot.

In cases where a pipe is circular and flowing full, Equation 10-8 can be rearranged in terms of the rate of flow, as illustrated in Equation 10-9, where:

$$Q = 0.006756CD^{2.63}H^{0.54} \qquad\qquad (10\text{-}9)$$

where:

Q = The discharge in gallons per minute.
C = The Hazen–Williams coefficient of roughness.
D = The inside diameter of the pipe in inches.
H = The headloss in feet per 1000 feet of pipe.

Another form of the basic Hazen–Williams equation used to solve directly for pipe frictional headloss is shown in Equation 10-10 where:

$$H_{\text{loss per 1000}} = \left[\frac{Q}{(0.006756)(C)(D^{2.63})} \right]^{1.85} \qquad (10\text{-}10)$$

where:

Q = The flow in gallons per minute.
C = The Hazen–Williams coefficient of roughness.
D = The pipe diameter in inches.

Using Equation 10-10, the corresponding friction loss per 1000 feet of eight inch pipe, using a Hazen–Williams C value of 140 for PVC, at the 500 gallons per minute rate, is 4.4 feet (or 0.0044 feet per foot of pipe).

D. Determine the Pump Suction Line Diameter.   The suction line should be sized for a velocity from four to eight feet per second using an average of six feet per second. At 500 gallons per minute a 6 inch diameter suction line yields a 5.67 feet per second velocity according to Equation 7-17. This velocity falls within the acceptable range. The corresponding frictional headloss (assuming ductile-iron piping with a Hazen–Williams C value of 120) using Equation 10-10 is 23.75 feet per 1000 feet of suction line (or 0.02375 feet per foot of pipe).

E. Compute the Total Dynamic Head (TDH) for the System Illustrated in Figure 10.21.   The TDH is comprised of the dynamic suction lift and the discharge head. To calculate the liquid lift heights, the elevations must now be established for pump controls. In most duplex pumping stations similar to the one included in this example, an electronic alternator is provided in the motor control panel to alternate pumps through each subsequent cycle of operation. If the left hand pump is labeled pump no. 1 and the right pump no. 2, the alternator would initially place pump no. 1 on line as the lead pump to pump down the wetwell when the inflow raises the liquid level to a predetermined normal

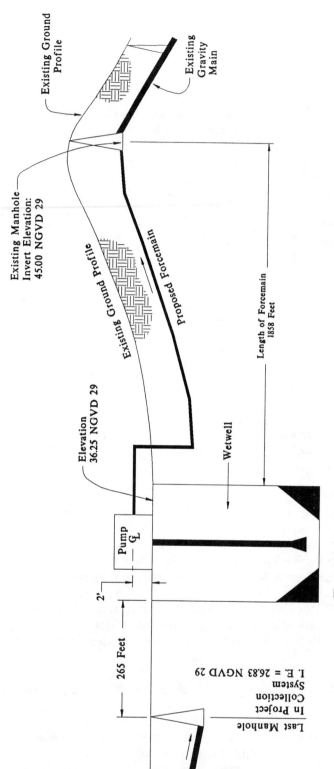

Figure 10.21 Site conditions for a sample sanitary sewer pump station design.

maximum height designated as the "lead pump on" elevation in the wetwell. Upon the removal of accumulated wastewater in the wetwell to a predetermined low level called "pumps off" using pump no. 1 (pump no. 2 on standby), the alternator would reverse pump roles. Pump no. 2 would then become the lead pump for dewatering for the subsequent cycle (pump no. 1 would be placed on standby) once the inflow raises the liquid level in the wetwell up to the "lead pump on" elevation. By alternating the pumps, they both receive equal wear and are both kept operational by continuous exercise. If one pump was constantly designated as the lead pump, the standby pump would probably freeze from nonuse and would not be capable of coming on line when required for backup.

The standby pump, called the lag pump, is available if the lead pump should malfunction and trip out. The standby pump also can be energized if the wetwell inflow exceeds the lead pump's capacity and liquid continues to rise in the wetwell. Both constant speed centrifugal pumps can be used simultaneously in this case to pump the wetwell down.

If local regulations require six inches of vertical distance between the influent invert elevation into the wetwell and the "lag pump on" elevation, and six inches of vertical distance between the "lag pump on" and the "lead pump on" elevation, then the wetwell elevations can be established. The influent invert elevation for this example is calculated using the invert elevation of the closest manhole away from the pumping station while providing acceptable slope (0.4 percent) to the connection line. For this example the manhole's invert elevation is 26.83 feet, as shown in Figure 10.21. The influent invert elevation is calculated as 26.83 − (0.004 feet per foot × 265 feet), or 25.77 feet. Figure 10.22 shows the completed pump control elevations in the wetwell.

(a) Compute:

(1) The dynamic suction head comprised of the suction lift distance from the low water level in the wetwell to the eye of the impeller that is two feet above the ground elevation.

(2) The frictional headloss in the suction line based on Step D as 0.0238 feet of headloss per foot of suction pipe (extending two feet below the pump's off elevation) at a 500 gallon per minute flow.

$$\text{static suction lift} = (36.25 - 17.97) + 2 = 20.28 \text{ feet}$$

$$\text{friction headloss: } (0.0238)(2 + 36.25 - 15.97) = 0.53 \text{ foot}$$

Minor headloss determination requires knowledge of the velocity head in the pipe, determined using Equation 10-11:

$$\text{velocity head} = V^2/(2G)$$

where:

V = Flow velocity in feet per second.
G = 32.2 feet per second squared.

Figure 10.23 shows various minor headloss coefficient values that can be multiplied by the velocity head to determine the minor headloss value. Minor losses assumed in this example in the suction line piping are as follows:

Minor Losses	Coefficient from Figure 10.23
2-each 90° ells	1.4
1-each gate valve	0.2
Bell entrance	0.23
Total	1.83

$$h_1 = K \ (V^2/64.4) = 1.83 \times (5.67^2/64.4) = 0.91 \text{ feet}$$

Total dynamic suction head then equals the sum of the component headlosses, or 21.72 feet.

(b) Compute the discharge head. The static discharge head is the difference between the elevation of the discharge point (also the high point in this case) and the elevation of

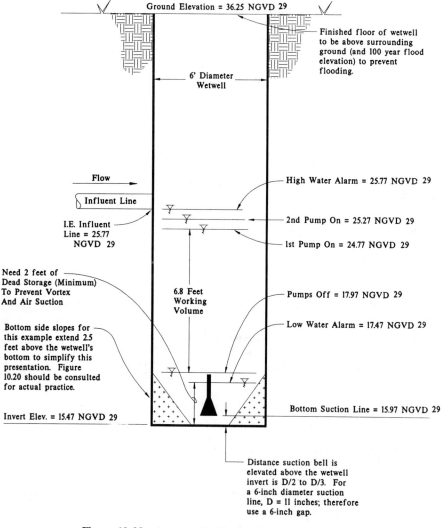

Ground Elevation = 36.25 NGVD 29

Finished floor of wetwell to be above surrounding ground (and 100 year flood elevation) to prevent flooding.

6' Diameter Wetwell

Flow

Influent Line

I.E. Influent Line = 25.77 NGVD 29

High Water Alarm = 25.77 NGVD 29

2nd Pump On = 25.27 NGVD 29

1st Pump On = 24.77 NGVD 29

Need 2 feet of Dead Storage (Minimum) To Prevent Vortex And Air Suction

6.8 Feet Working Volume

Pumps Off = 17.97 NGVD 29

Low Water Alarm = 17.47 NGVD 29

Bottom side slopes for this example extend 2.5 feet above the wetwell's bottom to simplify this presentation. Figure 10.20 should be consulted for actual practice.

Bottom Suction Line = 15.97 NGVD 29

Invert Elev. = 15.47 NGVD 29

Distance suction bell is elevated above the wetwell invert is D/2 to D/3. For a 6-inch diameter suction line, D = 11 inches; therefore use a 6-inch gap.

Figure 10.22 An example of wetwell elevation settings.

Fitting or Condition	Minor Headloss Coefficient, K
Gate Valve	
3/4 closed	24.00
1/2 closed	5.60
1/4 closed	1.20
Full Open	0.20
Angle Valve Open	2.50
Globe Valve Open	10.00
Swing Check Valve	0.6-2.5
Elbows	
90-deg standard	0.70
90-deg long radius	0.60
45-deg standard	0.50
Tee Flow Through Run	1.80
Sudden Contraction*	
d/D = 1/4	0.40
d/D = 1/2	0.30
d/D = 3/4	0.20
Sudden Enlargement*	
d/D = 1/4	0.90
d/D = 1/2	0.60
d/D = 3/4	0.20
Flow From Storage Tank	
Pipe Projecting Into Tank	0.78
Pipe Flush With Tank Wall	0.50
Slightly Rounded Entrance	0.23
Well Rounded Entrance	0.04
Flow Into Storage Tank	
All Conditions	1.00

* For Contractions and Enlargements, coefficients
are applied to the conditions in the smaller pipe
diameter.

Figure 10.23   Minor headloss factors. Source: The American Society of Civil Engineers. *Pipeline Design for Water and Wastewater,* 1975. Reproduced by permission of the American Society of Civil Engineers.

the eye of the impeller. The frictional headloss attributed to the forcemain is found by utilizing the headloss per foot of forcemain found previously in Step C as 0.0044 feet of headloss per foot of forcemain when conveying 500 gallons per minute.

The static discharge head equals 45.00 − 38.25, or 6.75 feet, while the forcemain friction loss equals 1858 linear feet of forcemain × 0.0044 feet per foot, or 8.18 feet.

Minor headlosses assumed for this example are given below:

Minor Headlosses	Coefficient from Figure 10.23
Discharge into manhole	1.0
6-each 90° ells	4.2
1-each check valve	2.5
1-each gate valve	0.2
2-each 45° ells	1.0
Total	8.9

$$h_1 = K \ (V^2/64.4) = 8.9 \times (3.19^2/64.4) = 1.41 \text{ feet}$$

The total discharge headloss equals $(6.75 + 8.18 + 1.41)$, or 16.34 feet.

The TDH equals the total dynamic suction head plus the total discharge head: $(21.72 + 16.34)$, or 38.06 feet.

A pump capable of pumping 500 gallons per minute at 38.06 feet TDH is required. According to Figure 10.24, the Burks Pumps performance curve, a model 4Cϕ4D (Imp. Dia. 8D), 10 horsepower at 1650 revolutions per minute, will be sufficient with an efficiency of 55 percent.

F. Compute the Required Brake Horsepower. The brake horsepower required to pump under design conditions using a pump of known efficiency can be determined using Equation 10-12:

$$\text{HP}_{\text{brake}} = \frac{\text{GPM} \times \text{TDH}}{3960 \times \text{Eff}} \tag{10-12}$$

where:

GPM = Flow rate in gallons per minute.
TDH = Total dynamic head of the system in feet when delivering the required flow rate.
3960 = A constant.
Eff = Pump efficiency expressed in decimal form.

For this example the required motor size is:

$$\text{HP}_{\text{brake}} = \frac{\text{GPM} \times \text{TDH}}{3960 \times \text{Eff}} = \frac{500 \times 38.06}{3960 \times 0.55} = 8.74 \text{ HP}$$

Ten horsepower is therefore adequate. For final motor selection there may be a requirement for the motor to be "nonoverloading" at every point along the performance curve. If this is a requirement, the engineer should check to see that the motor size is greater than the maximum horsepower required to drive the pump at all rates encompassed by the pump rating curve.

G. Check the Net Positive Suction Head (NPSH). The NPSH is the head required to lift the liquid into the eye of the pump's impeller so that the pump will remain primed during operation and help to eliminate cavitation. This relationship is shown in Figure 10.25.

available air pressure =	33.86 feet
− dynamic suction lift =	21.72 feet
− vapor pressure at 74°F =	1.00 foot
− factor of safety =	2.00 feet
total =	− 24.72 feet
NPSH available =	9.14 feet

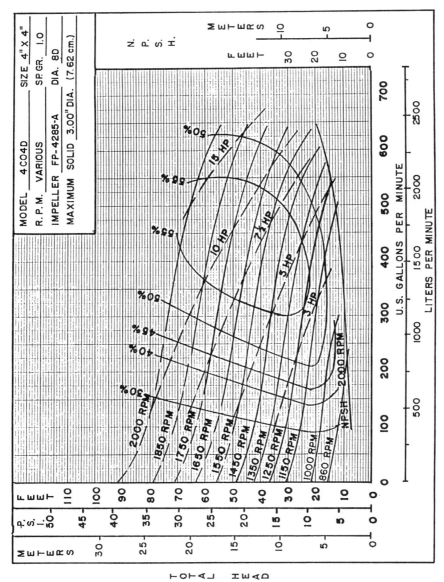

Figure 10.24 An example of a self-priming pump performance curve. Courtesy of Burks Pumps, Inc., 420 Third Street, Piqua, Ohio.

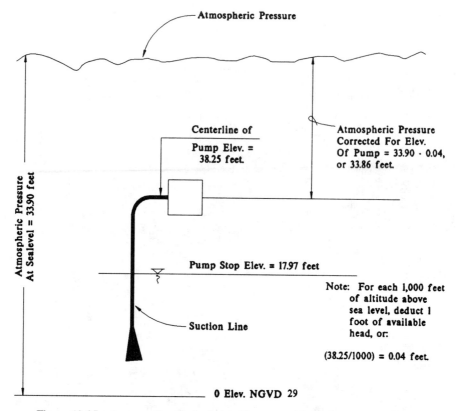

Figure 10.25 An example of calculations for net positive suction head (NPSH).

The selected Burks pump, a Model 4Cφ4D (Imp. Dia. 8D), requires 13 feet of NPSH. There is not enough NPSH available to allow proper operation of this model pump; therefore, another pump must be chosen.

The curve for an alternative pump, Burks 6Cφ5D (not shown here), is checked and found to be capable of delivering 500 gallons per minute at 38 feet TDH using 1000 revolutions per minute. The efficiency again is 55 percent; therefore, the 10 horsepower is adequate. The NPSH required for this particular model is approximately 4 feet, which is much less than the available 9.14 feet; therefore, this pump can be utilized.

H. Compute the Buoyant Forces. The buoyant forces affecting the pump station must be computed to insure that it will not float out of the ground as a result of upward pressure from high groundwater.

Upward forces attributed to groundwater, using the worst case situation, occur when the soil is saturated to the ground's surface. The total upward forces would equal the weight of displaced water, which is the wetwell's volume times the unit weight of water. For this example, the upward force is equal to the product of PI, the wetwell's radius squared, the wetwell height, and the unit weight of water. Assuming the 6 foot diameter wetwell has a 7 inch thick wall and bottom, the outside diameter of the wetwell is 7.17 feet. Numerically the upward force is equal to $3.14 \times 3.58^2 \times (36.25 \text{ feet} - 15.47 \text{ feet}) \times 62.4$ pounds per cubic foot, or approximately 52,210 pounds.

The downward forces contributed by the wetwell wall weight and bottom, using for the unit weight of concrete 150 pounds per cubic foot, are:

(a) The downward force contributed by the wetwell's wall is approximately equal to the product of PI, the wetwell's interior diameter, the wall thickness, the wetwell height, and the unit weight of concrete. Numerically the wall's downward force is 3.14 × 6 feet × 0.58 foot × 20.78 feet × 150 pounds per cubic foot, or 34,077 pounds.

(b) The downward force contributed by the wetwell's bottom (assuming it to be eight inches thick) is $3.14 \times 3.58^2 \times$ (8 inches/12 inches per foot) × 150 pounds per cubic foot, or approximately 4026 pounds. The total downward force estimated for the wetwell and bottom is 34,077 pounds plus 4026 pounds, or 38,103 pounds. To insure that this wetwell will not float, at least an additional 14,107 pounds must be provided by the pump station's floor slab, walls, and roof.

Several general considerations about the design of pump stations follow:

(a) The station should be sized for the ultimate "built out flow" expected during the life of the station if economically feasible. The wetwell may be oversized initially; however, pump control elevations can be adjusted to accommodate the actual flows encountered. These elevations are sensed by either mercury float switches or a pressure sensing system operated by air bubbles.

(b) The shutoff head (the head with the pump having its discharge closed off) of a pump should exceed the static lift from the low water level in the wetwell to the high point in the forcemain plus all upward gradient heads in the forcemain. On system start-up the pump must overcome all of these accumulated heads to charge the forcemain. Figure 10.26 illustrates this procedure.

(c) A system head curve must be made to check the performance of the selected pumps under various conditions. For example, to determine the pumping rate with both pumps on in parallel operation, a system curve first must be constructed. The static head needs to be plotted for 0 flow. This can be based on the maximum static lift in the system. In this example it is equal to 45.00 feet − 17.97 feet, or 27.03 feet. The TDH values must now be computed for various pump flows based on this example's system. A summary of these calculation results follows, based on the low water level in the wetwell for static lift determination:

Q, gpm	TDH, feet
0	27.03
100	27.57
200	29.01
300	31.25
400	34.28
500	38.07
600	42.60
700	48.11
800	53.84
900	60.47
1000	67.83

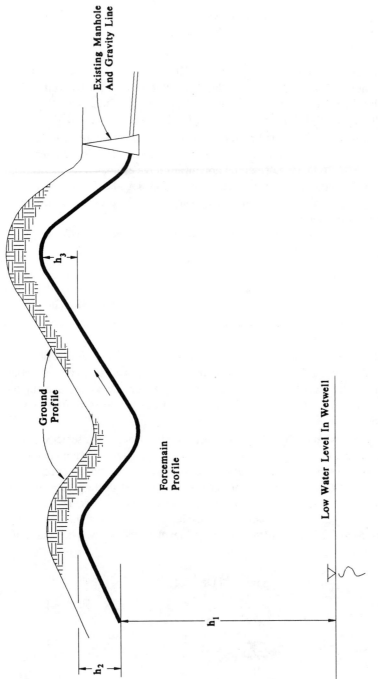

Figure 10.26 An example of minimum shutoff head determination.

A similar curve using the high water level in the wetwell for static head determination can also be prepared but is not included here. Both curves provide a range of values within which the pumps can be expected to operate, depending on where the wetwell liquid is located.

The pump performance curve for the chosen pump can now be superimposed on the graph. The intersection of the two curves yields the rated capacity of one pump and the system as shown in Figure 10.27. If both pumps are energized due to inflows greater than the capacity of one pump, the combined pumping capacity can be determined by using the system curve. For pumps in parallel, as illustrated by example in Figure 10.28, the flows are added while the corresponding heads remain the same. Figure 10.27 illustrates that the pumping rate does not double since it increases from 500 gallons per minute to approximately 615 gallons per minute. If the pumps are placed in series operation, as illustrated by the example in Figure 10.29 (not usually the case in a normal duplex pumping station configuration), then the combined pumping curve is obtained by adding the heads while the flow rates remain the same. In this case the pumping rate is increased from 500 gallons per minute to approximately 950 gallons per minute.

(d) When flooded suction pumps are used, the low water level of the wetwell should be kept above the eye of the pump's impeller to provide positive internal pressure. This helps eliminate vapor locks and prevent cavitation.

(e) If the station's interior piping is comprised of various pipe diameters, then any horizontal eccentric fittings must be oriented with the flat side up (eccentric down) to eliminate pockets that can trap vapor.

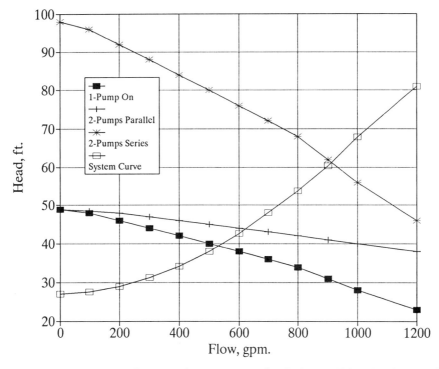

Figure 10.27 An example of pump and system curves for single, parallel, and series pumping.

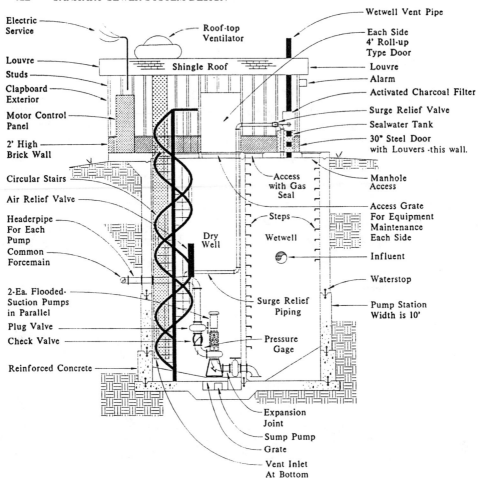

Figure 10.28   Elevation schematic of a flooded suction wastewater pump station using a parallel configuration. Source: ITT A-C Pump. *Pick A Pump,* 1981. Pump reproduced by permission of ITT A-C Pump, Cincinnati, Ohio.

(f) Discharge piping should be arranged to eliminate solids from being deposited into adjoining pipe segments where blockage could result. Figure 10.30 illustrates a condition that should be avoided.

(g) Surge control devices should be provided where long forcemains exist, or on forcemains that have unusual flows or geometry. One method is to provide an air release/vacuum breaker valve at all high points in the forcemain and a surge relief valve (valve that vents excessive internal pressure from the forcemain) at the pumping station. Another method is to provide pump control valves that regulate the rate of valve closure. Other methods are also available. Tullis (1989) provides a more detailed discussion on hydraulic transients.

(h) To vent accumulated gas, air release valves must be installed along the forcemain where high points occur. Vacuum breaker valves may also be required at high points to relieve negative pressures in the forcemain that could collapse the pipe walls or create leaks around joint seals.

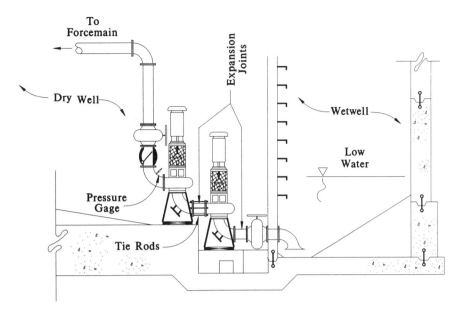

Figure 10.29 Two flooded suction wastewater pumps in series. Source: ITT A-C Pump. *Pick A Pump,* 1981. Pump reproduced by permission of ITT A-C Pump, Cincinnati, Ohio.

In addition forcemains require thrust blocking at bends and junctions similar to that required on potable water distribution mains (Section 11.7.2) in order to maintain watertight integrity of the forcemain.

(i) Proper interior ventilation in all pumping station drywells requires a complete change of air at least every two minutes. The wetwell must be vented to the atmosphere. An activated carbon air filter can be installed on the vent line to reduce complaints of foul odor from residents located close to the pumping station.

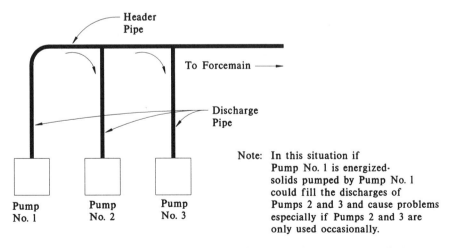

Figure 10.30 An elevation schematic of a parallel flooded suction pump station discharge piping system that should be avoided.

(j) Dehumidifier devices are required to protect equipment from moisture.

(k) Sump pumps are required to dewater pumping stations with drywells.

(l) Adequate access for equipment maintenance must be considered in the design process. Some pumping stations require the removal of roofs or hatches to allow access to installed equipment. A crane may also be required for equipment handling. Considerations for equipment repair and removal are paramount during the design process, and provisions for ease of access are important. Figure 10.28 shows a pumping station where the installed pumps can be easily removed by a truck boom hoist by simply opening an exterior roll-up door and removing a cast-iron grate in the ground level floor slab of the station.

Due to long term pump station operating expenses associated with labor and power, a cost effective design is imperative. One method to analyze a pump station design is to perform a cost analysis on several combinations of wetwell and forcemain diameters. Each variation yields different system headlosses, thus requiring different horsepower motors. The cost per gallon of pumping in dollars can be obtained using the following equation:

$$\text{cost per gallon of pumping} = \frac{(\text{power cost})(\text{TDH})}{(317,460)(E_p)(E_m)} \qquad (10\text{-}13)$$

where:

$$\text{power cost} = \text{Cost in dollars for electricity per kilowatt hour.}$$
$$\text{TDH} = \text{Total dynamic head in feet.}$$
$$E_p = \text{Pump efficiency.}$$
$$E_m = \text{Motor efficiency.}$$
$$317,460 = \text{A constant.}$$

A graph can be constructed to show the various combined costs attributed to the alternatives; from this graph a desirable choice of system relationships can be made. In the example provided an eight inch forcemain and a six foot diameter wetwell yield the most economical solution for the analysis period used. The calculations utilizing a 6 inch and 10 inch forcemain are not included. Figure 10.31 illustrates this process.

Although the four foot diameter wetwell curve is included in Figure 10.31, the NPSH available at the site (0.37 foot) would negate its use.

Local regulations may set forth other requirements; however, the general design process for a pumping station has been presented.

As design calculations are usually subject to outside review, it is imperative that the engineer conduct the analysis in a clear and logical manner so the reviewing party can follow the calculations without difficulty.

10.7   PUMPING STATION DISCHARGE

Discharge from a proposed pressure forcemain can be placed into another pumping station's wetwell, another forcemain, a gravity line, or directly into a wastewater treatment plant. The receiving facility should be capable of handling the proposed forcemain discharge along with other system flows.

Self Priming Pumping Station			
Job:			
Date:			
I. Site Conditions			
A. Finished Floor Elevation of Pumping Station, ft. =			36.25
B. I.E. Influent line, ft. =			25.77
C. Distance I.E. Influent to High Water Alarm, ft. =			0
D. Distance High Water Alarm to 2nd Pump On =			0.5
E. Distance 2nd Pump to 1st Pump On =			0.5
F. Distance Pumps Off to Low Water Alarm=			0.5
G. Average Daily Flow, GPM =			200
H. 'C' Value for Suction Line =			120
I. 'C' Value for Forcemain =			140
J. Hi Point Elev, in Forcemain, ft. =			45
K. Elevation of Discharge Point, ft., =			45
L. Length of Forcemain, ft. =			1858
M. Cost of Electricity, $kwh =			0.06
N. Height Pump Above Finished Floor, ft. =			2

II. Line Sizing

(Peak flow, gpm=) 500

A. Compute Suction Line Sizes

Pipe Dia	Velocity		Pump Station Cost Schedule		
(in.)	feet per second	Diameter	4	6	8
4"	12.77	Depth			
6"	5.67	8	$10,000.00	$14,000.00	$16,000.00
8"	3.19	12	$20,000.00	$26,000.00	$30,000.00
10"	2.04	16	$28,000.00	$32,000.00	$40,000.00
Use This, In. = 6		20	$32,000.00	$38,000.00	$46,000.00
		24	$37,000.00	$42,000.00	$50,000.00
B. Compute Forcemain Sizes		28	$44,000.00	$48,000.00	$55,000.00

Pipe Dia	Velocity,	Cost
(in.)	(FPS)	$
4"	12.77	$10.00
6"	5.67	$15.00
8"	3.19	$20.00
10"	2.04	$25.00
12"	1.42	$45.00
14"	1.04	$55.00
16"	0.80	$60.00

Use This Size Forcemain
For These Calculations=
8

III. Calculations For Wetwell Sizing

Wetwell Diameters, feet.	4	6	8
Working Volumes, gallons.	1440.00	1440.00	1440.00
Working Depths, feet.	15.32	6.79	3.83

Figure 10.31 Wastewater pump station design alternatives.

Summary of Wetwell Elevations				
F.F. Elevation, feet.		36.25	36.25	36.25
I.E. Influent Line, feet.		25.77	25.77	25.77
H.W. Alarm, feet.		25.77	25.77	25.77
2nd Pump On Elevation, feet.		25.27	25.27	25.27
1st Pump On Elevation, feet.		24.77	24.77	24.77
Pumps Off Elevation, feet.		9.45	17.98	20.94
Low Water Alarm Elevation, feet.		8.95	17.48	20.44
I.E. Suction Bell, feet.		6.95	15.48	18.44
I.E. Bottom Wetwell, feet.		6.45	14.98	17.94
Total Depth, feet.		29.80	21.27	18.31
IV. Compute Total Dynamic Head				
A. Dynamic Suction Losses				
1 Static Suction Lift, feet.		28.80	20.27	17.31
2. Friction HL,Suct. Pipe, feet.		0.74	0.54	0.47
3. Hl in Fittings.	No. Fit.			
a. No. 90 Deg. Ells	2	0.70	0.70	0.70
b. No. 45 Deg Ells	0	0.00	0.00	0.00
c. No. Gate Valves	1	0.10	0.10	0.10
d. No. Check Valves	0	0.00	0.00	0.00
e. No. Plug Valves	0	0.00	0.00	0.00
f. Inlet Losses	1	0.11	0.11	0.11
Total Dynamic Suction Headloss		30.45	21.72	18.69
B. Discharge Piping Losses, feet.				
1 Static Discharge Lift, feet.		6.75	6.75	6.75
2. Friction HL in Forcemain, feet.		8.18	8.18	8.18
3. Hl in Fittings.	No. Fit.			
a. No. 90 Deg. Ells	6	0.66	0.66	0.66
b. No. 45 Deg. Ells	2	0.16	0.16	0.16
c. No. Gate Valves	1	0.03	0.03	0.03
d. No. Check Valves	1	0.40	0.40	0.40
e. No. Plug Valves	0	0.00	0.00	0.00
f. Discharge Headloss-Subm.	1	0.16	0.16	0.16
Total Discharge Piping Losses, feet.		16.34	16.34	16.34
Total Dynamic Head (TDH), feet.		46.79	38.06	35.03
V. Compute NPSH Available At Site				
Wetwell Diameter, feet.		4	6	8
a. Available Pressure, feet.		33.90	33.90	33.90
b. Dynamic Suct. Lift, feet.		30.45	21.72	18.69
C. Vapor Press 74 degF, feet.		1.00	1.00	1.00
d. Factor of Safety, feet.		2.00	2.00	2.00
e. Loss to Site Elevation, feet.		0.04	0.04	0.04
NPSH Available At Site, feet.		0.41	9.14	12.17
VI. Horsepower Requirements				
a. Pump Efficiency		0.55	0.55	0.55
b. Motor Efficiency		0.95	0.95	0.95
Brake Horsepower Needed		10.74	8.74	8.04
VII. Cost Analysis Using 25 Year Analysis Period				
a. Assumed Interest Rate =	0.08			

Figure 10.31   (continued)

b. Annual % Increase in Cost		0		
c. Annual Pumping Cost		$1,779.24	$1,447.32	$1,331.99
Year	PW Factor	Cost 4' Dia.	Cost 6' Dia.	Cost 8' Dia.
1	0.92593	$1,647.45	$1,340.11	$1,233.33
2	0.85734	$1,525.41	$1,240.84	$1,141.97
3	0.79383	$1,412.42	$1,148.93	$1,057.38
4	0.73503	$1,307.80	$1,063.82	$979.05
5	0.68058	$1,210.92	$985.02	$906.53
6	0.63017	$1,121.22	$912.06	$839.38
7	0.58349	$1,038.17	$844.50	$777.20
8	0.54027	$961.27	$781.94	$719.63
9	0.50025	$890.06	$724.02	$666.33
10	0.46319	$824.13	$670.39	$616.97
11	0.42888	$763.09	$620.73	$571.27
12	0.39711	$706.56	$574.75	$528.95
13	0.36770	$654.22	$532.18	$489.77
14	0.34046	$605.76	$492.76	$453.49
15	0.31524	$560.89	$456.26	$419.90
16	0.29189	$519.34	$422.46	$388.80
17	0.27027	$480.87	$391.17	$360.00
18	0.25025	$445.25	$362.19	$333.33
19	0.23171	$412.27	$335.36	$308.64
20	0.21455	$381.73	$310.52	$285.78
21	0.19866	$353.46	$287.52	$264.61
22	0.18394	$327.27	$266.22	$245.01
23	0.17032	$303.03	$246.50	$226.86
24	0.15770	$280.59	$228.24	$210.05
25	0.14602	$259.80	$211.33	$194.49
d. Totals		$18,993.00	$15,449.82	$14,218.72
e. Forcemain Cost		$37,160.00	$37,160.00	$37,160.00
f. P.S. Const. Cost		$44,000.00	$42,000.00	$46,000.00
g. Total Present Worth of Costs		$100,153.00	$94,609.82	$97,378.72

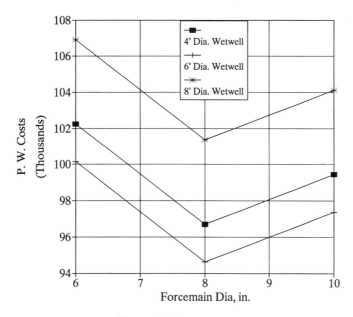

Figure 10.31 (continued)

10.7.1   Discharge into Another Pumping Station

If the pumping station under design is to discharge into another pumping station's wet-well, several factors must be investigated including:

(a) The determination of whether sufficient working storage capacity is available in the receiving pumping station's wetwell to accommodate the proposed additional flows.

(b) The determination of whether the installed pumps in the existing pumping station are capable of handling the additional flows. Alternative impeller sizes, motor sizes, or speeds may be considered; otherwise, new pumps and piping may be required. Should new motors be required, the electrical entrance, motor control panel, starters, and distribution wiring must be checked for adequacy.

(c) The capacity of the receiving pumping station's forcemain must be checked to determine its adequacy. Should the existing station's forcemain be undersize, a forcemain upgrade to a larger diameter might be required.

10.7.2   Discharge into Another Forcemain

If the pumping station under design is to discharge into an existing forcemain of sufficient capacity, the results will be a pressure transport system comprised of multiple pumping stations. System headlosses attributed to flows from various operating conditions necessitate that the designer analyze the forcemain under all possible conditions. These conditions include flows resulting from the various on-line combinations of pumping stations included in the network. For example, Figure 10.32 shows an existing forcemain and pumping station that are designated by points *A*, *B*, and *C*. A proposed new pumping station at *D* is to be tied into the existing forcemain at point *B*. Segment *A–B* will carry flows from either pumping station *C* or *D*, or both. Segment *B–C* carries only flows from pumping station *C*, while segment *B–D* carries only flows from pumping station *D*.

Because the existing system *A–B–C* is in place and currently handles only the flows from pumping station *C*, the design engineer should be concerned with the reduced pumping rate that will occur when pumping station *D* is on-line. To determine the TDH for pumping station *C* when station *D* is also pumping, the frictional headloss in forcemain segment *A–B* is based on the combined flows of pumping stations *C* and *D*. The frictional headloss for forcemain segment *B–C* is based only on the flow from pumping station *C*. The increased frictional headloss attributed to the increased flow in segment *A–B* will alter the system curve and cause the pump performance curve for the pumps in station *C* to intersect at a lower pumping rate.

Likewise, the engineer must determine the TDH for pumping station *D* when pumping station *C* is on-line as well as when it is off-line. The frictional headloss component for the forcemain segment *A–B* when pumping station *C* is on-line at the same time as the pumps in station *D* is attributed to the sum of the flows from the two stations. If pump station *C* is not on-line, the flow through segment *A–B* is attributed to the flow from pump station *B*. For forcemain segment *B–D*, the frictional headloss is attributed only to the flow from pumping station *D*. Data obtained from these headloss analyses can be used to prepare a series of system curves reflecting various operational characteristics attributed to one or

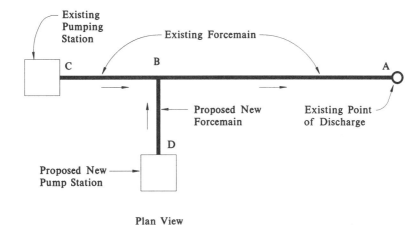

Plan View

Figure 10.32   Multiple wastewater pump stations discharging into a common forcemain.

both stations being on-line. From these data, both pumping stations can be analyzed to insure that the proper size pumps, motors, and impellers are provided.

High system headlosses resulting from combined flows may prohibit multiple pumping stations from being connected to a common forcemain; however, an alternative would be to upgrade portions of the forcemain to larger pipe diameters, thus reducing frictional headlosses in the system.

Another problem associated with multiple pumping stations in a common forcemain is that each pump station's performance must be analyzed where increased heads caused by the proposed new flow will alter previous operating points on their system curve. As a result, increased impeller diameter, motor size, and/or motor speeds may be required to overcome the new operating pressures. Analysis of waterhammer effects can be complicated, especially when a multiple station system on a common forcemain is involved.

10.7.3   Discharge into a Gravity Main

Forcemain discharges often are made into gravity sewer mains that flow to a wastewater treatment plant. The velocity of discharge from the forcemain can be significantly greater than the flow velocity that exists in the gravity sewer at the point of discharge. An expansion chamber can be used to reduce the discharge velocity from the forcemain and also allow for gas to escape from the system. Figure 10.33 illustrates a typical expansion chamber configuration.

The downstream capacity of the gravity lines must be checked to insure that the proposed flow can be properly accommodated all the way to the treatment plant.

10.7.4   Discharge into a Wastewater Treatment Plant

Forcemain discharges directly into a wastewater treatment plant can shock the treatment process if care is not exercised. Equalization facilities can be used to temporarily store the wastewater, so that it can be pumped into the plant at a lower than peak rate of delivery (over a sustained period of time). This helps provide more uniform hydraulic loadings to the plant.

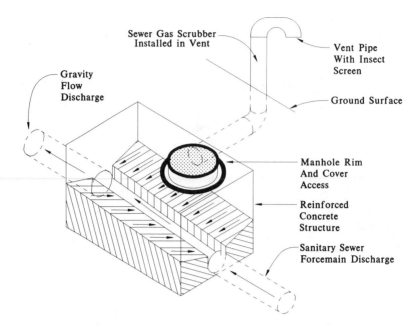

Figure 10.33   Forcemain discharge expansion chamber.

10.8   SPECIAL FORCEMAIN CONSIDERATIONS

Sanitary sewer forcemains are not always located in level terrain or in areas where their discharge points are positioned at elevations equal to or above the elevation of the pumps. This makes system head calculations less simple. To further complicate matters, the installation of combination air release/vacuum breaker valves (valves that allow air into the forcemain when negative internal pressures occur and accumulated air out of the forcemain when positive internal pressures occur) at high points can alter the hydraulic performance characteristics of the forcemain, making it possible for some segments to promote gravity flow conditions at times. This is especially true if the forcemain is sized for large ultimate flows while smaller initial flows produce only minimum self scouring velocities. The engineer must carefully determine the forcemain's system head at the desired pumping rate (or rates) because the flow might produce a transition from non-pressurized (open channel) to pressurized flow at a forcemain high point. Changes in flow conditions will result in the system's static and friction heads being altered. This requires the use of good engineering judgment when analyzing a specific forcemain configuration.

The system head of the forcemain shown in Figure 10.34 can be estimated when the rate of flow is Q_x by using the following procedure. The forcemain's diameter and roughness coefficient are assumed to be uniform between nodes and air release/vacuum breaker valves are assumed to be provided at each high point. Node 1 is the beginning of the forcemain.

Step A: Separately calculate the friction head (attributed to Q_x) using Equation 10-10 and minor losses between node 1 and every downstream node including the discharge node.

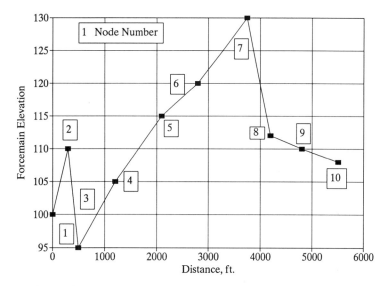

Figure 10.34 An example of an up-hill/down-hill forcemain profile.

Step B: For each friction head obtained in step A, calculate the static head by subtracting the elevation of node 1 from each downstream node's elevation.

Step C: Calculate the total dynamic head that is required to deliver Q_x from node 1 to each node in the forcemain by adding appropriate static and friction heads together.

Step D: Select the largest total dynamic head obtained from step C. This represents the head that the pump must deliver to the forcemain to provide a flow of Q_x.

To obtain enough values to construct the forcemain's system head curve, steps A through D must be repeated for each flow (Q_x) used to make the curve. Figure 10.35 illustrates the use of steps A through D when Q_x is 500 gallons per minute. The required forcemain TDH is 46.52 feet and is shown plotted in Figure 10.35*b* where the flow is 500 gallons per minute. Calculations for the other flows between 0 and 1000 gallons per minute are not included but follow the same procedure.

To obtain the total system head curve, the forcemain's system head curve must be adjusted to account for the level of the wastewater in the wetwell and all head losses in the pumping station between the wetwell and node 1.

Should the forcemain have a net negative slope (go downhill), a vacuum breaker valve can be installed at the beginning of the forcemain.

10.9 PARALLEL-SERIES PUMPING SYSTEMS

U.S. Patent 5,135,361—other patents applied for.

Pumping stations are normally configured to operate in either parallel or series modes. Nonclog centrifugal pumps customarily used in wastewater applications can be somewhat limited in the head that they can pump against. For this reason, the series application of

Flow (Qx)=		500	Gallons per Minute			Static	Friction	Cumulative	
Node	Elevation	Station	Segment	Diameter	Hazen-Williams	Head	Head	Frict. Head	TDH
	(feet)	(feet)	Nodes	(inches)	Coefficient	(feet)	(feet)	(feet)	(feet)
1	100	0	1-2	8	140	10	1.32	1.32	11.32
2	110	300	2-3	8	140	-5	0.88	2.20	-2.80
3	95	500	3-4	8	140	5	3.08	5.28	10.28
4	105	1200	4-5	8	140	15	3.96	9.25	24.25
5	115	2100	5-6	8	140	20	3.08	12.33	32.33
6	120	2800	6-7	8	140	30	4.18	16.52	46.52
7	130	3750	7-8	8	140	12	1.98	18.50	30.50
8	112	4200	8-9	8	140	10	2.64	21.14	31.14
9	110	4800	9-10	8	140	8	3.08	24.22	32.22
10	108	5500							

(a)

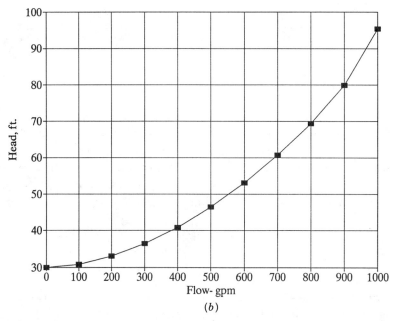

(b)

Figure 10.35 (a) An example of TDH calculations for an up-hill/down-hill forcemain when flow is 500 gallons per minute. (b) The completed system curve.

two standard sewage pumps occasionally has been employed to meet the requirement of some installations having higher heads. The design engineer should examine closely with the manufacturer all factors relating to such installations, including the pump's ability to operate in series configuration without damaging its casing, packing, and/or mechanical parts. Provisions have not been made in the past to allow an interchange of pumping modes in a station; however, parallel-series pumping systems allow both modes. As a result, smaller pump sizes often can be used or more pumping capacity usually can be attained.

Occasionally pumping stations must accommodate inflows that exceed the design capacity of the installed pumps. This can be attributed to infiltration, inflow, and/or undeterminable wastewater flows. In many cases parallel pumping in small duplex stations produces little increase in the discharge because of system curve limitations; therefore, little benefit is realized from activating the second pump. This large inflow could result in the wastewater surcharging the collection lines, causing it to back up into dwellings or overflow into streets and low lying areas.

Figure 10.27 demonstrates this situation where the base pumping rate is 500 gallons per minute and the parallel rate was found to be 615 gallons per minute. If the pumps were temporarily placed in series operation, using a parallel-series pumping system, the combined flow rate would be 950 gallons per minute.

10.9.1 Parallel-Series Pumping Technology

U.S. Patent 5,135,361—other patents applied for.

In cases where parallel pumping does not appreciably increase pumping capacity, the temporary reconfiguration of the pumping system into series pumping could produce additional capacity that heretofore was unavailable due to the design constraints of a duplex pumping system. Upon completion of series pumping, the pumps would be returned to the conventional parallel mode of operation. This is contrary to current practice, which suggests unless wastewater pumps need to be configured in an established dedicated series service attributed to high system heads, parallel configurations are normally utilized for multiple pump installations. Figure 10.36 illustrates a typical parallel-series piping configuration for self-priming pumps.

Other parallel-series pumping configurations can include the use of plug valves, piston valves, and other valve types to promote temporary reconfiguration. Submersible pumps are also capable of adapting to parallel-series pumping, as illustrated in Figure 10.37.

10.9.2 Parallel-Series Pumping Applications

U.S. Patent 5,135,361—other patents applied for.

In addition to being able to appreciably increase the discharge from a duplex constant speed centrifugal pumping station (if its system head curve is not relatively flat), parallel-series pumping can also be beneficially utilized in many other applications.

In some sanitary sewer pumping stations where more than two constant speed centrifugal pumps are to be utilized (based on the system's capacity and expected range of flows), a parallel-series pumping system can be considered. In smaller pump stations, 100 percent pumping standby is normally required; however, in larger stations the size and number of pumping units are selected so that the expected range of inflows can be met without starting and stopping the pumps too frequently and without requiring excessive wetwell storage capacity.

It is customary to provide a total pumping capacity equal to the maximum expected inflow with at least one of the largest pumping units out of service. Two units are sometimes considered out of service during emergencies or maintenance downtime in large stations. Variable speed drives can be utilized in stations such as these to match pumping rate with inflow rate.

Sometimes the capacity and depth of the wetwell can be coordinated with constant speed pumping units so that the rise and fall of liquid levels in the wetwell will result in a variable pumping capacity that might approximate inflow rates. Where the pumping head

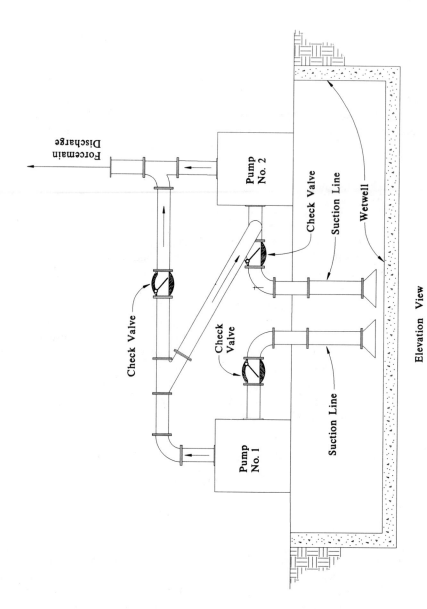

Figure 10.36 Parallel-series pumping schematic using self-priming pumps. Source: U.S. Patent 5,135,361, 1992. Other patents applied for.

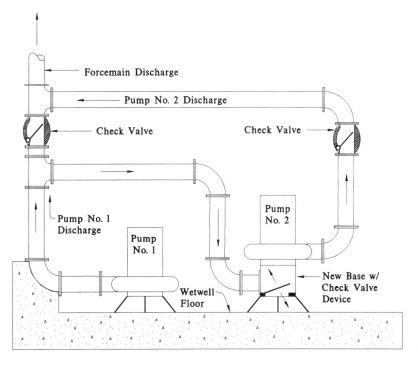

Elevation View

Figure 10.37 Parallel-series pumping schematic using submersible pumps. Source: U.S. Patent 5,135,361, 1992. Other patents applied for.

is low, normal level variations will change the pumping head so that a wide range of pumping capacities might be achieved.

Current practice suggests that the engineer select various combinations of pump characteristic curves that are plotted over the system head curve for high and low wetwell levels using parallel pumping configurations. Figure 10.38 illustrates the use of three

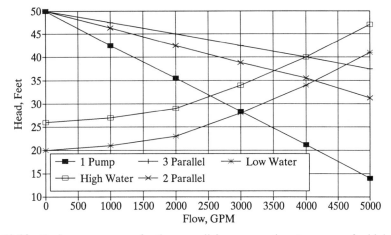

Figure 10.38 Performance curves for three parallel pumps, and system curves for high and low wetwell water levels.

similar parallel constant speed pumps to achieve a pumping range from 2500 gallons per minute at 32 feet of head to 4000 gallons per minute at 40 feet of head.

Based on BEP, these three similar pumps are selected where each has a shutoff head of 48 feet. For backup, a fourth pump must also be included.

Using a parallel-series pumping system, only two similar pumps are required to achieve the same approximate pumping rates. Each possesses a lesser shutoff head than is required in the three pump application, which usually means the pump's operating speed (wear-out factor) is reduced. In addition, a savings in construction and maintenance costs can be realized as a result of space requirements being reduced since less piping and equipment is needed when the total number of pumps is reduced from four to three. Figure 10.39 illustrates the use of parallel-series pumping to achieve the desired results with two pumps.

Another application for parallel-series pumping is to improve the performance of an existing pumping system that has deteriorated over time in service as a result of increased system head (possibly attributable to increased pipe friction, undetectable partially closed or malfunctioning valves, or even newer pumping stations being tied into the forcemain). This increased system head results in a modified system curve where the intended design delivery rate cannot be achieved. Figure 10.40 illustrates this situation: originally one pump delivered approximately 450 gallons per minute but it now delivers only 350 gallons per minute. Assuming the station is a constant speed duplex station in parallel configuration, originally the two pumps delivered 750 gallons per minute, but now only deliver 550 gallons per minute. Current methods to improve this system's performance include altering impeller diameters, changing impeller speeds, and providing increased motor sizes. If these alternatives are unsuccessful, other more expensive means can be considered including pump replacement, piping replacement, or even replacement of portions of the forcemain.

Using only a parallel-series pumping system, however, the maximum discharge from the station can be increased to approximately 760 gallons per minute, as illustrated in Figure 10.41.

Other applications for parallel-series pumping include the use of variable speed pumps.

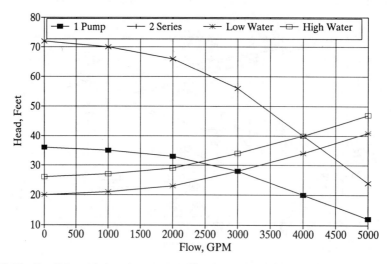

Figure 10.39 Parallel-series pumping curves using two pumps and system curves for high and low wetwell water levels.

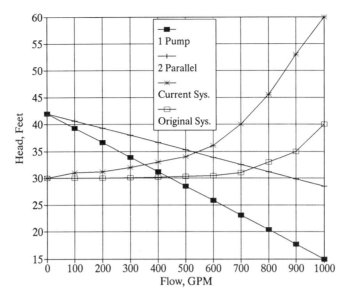

Figure 10.40   Examples of pump and system curves for the initial installation and after usage.

10.10   CENTRALIZED WASTEWATER TREATMENT PLANT CONSIDERATIONS

A centralized wastewater treatment plant may be required to service a proposed land development if there are sufficient dwelling units to warrant a facility. This decision is usually a function of regulatory review agency actions and how the proposed facility interfaces with the area's 208 Water Quality Management Plan. If a facility is needed, its function would be to treat wastewater to a predetermined level of purity where the

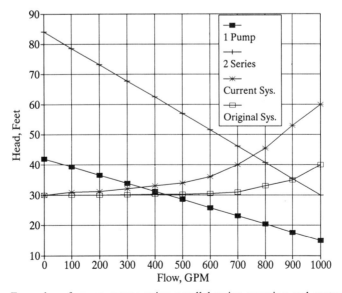

Figure 10.41   Examples of pump curves using parallel-series pumping and system curves for initial installation and after usage.

resulting treated water will not adversely affect the environment. Often field erected packaged wastewater treatment plants can be used to provide the necessary treatment for the development.

10.10.1   Wastewater Composition

Wastewater includes human waste products and household waste products such as wash water, detergents, and food preparation wastes (especially if garbage disposal units are present). Water attributed to inflow and infiltration can also be included. Typical characteristics of wastewater are shown in Figure 10.42.

Wastewater has various types of solids present, including organic and inorganic solids. Total solids include all solids present in the wastewater, of which approximately 50 percent are organic. Organic solids usually originate from animals or plants although synthetic organic compounds exist. Organic solids contain carbon, hydrogen, and oxygen along with possibly nitrogen, phosphorous, and/or sulfur. Organic solids can decay or decompose under the activity of bacteria or other living organisms and are capable of being burned.

Contaminants	Unit	Weak	Medium	Strong
		Concentration		
Solid, total (TS)	mg/L	350	720	1200
Dissolved, total(TDS)	mg/L	250	500	850
Fixed	mg/L	145	300	525
Volatile	mg/L	105	200	325
Suspended solids (SS)	mg/L	100	220	350
Fixed	mg/L	20	55	75
Volatile	mg/L	80	165	275
Settleable solids	mL/L	5	10	20
Biochemical oxygen demand, mg/L:				
5-day, 20-degC(BOD , 20-degC)	mg/L	110	220	400
Total organic carbon (TOC)	mg/L	80	160	290
Chemical oxygen demand (COD)	mg/L	250	500	1000
Nitrogen (total as N)	mg/L	20	40	85
Organic	mg/L	8	15	35
Free Ammonia	mg/L	12	25	50
Nitrites	mg/L	0	0	0
Nitrates	mg/L	0	0	0
Phosphorus (total as P)	mg/L	4	8	15
Organic	mg/L	1	3	5
Inorganic	mg/L	3	5	10
Chlorides	mg/L	30	50	100
Sulfate	mg/L	20	30	50
Alkalinity (as CaCO)	mg/L	50	100	200
Grease	mg/L	50	100	150
Total coliform	no/100 mL	10 - 10	10 - 10	10 - 10
Volatile organic compounds (VOCs)	ug/L	<100	100-400	>400

Source: Tchobanoglous et al, 1991.

Figure 10.42   Typical composition of untreated domestic wastewater. Source: Metcalf and Eddy, Inc, et al. *Wastewater Engineering: Treatment, Disposal, and Reuse,* 3rd ed. Copyright © 1991 by McGraw-Hill, Inc. Reprinted by permission of McGraw-Hill, Inc.

Inorganic solids are solids not subject to decay including sand, silt, and mineral salts contained in the water supply that produce water hardness. Usually inorganic solids are not combustible.

Suspended solids are solids that can be seen in suspension in wastewater. (Suspended solids can be defined as the solids retained by a filter mat in a Gooch crucible.) This type of solids can be removed physically or mechanically with sedimentation or filtration. Suspended particles can be composed of approximately 30 percent inorganic materials such as grit and sand. Some of the suspended solids can be settled, if of sufficient size and weight, in an Imhoff cone during a one hour period. The amount of settled material is usually quantified as millimeters of solids per liter of sewage or parts per million by weight. Solids remaining in suspension after the one hour period are classified as colloidal solids. Colloidal solids can be highly organic and subject to rapid decay.

Dissolved solids can be defined as those solids that are not retained by a filter mat in a Gooch crucible, although some of the very fine solids passing through the filter mat are not dissolved (about 10 percent).

Wastewater can include various levels of dissolved gasses including oxygen, carbon dioxide, nitrogen, hydrogen sulfide, and possibly some inorganic sulfur compounds. In addition, aerobic (needing oxygen) bacteria and anaerobic (no oxygen) bacteria can be present, along with facultative bacteria that can live in either environment. Water temperature can greatly affect the rate of bacteria growth and reproduction, which is of importance when treating wastewater. Parasitic or saprophytic bacteria can also be present, along with various other organisms including viruses.

10.10.2 Methods to Treat Wastewater

Wastewater can be treated using many methods including the following:

(a) The removal of suspended inorganic material can be accomplished using sedimentation or clarification, flocculation, and filtration. To remove dissolved solids reverse osmosis or electrodialysis can be used.

(b) Organic materials can be treated using activated sludge methods including extended aeration, contact stabilization, or other associative methods. Aerobic treatment can include trickling filters, lagoons, land treatment, and other methods. Anaerobic methods include mixed reactors or packed bed reactors.

During the treatment of wastewater, gravity can be used to reduce the amount of suspended particles present. Biological and chemical changes allow colloidal particles to flocculate and settle.

During aerobic decomposition oxygen is combined with organic materials contained in the wastewater to produce carbon dioxide, water, nitrates, sulfates, phosphates, and other substances. Anaerobic decomposition includes the removal of oxygen from complex compounds to form simpler ones. The process continues in steps until complex compounds are broken down into stable substances of organic or inorganic materials.

The activated sludge process is an aerobic biological process that incorporates an aeration tank and a solids liquid separator where the concentrated sludge can be recycled back to the aeration tank. This process is dependent upon the production of flocculent microorganisms that can be removed by gravity sedimentation. There are several varia-

Process	BOD Loading $\left(\frac{\text{lb BOD/day}}{1000 \text{ cu ft}}\right)^a$	F/M Ratio $\left(\frac{\text{lb BOD/day}}{\text{lb MLSS}}\right)^b$	Sludge Age (days)	Aeration Period (hr)	Return Sludge Rates (percent)	BOD Removal Efficiency (percent)
Conventional	30–40	0.2–0.5	5–15	6.0–7.5	20–40	80–90
Step aeration	30–50	0.2–0.5	5–15	5.0–7.0	30–50	80–90
Contact stabilization	30–50	0.2–0.5	5–15	6.0–9.0	50–100	75–90
High rate	80 up	0.5–1.0	3–10	2.5–3.5	50–100	70–85
High-purity oxygen	120 up	0.6–1.5	3–10	1.0–3.0	30–50	80–90
Extended aeration	10–20	0.05–0.2	20 up	20–30	50–100	85–95

^{a} 1.0 lb/1000 cu ft day = 16.0 g m^3·d
^{b} 1.0 lb/day·lb = 1.0 g d·g

Figure 10.43 General loading and operational parameters for activated sludge processes. F/m = food to microorganism ratio; MLSS = mixed liquor suspended solids. Source: Mark J. Hammer. *Water and Wastewater Technology,* 2nd ed. Copyright © 1986, p. 413. Reprinted by permission of Prentice-Hall, Englewood Cliffs, NJ.

tions of the activated sludge process including extended aeration, contact stabilization, step aeration, and high rate processes. Figure 10.43 contains various general loading and operational parameters that are used in various activated sludge processes. Figure 10.44 illustrates alternative activated sludge process flow diagrams that can be considered as a basis for wastewater treatment plant design.

Filters can be included to remove suspended particles as part of the treatment process. Nitrogen removal from the wastewater is also important. Most nitrogen present in the waste is in the form of ammonia. Nitrite nitrogen can be a product of oxidation of ammonia, while nitrate nitrogen is produced by the decomposition of nitrogenous materials. In wastewater "total Kjeldahl nitrogen" refers to ammonia and organic nitrogen whereas "total nitrogen" refers to ammonia, organic, nitrite, and nitrate nitrogen. Removal systems, such as a biological nitrification/denitrification system, can be used where ammonia nitrogen and nitrite nitrogen are converted into nitrates during the nitrification step. The denitrification step occurs where nitrates are converted into nitrogen gas.

Depending on the strength and composition of the waste, varying degrees or methods of treatment may be required to produce effluent that conforms to the waste load allocation set forth in issued permits. Other references on wastewater treatment plant design are available for the designer to consult. (Novotny et al, 1989) (Metcalf and Eddy, Inc. et al, 1991)

10.10.3 Waste Production from Treatment Processes

Sludge is produced as a by-product of the wastewater treatment process and must be handled properly. Some wastewater treatment plants use sludge drying beds to dewater the sludge prior to final disposal. Other facilities rely on dewatering equipment such as centrifuge and incineration for disposal. Regulatory agencies may require minimum set back distances from treatment plant facilities (especially chlorination storage and disinfection areas and sludge drying facilities) to the closest dwelling unit. Spatial requirements such as these place additional demands on the design engineer to allocate adequate land for treatment facilities.

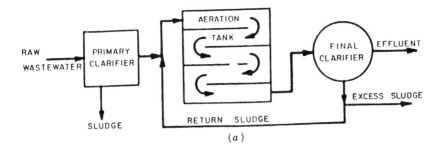

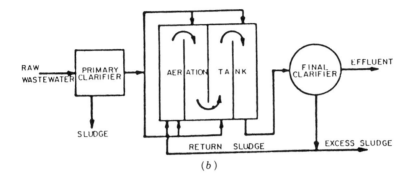

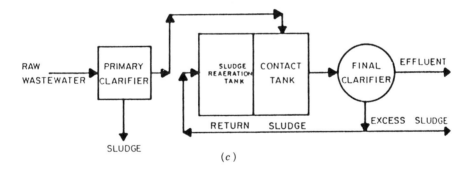

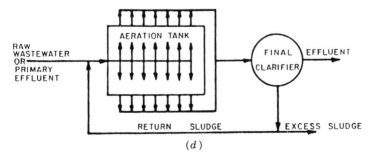

Figure 10.44 Flow diagrams for common activated sludge processes. (*a*) Conventional plug flow activated sludge plant. (*b*) Step aeration plant. (*c*) Contact stabilization activated sludge system. (*d*) Complete mix plant. Source: Karl Imhoff (Novotny et al.). *Karl Imhoff's Handbook of Urban Drainage and Wastewater Disposal.* Copyright © 1989 by John Wiley & Sons, Inc. Reprinted by permission of John Wiley & Sons, Inc.

10.11   WASTEWATER SYSTEM APPURTENANCES

Sanitary sewage collection systems are comprised of piping, fittings, laterals (or services), manholes and drops, and pipe bedding. Each of these components provides an important feature to the system.

10.11.1   Piping and Fittings

Most sanitary sewage piping and fittings utilized in land developments include small diameter circular conduits. Typical diameters, however, can range from 4 to 144 inches, depending on the material of construction. Pipes used in sanitary sewage systems must be strong, durable, abrasion resistant, impervious, and resistant to corrosive action. Pipe joints must be watertight to prevent infiltration of groundwater, as well as exfiltration of wastewater. Sewage pipes can be either gravity flow nonpressurized pipes or pressurized pipes similar to those used in forcemains.

Gravity flow pipes most commonly are made of vitrified clay, polyvinyl chloride (PVC), polyethylene (PE), precast concrete, and ductile-iron. Most pipe joints are of the slip-type, where the barrel of one joint "slips" into a gasket contained in the bell of the adjoining pipe. Ductile-iron pipe can be made with mechanical joints where the gasket is mechanically held into place, flange joints where pipe sections are bolted together using end flanges, and restrained jointed pipe (river crossing pipe) where each joint provides positive restraint of the joined pipes thus reducing the need for thrust blocking (Section 11.7.2). On occasion smooth walled steel pipe, corrugated steel pipe, or bituminous fiber pipes have been utilized.

Pipes made of materials susceptible to corrosion must be protected with coatings or linings. This corrosion occurs when released hydrogen sulfide gas (H_2S) in the wastewater condenses on the interior pipe's crown, allowing sulfuric acid to form. This acid can corrode certain pipe materials. Epoxy, plastic, or bituminous coatings help to protect susceptible materials.

10.11.2   Manholes and Drops

Sanitary sewage manholes allow system access for maintenance during the life of the system. The minimum manhole size used on most small diameter sanitary sewer mains is four feet inside; however, if acute angles of intersection of the mains joining the manhole exist, a larger diameter manhole can be used to allow more space inside to build an adequate invert to promote efficient hydraulic flow.

Manholes are often constructed of precast concrete manhole sections, which give superior water tightness over manholes constructed from masonry units.

10.11.3   Pipe Bedding

Proper installation of sanitary sewer mains requires that pipes be bedded properly in order to promote adequate performance. Bedding requirements are usually included in the construction specification (Chapter 13).

A properly bedded pipe will maintain its watertight integrity. Improperly bedded pipes can deflect at the joints, causing the joint seal to be lost. Also, properly bedded pipes can handle intended design live and dead loads without experiencing problems. Pipe strength should be investigated using methods similar to those included in Section 7.7.

REFERENCES

American Society of Civil Engineers (ASCE). *Pipeline Design for Water and Wastewater.* New York: ASCE, 1975, pp. 21, 33, 43. Referenced material is reproduced by permission of the American Society of Civil Engineers.

American Society of Civil Engineers (ASCE) and Water Pollution Control Federation (WPCF). *Design and Construction of Sanitary and Storm Sewers—ASCE Manual No. 37, MOP-9.* New York: ASCE, 1969.

American Society of Civil Engineers (ASCE) and Water Pollution Control Federation (WPCF). *Gravity Sanitary Sewer Design and Construction—ASCE Manual No. 60 WPCF No. FD-5.* New York: ASCE, 1982, pp. 36–37. Referenced material is reproduced by permission of the American Society of Civil Engineers.

Burks Pumps, Inc. *Pump Curve—4C⌀4D (Imp. Dia. 8D).* 420 Third St., Piqua, Ohio 45356-0603.

DAVCO—Division of Davis Water and Waste Industries, Inc. *DAVCO Equipment—Engineering: Pumping Station Design,* 1974. 1828 Metcalf Ave., Thomasville, Ga. 31792.

Dion, Thomas R. and William W. Gotherman. *U.S. Patent No. 5,135,361—Pumping Station in a Water Flow System.* U.S. Patent Office, 1992.

Fair, Gordon M., J. C. Geyer, and David A. Okun. *Water and Wastewater Engineering, Vol. I.* New York: Wiley, 1966, pp. 5–24.

Geyer, John C., Jerome B. Wolff, F. P. Linaweaver, Jr., and Acheson J. Duncan. Report on Phase One, Residential Water Use Project, *Final and Summary Report on Phase I of the Residential Water Use Research Project* January 1, 1959 to January 1, 1963. An unpublished report. Baltimore: Johns Hopkins University and Federal Housing Administration, October 1963, pp. 6-7, 6-8.

Hammer, Mark J. *Water and Wastewater Technology,* 2nd ed. Englewood Cliffs, N.J.: Prentice-Hall, 1986, p. 413.

ITT A-C Pump. *Pick-A-Pump,* 1981, p. 334. 1150 Tennessee Ave., Cincinatti, Ohio 45229.

J-M Manufacturing Company, Inc. *Green-Tite PVC Gravity Sewer Pipe—Installation Guide.* An unpublished manual. Stockton, Calif.: J-M Manufacturing Company, Inc., 1987.

Kaplan, O. Benjamin. *Septic Systems Handbook.* Chelsea, Mich.: Lewis, 1987.

Metcalf and Eddy, Inc. and George Tchobanoglous. *Wastewater Engineering: Collection and Pumping of Wastewater.* New York: McGraw-Hill, 1981, pp. 109, 111.

Metcalf and Eddy, Inc., George Tchobanoglous, and Frank L. Burton. *Wastewater Engineering: Treatment, Disposal, and Reuse,* 3rd ed. New York: McGraw-Hill, 1991, p. 109.

Novotny, V., Karl Imhoff, M. Olthof, and P. Krenkel, Eds. *Karl Imhoff's Handbook of Urban Drainage and Wastewater Disposal.* New York: Wiley, 1989, pp. 21, 162–168. Copyright © 1989 by John Wiley & Sons, Inc. Reprinted by permission of John Wiley & Sons, Inc.

Salvato, Joseph A. *Environmental Engineering and Sanitation,* 4th ed. New York: Wiley, 1992, p. 508.

South Carolina Department of Health and Environmental Control (SCDHEC). *Guidelines for Unit Contributory Loadings to Wastewater Treatment Facilities.* Columbia, S.C.: Water Pollution Control Division, SCDHEC, 1972.

South Carolina Department of Health and Environmental Control (SCDHEC). *Experimental Standard Elevated Infiltration System.* An unpublished memorandum. Columbia, S.C.: SCDHEC, 1987.

Submersible Waste Water Pump Association. *Submersible Sewage Pumping Systems Handbook.* Chelsea, Mich.: Lewis, 1985.

Tindall Concrete Products. *Underground Structures.* An unpublished report. Spartanburg, S.C.: Tindall Concrete Products, 1979.

Tullis, J. Paul. *Hydraulics of Pipelines.* New York: Wiley, 1989.

United States Environmental Protection Agency (U.S. EPA). *Small Wastewater Systems— Alternative Systems for Small Communities and Rural Areas. 830/F—92/001.* May 1992.

11 Potable Water System Design

The last system to be designed in a land development is usually potable water because water mains are pressurized and placed where conflicts with other designed systems are minimal.

To properly design this system, the engineer must determine anticipated demands of flows and pressures. The source of supply along with storage and fire suppression requirements are of importance as well as the configuration and sizing of the distribution system. This chapter centers on the design of small domestic systems normally found in residential land developments.

11.1 ENGINEER'S REPORT

The design engineer may be required as part of the regulatory review process to prepare a written report for a new water system, or for the extension of an existing system, prior to obtaining a permit for construction and operation. This report usually contains general project information (including phasing plans if an incremental approach is to be utilized), surface water and groundwater source information, a discussion of the proposed treatment process, and a description of the proposed distribution system.

11.1.1 General Information

The name and address of the proposed system's owner and the name and address of the design engineer normally are included in the engineer's report. A general description of the area being served, along with proposed land uses, terrain, location, and possibly consumer socioeconomic characteristics are included. The estimated number and type of potential customers are also described, and documentation of preliminary approvals from local regulatory agencies for the proposed development is provided. Should an existing public water supply be present, the distance to the project site is usually reported.

11.1.2 Surface Water Sources

The engineer's report may indicate the latitude and longitude of proposed raw water intake structures along with the name of the proposed water source and its classification. If a potential contributory watershed is involved, then its limits are usually described along with the expected low flow stream characteristics. The engineer should include the names of all contributory discharges that may affect the quality of the water source for possibly as far as 10 miles upstream, indicating classifications of each type of discharge as agricultural, municipal, industrial, and so on.

This report could include a description of the proposed intake and pumping facilities (pumping rates and the proposed pumping distance) as well as alternate sources of water that might be available.

11.1.3   Groundwater Sources

Should groundwater sources be proposed, the engineer's report could indicate the latitude and longitude of each proposed well site and the type of proposed well along with its diameter and depth. The report could describe the proposed positioning of each well on the well site along with the location of the 100 year flood boundary, if applicable. In addition, potential groundwater contamination sources would be delineated along with directions and distances to each potential source.

11.1.4   Water Treatment Plant

If a water treatment plant is proposed, the engineer's report may include the maximum anticipated plant capacity at project completion, along with an estimated completion date and interim plant phasing requirements. The location of the proposed plant and its proximity to the 100 year floodplain are of importance to most regulatory authorities. Also, the proposed treatment process is usually described along with proposed methods to handle waste by-products produced when treating water. Should the name of the operator of the proposed plant be known, it may be included along with the operator's qualifications.

11.1.5   Water Distribution System

The engineer's report for a residential land development includes anticipated average daily demands and peak requirements for domestic water use. Fire suppression flows using rating criteria formulated by the ISO Commercial Risk Services, Inc. (CRS) are usually considered along with types and capacities of proposed water storage facilities.

11.2   WATER SYSTEM DEMANDS

Rates of flow and volume demands in water systems vary based on geographical area, seasonal change, and user needs. Domestic, commercial, industrial, and fire protection requirements contribute to the required total system demands that must be met. According to Sweitzer and Flentje (1972), one month (during the annual seasonal cycle when the highest water usage occurs) can be designated as the maximum (water usage) month for that year. The maximum day for that year will likely be one of the days in the designated maximum month. A daily usage cycle that includes peak flow periods of about four hours duration and even extreme peaks of an hour or less can be superimposed upon the seasonal cycle. Typical peak flow ratios are given in Figure 11.1, where the range of values reflects possible differences in system users and sizes. In smaller built-up areas, water users tend to have similar habits, as indicated in Section 11.2.1, while water use patterns in larger areas tend to be more diversified, so that peak flow demands are reduced.

According to the American Water Works Association's (AWWA's) *Manual M 31*

Typical System Peak Flow Characteristics As Ratios Of The Average Daily Flow	
Average Daily Flow =	1.0
Average Day During The Maximum Month =	1.3-1.8
Average Rate On The Maximum Day =	1.6-2.2
Average Rate During The Maximum Four Hour Period =	2.0-3.0
Maximum Rate Sustained For A Period Of One Hour =	2.5-6.0

Figure 11.1 Typical water system peak flows. Source: Robert J. Sweitzer and Martin E. Flentje. *Basic Waterworks Management.* Copyright © 1972 by American Concrete Pressure Pipe Association. Reproduced with permission.

(1989),* the demands or rates of water use placed on water systems are usually classified into three general categories, as follows:

(a) *Average Daily Demand:* The average amount of water used each day during a one year period for the entire system.

(b) *Maximum Daily Demand:* The maximum amount of water used during one 24 hour period occurring during the latest three year increment. These data can be estimated from local water system data when new developments are being designed; however, such estimates are usually required to be at least 1.5 times larger than the average daily demand (CRS, 1980).

(c) *Maximum Hourly Demand:* The maximum amount of water used in a single hour, of any day, in a three year period, normally expressed as gallons per day by multiplying the hourly peak by 24.

Generally, the flows used for distribution system design purposes in land developments are either the peak hourly flow, or the needed fire flow plus the maximum daily demand, whichever is greater. A water system could be considered adequate by CRS if it can deliver the required fire flows to all points in the distribution system with the consumption at the maximum daily rate (average rate on maximum day of a normal year).

11.2.1 Domestic Water Use

Domestic water use includes water supplied to support residential activities such as flushing toilets, cleansing, drinking, cooking, and lawn/garden sprinkling. The average daily demand per capita ranges from approximately 50 to 150 gallons for metered systems and up to 250 gallons if unmetered. Domestic water rate of flow demands vary throughout the day and from day to day. According to a study by Wolff and Loos (1956) of the Baltimore area, two peak water use periods exist: one between 7 A.M. and 1 P.M., and a second between 5 P.M. and 9 P.M. Peak hourly flow requirements during these periods are reported to extend as much as 1500 percent above the average daily demand in residential areas with lots from one-half to three acres in new and old neighborhoods and from 500 to 600 percent in older neighborhoods comprised of well-settled small lots. New neighbor-

*Reprinted from AWWA *Manual M-31* (1989) by permission.

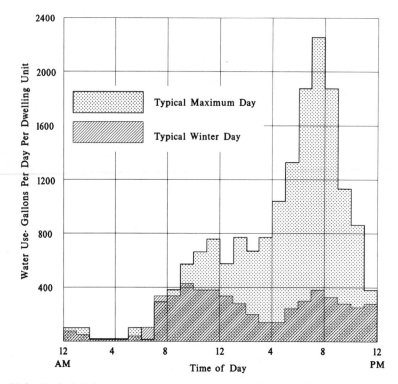

Figure 11.2   Typical daily water use patterns: maximum day and winter day. Source: John C. Geyer, Jerome B. Wolff, F. P. Linaweaver, Jr., and Acheson J. Duncan. Report on Phase One, Residential Water Use Project, *Final and Summary Report on Phase I of the Residential Water Use Project, January 1, 1959 to January 1, 1963.* October 1963. Reproduced with permission.

hoods studied, consisting of lots ranging from a quarter to a half acre, are found to reach hourly peaks as high as 900 percent. Lawn sprinkling is identified as a major factor that contributes to a high percentage of the peak flows. Many water utilities today attempt to limit or prohibit consumers from irrigating their lawns during certain periods, hoping to reduce demands on the system.

In a first phase study of residential water use in the Baltimore metropolitan area between 1959 and 1962, four topics were investigated: residential water use, peak hourly design criteria, commercial water use, and coincident water and sanitary sewage flows (Geyer et al., 1963). Typical daily water use patterns obtained from this study are shown in Figure 11.2.

This initial study was later expanded into a second phase to develop improved methods for the design of water distribution systems that would be applicable to most locations in the United States (Linaweaver et al., 1966). This included 41 residential areas ranging from 44 to 410 dwelling units in various climatic regions in the United States, as indicated in Figure 11.3.

The study results provide evidence that residential water demands depend upon:

(a) The number of customers and their economic level.

(b) The average irrigable lawn area.

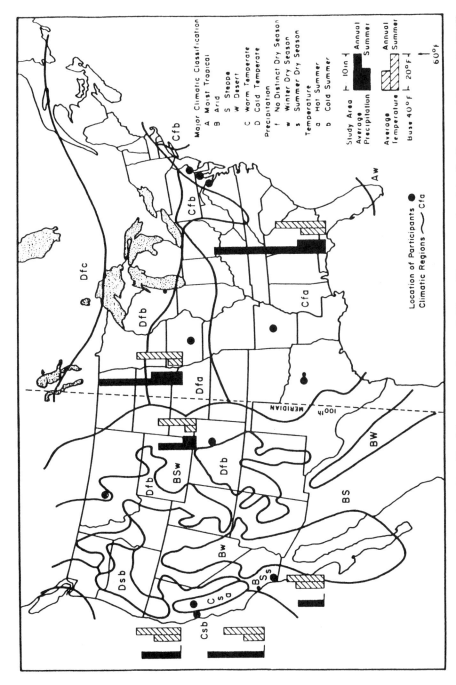

Figure 11.3 Study areas for Phase II residential water use. Source: F. Pierce Linaweaver, Jr., John G. Geyer, and Jerome B. Wolff. *Report V on Phase II—Residential Water Use Research Project,* 1966. Reproduced with permission.

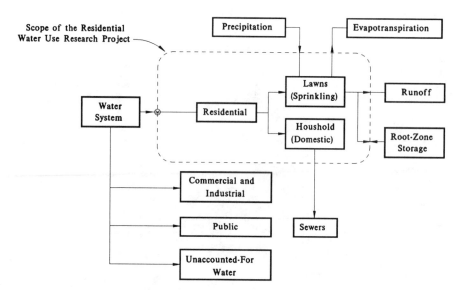

Figure 11.4 Schematic diagram of residential water use. Source: F. Pierce Linaweaver, Jr., John C. Geyer, and Jerome B. Wolff. *Report V on Phase II—Residential Water Use Research Project*, 1966. Reproduced with permission.

(c) The rate of transpiration.

(d) The amount of precipitation effective in reducing sprinkling.

(e) Whether customers were metered or unmetered.

Figure 11.4 shows residential water use activities schematically.

It was further determined that water systems serving residential areas with gross densities of about seven homes per acre or less should be capable of supplying lawn sprinkling water approximately equal to evapotranspiration from lawns covering the entire area served.

Traditionally water system demand estimates have been performed by making population estimates and multiplying those estimates by an average per capita daily use to estimate the total average use. Peaking factors based on entire service areas have then been used to estimate peak demands. According to the phase II study, "The peak to average ratios selected have been generally too low." In 1965 the Federal Housing Administration adopted as part of the Minimum Design Standards a policy whereby:

(a) If system operating data are available, the annual average demand should be ascertained on the basis of records or records of existing systems of similar nature in the area.

(b) In the absence of reliable experience records, an average demand of 100 gallons per capita per day should be used along with four persons being allocated to each dwelling unit. This provides an average demand of 400 gallons per day per dwelling unit.

(c) A maximum daily demand of 200 percent of the average demand was recommended.

(d) A maximum hourly demand of 500 percent of the average demand was suggested, except in areas where "extensive" lawn irrigation is commonly practiced. Then a rate of 700 percent or more should be used.

According to Linaweaver et al. (1966), "These criteria are an improvement, but there are situations when the use could lead to under design and in other situations to over design of the water distribution system." The results of the Phase II study are reported to provide a more equitable design approach.

A relationship for the design maximum daily demand and the design peak hourly demand are included in the Phase II study. The basic equation for predicting demands in metered and sewered residential areas are:

$$\overline{Q} = \overline{Q_d} + 0.6ca\,\overline{L_s}\,(\overline{E_{pot}} - \overline{P_{eff}}) \qquad \text{with } \overline{Q} \geq \overline{Q_d} \qquad (11\text{-}1)$$

where:

$\overline{Q}$ = The expected average demand for any period expressed in gallons per day.

$\overline{Q_d}$ = The expected average household use that equals the amount that is essentially returned to the wastewater system in gallons per day for all periods of a day or longer. This can be estimated as a function of the average market value of dwelling.

0.6 = Coefficient to adjust for the difference between actual and potential evapotranspiration from lawns.

c = A coefficient to adjust units, 2.72×10^4 gallons per acre-inch of water.

a = The number of dwelling units contributing to the demand.

$\overline{L_s}$ = The average irrigable area in acres per dwelling unit found by: $L_s = 0.803W^{-1.26}$ where W = the gross density in dwelling units per acre.

$\overline{E_{pot}}$ = The estimated average potential evapotranspiration for the period of demand in inches of water per day, determined from climatological data. (In the case of the Phase II study, the overall mean value for the study areas was found to be 0.28 inch of water.)

$\overline{P_{eff}}$ = The amount of natural precipitation effective in satisfying evapotranspiration for the period and thereby reducing the requirements for lawn sprinkling in inches of water per day.

During periods of high demands, precipitation becomes negligible and the equation takes the form of:

$$\overline{Q} = \overline{Q_d} + 0.6ca\overline{L_s}\,\overline{E_{pot}}$$

The domestic use in gallons per day per dwelling unit, Q_a / a, can be estimated from $Q_a / a = 157 + 3.46V$, where V is the average market value in $1000 per dwelling unit. Housing market values were correlated to the 1965 index.

Design curves that relate a 95 percent confidence interval were developed from Equation 11-1 and other data to show the relationship between numbers of dwelling units to peak system demands. Expected and design values for maximum daily and peak hourly demands are shown in Figures 11.5 and 11.6.

Peak flow requirements have also been expressed in terms of fixture units demands, in

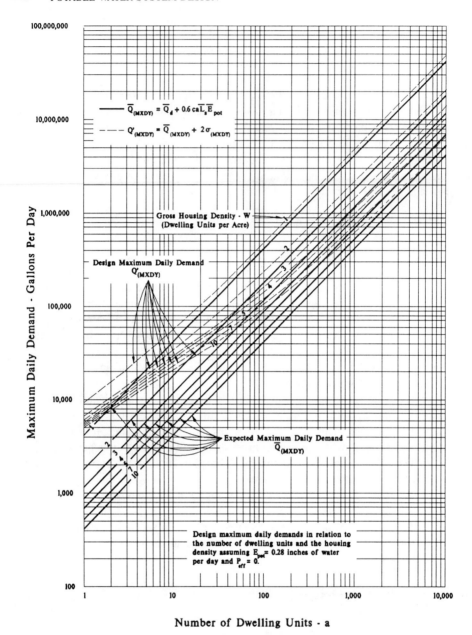

Figure 11.5   Maximum daily demands. Source: F. Pierce Linaweaver, Jr., John C. Geyer, and Jerome B. Wolff. *Report V on Phase II—Residential Water Use Research Project,* 1966. Reproduced with permission.

which a fixture unit represents a flow of 7.5 gallons per minute. For design using this criterion, the reader is referred to other sources (Konen and Chan, 1979).

11.2.2   Commercial and Industrial Consumption

Although water demands for commercial type residential rental units, such as apartments, can be included in domestic water estimates, other commercial establishments, such as

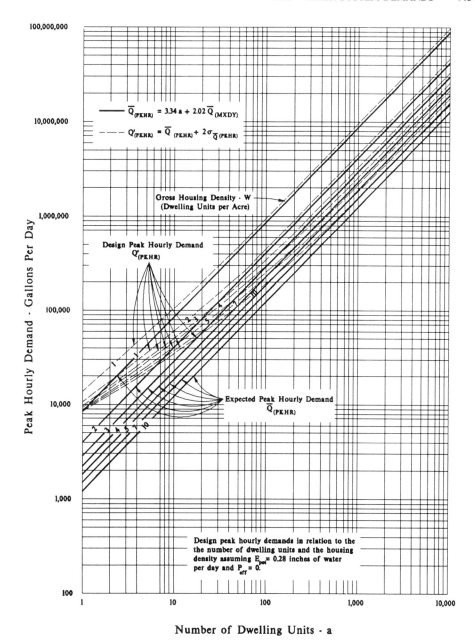

Figure 11.6 Design peak hourly demands. Source: F. Pierce Linaweaver, Jr., John C. Geyer, and Jerome B. Wolff. *Report V on Phase II—Residential Water Use Research Project*, 1966. Reproduced with permission.

offices, shopping facilities, and service stations, must be quantified. Some typical values are included in Figure 11.7.

Industrial water use varies based on makeup, process, and heating/cooling water needs. These demands vary greatly and can exceed other system demands; therefore, an accurate accounting of demand rates is essential.

Facility	Maximum Estimated Demands
Churches	0.4 Gallons per minute per member
Country Clubs	0.6 Gallons per minute per member
Beauty Shops	3.0 Gallons per minute per chair
Drug Store (No Fountain)	5.0 Gallons per minute
Launderette	8.0 Gallons per minute per unit
Motel	4.0 Gallons per minute per unit
Office Building	0.5 Gallons per minute per 100 s.f., or
	2.0 Gallons per minute per employee
Service Station	10.0 Gallons per minute per washrack

Figure 11.7 Nonresidential peak water demands. Source: Joseph S. Ameen. *Community Water Systems—Source Book,* 1971. Reproduced with permission.

11.2.3 Fire Protection Requirements

Today most potential buyers rely on the availability of fire protection when considering buying a dwelling in a development. Adequate water system design and performance are necessary to have fire fighting flows available of at least 500 gallons per minute with a minimum residual system pressure of not less than 20 pounds per square inch (AWWA, *M-31,* 1989).*

ISO Commercial Risk Services, Inc. (CRS) has prepared a schedule to rate public fire suppression facilities for insurance purposes on a basis from 1 to 10, with 1 being the best. This schedule, *Fire Suppression Rating Schedule—1980,* includes the evaluation of fire flow need, alarm handling, fire department equipment, personnel, training, and water supply. Although the schedule is "a fire insurance rating tool, and is not intended to analyze all aspects of a comprehensive public fire protection program" (CRS, 1980), it does provide broad guidelines that generally can be incorporated into the water system design process where the installed system will enhance fire suppression capabilities.

According to CRS the needed fire flow (NFF) for selected locations throughout the protected area in gallons per minute is the rate of flow required to suppress a specific fire under certain conditions. For one and two family dwellings without wood shingles and not exceeding two stories in height, the NFF is given in Figure 11.8 along with the minimum durations.

Needed Fire Flows For 1 And 2 Family Dwelling Not Exceeding 2 Stories In Height	
Distance Between Buildings	Needed Fire Flow
Over 100 feet	500 Gallons per minute
31 feet to 100 feet	750 Gallons per minute
11 feet to 30 feet	1,000 Gallons per minute
10 feet or less	1,500 Gallons per minute
Needed Fire Flow Duration	
Needed Fire Flow	Duration In Hours
NFF $\leq$ 2,500 Gallons per minute	2 Hours
3,000 GPM $\geq$ Needed Fire Flow $\leq$ 3,500 GPM	3 Hours

Figure 11.8 Needed fire flows (NFFs) for one and two family dwellings. Source: ISO Commercial Risk Services, Inc. *Fire Suppression Rating Schedule.* Copyright © 1980, ISO Commercial Risk Services, Inc. Reprinted with permission.

*Reprinted from AWWA Manual M-31 (1989) by permission.

For other types of residential units, or structures not fitting the above category, the NFF must be calculated from parameters addressing construction class, effective building area, occupancy, exposure, and communication factors associated with a building or within a fire district. In any case the NFF will not be less than 500 gallons per minute.

A public water system in a residential development that provides fire protection should be able to:

(a) Supply to the distribution system the NFF simultaneously with the maximum daily demand.

(b) Deliver throughout the distribution system the required NFF with a residual pressure of at least 20 pounds per square inch gauge (psig).

(c) Include three-way fire hydrants with standard fittings that are no more than 300 feet (closer in high value districts) away from any dwelling measured by hose length.

Fire suppression capabilities of a water system are site specific and therefore require an in-depth analysis by CRS. Questions pertaining to system adequacy can *only be resolved on a case by case basis by CRS, and the results of this analysis are governed solely by CRS's independent application and interpretation of current CRS produced documents.*

If the development includes total development activities, the design professional should incorporate adequate fire protection measures for all buildings. Various considerations of this nature are discussed by the National Fire Protection Association (NFPA) in *Fire Protection in Planned Building Groups—NFPA 1141.*

11.3 WAYS TO PROVIDE WATER TO A DEVELOPMENT

Potable water for the development should be obtained from an existing reliable municipal or public source, if available; otherwise, a well system must be installed or a suitable surface water source must be found. Wells or surface water sources also require treatment and storage facilities. Provisions must be made for the operation, maintenance, and control of these systems. Requirements are set forth in the Safe Drinking Water Act (PL 93-523) as amended, the National Primary Drinking Water Regulations (40 CFR 141), and the National Secondary Drinking Water Regulations (40 CFR 143) as amended. These requirements make the option of providing an extension to an existing municipal system very attractive.

Should the developer rely either on wells or surface water for supply, provisions must be made to obtain the right of use for that source in accordance with regulatory permit requirements and to secure legal water rights for the source's use. Reliance on these types of supplies requires the absence of pollutant sources such as sewage lines, septic tanks and drain fields, sewer mains, abandoned unsealed wells, wastewater treatment and storage works, feedlots for animals, chemical handling and storage facilities, petroleum storage facilities (including those used for herbicides, pesticides, and fertilizers), waste disposal sites (including hazardous and radioactive wastes), and mining operations.

11.3.1 Extension to an Existing Water System

If the project is located close enough for the developer to economically provide an extension to an existing system, permission must be obtained from that system's owner.

Because this extension becomes part of the existing system once installed and accepted, all proposed design methods, materials, and provisions must be approved. Often the existing water utility assumes the responsibility of providing an adequate supply of water and storage volume, although impact fees may be assessed for such purposes.

A field hydrant flow test must be conducted to inventory the existing system's capability. If the hydrant test is to check fire suppression flows, the test should be conducted during periods of ordinary system demands according to the National Fire Protection Association (NFPA, 1988).* The hydrant flow test requires a minimum of two hydrants located close to the point of connection for the new extension. The location should be free from traffic, pedestrians, and other interferences. One hydrant is designated as the residual hydrant where normal pressure is initially observed with all other hydrants being closed (and later pressure is measured at the residual hydrant when the other designated hydrants are opened for flow). The residual hydrant is generally located away from the major source of supply. Flowing hydrants should be located between the residual hydrant and the major supply point. Figure 11.9a through e shows several suggested test layouts for hydrant selection. The number of flowing hydrants utilized depends on the strength of the distribution system. The flow from one or several hydrants must produce a pressure drop of at least 25 percent in the residual hydrant for usable test results, or the observed rate of flow must be equal to the NFF (NFPA 1988).*

To generally describe the test procedure using Figure 11.9e as the example, initially the system's static pressure is observed at the residual hydrant with nothing flowing through hydrant 1. Next, hydrant 1 is opened for maximum flow. The rate of flow at hydrant 1 is determined using a Pitot tube to measure the pressure of the discharge stream, as illustrated in Figure 11.9f. According to NFPA 291* the corresponding flow can then be calculated using the Pitot reading by:

$$Q = 29.83Cd^2P^{0.5} \tag{11-3}$$

where:

Q = The discharge from the flowing hydrant in gallons per minute.
C = The coefficient of discharge for various outlet configurations shown in Figure 11.8g.
d = The diameter of the flowing outlet in inches.
P = The measured gauge pressure in pounds per square inch gauge (psig) from the Pitot.

The calculated flow is recorded at the time of day that the reading was obtained. Usually the Pitot reading is taken off the flowing hydrant's standard 2.5 inch outlet rather than a pumper outlet because better flow characteristics are obtained. While this peak flow continues, the system's dynamic (or residual) pressure is observed at the residual hydrant

---

*Reprinted with permission from NFPA 291, *Fire Flow Testing and Marking of Hydrants*. Copyright © 1988, National Fire Protection Association, Quincy, Mass. 02269. This reprinted material is not the complete and official position of the National Fire Protection Association on the referenced subject, which is represented only by the standard in its entirety.

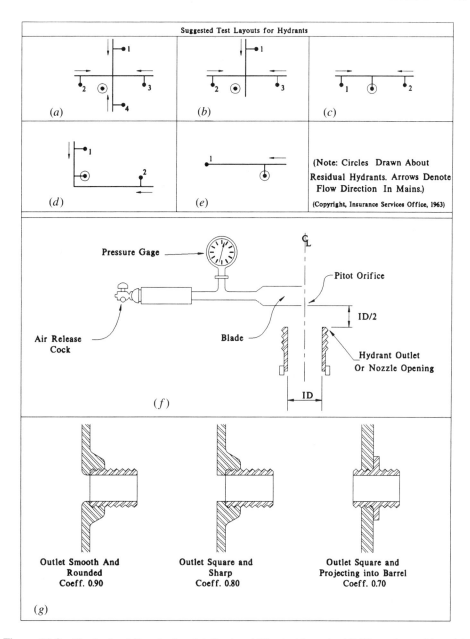

Figure 11.9 Fire hydrant flow testing details. (*a–e*) Flowtest layouts. (*f*) Pitot tube position. (*g*) Three general types of hydrant outlets and their coefficients of discharge. Reprinted with permission from NFPA 291, *Fire Flow Testing and Marking of Hydrants.* Copyright © 1988, National Fire Protection Association, Quincy, Mass. 02269. This reprinted material is not the complete and official position of the National Fire Protection Association on the referenced subject, which is represented only by the standard in its entirety. This figure includes copyrighted material of ISO Commercial Risk Services, Inc. with its permission. Copyright, ISO Commercial Risk Services Inc, 1963.

and recorded. With this field information, the corresponding residual pressure at any flow can be computed using Equation 11-4 (NFPA 291):*

$$Q_r = Q_f \frac{h_r^{0.54}}{h_f^{0.54}}$$ (11-4)

where:

Q_r = Flow available at the desired residual pressure.
Q_f = Maximum recorded flow at the flowing hydrant during the field test.
h_r = (Static pressure at no flow)—(desired residual pressure).
h_f = (Static pressure at no flow)—(residual pressure at maximum test flow).

Equation 11-4 is valuable for estimating available residual pressures at the connection point of an existing system when various design demands required by a potential development are delivered through the system.

Often a graph similar to Figure 11.10 is used instead of Equation 11-4. The static pressure is initially plotted where the flow equals zero. The maximum measured flow and corresponding observed residual pressure are then plotted. Residual pressures for any other flow can be obtained along the line that connects the plotted data.

Later the design professional can compute the headloss in the proposed water main extension from the proposed point of connection to the critical point (usually considered the location of the hydrant in the development that is most remote from the point of connection to the existing water system). If the difference between this calculated headloss in the proposed water main extension and the available system pressure when delivering the design flow is greater than the minimum residual pressure allowed (usually 20 pounds per square inch), the proposed system would be considered adequate. If this difference produces a residual pressure that is less than the allowable, the proposed system must be redesigned.

When conducting a flow test in the field, the following precautions must be followed:

(a) Hydrants must be slowly flushed prior to installing any gauges or Pitot tubes to prevent rust or particles within the hydrant from clogging the testing device.

(b) Hydrants must be opened and closed slowly to prevent waterhammer.

(c) Hydrant valve stems must be completely closed in order to open the hydrant's weep hole for interior hydrant drainage when tests are completed.

As an example, data collected from a field hydrant flow test are shown below under conditions similar to Figure 11.9e:

Observed Static Pressure:	61 pounds per square inch gauge
Observed Maximum Flow:	1256 gallons per minute corrected for outlet discharge loss C
Observed Dynamic Pressure:	25 pounds per square inch gauge

*Reprinted with permission from NFPA 291, *Fire Flow Testing and Marking of Hydrants.* Copyright © 1988, National Fire Protection Association, Quincy, Mass. 02269. This reprinted material is not the complete and official position of the National Fire Protection Association on the referenced subject, which is represented only by the standard in its entirety.

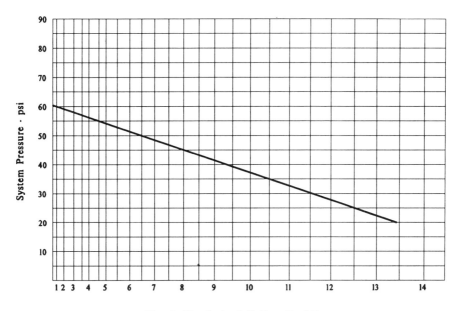

Figure 11.10 Pressure-flow graph for water supply analysis.

Consider an example of a proposed residential development of approximately 100 dwelling units with a gross density of three dwelling units per acre; it is to be served by an extension to the system tested above. The average daily demand based on 24 hours (assuming 150 gallons per person per day) would be (150 gallons per person) × (3.4 persons per unit) × (100 units) = 51,000 gallons flow per day. The corresponding maximum daily demand, using Figure 11.5, would be 170,000 gallons per day or 118 gallons per minute, while the peak hour demand would be estimated as 360,000 gallons per day (or 250 gallons per minute) from Figure 11.6.

If the NFF is assumed to be 750 gallons per minute in this example, an acceptable design flow to demonstrate system adequacy would be to provide at least 750 gallons per minute + 118 gallons per minute, or 868 gallons per minute (which is greater than the peak hour demand), at the most remote hydrant when testing after construction but before dwelling occupancy. The estimated residual pressure at the point of connection for that design flow can now be calculated using the flow test data and Equation 11-4 where:

$$868 = 1256 \, \frac{(61 - \text{residual})^{0.54}}{(61 - 25)^{0.54}}$$

estimated residual pressure = 42.8, or 43 pounds per square inch gauge

Using a minimum allowable system pressure of 20 pounds per square inch, the sum of all static (elevation differences) and frictional headlosses associated with the proposed extension and distribution network from the connection point to the most remote hydrant must not exceed 23 pounds per square inch. Otherwise the proposed system would be inadequate and would need to be redesigned by either increasing pipe sizes, adding a booster pump, or including an elevated tank to increase pressure.

11.3.2   Water Wells

A water well for use as a groundwater supply consists of the well itself, and the pump, fittings, and piping. The well is a hole in an aquifer that is bored, dug, punched, or drilled and that can be lined with a casing. Most deep wells are drilled and are the type most commonly associated with land development water supplies. During the drilling operation, a driller's log is maintained where all subsurface formations encountered are recorded, with the data including coloring, mineralogy, rock types, grain size distribution, and geophysical and mechanical log data. Openings through perforations or screens allow water to enter the well while providing lateral support to prevent the walls from collapsing. Water that enters is removed by pumping (unless it flows naturally from artesian pressure), and replacement water is then allowed to enter. Provided the rate of withdrawal does not exceed the rate of recharge, the well operates within what is classically known as its safe yield. As water enters the well, the surrounding soil resists the flow, resulting in headloss. This causes the water's surface to form a cone of depression, known as drawdown. Figure 11.11 illustrates this phenomena.

According to Hammer and MacKichan (1981), Darcy's law is generally used to define the velocity of water flowing through a granular soil by:

$$v = Ki \qquad (11-5)$$

where:

v = Velocity of moving water under laminar flow conditions in feet per second.
K = A coefficient having the same units as the velocity.
i = Hydraulic slope gradient in feet per foot.

More commonly, the equation takes the form of:

$$Q = AKi$$

where, in addition to the previously defined parameters:

Q = The flow rate in cubic feet per second.
A = The cross-sectional flow area in square feet.

Factors affecting the coefficient of permeability include soil parameters, water viscosity, and varying subsurface conditions. According to Fair et al. (1966), the U.S. Geological Survey has chosen as its "standard coefficient of permeability" the flow of water in gallons per day at 60°F through a cross-sectional area of one square foot under a unit gradient. If the coefficient of permeability is obtained at field temperatures other than 60°F, the "standard coefficient of permeability" can be determined with the following relationship (Cedergren, 1989):

$$K_{\text{at any temp.}} = \left[K_{60F} \right] \left[\frac{\mu_{60}}{\mu_{\text{at any temp.}}} \right] \qquad (11-7)$$

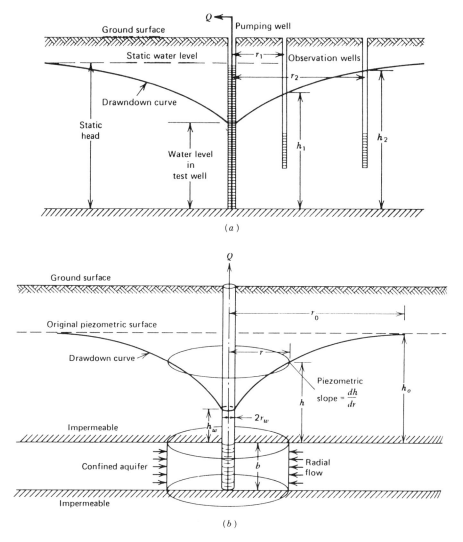

Figure 11.11 Aquifer pumping. (*a*) Flow of water toward a well during a pumping test in an unconfined coarse-grained aquifer. (*b*) Steady radial flow to a well completely penetrating an ideal confined aquifer. Source: Mark J. Hammer and K. A. MacKichan. *Hydrology and Quality of Water Resources.* Copyright © 1981 by John Wiley & Sons, Inc.

where:

$$K_{60F} = \text{Standard coefficient of permeability.}$$
$$K_{\text{at any temp.}} = \text{Coefficient of permeability obtained under field conditions.}$$
$$\mu_{60} = \text{Dynamic viscosity of water at 60°F}$$
$$\mu_{\text{field}} = \text{Dynamic viscosity of water at field temperature.}$$

According to Hammer and MacKichan (1981), for ideal unconfined flows (Figure 11.1*a*) the observed coefficient of permeability can be obtained after steady state pumping conditions have been attained by:

$$K = \frac{Q \, \log_e(r_2/r_1)}{3.14(h_2^2 - h_1^2)} \qquad (11\text{-}8)$$

where:

K = Coefficient of permeability in feet per second.
Q = Well's discharge in cubic feet per second.
r and h = Observed well data measured in feet, as indicated in Figure 11.11a.

For confined aquifer conditions (Figure 11.11b), the observed coefficient of permeability after steady state pumping conditions are attained is:

$$K = \frac{Q \, \log_e(r_2/r_1)}{2(3.14)(b)(h_2 - h_1)} \qquad (11\text{-}9)$$

where:

Q = Well's discharge in cubic feet per second.
K = Coefficient of permeability in feet per second.
b = Aquifer's thickness in feet.
r and h = Observed well data measured in feet as defined in Figure 11.11a.

Other factors must be considered if the orientation of the original hydraulic grade line is not horizontal. A more complete presentation on this topic can be found elsewhere (Cedergren, 1989; Fair et al, 1966; Hammer and MacKichan, 1981).

Due to a number of conditions that can cause well yields to vary (including encountering nonsteady state flows), the design professional should consult other reference sources, as well as local well drillers and regulatory agencies, to determine additional design information including well sizes. Generally the larger the diameter of the well, the more it can produce. Several well sizes relative to possible production rates are depicted in Figure 11.12 as proposed by Ameen (1974).

Well screens that allow water to enter are positioned at depths (extending through the water bearing strata) where good water is available. Information necessary to position screens comes from soil samples obtained during drilling and the logged information on the well. Precautions must be taken to protect these screens. They can clog if improperly sized or if the inflow velocity exceeds a maximum safe velocity of approximately one to two inches per minute. An experienced well driller is needed to properly develop the

Desired Daily Yield Of Well	Recommended Diameter Of Well
Flows up to 50,000 gallons per day	6-inch
50,000 to 100,000 gallons per day	8-inch
100,000 to 250,000 gallons per day	10-inch
250,000 to 500,000 gallons per day	12-inch
500,000 and above	Gravel packed wells

Figure 11.12 Typical well sizes for various yields. Source: Joseph S. Ameen. *Community Water Systems—Source Book,* 1971. Reproduced with permission.

well's production capability. A typical deep well with a vertical turbine pump installed is shown in Figure 11.13.

If several wells are to be installed, their locations should be dispersed to minimize the compounding effects of overlapping cones of depression resulting from drawdown. Wells also should be located where the possibility of surface water flooding, with its potential contamination, is remote and the intrusion of salt water is unlikely. Small diameter test wells are often drilled to provide initial subsurface information that can be used to design the permanent production well. The production well can be constructed by either reaming and developing the initial test well, or by drilling a new hole. If a new hole is made, the test well must be properly sealed prior to abandonment to prevent groundwater contamina-

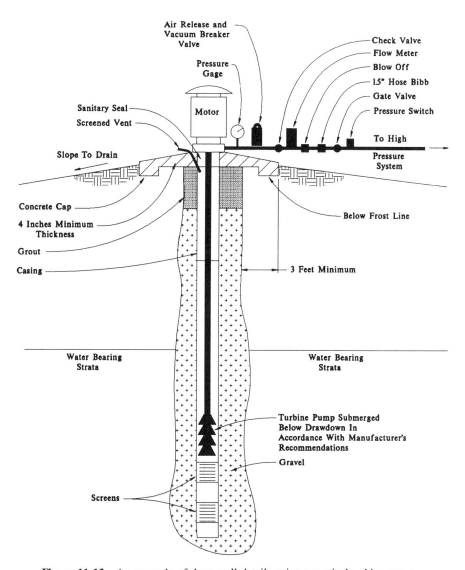

Figure 11.13 An example of deep well details using a vertical turbine pump.

tion. Once the production well has been developed, it can be tested by pumping to determine its yield and specific capacity. Upon completion, it can be disinfected.

To determine the pump size that will deliver the desired yield, field drawdown and permeability data must be utilized. For example, if a well is to withdraw 350 gallons per minute from an unconfined aquifer, an estimated 12 inch diameter well can be utilized, as shown in Figure 11.12. Once the test well or production well is installed and developed and steady state conditions exist, a field pumping test can be undertaken. The following sample drawdown data are obtained from such a test:

(1) Steady state pumping rate: 500 gallons per minute, or 1.114 cubic feet per second (0.002228×500 gallons per minute).

(2) $r_1 = 15$ feet; $h_1 = 415$ feet.

(3) $r_2 = 60$ feet; $h_2 = 440$ feet.

(4) Depth from ground surface to bottom of aquifer: 450 feet.

(5) Ground surface elevation: 50.0 feet (NGVD29).

(a) Determine Field permeability using Equation 11-8:

$$K = \frac{1.114 \text{ cubic feet per second} \times \log_e(60/15)}{3.14(440^2 - 415^2)} = 0.000023 \text{ feet per second}$$

(b) Determine the amount of drawdown (at one foot from the well) when the required 350 gallons per minute yield (or 0.78 cubic foot per second) is withdrawn, using Equation 11-8:

$$0.000023 = \frac{0.78 \text{ cubic foot per second} \times \log_e(60/1)}{3.14(440^2 - h_1^2)}$$

$$h_1 = 386.5 \text{ feet}$$

(c) Determine the elevation of drawdown when the yield is 350 gallons per minute:

Ground elevation:	+ 50 feet (NGVD29)
Depth to aquifer bottom:	−450 feet
Elevation of aquifer bottom:	−400 (NGVD29)
Height to drawdown at 350 gallons per minute:	+386.5 feet
Elevation of drawdown:	− 13.5 (NGVD29)

(d) Determine the TDH for the well pump. This example assumes the well pump will be submerged at least 35 feet and that the pump will discharge directly into a ground storage tank at the well site where the maximum height of water in the tank is estimated to be 24 feet.

The Hazen–Williams formula, Equation 10-10, is utilized to calculate the various frictional headlosses (per 1000 feet of pipe) as shown below. The 6 inch pipe column used to convey water to the ground surface is 98.5 feet with a Hazen–Williams C of 100 where:

$$H_{1\ 1000} = \left[\frac{350}{(0.006756)(100)(6^{2.63})} \right]^{1.85} = 17.2 \text{ feet}/1000$$

A summary of headlosses are:
Static system elevation head:
(50 NGVD29 + 24 foot tank − (−13.5) = 87.5 feet

Friction in 6 inch pipe column—98.5 feet long:	1.7 foot
Minor losses (assumed for this example):	0.5 feet
Total:	89.7 feet
Use:	90.00 feet

(e) Select possible vertical turbine pump sizes that can fit in the 12 inch well. Since various stages can be incorporated into the final pump configuration where each stage acts in series, the TDH can be incremented into various head segments that correspond to integer stage possibilities where the most efficient pump combination can be selected. To illustrate this point, the required 90 feet of head determined previously can be incremented into the following segments:

Number of Possible Stages	Corresponding Head/Stage
2	90/2 = 45 feet
3	90/3 = 30 feet
4	90/4 = 22.5 feet

The design professional can utilize various pump curves to select possible combinations of stages for various model pumps to obtain the most efficient system. Figure 11.14 illustrates one such typical set of curves where four stages, each having impellers of 5.469 inches and contributing 22.5 feet of head, can be satisfactorily used to pump the required 350 gallons per minute. The NPSH available must be at least 7 feet and the brake horsepower per stage is approximately 2.4 horsepower according to this figure.

Other factors relating to specific manufacturers' data must be evaluated including motor horsepower, thrust conditions, total efficiency based on possible corrections relating to the number of stages utilized, shaft sizing and stretch potential, and discharge size selection.

11.3.3 Surface Water Sources

Land developers might choose not to rely on surface water supply sources because substantial contributory drainage basins are required. If the development, however, does utilize this type of source, either a suitable surface water supply must be available with sufficient flow to meet system demands during dry periods, such as a river or lake, or an impoundment (or reservoir) must be constructed that has a storage capacity capable of accommodating withdrawal demands during dry periods based on sufficient inflows from streams, rainfall, and possibly melting ice and snow. The required storage volume of a reservoir is usually determined using a mass curve that shows the cumulative inflow; it must allow for loss of volume attributed to silting and the loss of water due to evaporation and percolation. It must also feature protection from flooding. The design professional

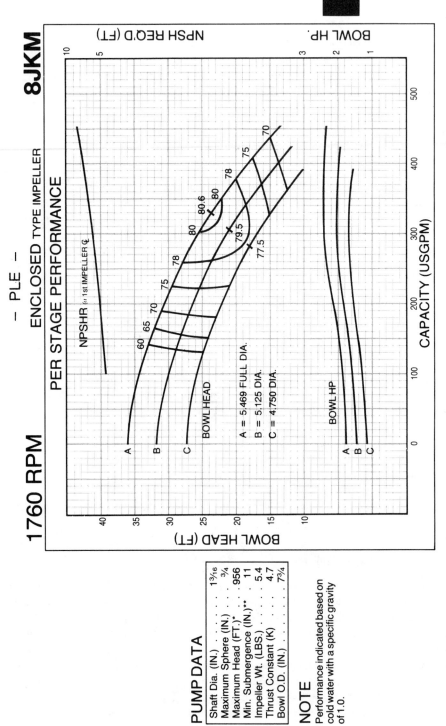

Figure 11.14 Examples of vertical turbine pump performance curves. Courtesy of Floway Pumps, Inc., Fresno, Calif. 93707.

456

must determine whether any wells exist within the area proposed for inundation to insure that they are sealed properly to prevent groundwater contamination.

According to Fair et al. (1971), the safe yield of a river or stream is its lowest dry weather flow, whereas with full development of storage, the safe yield approaches the mean annual flow. The attainable yield, which is of interest to the design professional when analyzing sources of supply, is affected by evaporation, bank storage, seepage, and silting. The reader is referred elsewhere for a more complete treatment of this subject (Fair et al., 1971).

11.4 TREATMENT CONSIDERATIONS

Potable water offered to the public for consumption in land developments must comply with the Safe Drinking Water Act of 1974 as amended (along with other federal and state regulations) and must be noncorrosive. Should the water meet all regulatory quality standards, which is possible with water produced from some wells, then the only treatment required is the addition of a disinfectant such as chlorine, chlorine dioxide, chloramines, or ozone. The production of trihalomethanes (THMs) must be minimized during the disinfection process especially when organic precursors are present. Raw water, however, can contain suspended and dissolved solids, dissolved gases, and organic compounds that make the water unusable without some type of treatment.

11.4.1 Suspended Solids

Organic or inorganic suspended solids are usually found in surface waters. They can be removed by physical or mechanical means using filtration or sedimentation. Organic suspended solids normally are products of plant or animal life, but may include synthetic organic compounds. These solids can decay or decompose or be burned. Primarily, organic solids consist of oxygen, hydrogen, and carbon in possible combination with sulfur, phosphorous, and nitrogen.

Inorganic suspended solids do not decay as they are inert and include mineral substances that produce hardness and mineral levels in the raw water.

Suspended solids may be of sufficient size and weight to allow settling with time; however, smaller suspended solids (usually less than one micron) are considered colloidal and normally will remain in suspension unless combined into flocs.

11.4.2 Dissolved Solids

Dissolved solids are present more often in groundwater than in surface water as a result of subsurface minerals being dissolved in the underground water. These solids are removed using chemical processes, such as oxidation and precipitation, or physical processes, such as reverse osmosis.

Iron and manganese often are found in groundwater and anaerobic surface waters as Fe^{2+} and Mn^{2+}. Upon exposure to oxygen, they are transformed into more stable forms (Fe^{3+}, Mn^{4+}). These ions yield unpleasant tastes and odors in water and can produce stains in laundry and porcelain plumbing fixtures.

Water hardness is caused primarily by the presence of calcium and magnesium cations in the water. Hard water produces plumbing scale and requires large quantities of soap for cleansing. These ions are associated with bicarbonate or sulfate anions; however, they may also be associated with chlorides, nitrates, and silicates. Carbonate hardness refers to the calcium and magnesium carbonates and bicarbonates present. Noncarbonate hardness refers to other calcium and magnesium compounds such as calcium or magnesium sulfates and chlorides. High fluoride concentrations are usually associated with groundwater supplies in several states. Two methods to lower fluoride concentrations are activated alumina absorption and reverse osmosis.

11.4.3   Dissolved Gases

Dissolved gasses can be present in raw water supplies. For example, nitrogen and oxygen are both present in surface waters. Also, carbon dioxide can be present in groundwater and will begin to ionize into carbonic acid if the water pH is above 4.5. Carbon dioxide, which can be present in groundwater, usually lowers the pH of the water, resulting in distribution system corrosion.

Another dissolved gas, hydrogen sulfide, produces a rotten egg smell that often is offensive to consumers. If the pH is low, air stripping may remove the odor; however, chemical oxidation may be required if the pH is high. Ammonia also may be found dissolved in some surface waters. This results in unpleasant odors and tastes and can be stripped by aeration.

11.4.4   Trace Organic Compounds

Trace organic compounds can be found in both surface water and groundwater. These compounds are usually man-made and can adversely affect consumer health. Volatile organic chemicals (VOC's) are essentially solvents, such as benzene and chloroform, that may have been used for industrial purposes and have escaped into the ground from underground storage tanks, leached from disposal sites, and been spilled. Other trace organic compounds include pesticides and herbicides that have been transported into surface waters by storm water runoff or into groundwater aquifers by percolation. Other contaminants may exist in various forms and from various sources, all of which may require removal.

11.4.5   Microbiological Contamination

Microbiological contamination may be encountered in surface waters and some groundwater sources; however, filtration and disinfection can eliminate most of these types of contamination.

11.4.6   Water Treatment Processes

Water treatment processes include aeration, chemical addition, mixing, coagulation, flocculation, sedimentation, filtration, adsorption, ion exchange, membrane separation, disinfection, and waste by-product handling. Most water treatment facilities utilize several of these processes to produce finished water. A typical treatment process is shown in Figure 11.15.

Fluorides, sodium, iron, manganese, magnesium, or calcium commonly are removed by the lime-soda ash process, reverse osmosis, or ion exchange. Chemical coagulation is

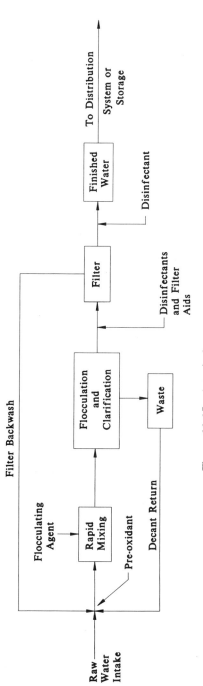

Figure 11.15 A typical water treatment process.

459

also used to form flocs of suspended colloidal particles that, once removed from the system, reduce the color, odor, taste, and turbidity of the water. Alum is probably the most common coagulant used, although many others are available. Once the flocs are formed, the water normally is clarified and filtered. The general effectiveness of water treatment processes for contaminant removal is shown in Figure 11.16.

Lack of space prohibits a more complete presentation of information pertaining to the purification of water for consumption as indicated in Figure 11.16. Other references on water treatment are available for the reader to consult (AWWA, 1990; Salvato, 1992; Weber, 1972).

11.5 TREATED WATER STORAGE FACILITIES

Treated water storage facilities hold water for emergency reserve, fire fighting, equalization for peak demands, and providing idle time for supply and treatment works. These facilities can be elevated tanks, standpipes, ground level high pressure tanks, or low pressure ground storage tanks. The most advantageous is the elevated tank because it is a ready source of water volume and pressure.

11.5.1 Elevated Water Tanks

Elevated water tanks are placed where the ground elevation is the highest in the surrounding service area and, if possible, near the center of areas of high demands or opposite the source of supply to help equalize pressures. Tanks can provide equalization storage ranging from approximately 15 to 35 percent of the maximum daily demand when constant pumping is utilized.

The engineer should obtain local information that reflects maximum daily demands and water use patterns for design purposes. If local information is not available, reasonable estimates should be made. In this example, Figure 11.5 is used to establish the maximum daily demand, and Figure 11.2 is used to allocate the demand throughout the maximum day.

To illustrate a method for sizing an elevated tank, consider an upscale development of 1000 persons with each person having an average demand of 150 gallons of water per day and a NFF requirement of 1000 gallons per minute for two hours. The required tank capacity can be determined by:

(a) *Required Fire Storage Reserve:*

1000 gallons per minute × 2 hours × 60 minutes per hour = 120,000 gallons

(b) *Required One Day Emergency Reserve:*

150 gallons per person per day × 1000 persons = 150,000 gallons

For this example two 75,000 ground storage tanks will be provided (each 23.5 feet in diameter × 24 feet tall), with one located at each of the two proposed well sites. This allows the one day emergency reserve to be turned over on a regular basis to keep it fresh.

Contaminant Categories	Aeration and Stripping	Coagulation processes, sedimentation, filtration	Lime softening	Ion Exchange Anion	Ion Exchange Cation	Reverse osmosis	Ultra filtration	Electrodialysis	Chemical oxidation, disinfection	GAC	PAC	Activated alumina
A. Primary Contaminants												
1. Microbial and Turbidity												
Total Coliforms	P	G-E	G-E	P	P	E	E	--	E	F	P	P-F
Giardia Lamblia	P	G-E	G-E	P	P	E	E	--	E	F	P	P-F
Viruses	P	G-E	G-E	P	P	E	E	--	E	F	P	P-F
Legionella	P	G-E	G-E	P	P	E	E	--	E	P	P	P-F
Turbidity	P	E	G	F	F	E	E	--	P	F	P	P-F
2. Inorganics												
Arsenic (+3)	P	F-G	F-G	G-E	P	F-G	--	F-G	P	F-G	P-F	G-E
Arsenic (+5)	P	G-E	G-E	G-E	P	G-E	--	G-E	P	F-G	P-F	E
Asbestos	P	G-E	--	--	--	--	--	--	P	--	--	--
Barium	P	P-F	G-E	P	E	E	--	G-E	P	P	P	P
Cadmium	P	G-E	E	P	E	E	--	E	P	P-F	P	P
Chromium (+3)	P	G-E	G-E	P	E	E	--	E	F	F-G	F	P
Chromium (+6)	P	P	P	E	P	G-E	--	G-E	P	F-G	F	P
Cyanide	P	--	--	--	--	G	--	G	E	--	--	--
Fluoride	P	F-G	P-F	P-F	P	E	--	E	P	G-E	P	E
Lead	P	E	E	P	F-G	E	--	E	P	F-G	P-F	P
Mercury (inorganic)	P	F-G	F-G	P	F-G	F-G	--	F-G	P	F-G	F	P
Nickel	P	F-G	E	P	E	E	--	E	P	F-G	P-F	P
Nitrate	P	P	P	G-E	P	G	--	G	P	P	P	P
Nitrite	F	P	P	G-E	P	G	--	G	G-E	P	P	P
Radium (226 and 228)	P	P-F	G-E	P	E	E	--	G-E	P	P-F	P	P-F
Selenium (+6)	P	P	P	G-E	P	E	--	E	P	P	P	G-E
Selenium (+4)	P	F-G	F	G-E	P	E	--	E	P	P	P	G-E
3. Organics												
VOCs	G-E	P	P-F	P	P	F-E	F-E	F-E	P-G	F-E	P-G	P
SOCs	P-F	P-G	P-F	P	P	F-E	F-E	F-E	P-G	F-E	P-E	P-G
Pesticides	P-F	P-G	P-F	P	P	F-E	F-E	F-E	P-G	G-E	G-E	P-G
THMs	G-E	P	P	P	P	F-G	F-G	F-G	P-G	F-E	P-F	P
THM precursors	P	F-G	P-F	F-G	--	G-E	F-E	G-E	F-G	F-E	P-F	P-F
B. Secondary contaminants												
Hardness	P	P	E	P	E	E	G-E	E	P	P	P	P
Iron	F-G	F-E	E	P	G-E	G-E	G	G-E	G-E	P	P	P
Manganese	P-F	F-E	E	P	G-E	G-E	G	G-E	F-E	P	P	P
Color	P	F-G	F-G	P-G	--	--	--	--	F-E	E	G-E	G
Taste and Odor	F-E	P-F	P-F	P-G	--	--	--	--	F-E	G-E	G-E	P-F
Total dissolved solids	P	P	P-F	P	P	G-E	P-F	G-E	P	P	P	P
Chloride	P	P	P	F-G	P	G-E	P	G-E	P	P	P	--
Copper	P	G	G-E	P	F-G	E	--	E	P-F	F-G	P	--
Sulfate	P	P	P	G-E	P	E	P	E	P	P	P	G-E
Zinc	P	F-G	G-E	P	G-E	E	--	E	P	--	--	--
TOC	F	P-F	G	--	G-E	G	G-E	P-G	G-E	F	F-G	--
Carbon Dioxide	G-E	P-F	E	P	P	P	P	P	P	P	P	P
Hydrogen sulfide	F-E	P	F-G	P	P	P	P	P	F-E	F-G	P	P
Methane	G-E	P-E	P	P	P	P	P	P	P	P	P	P
C. Proposed contaminants												
VOCs	G-E	P	P-F	P	P	F-E	F-E	F-E	P-G	F-E	P-G	P
SOCs	P-F	P-G	P-F	P	P	F-E	F-E	F-E	P-G	F-E	P-E	P-G
Disinfection by-products	--	P-E	P-F	P-F	--	P	F-G	F-G	F-G	F-E	P-G	--
Radon	G-E	P	P	P	P	P	P	P	P	E	P-F	P
Uranium	P	G-E	G-E	E	G-E	E	--	E	P	F	P-F	G-E
Aluminum	P	F	F-G	P	G-E	E	--	E	P	--	--	--
Silver	F-G	G-E	P	G	--	--	--	P	F-G	P-F	--	--

P-poor (0 to 20 percent removal); F-fair(20 to 60 percent removal); G-good (60-90 percent removal);
E-excellent (90-100 percent removal); "----" not applicable/insufficient data
Note: Costs and local conditions may alter a process applicability.

Figure 11.16 General effectiveness of water treatment processes for contaminant removal. Source: Frederick W. Pontius, Ed. *AWWA Water Quality and Treatment*, 4th ed. Copyright © 1990 by McGraw-Hill, Inc. Reprinted by permission of McGraw-Hill, Inc.

(c) *Required Equalization Volume Using a Graphical Approach:*

(1) For this example it is assumed that two wells will be at opposite sides of the development to provide two sources of water for 100 percent backup capability. If each well pumps 12 hours, providing effectively 24 hours of pumping, the required maximum daily demand will be provided as the pumping rate. The maximum daily demand can be determined using local data; however, if local data are lacking, the demand can be determined from Figure 11.5 unless the parameters used to construct Figure 11.5 are not germane. By entering Figure 11.5 with 300 dwelling units at a 3 dwelling unit per acre density (1000 persons/approximately 3.4 people per dwelling yields approximately 300 dwellings), about 500,000 gallons per day for the maximum daily demand is conservatively estimated. The required 24 hour pumping rate is calculated by:

$$\frac{500{,}000 \text{ gallons per day}}{24 \text{ hours} \times 60 \text{ minutes per hour}} = 347 \text{ gallons per minute}$$

which can be rounded to 350 gallons per minute.

(2) To obtain mass ordinates for constructing the diagram, a water use hydrograph based on local data must be obtained. For this example percentages of water use have been extrapolated from Figure 11.2 for the maximum day. Figure 11.17 contains the calculations.

(3) The calculated ordinates can now be used to construct the mass curve shown in Figure 11.18. Using the slope of the corresponding 24 hour pumping supply curve to

Maximum Day Demands				
Time Interval-%	% of Total	Volume, gal	Time of Day	Cumulative Volume, gal.
12-1 A.M.	0.61	3050	Midnight	0
1-2	0.61	3050	1	3050
2-3	0.12	600	2	6100
3-4	0.12	600	3	6700
4-5	0.12	600	4	7300
5-6	0.61	3050	5	7900
6-7	0.12	600	6	10950
7-8	1.70	8500	7	11550
8-9	2.31	11550	8	20050
9-10	3.52	17600	9	31600
10-11	3.95	19750	10	49200
11-12	4.55	22750	11	68950
12-1 P.M.	3.46	17300	12 Noon	91700
1-2	4.55	22750	1	109000
2-3	4.01	20050	2	131750
3-4	4.63	23150	3	151800
4-5	6.25	31250	4	174950
5-6	8.07	40350	5	206200
6-7	11.41	57050	6	246550
7-8	13.66	68300	7	303600
8-9	11.41	57050	8	371900
9-10	6.80	34000	9	428950
10-11	5.10	25500	10	462950
11-12 P.M.	2.31	11550	11	488450
		500000	Midnight	500000

Figure 11.17   Sample calculations for mass ordinates using maximum day demands.

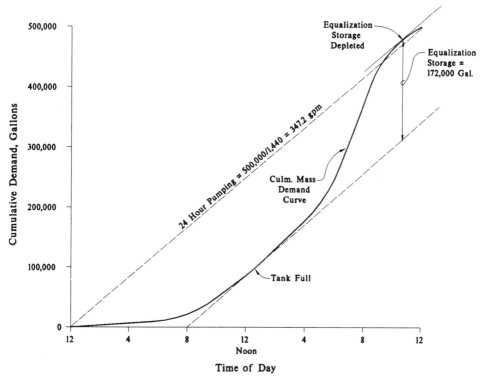

Figure 11.18   An example of a mass diagram for equalization storage based on maximum day demands.

construct tangents at the full tank and the depleted conditions, an equalization storage is determined graphically to be approximately 172,000 gallons.

(d) The total storage requirement is then:

Fire Reserve:	120,000 gallons
Emergency Reserve:	150,000 gallons
Equalization Storage:	<u>172,000 gallons</u>
Total:	442,000 gallons

The elevated tank should include fire reserve and equalization storage of 292,000 gallons, with the one day emergency reserve being placed in ground storage tanks in this example.

If the contribution of the well is included during fire flow, the required tank's capacity can be reduced by 350 gallons per minute × 2 hours × 60 minutes per hour, or 42,000 gallons. The net elevated tank size would then be 292,000 − 42,000 = 250,000 gallons. Therefore, a standard 250,000 gallon tank would be used in conjunction with two 75,000 gallon ground storage tanks.

The design professional must determine the nominal water system operating pressure in order to establish the elevation of the tank's bottom, E_{tank}. Using for this example 40 pounds per square inch as the normal system operating pressure and an estimated elevation of the highest fixture in the proposed system of 70 feet NGVD, the tank's bottom

must have an elevation of (40 pounds per square inch gauge design working pressure) ×
(2.31 feet of water per pound per square inch) + 70 feet = 162.4 or 162 feet NGVD 29. If
the ground elevation at the tank site is 50 feet NGVD 29, the height of the tank will be 112
feet. Figure 11.19 depicts the general features of an elevated tank.

The high pressure service pumps delivering finished water into the system from the two
75,000 ground storage tanks are sized in this example to deliver 350 gallons per minute at
a head equal to the system head from the pump to the elevated tank at high water (194
feet).

The proposed supply and storage system is shown in Figure 11.20.

11.5.2   Water Storage Standpipes

A standpipe for water storage is only a cylindrical water reservoir that has a height greater
than its diameter. A standpipe is advantageous if the site's terrain includes an area where it
can be located above the surrounding service area to provide adequate system working
pressures.

The use of a standpipe can be considered instead of the elevated tank illustrated in
Section 11.5.1 if the proposed tank site's elevation is 130 feet NGVD 29. A tank 32 feet,
10 inches in diameter × 64 feet tall, containing 400,000 gallons of storage (sum of all
storage requirements equaling 442,000 gallons minus pump contributions during fire flow
of 42,000 gallons), could be used, as shown in Figure 11.21.

If the highest plumbing elevation in the service area is the same 70 feet NGVD 29, the
standpipe will approximately maintain the minimum system operating pressure of 40
pounds per square inch gauge using: (165.1 low equalization level - 70.00 feet of highest
fixtures) / 2.31 feet of water per pound per square inch). A high capacity pump would
have to be installed to increase the system pressure for flows below the equalization
volume if the pressure of 40 pounds per square foot is to be maintained for the 1000
gallons per minute fire flow.

11.5.3   High Pressure Service Tanks

For small land developments without fire protection, high pressure service tanks at ground
level can be incorporated into the system to meet demands. Various approaches are used to
design high pressure service tanks. Some tanks (hydro-pneumatic) are designed to "float"
on the system (Rothrock, 1974) and therefore add little storage capacity to the system.
The value of this type of "floating" tank is that it allows the intermittent use of constant
speed electric pump motors and simple controls to respond to various system demands. A
typical "floating" tank is shown in Figure 11.22.

This system requires the selected pump to be capable of delivering approximately 125
percent of the peak system demand at the high operating pressure. Use of this type of
system requires high capacity supply wells or the incorporation of low pressure storage
facilities to provide adequate standby reserve for pumping demands.

Some regulatory agencies require high pressure service tanks to have adequate storage
to accommodate peak system demands for 20 to 30 minutes when incorporated into small
domestic water systems. According to Ameen (1974), each tank should be capable of
storing the maximum flow demand for a 20 minute duration while allowing a reduction
in storage volume for any contributions made by other sources (flows from low pressure
storage facilities) during a 20 minute period. This volume based on flow demands occu-

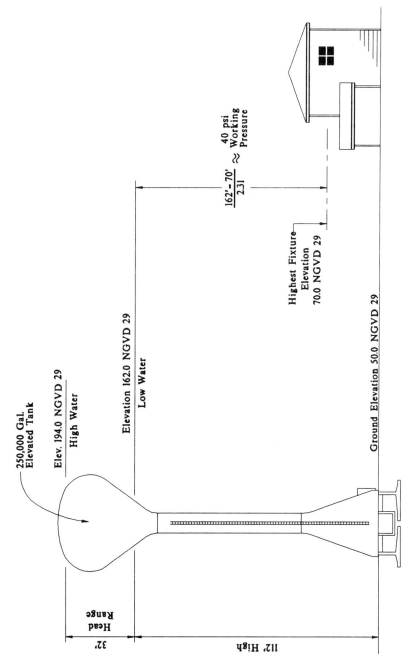

Figure 11.19 Typical 250,000 gallon elevated storage tank. Courtesy of CBI Na-Con (Chicago Bridge and Iron Company), Norcross, Ga.

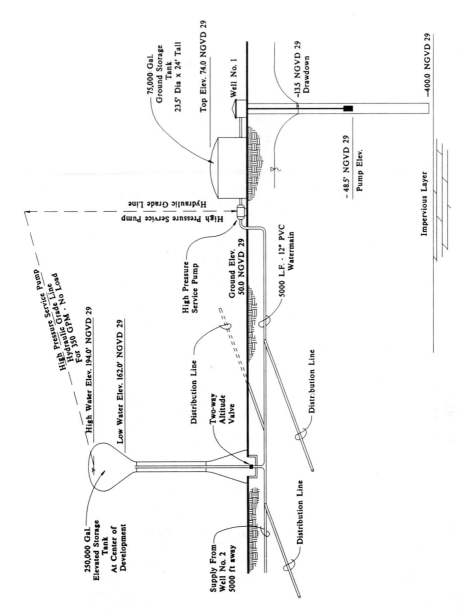

Figure 11.20 An example of a water supply system feeder configuration. Tank courtesy of CBI Na-Con (Chicago Bridge and Iron Company), Norcross, Ga.

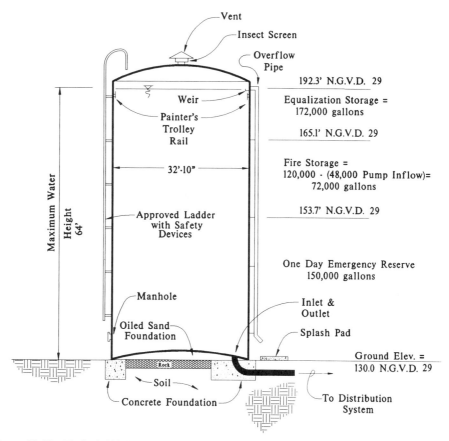

Figure 11.21 Typical 400,000 gallon potable water standpipe. Tank courtesy of CBI Na-Con (Chicago Bridge and Iron Company), Norcross, Ga.

pies approximately 25 percent of the total storage volume since adequate pressurized air must be included. In any case this storage should not be less than 20 times the pump's yield, with the pump's capacity being between 30 and 40 percent of the maximum flow demand (Ameen, 1974).

It is important to prevent air from escaping the high pressure storage tank into the distribution system at low water level and to minimize vortex action. An adequate dead water storage level must be maintained and can be determined using Figure 11.23.

A typical high pressure storage facility is shown in Figure 11.24.

11.5.4 Low Pressure Storage

When the rate of water being supplied to the system is low, or when a large system demand is imposed for short periods of time, such as that generated by lawn sprinkling or schools, a low pressure storage facility (at atmospheric pressure) can be provided close to the supply point. This facility can be designed to accommodate some equalization flows along with emergency reserve capacity to allow for maintenance downtime. This use has been illustrated in Section 11.5.1 in conjunction with elevated tanks.

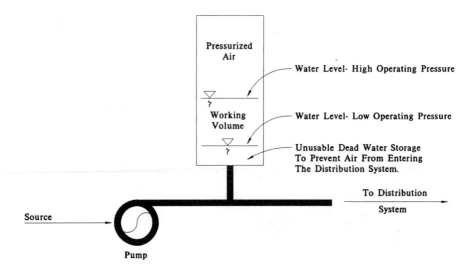

Figure 11.22 Floating hydro-pneumatic high pressure tank. Source: Donald F. Rothrock. Red Coat Reports, 1074 East Pueblo Road, Phoenix, Ariz. Copyright © 1974 by Donald F. Rothrock.

Low pressure storage facilities also can be used in conjunction with high pressure storage facilities that are equipped with high service pumps capable of pumping at 30 to 40 percent of the peak demand to deliver the reserve directly into a high service storage tank (Ameen, 1974). These pumps are usually chosen with run cycle times that extend from 15 minutes to 1 hour.

11.6 WATER DISTRIBUTION SYSTEMS

Water distribution systems consist of water lines and fittings, valves, service lines, meters, thrust blocks or restraining devices, and fire hydrants. The most desirable system provides adequate water service to all areas while minimizing unnecessary pipe length

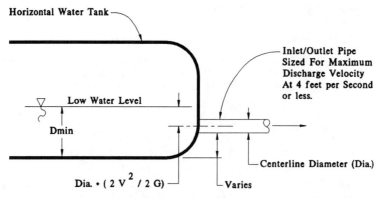

Figure 11.23 Dead storage determination for a high pressure water tank. Source: Donald F. Rothrock. Red Coat Reports, 1074 East Pueblo Road, Phoenix, Ariz. Copyright © 1974 by Donald F. Rothrock.

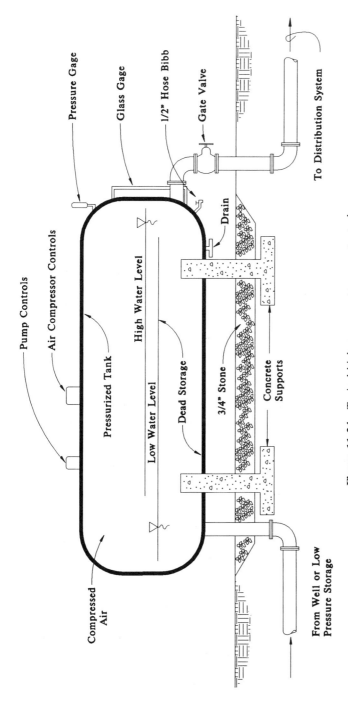

Figure 11.24 Typical high pressure storage water tank.

469

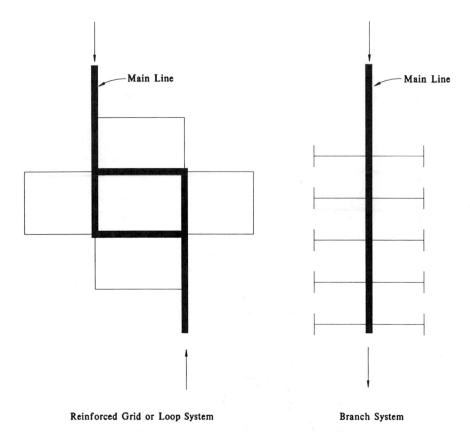

Reinforced Grid or Loop System   Branch System

Figure 11.25   Typical water distribution system configurations.

consistent with the incorporation of adequate redundant loops to reinforce the system. The most desirable network is a grid with a loop feeder to allow flexible system operation. Should a line rupture occur in a looped system, the whole system does not have to be secured since isolation valves and looped connections allow the system to continue operation outside the isolated problem area. Reinforced (or looped) systems also provide flow in more than one direction, resulting in pressures being more uniformly maintained. Branched systems do not exhibit these qualities and should be avoided if possible. Figure 11.25 illustrates these systems.

Gate valves are usually placed between the water main and fire hydrants and at water main intersections. AWWA recommends placing four isolation gate valves at crossings (where a four way intersection of water mains occurs) and three at tee type intersections. Other valves can be included along water mains, usually not more than 800 feet apart on long branches and 500 feet apart along mains in high value districts (AWWA, M-31, 1989).*

The installed distribution system must be adequately protected from freezing during cold weather by using burial depths greater than the frost penetration depths experienced in the area. Figure 11.26 indicates the minimum recommended pipe burial depths where flows are low such as in a private service main. Municipal water mains, however, can be buried at depths that are approximately six inches less than those shown.

*Reprinted from AWWA Manual M-31, by permission.

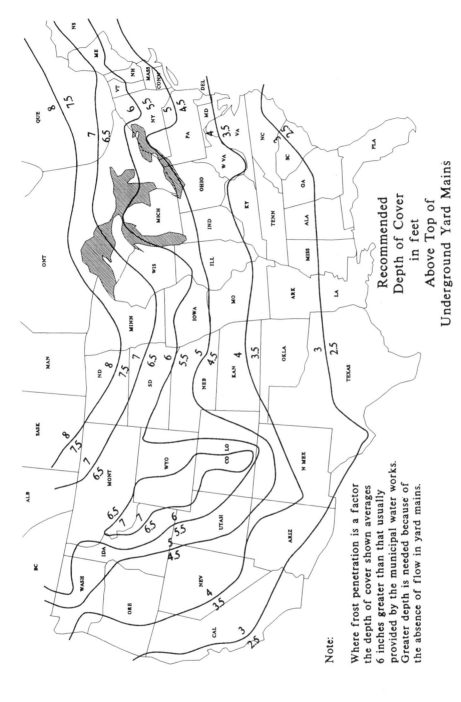

Note:

Where frost penetration is a factor the depth of cover shown averages 6 inches greater than that usually provided by the municipal water works. Greater depth is needed because of the absence of flow in yard mains.

Recommended
Depth of Cover
in feet
Above Top of
Underground Yard Mains

Figure 11.26 Water main burial depths. Reprinted with permission from NFPA 24, *Installation of Private Fire Service Mains*. Copyright © 1992, National Fire Protection Association, Quincy, Mass. 02269. This reprinted material is not the complete and official position of the National Fire Protection Association on the referenced subject, which is represented only by the standard in its entirety.

471

Other concerns associated with water main locations include:

(a) That water mains have at least 36 inches of cover.

(b) That potable water mains are horizontally separated from sanitary sewer mains to avoid possible contamination of the water system in case of line rupture.

(c) That proper vertical separation is maintained where water mains cross sanitary sewers.

These concerns are addressed in part 3.04 of the Project Manual's Section 02600, located in Chapter 13 (Section 13.3).

11.6.1   Water Distribution System Design Criteria

AWWA recommends that water distribution systems with fire suppression capability include the following (AWWA 1989):*

Minimum line size in the network:	6 inches
Smallest branching pipe for dead ends:	8 inches
Largest spacing of 6 inch grid:	600 feet
Smallest pipes in high value district:	8 inches
Smallest pipes on principal streets in central district:	12 inches
Largest spacing of supply mains or feeders:	3000 feet

To properly design a water distribution system, the design professional must have:

(a) The peak system demands.

(b) Fire flow requirements if applicable.

(c) The nominal water system operating pressure.

(d) The flow capacity of various pipe sizes.

(e) A scale drawing (preliminary development plan) of the proposed development.

Information about peak system demands should be used from local sources if available. If local data are not available, the peak hourly demands shown in Figure 11.6 can be utilized, unless the criteria used to construct the figure are not applicable. Regardless of the values obtained from Figure 11.6 Ameen (1974) recommends 15 gallons per minute for one dwelling without reference to density; thus design demands should be at least:

Number of Dwelling Units	Design Peak Hour Demand (gallons per minute)
1	15
2	16
3	17
4	18
5	19
6	20

*Reprinted from AWWA, Manual *M-31*, by permission.

A table containing the peak hourly demands is shown in Figure 11.27 for developments having a gross density of 3 dwelling units per acre and containing from 1 to 300 dwellings. The design peak hour values for units numbering four and under reflect minimum values given above. Tables can be constructed for other dwelling unit densities. These tabular peak hourly flows coupled with the additional NFF provide a basis for designing water mains in a small residential system.

Before designing the system, the engineer must also determine the nominal water system operating pressure, as illustrated in Figure 11.19. For the following example, a 40 pound per square inch working pressure is utilized.

Another table can be constructed using a spreadsheet for various pipe flows that can also aid in the design process. Figure 11.28 utilizes Equation 10-9 with a Hazen–Williams C of 130 as a design parameter, along with a working pressure of 40 pounds per square inch. A minimum residual pressure of 20 pounds per square inch is also used. Actual interior pipe diameters for various pipe lengths can be used to calculate flow delivery rates in gallons per minute when a corresponding 20 pounds per square inch pressure drop (40 pounds per square inch - 20 pounds per square inch) is experienced across each tabulated pipe length. Other tables can be constructed to represent different design parameters.

11.6.2 Water Distribution Design

Having the prerequisite information available, the designer can correlate peak hourly demands generated by each building site to the actual project layout. This is accomplished by placing a transparent paper overlay on top of the preliminary plan where the water mains are sketched, along with service lines to each building site. Often dual water services are used in single family detached developments because they are more economical to install, operate, and maintain than single services even though the larger 1 inch service line is provided instead of the normal $\frac{3}{4}$ inch diameter single service size. Figure 11.29 shows typical dual services for water and sewer.

To illustrate the design of a small residential development using single service taps (without fire protection), a method similar to that proposed by Ameen (1974) is utilized. This example uses the lot configuration shown in Figure 11.30a. The designer must sum the various cumulative maximum hourly demands using information contained in Figure 11.27 for each lot, beginning at the most remote point from the point of supply. Summation continues to the point of supply, with Figure 11.30b illustrating this process. If a nominal water system pressure of 40 pounds per square inch is chosen, information in Figure 11.28 can be utilized to match a pipe that can deliver the required maximum demand to all points along the line. The process begins with the smallest pipe diameter being initiated at the point of supply. By comparing demands with available capacities at specific distances from the supply point, the beginning point of the pipe is systematically moved away from the supply point until all pipe capacities are capable of meeting the demands at each location. The next larger pipe diameter that is available is utilized to repeat the process for servicing the distance between the point of supply and the point where the initial pipe was found to be satisfactory. This larger pipe must also be capable of delivering a capacity equal to or exceeding the capacity of the previous pipe, as illustrated by the 3 inch main supplying 298 gallons per minute at a point 200 feet from the supply point, which exceeds the 149 gallons per minute capacity of the 2 inch main at 300 feet. (Refer to Figures 11.28 and 11.30.) This process continues until the entire line is designed.

Unit	G.P.M.	Unit	G.P.M.	Unit	G.P.M.	Unit	G.P.M.	Unit	G.P.M.	Unit	G.P.M.
1	15	51	132	101	242	151	351	201	458	251	565
2	16	52	134	102	245	152	353	202	460	252	567
3	17	53	136	103	247	153	355	203	462	253	569
4	18	54	138	104	249	154	357	204	465	254	571
5	20	55	141	105	251	155	359	205	467	255	573
6	23	56	143	106	253	156	362	206	469	256	576
7	26	57	145	107	256	157	364	207	471	257	578
8	29	58	147	108	258	158	366	208	473	258	580
9	31	59	150	109	260	159	368	209	475	259	582
10	34	60	152	110	262	160	370	210	477	260	584
11	37	61	154	111	264	161	372	211	480	261	586
12	39	62	156	112	266	162	375	212	482	262	588
13	42	63	159	113	269	163	377	213	484	263	591
14	45	64	161	114	271	164	379	214	486	264	593
15	47	65	163	115	273	165	381	215	488	265	595
16	50	66	165	116	275	166	383	216	490	266	597
17	52	67	168	117	277	167	385	217	492	267	599
18	55	68	170	118	279	168	387	218	495	268	601
19	57	69	172	119	282	169	390	219	497	269	603
20	59	70	174	120	284	170	392	220	499	270	605
21	62	71	176	121	286	171	394	221	501	271	608
22	64	72	179	122	288	172	396	222	503	272	610
23	67	73	181	123	290	173	398	223	505	273	612
24	69	74	183	124	292	174	400	224	507	274	614
25	72	75	185	125	295	175	402	225	509	275	616
26	74	76	188	126	297	176	405	226	512	276	618
27	76	77	190	127	299	177	407	227	514	277	620
28	79	78	192	128	301	178	409	228	516	278	622
29	81	79	194	129	303	179	411	229	518	279	625
30	83	80	196	130	305	180	413	230	520	280	627
31	86	81	199	131	308	181	415	231	522	281	629
32	88	82	201	132	310	182	417	232	524	282	631
33	90	83	203	133	312	183	420	233	527	283	633
34	93	84	205	134	314	184	422	234	529	284	635
35	95	85	207	135	316	185	424	235	531	285	637
36	97	86	210	136	318	186	426	236	533	286	639
37	100	87	212	137	321	187	428	237	535	287	642
38	102	88	214	138	323	188	430	238	537	288	644
39	104	89	216	139	325	189	432	239	539	289	646
40	107	90	218	140	327	190	435	240	541	290	648
41	109	91	221	141	329	191	437	241	544	291	650
42	111	92	223	142	331	192	439	242	546	292	652
43	113	93	225	143	334	193	441	243	548	293	654
44	116	94	227	144	336	194	443	244	550	294	656
45	118	95	229	145	338	195	445	245	552	295	659
46	120	96	232	146	340	196	447	246	554	296	661
47	123	97	234	147	342	197	450	247	556	297	663
48	125	98	236	148	344	198	452	248	559	298	665
49	127	99	238	149	347	199	454	249	561	299	667
50	129	100	240	150	349	200	456	250	563	300	669

Figure 11.27   Peak hourly demands in gallons per minute—three dwelling units per acre density.

Hazen Williams C	130						
Design Pres., psig=	40						
Residual Pres, psig	20						
Pipe Size,inches=	2	3	4	6	8	10	12
Length, ft.							
100	149	434	925	2686	5724	10293	16626
200	103	298	636	1847	3936	7079	11435
300	83	240	511	1484	3162	5687	9186
400	71	205	437	1270	2707	4869	7865
500	63	182	388	1126	2400	4316	6972
600	57	165	351	1021	2175	3911	6318
700	52	152	323	939	2001	3599	5813
800	49	141	301	874	1862	3349	5409
900	46	132	282	820	1747	3142	5076
1000	43	125	267	775	1651	2969	4795
1100	41	119	253	736	1568	2820	4554
1200	39	113	242	702	1496	2690	4345
1300	37	109	231	672	1433	2576	4162
1400	36	104	222	646	1376	2475	3998
1500	35	101	214	622	1326	2385	3852
1600	33	97	207	601	1281	2303	3720
1700	32	94	200	582	1239	2229	3600
1800	31	91	194	564	1202	2161	3491
1900	30	88	189	548	1167	2099	3390
2000	30	86	183	533	1135	2042	3298
2100	29	84	179	519	1106	1989	3212
2200	28	82	174	506	1078	1939	3132
2300	27	80	170	494	1053	1893	3058
2400	27	78	166	483	1029	1850	2989
2500	26	76	163	472	1006	1810	2923
2600	26	75	159	462	985	1772	2862
2800	25	72	153	444	947	1702	2750
3000	24	69	147	428	912	1640	2649
3200	23	67	142	413	881	1584	2559
3400	22	65	138	400	852	1533	2476
3600	22	63	134	388	827	1486	2401
3800	21	61	130	377	803	1444	2332
4000	20	59	126	366	781	1404	2268
4200	20	58	123	357	761	1368	2209
4400	19	56	120	348	742	1334	2154
4600	19	55	117	340	724	1302	2103
4800	18	54	114	332	708	1273	2055
5000	18	52	112	325	692	1245	2011
5500	17	50	106	309	657	1182	1910
6000	16	48	101	294	627	1128	1822
6500	16	46	97	282	601	1080	1745
7000	15	44	93	271	577	1038	1677
7500	15	42	90	261	556	1000	1615
8000	14	41	87	252	537	966	1560
8500	14	39	84	244	520	935	1510
9000	13	38	81	236	504	906	1464
9500	13	37	79	230	489	880	1422
10000	12	36	77	223	476	856	1383

Figure 11.28 Pipe capacities in gallons per minute for various lengths for a 40 pound per square inch system operating pressure.

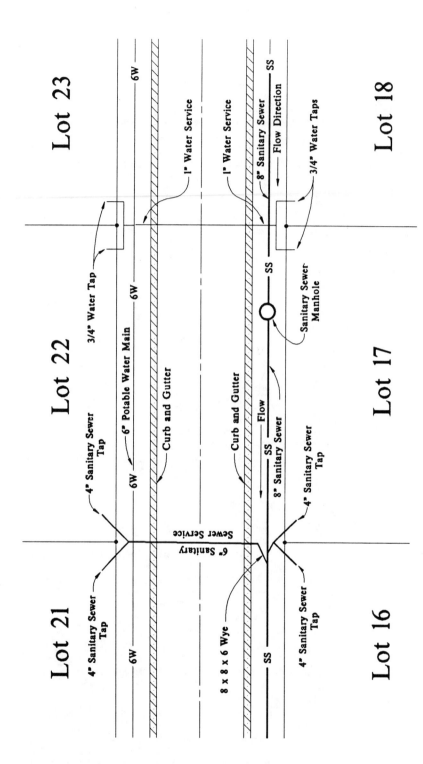

Figure 11.29 Typical double utility services in a single family residential development.

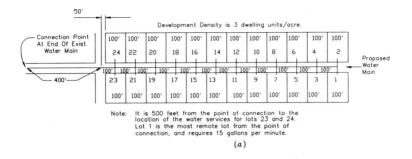

(a)

100	200	300	400	500	600	700	800	900	1000	1100	1200	1300	1400	1500	1600	Dist. From Connection, ft.
69	69	69	69	69	64	59	55	50	45	39	34	29	23	18	16	Cumulative Peak Hour Demand From Figure 11.27

(b)

100	200	300	400	500	600	700	800	900	1000	1100	1200	1300	1400	1500	1600	Dist. From Connection, ft.
149	103	83	71	63	57	52	49	46	43	41	39	37	36	35	33	2 inch line capacity from Figure 11.28.

∗

The capacity of the 2 inch watermain starting 500 feet from the connection is inadequate.

(c)

100	200	300	400	500	600	700	800	900	1000	1100	1200	1300	1400	1500	1600	Dist. From Connection, ft.
	149	103	83	71	63	57	52	49	46	43	41	39	37	36	35	2 inch line capacity from Figure 11.28.

∗

∗The capacity of a 2 inch watermain 600 feet from the connection point is inadequate because it can only deliver 63 gallons per minute when 64 are required.

(d)

100	200	300	400	500	600	700	800	900	1000	1100	1200	1300	1400	1500	1600	Dist. From Connection, ft.
		149	103	83	71	63	57	52	49	46	43	41	39	37	36	2 inch line capacity from Figure 11.28.

The capacity of a 2 inch watermain is satisfactory starting 200 feet from the connection point.

(e)

100	200	
		------------> A 2 inch main is adequate.
434	298	3 inch watermain capacity.

The 3 inch watermain is satisfactory from the connection point for a distance of 200 feet.

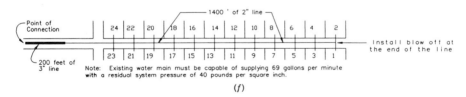

(f)

Figure 11.30 An example of the design procedure for a simple water distribution system. (a) Simple development layout for water system design. (b) Cumulative system demands determined in gallons per minute. (c) Using a working pressure of 40 pounds per square inch to design the water mains, the capacity of a two inch water main, starting at the connection point, is as shown. (d) The two inch water main will be analyzed using its beginning 100 feet from the connection point. (e) The two inch water main will be analyzed using its beginning 200 feet from the connection point. (f) A three inch line will be evaluated from the connection point to the beginning of the two inch lines. (g) The final system design would be as shown.

The design of a more typical residential water system is illustrated in Figure 11.31, where fire protection is included. Because this development will be serviced by extending an existing water system, local authorities are requiring that the completed system provide flows equal to the NFF plus the maximum daily demand at the remotest hydrant with a residual pressure of at least 20 pounds per square inch. The engineer may choose to design the proposed system using the peak hourly demands plus the NFF. The performance of the designed system is then analyzed by checking the static and frictional headlosses with a flow equal to the NFF plus the maximum daily demand at the critical hydrant using Hardy Cross (1936) methods, equivalent pipes, or digital computer analysis.

The design peak hourly demands and NFF at various points in the system are determined as follows:

(a) The design peak hourly demand at point B is the peak hourly demand for 6 lots, which is 23 gallons per minute.

(b) The design peak hourly demand and NFF at point C is the peak hourly demand for 16 lots plus 750 gallons per minute or 50 + 750, or 800 gallons per minute. It should be noted that this total demand can be supplied to point C with flows coming through either point D or E, or both. Therefore, line C–D–F must be sized to handle this flow as well as C–E–F.

(c) At point D the peak hourly demand and NFF for 58 lots is 147 gallons per minute plus 750 gallons per minute, or 897 gallons per minute.

(d) At point E the peak hourly demand and NFF for 58 lots is 147 gallons per minute plus the NFF, or 897 gallons per minute.

(d) At point F the peak hourly demand for 100 lots (peak hourly demand of 240 gallons per minute) and the NFF is 990 gallons per minute.

Figure 11.32 contains a method to design the various water main segments.

11.6.3   Distribution System Analysis

For the aggregate 100 lots, the critical flow (Section 11.2.3) of 868 gallons per minute would have to be analyzed (750 gallons per minute NFF plus 118 gallons per minute maximum daily demand from Figure 11.5). If the available pressure at point G in Figure 11.31 is 43 pounds per square inch gauge when delivering 868 gallons per minute (using data from the example in Section 11.3.1), the amount of allowable system loss from point G to B would be (43 pounds per square inch gauge - 20 pounds per square inch), or 23 pounds per square inch, assuming that the ground elevation was relatively flat. Although analysis methods proposed by Hardy Cross (1936) or computer software can be utilized to analyze this system, the equivalent pipe approach is presented here. All pipes from B to G will be converted to an equivalent eight inch single pipe using the following steps:

(a) Segment F–G is already in terms of an 8 inch pipe so the actual 1200 linear feet will be used for this segment.

(b) Segment C–F is comprised of two parallel 8 inch mains of 2000 feet each, which must be replaced with a single 8 inch main. For parallel line replacement, a headloss is assumed between the terminal ends of the parallel pipes. In this example five feet will be

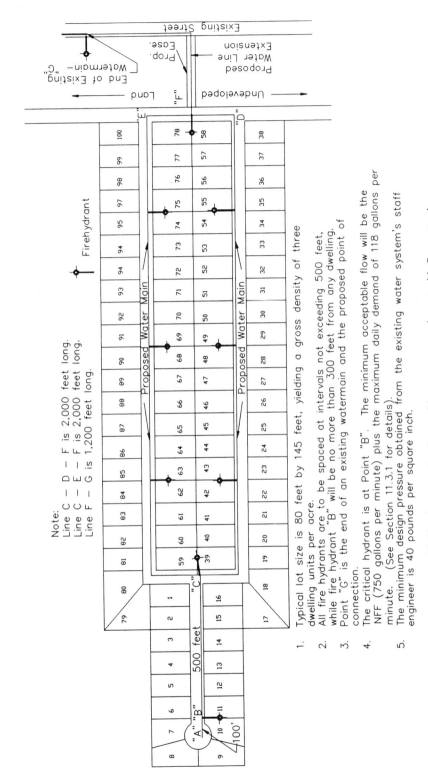

Note:
Line C – D – F is 2,000 feet long.
Line C – E – F is 2,000 feet long.
Line F – G is 1,200 feet long.

1. Typical lot size is 80 feet by 145 feet, yielding a gross density of three dwelling units per acre.
2. All fire hydrants are to be spaced at intervals not exceeding 500 feet, while fire hydrant "B" will be no more than 300 feet from any dwelling.
3. Point "G" is the end of an existing watermain and the proposed point of connection.
4. The critical hydrant is at Point "B". The minimum acceptable flow will be the NFF (750 gallons per minute) plus the maximum daily demand of 118 gallons per minute. (See Section 11.3.1 for details).
5. The minimum design pressure obtained from the existing water system's staff engineer is 40 pounds per square inch.

Figure 11.31 An example of a potable water system layout with fire protection.

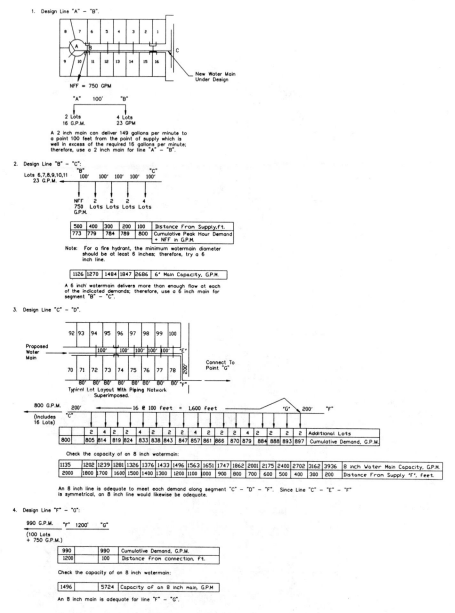

Figure 11.32 Design of a typical water distribution system.

utilized. This amounts to 5 feet/2000 feet of 8 inch line, which is 2.5 feet/1000 feet for both pipes. Using Equation 10-9, each pipe can deliver:

$$Q = 0.006756 \times 130 \times 8^{2.63} \times 2.5^{0.54}$$

or 341.7 gallons per minute at the assumed headloss of 5 feet. The corresponding total flow of both pipes is 683.4 gallons per minute.

A single 8 inch pipe will have a corresponding headloss per 1000 feet of pipe for the 683.4 gallons per minute flow of:

$$H_{1\,1000} = \left[\frac{683.4}{(0.006756)(130)(8^{2.63})} \right]^{1.85}$$

or 9 feet/1000 feet of 8 inch pipe, which yields an equivalent pipe (5/9) × 1000, or 555 feet.

(c) For the replacement of the 6 inch main from B to C (which constitutes series replacement), an arbitrary flow of 200 gallons per minute is assumed through B–C. For a 6 inch line with a Hazen–Williams C of 130, the corresponding headloss per 1000 feet of pipe is:

$$H_{1\,1000} = \left[\frac{200}{(0.006756)(130)(6^{2.63})} \right]^{1.85}$$

or 3.76 feet per 1000 feet, which gives a total headloss for B–C of (3.67/1000) × 500 feet = 1.88 feet. For the equivalent 8 inch pipe at 200 gallons per minute, the headloss per 1000 would be:

$$H_{1\,1000} = \left[\frac{200}{(0.006756)(130)(8^{2.63})} \right]^{1.85}$$

or 0.93 foot per 1000 feet of pipe. Therefore, the equivalent 8 inch line is (1.88/0.93) × 1000 or 2022 feet.

(d) The equivalent 8 inch replacement pipe is equal to 1200 + 555 + 2022, or 3777 feet.

(e) For 868 gallons per minute flowing through an 8 inch main with a Hazen–Williams C of 130, the headloss per 1000 feet of pipe is:

$$H_{1\,1000} = \left[\frac{868}{(0.006756)(130)(8^{2.63})} \right]^{1.85}$$

or 14 feet per 1000 feet of pipe. The total headloss from G to B would be 14 × 3.777 or 52.88 feet of head (22.9 pounds per square inch).

(f) The calculated residual pressure at B while 868 gallons per minute are flowing would be 43.0 - 22.9 or 20.1 pounds per square inch, which is acceptable since the assumption for this example was that the terrain was relatively flat. If the resulting residual pressure was below 20 pounds per square inch, various pipe sizes would have to be increased until an acceptable residual pressure was attained.

11.7 WATER SYSTEM COMPONENTS

Potable water distribution systems are comprised of pipes, fittings, thrust blocks, restraining joints, valves, blow offs, laterals, meters, and back flow preventers. Each plays an important role when incorporated into a water system.

11.7.1   Piping and Fittings

Most new residential water distribution systems rely on thermoplastic or ductile-iron pipe and fittings, although reinforced concrete pressure pipe is used in larger size mains. Asbestos cement pipe and cast-iron pipe have been widely used in the past and may be the type of pipe encountered when making extensions to existing water systems. Asbestos cement pipe is often present in areas where corrosive soil exists. Most smaller diameter pipes come in 18 to 20 foot lengths and are made to conform to various industrial standards.

Pipes must have adequate tensile and flexural strength to withstand internal and external loads resulting from backfill and other imposed loads. The interior must be smooth and noncorrosive to provide long service life and should provide satisfactory characteristics for the placement of field taps when installing service lines. Ductile-iron pipe and cast-iron pipe may exhibit increased interior wall friction as a result of tuberculation from aggressive water. Cement linings and coatings are effective in combating this problem. Pipe wall thicknesses are selected using the internal system pressure (including surge), external loadings, bedding conditions (or support conditions when the pipe is on piers or held by pipe hangers), corrosive thickness allowances, and factors of safety.

11.7.2   Thrust Blocking and Restraining Joints

Water systems consisting of slip joint piping and fittings require restraining systems called thrust blocks to transfer forces produced by moving water within the conduit to the soil. Failure to provide adequate bearing on the surrounding soil may result in ruptured lines caused by the separation of pipe segments. Thrust blocking is used to provide the necessary bearing. According to King et al. (1948), Figure 11.33 shows in plan view typical forces resulting at a bend having a deflection angle of Θ and a diameter that decreases from AB to CD. If p_1, A_1, and V_1 represent the pressure, area, and mean velocity at AB, then p_2, A_2, and V_2 can be used to represent similar variables at CD. The fluid is accelerated from V_1 to V_2 when flowing from AB to CD and the force producing this acceleration is the result of all forces acting on the mass $ABCD$. P_X and P_Y represent the X and Y components of the forces exerted by the closed channel upon the fluid and can be further defined by:

$$P_X = A_1 p_1 - A_2 p_2 \cos \Theta + \frac{Qw}{g} (V_1 - V_2 \cos \Theta)$$

$$P_Y = A_2 p_2 \sin \Theta + \frac{Qw}{g} V_2 \sin \Theta$$

$$\text{total force} = P = (P_X^2 + P_Y^2)^{0.5}$$

where:

p = Pressure in pounds per square inch.
A = Cross-sectional area in square inches.
Q = Flow rate in cubic feet per second.
V = Velocity of flow in feet per second.
w = Unit weight of water, 62.4 pounds per cubic foot.
g = Acceleration due to gravity, 32.2 feet per second.

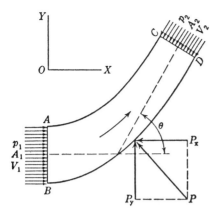

Figure 11.33 Thrust block analysis. Source: Horace W. King, Chester O. Wisler, and James G. Woodburn. *Hydraulics,* 5th ed. Copyright © 1948 by John Wiley & Sons, Inc. Reprinted by permission of John Wiley & Sons, Inc.

To illustrate the design of a thrust block for a 45° bend on a 12 inch water main where the operating pressure and surge pressure are 100 pounds per square inch and the flow rate is 9 cubic feet per second, the following example is given:

(a) The corresponding velocity for a nine cubic foot per second flow is calculated using Equation 7-18 rearranged as:

$$V = Q/A = 9/[(3.14 \times 6^2)/(144 \text{ square inches per square foot})]$$

or 11.46 feet per second.

(b) $P_X = (3.14 \times 36)(100) - (3.14 \times 36)(100)(0.707)$

$$+ \frac{9 \times 62.4}{32.2} (11.49 - 11.49 \times 0.707)$$

$$= 3371 \text{ pounds}$$

(c) $P_Y = (3.14 \times 36)(100)(0.707) + \frac{9 \times 62.4}{32.2} (11.49)(0.707)$

$$= 8134 \text{ pounds}$$

(d) $P_t = (3371^2 + 8134^2)^{0.5}$

$$= 8805 \text{ pounds}$$

(e) If a factor of safety of 1.25 is applied, the design load would be approximately 11,000 pounds.

(f) The bearing area required by the thrust block would be 11,000/1000 pounds per square foot allowable soil pressure, or 11 square feet of bearing surface. If a 3 foot × 4 foot bearing area is used for ease of construction, the installed thrust block should look like Figure 11.34.

Should field conditions prohibit the thrust block from being placed on the outside of the pipe bend (this often occurs when the pipe is close to a ditch bank), the thrust block must be located on the inside of the bend. The 45° fitting must then be secured to the thrust

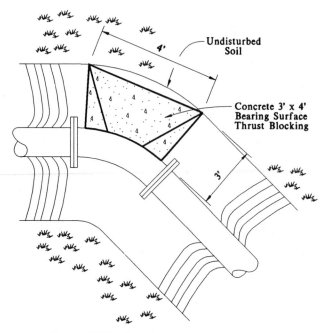

Figure 11.34 Completed thrust block design.

block by tie rods. If portland cement concrete is used to anchor the fitting from the inside, the number of cubic yards required would be (11,000/150 pounds per cubic foot of concrete)/27 cubic feet per cubic yard, or 2.7 cubic yards.

Another way to provide thrust restraint is to use either pipe restraining joints or tie rods that clamp and hold segments together.

11.7.3 Valves

Valves are used to control the direction or amount of flow in a water system, with globe, swing, slide, and rotating types being commonly available. Valves can be classified by use such as:

(a) *Shutoff Valves:* Normally gate type valves or butterfly types are placed within the distribution system for repair and emergency purposes.

(b) *Check Valves:* These are used to control the direction of water flow in only one direction. The most common types of check valves are the swing check and lift check valves, which often incorporate a nonslamming feature.

(c) *Pressure Reducing Valves:* These are used to automatically reduce inlet pressures to a controlled lower outlet pressure.

(d) *Altitude Control Valves:* These valves are used to automatically control the flow into and out of elevated storage tanks, standpipes, or ground storage tanks. These valves normally are located in a pit close to the storage reservoir. Altitude control valves can be classified as either single acting or double acting.

A single acting type closes on tank filling when water reaches the highest water level. Water must discharge through a separate discharge line. A double acting valve closes on tank filling when the water reaches the high water level, but opens for return flow when pressure on the valve inlet is less than that of the storage tank.

(e) *Air Release Valves:* These are used to release trapped air within the water system. Pockets of air that restrict water flow accumulate at high points in water mains. Air release valves vent the accumulated air to the atmosphere and maintain use of the conduit's full flow area.

(f) *Back Flow Preventers:* These are antisiphon valves used to protect safe water systems from contamination as a result of cross connections made with systems containing contaminated or hazardous wastes or from siphoning action. They normally take the form of double check valves, air gap devices, vacuum breaker valves, or reduced pressure back flow preventers. Siphoning risk is minimized by maintaining normal system operating pressures within the distribution system; however, low pressures or ruptured water mains may cause siphoning action to attempt to take place. These valves will prohibit the action.

11.7.4 Blow Offs

These are outlets that are used periodically to flush the distribution system of accumulated sediment, rust, scale, and other undesirable products. They usually consist of a gate valve and an orifice in a vault placed at the end of a dead end line or at low points in the main distribution lines. Blow offs should be sized to allow at least a 2 foot per second flushing velocity in the flushed pipe.

11.7.5 Laterals and Meters

Laterals (service lines) are pipes that are used to convey water from the water main to the dwelling unit property line, with the minimum size for a single family dwelling being $\frac{5}{8}$ inch, although a $\frac{3}{4}$ inch is recommended. The lateral is connected to the main using a saddle or screwed fitting that is threaded through the wall, although other types of connections are available. Laterals usually terminate at a service meter box that is located on the edge of the road right of way. If the system is located in cold weather areas, however, the meter is positioned inside the dwelling. The connection between the meter and the dwelling unit is installed and maintained by the dwelling unit owner. Meters are usually purchased and installed by the organization operating the water utility so that control over meter maintenance, spare parts, calibration, and the end user (meter locking for bill nonpayment) can be attained.

11.8 WATER SYSTEM PROTECTION

Water systems require protection from cold weather and corrosion. Cold weather protection for piping utilizes burial depths discussed in Section 11.6; however, tanks and equipment must also be protected. Water heaters are often used to heat water in elevated tanks and standpipes, while space heaters can provide equipment protection from cold weather damage.

Water systems also must be protected from corrosion attributed to the water itself, to the atmosphere, or to aggressive soil conditions. Lining water mains with cement or other coatings can protect the pipes from deterioration. Corrosion attributed to aggressive soils can be minimized by burying the mains in a protective barrier to prevent the soil from coming into contact with the mains. Tanks and equipment can be protected from corrosion by painting and by providing sacrificial metal sources to minimize corrosion. These precautions must be considered when corrosion is expected to pose a problem.

REFERENCES

Ameen, Joseph S. *Community Water Systems—Source Book* 5th ed. High Point, N.C.: Technical Proceedings, 1974, pp. 19, 52–57, 62, 63, 70–75.

American Water Works Association (AWWA). *Distribution System Requirements for Fire Protection, AWWA Manual M 31.* Denver: AWWA, 1989, pp. 16–21. Copyright © 1989 by American Water Works Association. Referenced material reprinted by permission of American Water Works Association.

American Water Works Association (AWWA). *Water Quality and Treatment,* 4th ed. New York: McGraw-Hill, 1990, pp. 184–185.

CBI Na-Con (Chicago Bridge and Iron Co.). PO Box 5650, Norcross, Ga.

Cedergren, Harry R. *Seepage, Drainage, and Flow Nets,* 3rd ed. New York: Wiley, 1989, p. 41. Copyright © 1989 by John Wiley & Sons, Inc. Reprinted by Permission of John Wiley & Sons, Inc.

Fair, Gordon M., John C. Geyer, and Daniel A. Okun. *Elements of Water Supply and Wastewater Disposal, Vol. I.* New York: Wiley, 1966, pp. 9–12. Copyright © 1966 by John Wiley & Sons, Inc. Reprinted by permission of John Wiley & Sons, Inc.

Fair, Gordon M., John C. Geyer, and Daniel A. Okun. *Elements of Water Supply and Wastewater Disposal,* 2nd ed. New York: Wiley, 1971, p. 74. Copyright © 1971 by John Wiley & Sons, Inc. Reprinted by permission of John Wiley & Sons, Inc.

Floway Pumps, Inc. *Peabody Floway Vertical Turbine Data Handbook Curve # 8JKM.* Floway Pumps, Inc., P.O. Box 226, Fresno, Calif. 93707.

Geyer, J. C., Jerome B. Wolff, F. P. Linaweaver, Jr., and A. J. Duncan. Report on Phase One, Residential Water Use Project, Final and Summary *Report on Phase I of the Residential Water Use Research Project,* January 1, 1959 to January 1, 1963. An unpublished report. Baltimore: Johns Hopkins University and Federal Housing Administration, October 1963, pp. 1–7.

Hardy Cross. "Analysis of Flow in Networks of Conduits or Conductors," *University of Illinois Bulletin 286.* 1936.

Hammer, M. J. and K. A. MacKichan. *Hydrology and Quality of Water Resources.* New York: Wiley, 1981, pp. 45, 47, 59–60.

ISO Commercial Risk Services, Inc. (CRS). *Fire Suppression Rating Schedule.* New York: ISO Commercial Risk Services, Inc., 1980, pp. 1–7, 25–27.

King, Horace W., Chester O. Wisler, and James G. Woodburn. *Hydraulics,* 5th ed. New York: Wiley, 1948, pp. 301–302. Copyright © 1948 by John Wiley & Sons, Inc. Reprinted by permission of John Wiley & Sons, Inc.

Konen, Thomas P. and Wen-Yung W. Chan. "An Investigation of the Frequency and Duration of Use Parameters in Sizing Water Distribution Systems," An American Society of Plumbing Engineers (ASPE) Research Project. West Lane, Calif.: ASPE, 1979.

Linaweaver, F. P., Jr., J. C. Geyer, and Jerome B. Wolff. *Report V on Phase II—Residential Water Use Research Project.* An unpublished report. Baltimore: Johns Hopkins University and Federal Housing Administration, 1966.

National Fire Protection Association (NFPA). *Centrifugal Fire Pumps, NFPA 20.* Quincy, Mass.: NFPA, 1990.

National Fire Protection Association (NFPA). *Water Tanks for Private Fire Protection, NFPA 22.* Quincy, Mass.: NFPA, 1987.

National Fire Protection Association (NFPA). *Installation of Private Fire Service Mains and Their Appurtenances, NFPA 24.* Quincy, Mass.: NFPA, 1987, p. 24-20.

National Fire Protection Association (NFPA). *Fire Flow Testing and Marking of Hydrants, NFPA 291.* Quincy, Mass.: NFPA, 1988, pp. 291-4–291-9.

National Fire Protection Association (NFPA). *Fire Protection in Planned Building Groups, NFPA 1141.* Quincy, Mass.: NFPA, 1990.

National Fire Protection Association (NFPA). *Suburban and Rural Fire Fighting, NFPA 1231.* Quincy, Mass.: NFPA, 1989.

Pontius, Frederick W., Ed. *AWWA Water Quality and Treatment,* 4th ed. New York: McGraw-Hill, 1990, pp. 184–185.

Rothrock, D. F. *Hydro-Pneumatic Tanks for Water Systems.* Phoenix: Red Coat Reports, 1974, pp. 1–4, 6.

Salvato, Joseph A. *Environmental Engineering and Sanitation,* 4th ed. New York: Wiley, 1992.

Sweitzer, Robert J. and Martin E. Flentje. *Basic Water Works Management.* Arlington, Va.: American Concrete Pipe Association, 1972, pp. 111–113.

Walski, Thomas M. *Analysis of Water Distribution Systems.* New York: Van Nostrand Reinhold, 1984.

Weber, Walter J., Jr. *Physicochemical Processes for Water Quality Control.* New York: Wiley, 1972.

Wolff, Jerome B. and John F. Loos. "An Analysis of Public Water Demands," *Public Works.* Sept. 1956, pp. 111–115.

12 Miscellaneous Design Considerations

Several independent topics related to land developments have yet to be presented. These include recreational playing courts, swimming pools, pipe jacking and boring, and fencing.

12.1 RECREATIONAL PLAYING COURTS

There are a number of recreational courts from which to choose for any facility plan. Tennis, basketball, volleyball, and shuffleboard courts are presented here.

12.1.1 Surface Composition and Orientation

Most paved courts are composed of reinforced portland cement concrete, with some being constructed with hot mix asphaltic concrete. All courts require stable subgrades with good drainage features. Courts constructed of portland cement concrete are usually made with a minimum 4 inch thick slab using a 3000 pound per square inch concrete mix design along with temperature reinforcement being installed. The slab is usually formed over a polyethylene vapor barrier, and includes construction joints to control expansion, contraction, and possible warping stresses depending on the total facility size. Expansion joints are placed at right angles across the court's center (net line) in most cases. Contraction joints are usually located outside of the playing area.

Each court included in this discussion is normally positioned with the long court dimension being parallel to a north-south line. This will minimize adverse player sight conditions resulting from the sun.

12.1.2 Tennis Courts

Typical tennis courts are illustrated in Figure 12.1a. The surface is sloped to drain from end to end or from side to side, since a uniform playing surface is required. The nets are supported so that the tops are 3.5 feet above the playing surface.

12.1.3 Basketball Courts

A typical college basketball court is depicted in Figure 12.1b. The backboards are comprised of 4 foot × 6 foot surfaces and are mounted 10 feet above the playing surface at each end, 4 feet from the end lines.

12.1.4 Volleyball Courts

A typical volleyball court's dimensions are shown in Figure 12.1c. The center net is positioned where its top is 8.25 feet above the playing surface.

12.1.5 Shuffleboard

A typical shuffleboard court is illustrated in Figure 12.1d and requires a very smooth surface finish. The one inch striping used to mark the court must also provide a smooth riding surface.

12.2 SWIMMING POOL FACILITIES

Land developments often include swimming pools as part of the project's recreational facility plan. Such pools are normally considered "controlled public" because they are

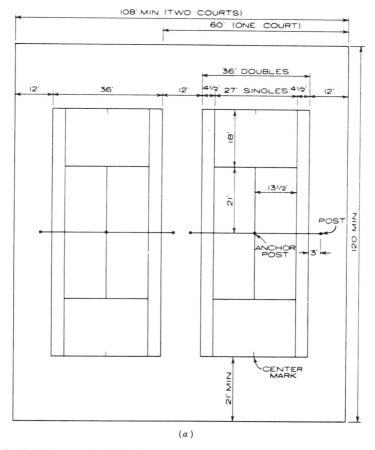

(*a*)

Figure 12.1 Typical recreational playing courts. (*a*) Tennis courts. (*b*) Basketball court. (*c*) Volleyball court. (*d*) Shuffleboard court. Source: Harvey M. Rubenstein. *A Guide to Site and Environmental Planning,* 3rd ed. Copyright © 1987 by John Wiley & Sons, Inc. Reprinted by permission of John Wiley & Sons, Inc.

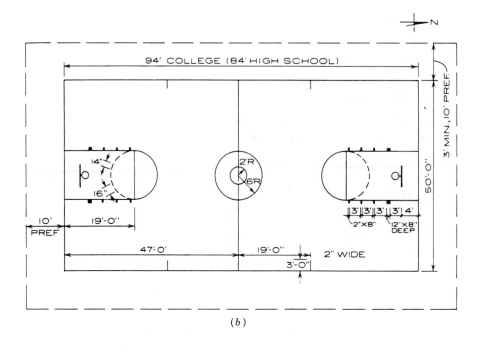

(*b*)

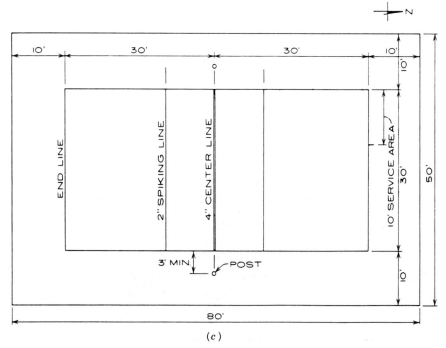

(*c*)

Figure 12.1   (continued)

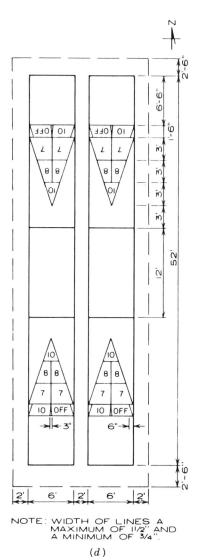

NOTE: WIDTH OF LINES A
MAXIMUM OF 1½" AND
A MINIMUM OF ¾".

(*d*)   **Figure 12.1**   (continued)

utilized by more than one family, but are not open for general public use. These pools usually fall under regulatory controls that necessitate plan approval prior to construction (SC DHEC, 1983). During construction these pools are often subjected to regulatory inspections.

Controlled public pools are normally constructed of reinforced concrete, shotcrete (gunite), or similar materials and include National Sanitation Foundation approved materials. Pools exist in a variety of styles and shapes, with the most common being rectangular, ell, or tee shaped. They should be protected from dust, smoke, and soot. Decks and coping provisions, water circulation and treatment equipment, fencing, lighting, bath house and toilet facilities, chlorine room, water supply lines, and wastewater discharge lines must all be provided.

12.2.1   Pool Features

Pools are usually comprised of vertical walls, with a minimum of six inch radius fillets being provided at all interior intersections. Wall tops are usually capped with coping tile that is suitable for gripping from within the pool, unless a perimeter guttering system is utilized. The pool's normal water level is usually delineated by installing six inch frost free glazed tile under the coping tile. The interior is usually finished with a smooth white (or light colored) plaster or nontoxic paint that is durable and water resistant. Depth markers are placed on the walls and deck at maximum and minimum points and at 2 foot depth increments (with a maximum spacing being 15 feet apart). Typical water depths and spatial requirements are shown in Figure 12.2.

Other considerations include the positioning and sizing of auxiliary equipment including:

(a) Ladders and steps.

(b) Diving structures.

(c) Surface skimmers or gutters.

(d) Lighting.

(e) Fittings.

(f) Drains and inlets.

(g) Pipe sizes and configuration.

(h) Fill spout and valve.

(i) Surge chamber (if required).

The design professional usually includes in the plans the pool's volume, surface area, perimeter, deck area, turnover flow rate, and maximum user capacity.

12.2.2   Deck Area Features

The pool's apron, or deck area, is shown on the construction plans with dimensions. The deck is normally a minimum of 4 feet wide, but often extends at least 10 feet. Its surface is made of nonslip material (often portland cement concrete with a brush finish) and slopes away from the pool's coping using approximately $\frac{1}{2}$ inch fall per horizontal foot. Water depths are normally shown adjacent to corresponding wall marks. The position of area lighting fixtures and wiring is included. Piping for hose bibbs fitted with vacuum breaker valves, a foot shower, and a drinking fountain are also shown. Care should be exercised by the design professional to maintain clear air space for at least ten feet horizontally from the edge of the pool to prevent adjacent balcony or rooftop access for diving; otherwise protective barriers need to be installed. Areas next to the deck should be paved or consist of grass to minimize potential pool water contamination.

12.2.3   Bath House Facilities

Local regulations can require separate facilities for each gender, facilities that are well drained, and that have adequate light and ventilation (SC DHEC, 1983). Unbreakable mirrors, soap dishes, tissue dispensers, a service sink, and hot water are included. To accommodate approximately 100 male swimmers (and every increment of 200 thereafter),

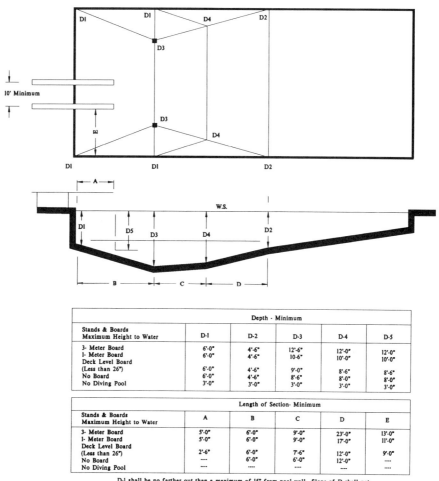

	Depth · Minimum				
Stands & Boards **Maximum Height to Water**	D-1	D-2	D-3	D-4	D-5
3- Meter Board	6'-0"	4'-6"	12'-6"	12'-0"	12'-0"
1- Meter Board	6'-0"	4'-6"	10-6"	10'-0"	10'-0"
Deck Level Board					
(Less than 26")	6'-0"	4'-6"	9'-0"	8'-6"	8'-6"
No Board	6'-0"	4'-6"	8'-6"	8'-0"	8'-0"
No Diving Pool	3'-0"	3'-0"	3'-0"	3'-0"	3'-0"

	Length of Section· Minimum				
Stands & Boards **Maximum Height to Water**	A	B	C	D	E
3- Meter Board	5'-0"	6'-0"	9'-0"	23'-0"	13'-0"
1- Meter Board	5'-0"	6'-0"	9'-0"	17'-0"	11'-0"
Deck Level Board					
(Less than 26")	2'-6"	6'-0"	7'-6"	12'-0"	9'-0"
No Board	....	6'-0"	6'-0"	12'-0"	....
No Diving Pool	....	....	....	....	....

D-1 shall be no farther out than a maximum of 15" from pool wall. Slope of D shall not
exceed 1' · 00" vertical to 3' · 00" horizontal. The maximum values of A are 6'-00" for
1-meter and 3-meter boards and 4'-00" for deck level boards. Clearance E must extend
the entire length of sections B, C, & D. Depth D-5 is measured at midpoint of section B
where a diving board is not provided. Where a diving board is provided D-5 shall be
measured from the tip of the board. The minimum distance between the diving well wall
on the deep end and any opposite wall shall not be less than six (6) feet greater than
the diving bowl dimensions (B, C, D).

Figure 12.2   Minimum pool safety setbacks and water depths. Source: South Carolina Department
of Health and Environmental Control, 1983.

one water closet, lavatory, and urinal are necessary. Female accommodations include two
water closets and lavatories for each 100 users. Every shower installed in each facility can
service approximately 50 swimmers. Privacy partitions are usually included in the female
facilities.

12.2.4   Equipment House

The equipment house provides a safe location for all circulating and purification equipment necessary to efficiently operate the pool. The floor is usually constructed of concrete
and sloped. This will allow positive drainage to protect both operators and equipment.
Adequate ventilation is required to allow proper cooling of operating equipment when the

door is secured. Adequate illumination is also desirable. All electrical circuits must be installed with grounds (and ground fault interrupters where required) to eliminate shock hazards. To promote safety, all circuits must comply with electrical code requirements throughout the entire facility area.

Diatomaceous earth or rapid sand filters are usually installed in conjunction with corrosion resistant centrifugal pumps to cleanse and circulate the pool's water. Normally piping and valves are labeled to show flow directions and include pressure gauges, a flow meter, a backwash sight glass, and connections for chemical feed dispensing. If PVC pipe and fittings are used, Schedule 40 or 80 classifications are satisfactory in most cases.

12.2.5   Chlorine Room Facility

There are several methods available to disinfect pool water; however if chlorine gas is to be utilized, the dispensing and storage facility should be clearly marked as a chlorine facility with prominently displayed signs. This installation should be located no nearer than 200 feet to the closest dwelling, but located close to the main circulation pumps and filters to reduce the gas feed line distance. The room includes a corrosion resistant exhaust fan located as close to the concrete floor as possible with its discharge to the out of doors. The fan should be able to provide at least one change of air per minute. Automated air intake louvers are needed to supply fresh air inside and are located as high as possible in the wall to allow circulation when the blower is energized. Switches to control interior lights and the exhaust fan are located outside the chlorine room for the operator's protection. A lockable door is provided that opens only to the outside. A viewing window is normally incorporated in this access door to allow observations of the interior without entry.

The installed dispensing equipment must be capable of maintaining a minimum of 1.5 parts per million concentration of free chlorine throughout the pool. A range of 1 to 3 parts per million must be maintained whenever the pool is in use. To insure an uninterrupted chlorine supply, if 150 pound chlorine cylinders are used, either an automatic changeover device capable of switching chlorine cylinders is required, or a scale to determine when the cylinder is empty must be utilized.

A gas mask for emergencies is provided at a location outside the chlorine room. Also a gas leak detector should be installed that has both audible and visual alarms, to alert the operator of potential problems.

12.2.6   Chemical Dispensing Facilities

The pH of the pool's water must be maintained between 7.2 and 7.8 for good results. The addition of soda ash or muriatic acid may be necessary to keep the pH in balance. Installed corrosionless tanks with metered pumps can be utilized to inject the necessary doses directly into the filtered water's piping. This facility should be separated from the pump and filter area but could be part of the equipment house.

12.2.7   Facility Security

To provide controlled access use, the pool facility should be enclosed with a chain link type fence at least six feet high that is lockable and nonscalable. This fence may be required by law and will help to lower owner liability, since facilities such as pools are considered "attractive nuisances." Fencing also keeps pets and animals from entering the premises. Security lighting to illuminate the facility after hours is also helpful in reducing the chances of unauthorized use.

12.2.8 Miscellaneous Provisions

To increase swimmer safety a lifeguard stand is usually provided for every 2000 square feet of pool surface area, along with life saving and first aid equipment. In addition, provisions for a telephone must be made where emergency phone numbers (including an ambulance) can be posted. The design must also include a place to conspicuously post and display the pool regulations.

12.2.9 An Example of Pool Design

For this example, it is assumed that the pool must be capable of serving a 50 family membership. Based on site constraints this pool's shape will be in the form of an "L" with a diving well.

A. Determine the Required Pool Surface Area.   If 3.4 persons are estimated in each membership based on the development's targeted market, and maximum utilization is estimated for this example at 85%, the pool must accommodate (50 memberships) × (3.4 persons/membership) × 85% = 144.4 or 145 maximum users.

Allowing a maximum of three persons to occupy the 20 foot side diving well where a deck level board (less than 26 inches) is to be installed, the required surface area of the well is estimated by assuming that the board's tip will project horizontally past the coping by three feet. Providing an additional 12 feet of horizontal distance for safety, the resulting diving well's surface area is (3 feet for board projection + 12 feet for clearance) × (20 feet wide) = 300 square feet.

If the percentage of users to occupy the nonswimming area is assumed to be 75%, and 10 square feet of surface area is allocated to each user, the corresponding nonswimming area is (145 maximum users) × (10 square feet per user) × 75% = 1087.5 square feet.

Since 2% of the users are allocated to diving in this example and 75% to nonswimming, the swimming area must accommodate the remaining 23%. Allowing 25 square feet of surface area for each swimmer, the resulting area would be (145 maximum users) × (25 square feet per user) × 23% = 834 square feet.

The pool's interior dimensions for wall clearances are determined by:

(a) The diving well's width is 20 feet and the length is 15 feet to provide the required 300 square feet of area.

(b) The swimming area that adjoins the diving well utilizes the same 20 foot width and would need a corresponding length of (834 square feet required)/(20 feet wide)= 41.7 feet.

(c) The nonswimming area that adjoins the swimming area is assumed to have a width of 28 feet. The corresponding length is (1087.5 square feet required) / (28 feet wide) = 38.8 feet long.

If these values are rounded, an adjusted layout is obtained. Figure 12.3*a* shows these adjusted dimensions, which are used in this design. The configuration provides 2280 square feet of surface area with a corresponding perimeter of 236 linear feet.

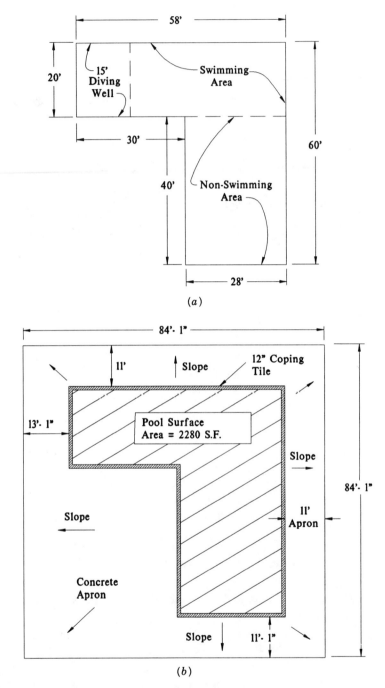

Figure 12.3 An example of pool dimensions. (*a*) Adjusted dimensions for pool's interior. (*b*) Adjusted deck dimensions.

B. Determine the Corresponding Deck Area.   If the deck area adjoining the pool is to be comprised of coping tile that is 1 foot wide and a concrete apron that is 11 feet wide (in addition to the 30 foot × 40 foot area contiguous to the pool) the resulting area is 4608 square feet. The minimum deck area required, however, allowing a 33 square foot area for each swimmer, is 4785 square feet. This necessitates an increase in the width of the apron adjoining the diving well to 13 feet, 1 inch, and of the apron contiguous to the shallow area to 11 feet, 1 inch. The final deck configuration is shown in Figure 12.3*b;* it contains 4790 square feet.

C. Establish the Pool Depths.   The shallow depth of the pool in the nonswimming area usually ranges from 3 to 3.5 feet, with 3 feet being utilized in this example. A depth of 4.5 feet is normally used to separate swimming from nonswimming activities with the pool's bottom sloping between these two limits. If the diving board is to be located where its surface is approximately 20 inches above the deck, the swimming pool depths can be determined by using information similar to that shown in Figure 12.2.

Figure 12.4 shows the depths that are incorporated in this sample pool. They provide a corresponding volume of 84,300 gallons.

D. Determine Pumping Rate.   Using a turnover rate for the pool's water of once every 6 hours, the corresponding pumping rate is (84,300 gallons pool capacity)/[(6 hour turnover) × 60 minutes per hour)] = 235 gallons per minute.

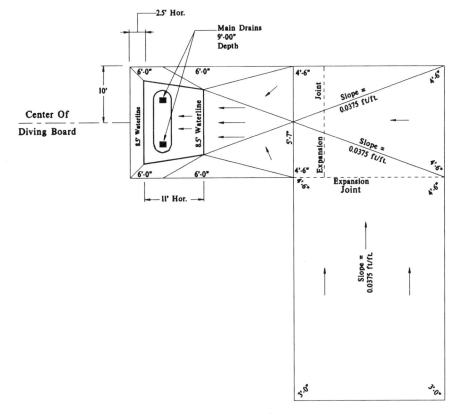

Figure 12.4   An example of pool depths.

E. Determine Filter Bed Area.   By limiting the maximum flow rate through rapid sand filters to 15 gallons per minute per square foot of filter surface area, the corresponding minimum filter bed surface area for this example is (235 gallons per minute pumping rate) / (15 gallons per minute per square foot of filter area) = 16 square feet, rounded.

F. Determine Intake Skimmer Requirements.   If the maximum flow rate through each skimmer is limited to 30 gallons per minute according to the manufacturer, the minimum number of required skimmers is (235 gallons per minute pumping rate) / (30 gallons per minute limit per skimmer) = 7.8 or 8 skimmers. Each skimmer includes a basket strainer, weir, and equalizer line with an intake that must be submerged more than a foot below the normal water surface level to be effective. These skimmers would be strategically located to allow complete pool surface coverage.

G. Determine Pool Inflow Inlet Requirements.   Since each wall inlet orifice is adjustable for both flow and direction, the spacing interval to be used will be not greater than 15 feet to provide complete circulation. A total of 17 inlet nozzles are provided.

H. Determine Main Drain Requirements.   Most pools have at least two main drains that are innerconnected in case one clogs. If the maximum intake velocity is limited to 1.5 feet per second through each drain to minimize the effects of vortex when the desired pumping rate enters, the minimum clear way area of the drain grate can be calculated. First the flow rate must be converted from gallons per minute to cubic feet per second by (235 gallons per minute pumping rate) × (0.002228 cubic feet per second per gallon per minute) = 0.52 cubic foot per second equivalent flow rate. The corresponding minimum clear way area for each grate is determined by (0.52 cubic foot per second flow) / (1.5 feet per second limiting velocity) = 0.35 square foot of open grate area; therefore, each grate chosen from manufacturers' literature must have this area available.

Unless the pool's weight is sufficient to overcome buoyant forces resulting from the presence of groundwater, the pool could float. This can occur if the water level in the pool is sufficiently lower than that of the surrounding groundwater. Since most pools do not have enough weight to counter this buoyant force, care must be taken to keep the groundwater table low. This can be accomplished by installing a well point (or points) that can be pumped throughout construction. When construction is completed and the pool is filled, the weight of the pool and included water prevent flotation. The well point system can be left in place for dewatering that might be required in the future for pool maintenance.

As a precaution, a hydrostatic relief (check type) valve is installed in the drain sump that allows groundwater to enter the pool until the levels are equalized if the pool's water surface drops below the surrounding groundwater level. Under normal operating conditions this valve remains closed.

I. Determine Necessary Pump Size.   Once the number and arrangement of intake and discharge outlets are known, the pipe sizes are designed. Normally the Hazen–Williams formula is used in conjunction with the following maximum headloss parameters:

(a) *Suction Lines:* 6 feet/100 feet of pipe.
(b) *Discharge Lines:* 12 feet/100 feet of pipe.

The pump's TDH is calculated by including static, friction, filter media, and minor system headlosses. This procedure is similar to that presented in Chapter 10. An efficient pumping unit can then be chosen from manufacturer's data.

It has proven beneficial to install a backup circulating pump and filter system that allows continuous operation when repairs are necessary. This keeps the membership satisfied throughout the swimming season for nominal additional capital expenditures. A typical piping schematic for this setup is shown in Figure 12.5.

J. Determine Necessary Gas Chlorinator Size.   An adequate chlorinator must be installed to maintain the proper chlorine residual. Chlorinators are adjustable to allow operator manipulation according to demands. If an average chlorine dose rate of eight parts per million is used, the number of pounds of chlorine required per day is calculated by first determining the total number of pounds of water circulated each day. This is obtained by (235 gallons per minute) $\times$ (1440 minutes per day) $\times$ 10^{-6} = 0.338400 million gallons per day. The equivalent number of million pounds of purified water each day is (0.3384 million gallons per day) $\times$ (8.34 pounds per gallon of water) = 2.822 million pounds. Applying eight parts per million, the daily chlorine demand is (2.822 million pounds of water per day) $\times$ (8 parts per million chlorine) = 22.6 pounds per day. A chlorinator capable of delivering from 0 to 100 pounds per day would be satisfactory and would utilize one 150 pound cylinder in about a week.

K. Determine Pool Lighting Requirements.   Adequate pool lighting is obtained when each installed fixture is allowed to illuminate no more than 2000 square feet of the pool's subsurface area. In the case of this example, the minimum required number of underwater light fixtures is (2280 square feet of pool surface) / (2000 square feet effective area) = 1.14 fixtures; however, due to the configuration of this pool, 3 are provided. The amount of wattage required for each fixture based on 0.5 watt per square foot of pool surface area is (2280 square feet) $\times$ (0.5 watt per square foot) = 1140 watts total. Therefore, each fixture will be outfitted with a 400 watt bulb.

L. Consider Miscellaneous Provisions.

(a) *Steps and Ladders:* Usually two sets of steps and ladders are provided in each pool as a minimum, with additional ones being located approximately 75 feet apart. One set of steps in the shallow end and three sets of ladders are included in this example.

(b) *Fill Spout:* The pool's fill spout is usually located in a safe place to prevent accidents and to make effective use of the area under the diving board. The pipe's outlet elevation is usually two pipe diameters or more above the pool's coping to prevent possible siphonage.

(c) *Surge Tanks:* Water is displaced by each swimmer's body volume and by dynamic displacement from swimmer turbulence. Some pools require surge tanks to equalize this displaced water. This temporary excess of water volume is allowed to be stored until it can be circulated back into the pool without flooding. This example does not include a surge tank because the pool is relatively small and skimmer equalizer lines are provided to prevent loss of pump prime during peak pool usage.

(d) *Overflow Protection:* The pool must be protected from flooding attributed to storm water or other sources. This requires that proper drainage be provided to eliminate this potential problem.

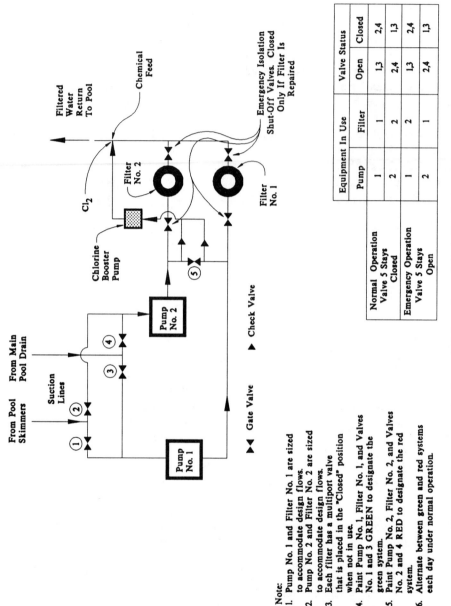

Note:
1. Pump No. 1 and Filter No. 1 are sized to accommodate design flows.
2. Pump No. 2 and Filter No. 2 are sized to accommodate design flows.
3. Each filter has a multiport valve that is placed in the "Closed" position when not in use.
4. Paint Pump No. 1, Filter No. 1, and Valves No. 1 and 3 GREEN to designate the green system.
5. Paint Pump No. 2, Filter No. 2, and Valves No. 2 and 4 RED to designate the red system.
6. Alternate between green and red systems each day under normal operation.

▲▼ Gate Valve   ▲ Check Valve

	Equipment In Use		Valve Status	
	Pump	Filter	Open	Closed
Normal Operation Valve 5 Stays Closed	1	1	1,3	2,4
	2	2	2,4	1,3
Emergency Operation Valve 5 Stays Open	1	2	1,3	2,4
	2	1	2,4	1,3

Figure 12.5 An example of a pool piping schematic including backup pumps and filters.

M. Finalize Pool Layout. The final pool layout is shown in Figure 12.6, including the locations of lights, ladders and stairs, skimmers, inflow inlets, main drains, filling spout, and diving board.

N. Determine Structural Design. The bottom of a swimming pool is usually made at least six inches thick when constructed of reinforced concrete or similar materials. The magnitude of earth pressures acting on the outside of the pool's buried surface and hydrostatic forces acting inside or out of the pool are used as criteria for structural design. The design must include adequate expansion joints that incorporate the use of waterstops to prohibit leaks. Structural design sources are available elsewhere that address this topic.

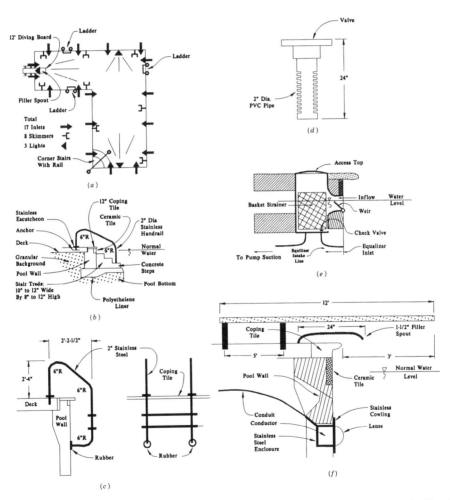

Figure 12.6 Final layout for the sample pool, with typical pool details. (*a*) Layout details of sample pool. (*b*) Stair details. (*c*) Ladder details. (*d*) Automatic hydrostatic relief valve mounted in main drain sump. (*e*) Skimmer details. (*f*) Diving Board area details.

12.3   FENCING

Chain link type fences are commonly used to protect development property, provide safety, and act as enclosures. Applications include protecting utility facilities (sanitary sewer pumping station or wastewater treatment plant), providing security around recreational areas such as swimming pools, and protecting hazardous areas such as deep storm water detention facilities. Some fencing systems installed for security purposes include barbed wire strands along the fence top using special post caps. Figure 12.7 shows details of some typical installations.

12.4   PIPE JACKING AND BORING

Today many regulatory authorities require underground pipelines to be installed under existing improvements by means other than using open trenching and backfilling. A commonly used procedure is the process of boring a pilot hole through the soil into which the pipe is forced under the improvement using a hydraulic jack.

12.4.1   The Installation Process

Initially a boring pit is excavated to a depth that allows the boring machine to be properly positioned. The machine is secured in place once it has been oriented to line and grade. The boring machine forces a rotating auger forward to pre-bore the casing's hole, if a protective casing is to be installed, otherwise the carrier pipe's hole. Periodically boring ceases so that sections of casing pipe can be welded to previous sections to form a continuous conduit and then be jacked forward. On completion, the casing (or uncased carrier pipe) extends to the desired location, at which point open trenching is continued.

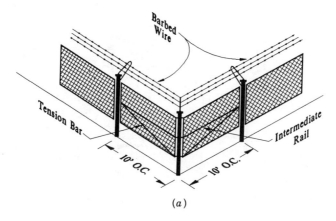

(a)

Figure 12.7 Typical fencing details. (a) An example of a security fence. (b) An example of a single gate. Source: Harvey M. Rubenstein. *A Guide to Site and Environmental Planning*, 3rd ed. Copyright © 1987 by John Wiley & Sons, Inc. Reprinted by permission of John Wiley & Sons, Inc.

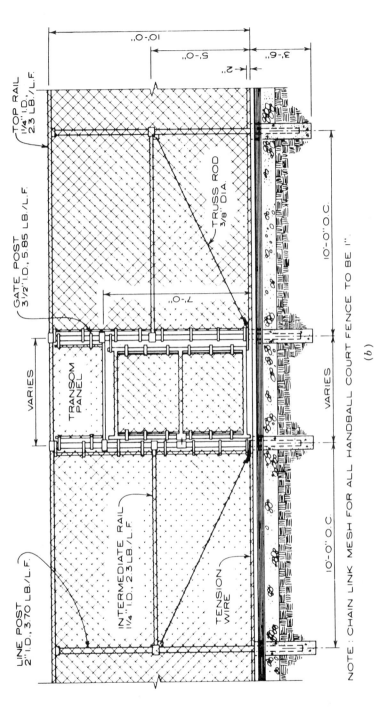

Figure 12.7 (continued)

12.4.2 Jacking and Boring Requirements

Depending on the situation, regulatory agencies have criteria for these activities. To illustrate an example using the American Railway Engineering Association's guidelines (AREA, 1991), a pressurized pipeline carrying nonflammable substances that crosses under a railroad may be required to horizontally intersect the track at an angle between 45° and 90°. Installation of the new pipe is usually not permitted through existing culverts or under railroad bridges if a reduction in flow capacity is realized or if any foundation has the potential of being weakened. Pressurized carrier pipe joints can be welded or of the mechanical type, with the pipe being installed so that it is not in tension. Casing pipe and joints are leak-proof and capable of withstanding imposed railroad loads. Minimum 35,000 pounds per square inch yield strength steel casing that is subjected to E-80 railroad loads must have a wall thickness of at least that shown in Figure 12.8.

If the casing does not have a protective coating and is not cathodically protected, these thicknesses are increased by at least 0.063 inch. The casing pipe's inside diameter must be at least two inches larger than the maximum outside diameter of the carrier pipe, including any joints or couplings, for carrier pipes less than six inches in diameter. For larger pipes a four inch minimum differential is required.

The top crown of most casings must be at least 5.5 feet below the main track rail and may be required to extend 2 feet beyond the toe of slope, 3 feet beyond ditches, otherwise to extend 25 feet from the track centerline. Figure 12.9b illustrates a typical pipe installation under a railroad.

For a more complete presentation of the requirements for pipeline installations within railroad rights of way, the reader is referred to Chapter 1, Parts 4 and 5 of the current AREA's *Manual for Railroad Engineering* (1991).

12.4.3 Potential Problems Encountered During Boring

Many factors contribute to problems associated with boring misalignment and progress, including groundwater interference, varying soil conditions, and obstacles (including rocks, roots, and buried debris). If the auger is unable to maintain line and grade, often the

Nominal Diameter (in.)	Nominal Thickness (in.)
14 and Under	0.188
16	0.219
18	0.250
20 - 22	0.281
24	0.312
26	0.344
28	0.375
30	0.406
32	0.438
34 - 36	0.469
38	0.500
40	0.531
42	0.563

Figure 12.8 Minimum wall thicknesses for casing pipe—E 80 loading. Taken from *Manual for Railway Engineering,* Chapter 1, Roadway and Ballast, Part 5, Pipelines. Copyright © 1991 by the American Railway Engineering Association. Reprinted by permission of the American Railway Engineering Association.

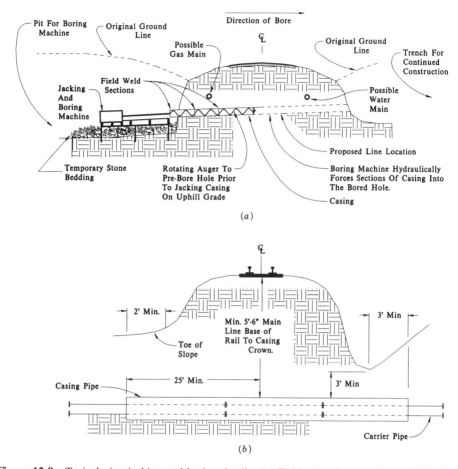

Figure 12.9 Typical pipe jacking and boring details. (*a*) Field setup for street bore. (*b*) Typical pipeline installation under a railroad. Taken from *Manual for Railway Engineering,* Chapter 1, Roadway and Ballast, Part 5, Pipelines. Copyright © 1991 by the American Railway Engineering Association. Reprinted by permission of the American Railway Engineering Association.

boring machine is located, with regulatory approval, at another position close by for another try. This minor relocation does not usually create operational problems if the line is pressurized. However, a gravity flow carrier pipe may require additional provisions to accommodate the pipe's grade.

To allow more flexibility during construction, for example, consideration should be given to the inclusion of a manhole at each end of an amply oversized casing pipe when gravity wastewater mains are being installed. A flange ductile-iron gravity carrier pipe can be inserted on intermediate supports and afforded room for position adjustments (to attain the desired grade) prior to being permanently secured in place.

On occasion poor boring conditions aggravated by groundwater and soft soil may cause the augered hole to collapse. Success under these conditions has been attained by using well points to dewater, or by freezing the soil and then boring.

REFERENCES

American National Standards Institute (ANSI) and National Spa and Pool Institute (NSPI). *1991 Standard for Public Swimming Pools, ANSI/NSPI-1*. Alexandria, Va.: ANSI, 1991.

American Railway Engineering Association (AREA). *AREA Manual for Railway Engineering*. Washington: AREA, 1989, Chapter 1, Part 4, Culverts.

American Railway Engineering Association (AREA). *AREA Manual for Railway Engineering*. Washington: AREA, 1991, Chapter 1, Part 5, Pipelines, pp. 1-5-7, 1-5-8, and 1-5-11. Referenced material is reprinted by permission of the American Railway Engineering Association.

Rubenstein, Harvey M. *Site and Environmental Planning*, 3rd ed. New York: Wiley, 1987, pp. 324, 328, 329, 331, 377–397.

South Carolina Department of Health and Environmental Control (SC DHEC). *Public Swimming Pools—Regulation 61-51*. Columbia, S.C.: SC DHEC, 1983.

13 Project Manual Including Specifications

There are several recognized methods of organizing and preparing construction documents and specifications that are used by civil engineers today. The Project Manual approach is presented here.

According to the Construction Specifications Institute's (CSI) *Manual of Practice* (CSI, 1985), the term "Project Manual" originated with the American Institute of Architects (AIA) in 1964. The Project Manual includes bid information, contractual conditions, and construction specifications. The Project Manual's construction specifications section is jointly categorized into 16 construction methods and materials divisions by CSI in the United States, and by Construction Specifications Canada (CSC) there, in an effort to foster better specification writing. The jointly prepared *MASTERFORMAT* (CSI and CSC, 1988) contains a master listing of construction information that has been sequenced into groups comprised of bidding requirements, contract forms, conditions of the contrast, and specification divisions 1 through 16. Although this system evolved from architectural project applications, civil engineers have increased interest in the use of this format for land development projects since 1981 when CSI became a member of the Engineers Joint Contract Documents Committee (EJCDC). The American Society of Civil Engineers (ASCE) and the National Society of Professional Engineers (NSPE) are two of the EJCDC's members.

13.1 CONSTRUCTION AND CONTRACTUAL DOCUMENTS

Construction and contract documents include all graphic and written documents that communicate the project's design intentions and construction administration activities. These documents, according to CSI's *Manual of Practice* (1985), include the following:

(a) Bidding requirements:
   (1) Invitation to bid.
   (2) Instruction to bidders.
   (3) Information available to bidders.
   (4) Bid form.
   (5) Bid bond form.
(b) Contract forms:
   (1) Agreement.
   (2) Performance bond.
   (3) Payment bond.
   (4) Certificates.

(c) Conditions of the contract:
   (1) General conditions.
   (2) Supplementary conditions.
(d) Specifications.
(e) Drawings.
(f) Addenda.
(g) Contract modifications:
   (1) Change orders.
   (2) Field orders.
   (3) Supplemental instructions.

The relationships of these elements are illustrated in Figure 13.1.

The bidding requirements, contract forms, and conditions of the contract are not considered specifications and usually require the services of the developer's attorney and insurer when they are being prepared. Depending on the developer's approach to achieving project completion, various EJCDC standard forms can be incorporated as part of the contract documents. For example, Section 00700—General Conditions could be comprised of EJCDC's *Standard General Conditions of the Construction Contract* (No. 1910-8). According to Smith (1991), the 1990 EJCDC General Conditions, Instructions to Bidders, Bid Form, and Supplementary Conditions, "Are all written assuming this schedule will be used." He concludes with "A change in one (document) will require a change in all."

The appropriate location of subject matter within the Project Manual can be determined by using the EJCDC (and the American Institute of Architects) prepared document designated as the "Uniform Location of Subject Matter—Information in Construction Documents" (No. 1910-16). Utilization of this guideline will aid the specifier in coordinating the various elements that are included in the Project Manual.

The CSI/CSC (1988) format provides a standard set of rules for the arrangement, grammar, punctuation, and sequencing of material. The standard location for specific specification information is based primarily on 16 broad divisions, with each category being assigned a 5 digit identification number. This system lends itself to computer adaption through the use of this numbering system. The first two digits refer to the division within which the category is grouped. Each part is sequentially numbered beginning with the cover sheet labeled 00001. Preferred section numbers and titles are contained in CSI and CSC's periodically updated *MASTERFORMAT* document, which specification writers can use to choose applicable entries when preparing specific Project Manuals.

MASTERFORMAT contains classification numbers for all broadscope and mediumscope categories. Broadscope titles are used for broad categories of work, while the medium scope titles are more limited. Narrowscope titles can be used to focus on specific work elements. Narrowscope sections are numbered by the specifier using uncommitted numbers within the mediumscope sections; this allows a flexible system of writing specifications. For example, under the broadscope category of Earthwork (02200) several mediumscope categories are included such as Grading (02210), Base Courses (02230), and Soil Stabilization (02240). The specifier has the option of using only the broadscope category (02200) by incorporating all earthwork information together. However, if the specifier's approach is to address more categorically various aspects of earthwork, then several

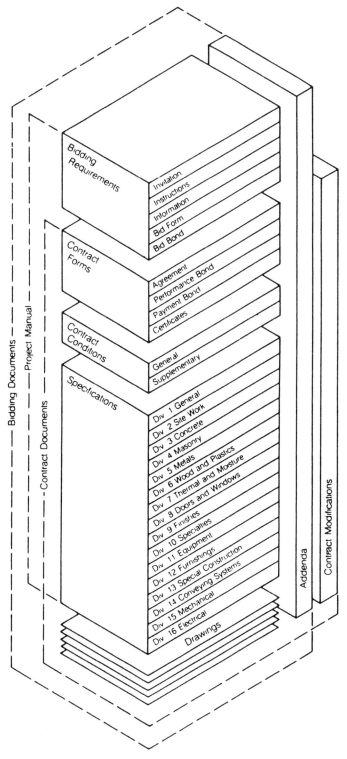

Figure 13.1 Construction document relationships. Taken from the Construction Specifications Institute's *Manual of Practice*. Copyright © 1985 by the Construction Specifications Institute. Reprinted by permission of the Construction Specifications Institute, Alexandria, Va.

applicable mediumscope categories can be utilized. The specifier also has the option of further refining the presentation of material by using a narrowscope approach, in which many applicable aspects of earthwork are narrowly defined while unused section numbers are assigned for their identification. The result is a document that is well organized and can be conveniently used to monitor construction activities and costs.

13.2 SPECIFICATIONS

Construction specifications are written instructions that provide information not found on the construction drawings. They convey to the contractor what final quality (performance) is expected in materials and workmanship used on the job. Specifications can be written using various formats including:

(a) *Descriptive:* Describing methods and materials without the inclusion of propri-etary names.

(b) *Performance:* Requiring specified end results that can be verified.

(c) *References:* Requiring a product, procedure, or process that conforms with an established standard.

(d) *Proprietary:* Requiring a brand name or product.

Each of the 16 CSI/CSC divisions being subdivided into broad, medium, and narrow scope sections addressing specific topics; each of these sections are further categorized by CSI and CSC into a standard "Section Format" containing three parts:

> Part 1   General
> Part 2   Products
> Part 3   Execution

Part 1 includes the scope of work involved in the section, including requirements for inspections and submittals. Part 2 contains technical provisions for all materials and equipment used in this section. Part 3 includes installation and standards relating to workmanship. If one of the parts is not germane, then it is excluded. Figure 13.2 contains possible elements that can be included in the various parts.

The engineer who prepares the construction specifications must have a thorough knowledge of the work involved with the job, the kinds of materials and the methods to be used in construction, and an ability to write clearly. The legal importance of specifications is that they usually govern and take precedence over drawings, unless the specifications specifically state that the drawings govern. The specifier must select appropriate words to precisely impart the intended meaning to the reader. In addition, capitalization, punctua-tion, and grammar are important. Often the sentence structure takes the form of the indicative mood in passive voice, which traditionally has been incorporated into specifica-tion writing. This format requires the use of "shall." The imperative mood, however, can be used by placing the verb at the beginning of the sentence. Examples of these methods are:

Indicative Mood: Soil shall be spread in six inch lifts prior to compacting.
Imperative Mood: Compact soil in six inch lifts.

SECTION FORMAT OUTLINE

PART 1 GENERAL

SUMMARY
Section Includes
Products Furnished but not
  Installed Under this Section
Products Installed but not
  Furnished Under this Section
Related Sections
Allowances
Unit Prices
Alternates/Alternatives*

REFERENCES

DEFINITIONS

SYSTEM DESCRIPTION
Design Requirements
Performance Requirements

SUBMITTALS
Product Data
Shop Drawings
Samples
Quality Control Submittals
  Design Data
  Test Reports
  Certificates
  Manufacturer's Instructions
  Manufacturer's Field Reports
Contract Closeout Submittals
  Project Record Documents
  Operation and Maintenance
  Data Warranty

QUALITY ASSURANCE
Qualifications
Regulatory Requirements
Certifications
Field Samples
Mock-Ups
Pre-Installation Conference

DELIVERY, STORAGE, AND HANDLING
Packing and Shipping
Acceptance at Site
Storage and Protection

PROJECT/SITE* CONDITIONS
Environmental Requirements
Existing Conditions
Field Measurements

SEQUENCING AND SCHEDULING

WARRANTY
Special Warranty

MAINTENANCE
Maintenance Service
Extra Materials

PART 2 PRODUCTS

MANUFACTURERS

MATERIALS

MANUFACTURED UNITS

EQUIPMENT

COMPONENTS

ACCESSORIES

MIXES

FABRICATION
Shop Assembly
Shop/Factory/Finishing
Tolerances

SOURCE QUALITY CONTROL
Tests
Inspection
Verification of Performance

PART 3 EXECUTION

EXAMINATION
Verification of Conditions

PREPARATION
Protection
Surface Preparation

ERECTION
INSTALLATION
APPLICATION

Special Techniques
Interface with Other Products
Tolerances

FIELD QUALITY CONTROL
Tests
Inspection
Manufacturer's Field Service

ADJUSTING

CLEANING

DEMONSTRATION

PROTECTION

SCHEDULES

Figure 13.2 Construction Specification Institute's section format. Taken from the Construction Specifications Institute's *Manual of Practice*. Copyright © 1985 by the Construction Specifications Institute. Reprinted by permission of the Construction Specifications Institute, Alexandria, Va.

An excellent presentation on the preparation of a Project Manual is contained in CSI's (1985) *Manual of Practice*.

Development projects using public monies may be required to specify multiple source suppliers for material and products proposed for use in construction. Such provisions normally include at least two manufacturer suppliers and the phrase "or equal" when specifying products. "Or equal" normally refers to dimensional, mechanical, and material aspects of the desired product and can also include performance aspects such as operational speed, efficiency, and horsepower.

Because materials, technology, and construction methods are changing rapidly, the specifier can obtain from CSI the current master guide specifications designated as "SPECTEXT" or "SPECTEXT II." These master specifications are produced by the Construction Sciences Research Foundation (CSRF), and are available in either hardcopy or magnetic media formats (for computer word processing).

13.3   AN EXAMPLE OF A PROJECT MANUAL

Each Project Manual must be individually prepared to meet a specific development's needs. Examples of various typical sections are shown here to illustrate the general requirements of a typical privately funded improved land development. The following table of contents is applicable to projects that consist of the construction of roads, storm drainage facilities, potable water distribution facilities, and wastewater collection and flooded suction pumping systems similar to the one shown in Figure 10.28. CSI/CSC *MASTERFORMAT* section designations used in this example follow:

TYPICAL PROJECT MANUAL

TABLE OF CONTENTS

INTRODUCTORY INFORMATION

00001   Cover
00002   Certification Page
00003   Table of Contents
00004   List of Drawings, Tables, and Schedules
00005   Project Consultants

**BIDDING REQUIREMENTS, CONTRACT FORMS,
AND CONDITIONS OF THE CONTRACT**

00020   Invitation to Bid
00100   Instructions to Bidders
00220   Geotechnical Data
00300   Bid Forms
00410   Bid Security Forms
00420   Bidder's Qualification Forms

00430   Subcontractor Listing
00470   List of Estimated Quantities
00500   Agreement Forms
00610   Performance Bonds
00620   Payment Bonds
00650   Certificates of Insurance
00670   Affidavit of Prime Contractor*
00680   Release and Waiver of Claims*
00690   Contractor's General Warranty*
00700   General Conditions
00800   Supplementary Conditions
00900   Addenda*

SPECIFICATIONS

Division 1—General Requirements

01025   Measurement and Payment
01035   Modification Procedures
01040   Coordination
01050   Field Engineering
01060   Regulatory Requirements*
01200   Project Meetings
01310   Progress Schedules
01340   Shop Drawings, Product Data, and Samples
01380   Construction Photographs
01410   Testing Laboratory Services
01445   Manufacturer's Field Services
01500   Construction Facilities and Temporary Controls*
01610   Delivery, Storage, and Handling
01650   Facility Startup/Commissioning
01710   Final Cleaning
01720   Project Record Documents
01730   Operation and Maintenance Data*
01760   Warranty Inspection

Division 2—Sitework

02110   Site Clearing*
02140   Dewatering*
02150   Shoring and Underpinning*
02210   Grading*
02220   Excavating, Backfilling, and Compacting*

02230   Base Courses*
02240   Soil Stabilization*
02270   Slope Protection and Erosion Control*
02510   Asphaltic Concrete Paving*
02520   Portland Cement Concrete Paving
02600   Water Distribution*
02720   Storm Sewerage*
02730   Sanitary Sewerage*
02770   Ponds and Reservoirs
02800   Site Improvements
02900   Landscaping

Division 3—Concrete

03300   Cast in Place Concrete
03600   Grout

Division 4—Masonry

04210   Clay Unit Masonry

Division 5—Metals

05510   Metal Stairs
05520   Handrails and Railings
05540   Castings

Division 6—Wood and Plastics

06100   Rough Carpentry
06190   Wood Trusses

Division 7—Thermal and Moisture Protection

07110   Sheet Membrane Waterproofing
07300   Shingles and Roofing Tiles

Division 8—Doors and Windows

08100   Metal Doors and Frames
08330   Coiling Doors and Grilles

Division 9—Finishes

09900   Painting

Division 10—Specialties

10200   Louvers and Vents

Division 11—Equipment

11310   Sewage and Sludge Pumps

Division 12—Furnishings (Omit)

Division 13—Special Construction

13152   Swimming Pools

Division 14—Conveying Systems (Omit)

Division 15—Mechanical

15500   Heating, Ventilating, and Air Conditioning

Division 16—Electrical

16050   Basic Electrical Materials and Methods
16400   Service and Distribution
16500   Lighting
16900   Controls
16950   Testing

Because of space limitations, an example of a complete Project Manual cannot be included; however several illustrative sections (designated with asterisks above) have been included.

SECTION 00670

AFFIDAVIT OF PRIME CONTRACTOR

(PAYMENT STATUS—SUBCONTRACTORS AND MATERIAL SUPPLIERS)

Date: _____

We, _____Name of Prime Contractor_____ , do hereby certify that to the best of our knowledge and belief no claims or liens exist against any subcontractors or material suppliers who furnished labor and/or material for the construction of _____Project's Name and Location_____ . We shall hold harmless the Owner from any subsequent liens made of record. If any liens remain unsatisfied after final payment is made to the Contractor, the Owner will satisfy the lien, and all incurred costs including reasonable attorney's fees will be immediately due and paid to the Owner by the Contractor.

Sworn to and subscribed before me this _____ day of ____19____ .

*Example sections follow.

_____ Signature _____

By: _____

Title: _____

Date: _____

Notarized By: _____   Seal:

Date: _____

Information contained in this section is subject to change and is intended for educational purposes only. CSI, CSC, and CSRF did not prepare nor do they specifically endorse this material. The current format is contained in CSI/CSC's effective edition of MASTERFOR-MAT along with current CSI sponsored master guide specifications containing CSRF's SPECTEXT/SPECTEXT II.

End of Section 00670

SECTION 00680

RELEASE AND WAIVER OF CLAIM

On _____ , 19 _____ there personally appeared before me the undersigned authority in and for said County of _____ and State of _____, __Authorized Person's Name__, who, being duly sworn, states that all material bills, sales and other taxes, licenses, payrolls, state and federal unemployment insurance, and all other liabilities have been paid in full that were incurred in the performance of the contract for __Project's Name and Location__ , and waives any claims and releases the Owner from any rights or claims for debts due as a result of any material or supply liens.

Sworn to and subscribed before me this _____ day of _____ 19 ____ .

_____ Signature _____

By: _____

Title; _____

Date: _____

Notarized By: _____   Seal:

Date: _____

Information contained in this section is subject to change and is intended for educational purposes only. CSI, CSC, and CSRF did not prepare nor do they specifically endorse this material. The current format is contained in CSI/CSC's effective edition of MASTERFOR-MAT along with current CSI sponsored master guide specifications containing CSRF's SPECTEXT/SPECTEXT II.

End Section 00680

SECTION 00690

CONTRACTOR'S GENERAL WARRANTY

PROJECT'S NAME

PROJECT'S LOCATION

The undersigned Contractor does hereby warrant in accordance with the provisions and terms of the Contract Documents, all materials and workmanship incorporated in _____Project's Name and Location_, against any and all defects due to faulty materials, workmanship, or negligence for a period of 12 months for all streets and drainage facilities, and 12 months for all other improvements based on the effective date of this warranty. This Contractor further warrants that all work incorporated in this project will remain leakproof and watertight at all points for a period of 24 months from the effective date of this warranty.

This warranty shall be binding where defects occur due to normal use and does not cover wanton and malicious damages, damage caused by acts of God, or damage caused by casualties not controlled by the Contractor.

This warranty shall be in addition to all other warranties and guarantees set forth in the Contract Documents, and shall not act to constitute a waiver of additional protection of the Owner if afforded by consumer protection and product liability provisions of the law, and these stipulations shall not constitute waivers of any additional rights or remedies available to the Owner under the law.

Signature: _____

Name: _____

Title: _____

Date: _____

Subscribed and sworn before me this _____day of ____, 19____ .

Notarized By: _____   Seal

Date: _____

Information contained in this section is subject to change and is intended for educational purposes only. CSI, CSC, and CSRF did not prepare nor do they specifically endorse this material. The current format is contained in CSI/CSC's effective edition of MASTERFORMAT along with current CSI sponsored master guide specifications containing CSRF's SPECTEXT/SPECTEXT II.

End of Section 00690

SECTION 00900

ADDENDA

ADDENDUM NO. 1

PROJECT'S NAME

PROJECT SECTION OR PHASE

    1. Modification to the contract documents pertaining to the Project cited above is as follows:

        A. *Drawings:* [Cite sheets with modifications or changes.]

        B. *Project Manual:* [Cite applicable Sections and paragraphs requiring modifications or changes.]

        C. *Other:* [State applicable information.]

Engineer's Signature
_____
Engineer's Name

Date

Engineer's Seal

End Section 00900

SECTION 01060

REGULATORY REQUIREMENTS

Part 1—General

1.01 Summary:

    A. Section contains provisions for compliance with jurisdictional regulations affecting the Work.

    B. Related Sections include but are not necessarily limited to:

        1. General Conditions.

        2. Supplementary Conditions.

        3. Sections in Division 1 of this Project Manual.

1.02 References:

    A. Regulatory requirements required by the Contract Documents include, but are not necessarily limited to, obtaining, maintaining, or complying with:

        1. Owner Procured:

            a. Sedimentation and Erosion Control Permits.

            b. Highway Department Encroachment Permits.

            c. Potable Water System Construction Permit.

            d. Permit to Construct in Navigable Streams.

            e. Encroachment Permits for Public Utilities.

      f.   Permit to Construct Storm Drainage System.

      g. Coastal Zone Construction Permits.

      h. Permits to Withdraw Surface or Subsurface Waters.

      i.   Permit to Construct Public Swimming Facilities.

  2. Contractor Procured:

      a. Current business license.

      b. Current General Contractor's License for State in which the Project is located.

      c. Building permits not included in this Project Manual for temporary offices and buildings.

      d. Burn permits.

      e. Fire, hazardous waste, safety, and OSHA Standards.

      f.   Permit for Hauling Waste and Disposal.

      g. Permit for temporary utility services and construction coordination for Project utility connections.

      h. Permit to exceed load limits of certain bridges if they are on haul route and have very low load limits.

  3. Permits procured by the Owner and/or Contractor, if the Contractor undertakes temporary or independent activity.

      a. Filling Permits.

      b. Permits to Discharge Water.

      c. Sign Permits.

      d. Landscape and Tree Removal Permits.

      e. Permits to restrict work in temporary construction easements.

1.03 Submittals:

  A. Comply with germane portions of Section 01340.

  B. Provide required reports to regulatory agencies in accordance with provisions of issued permits.

1.04 Quality Assurance:

  A. Designate in writing to the Owner a member of the Contractor's organization to monitor compliance with provisions of issued permits.

  B. Conduct inspections of construction activities to insure compliance.

1.05 Site Conditions:

  A. Practice and be fully responsible for good housekeeping activities and procedures to prevent oil spills, hazardous waste contamination, unauthorized environmental pollution, and loss of safety.

  B. Maintain drainage features while incorporating sedimentation and erosion control measures.

  C. Maintain traffic using recognized traffic zone control measures.

1.06 Sequencing and Scheduling:

  A. Provide ample notification to regulatory agencies for scheduled construction activities if a regulatory inspector is required for observation.

[Include copies of owner procured permits in this section.]

Information contained in this section is subject to change and is intended for educational purposes only. CSI, CSC, and CSRF did not prepare nor do they specifically endorse this material. The current format is contained in CSI/CSC's effective edition of MASTERFOR-

MAT along with current CSI sponsored master guide specifications containing CSRF's SPECTEXT/SPECTEXT II.

End of Section 01060

SECTION 01500

CONSTRUCTION FACILITIES AND TEMPORARY CONTROLS

Part 1—General

1.01   Summary:
A.   Section includes provisions for construction facilities and temporary controls necessary for and as part of the Work.
B.   Related Sections include but are not necessarily limited to:
1.   General Conditions.
2.   Supplementary Conditions.
3.   Sections in Division 1 of this Project Manual.
4.   Section 02270.

1.02   Submittals:
A.   Comply with provisions of Section 01340.
B.   Submit temporary drainage plans along with sedimentation and erosion control plans to the Engineer for approval. Include schedule of activities required by NPDES General Permit for Storm Water, and keep schedule current.

1.03   Quality Assurance:
A.   Instruct personnel to maintain cleanliness throughout the construction area and prevent unauthorized pollution.
B.   Conduct daily inspections to insure that standards are being maintained.

1.04   Delivery, Storage, and Handling:
A.   Monitor and maintain construction facilities and temporary control in proper working order to promote safety while prosecuting the Work.

1.05   Site Conditions:
A.   Provide construction facilities and temporary controls necessary to conduct the Work, including but not necessarily be limited to:
1.   Field Office and Sheds.
2.   Temporary Storage Areas.
3.   Temporary Sanitary Facilities.
4.   Temporary Utilities.
5.   Temporary Fencing, Barriers, Enclosures, and Security Devices.
6.   Temporary Safety Provisions.
7.   Temporary Drainage Facilities.
8.   Temporary Signage for Identification and Traffic Control.
9.   Temporary Measures for Dust Control.
10.   Routine Cleaning and Waste Removal.

1.06   Maintenance:
A.   Maintain all areas, improvements, work areas and access roads for clear passage and safe functional use.

B. Locate field offices, temporary structures, and storage areas in accordance with the Engineer's directives.

Part 2—Products

2.01 Components:
   A. Provide temporary field offices and storage areas for use during construction.
   B. Provide temporary water supply to job site for use during construction.
      1. Arrange and pay for connection and fees.
      2. Install temporary piping and fittings.
      3. Pay for water used for construction.
      4. Remove temporary connection and piping at completion of job.
   C. Provide temporary electrical power to the job site during construction.
      1. Arrange and pay for connection and fees.
      2. Install temporary wiring and accessories.
      3. Pay for electricity used during construction.
      4. Remove temporary connection and wiring at completion of job.
   D. Provide temporary telephone service to the job site during construction.
      1. Arrange and pay for connection and fees.
      2. Install temporary wiring and accessories.
      3. Pay for telephone service used during construction.
      4. Remove temporary connection and wiring at completion of job.
   E. Provide and pay for temporary heat to the job site during construction.
   F. Provide warning signs, barricades, and temporary safety devices in compliance with regulatory requirements.
   G. Provide temporary fencing and security to prevent unauthorized entry by the public.
   H. Provide signs to identify the project site and to control traffic. Signs must be acceptable to the Engineer.
   I. Provide as part of the Contract Price all temporary drainage facilities necessary to prosecute the Work in a timely manner. The temporary drainage plan must conform to the overall drainage and sedimentation and erosion control plan for the site and be approved by the Engineer and any required regulatory authorities and provisions of Section 02270. This contractor prepared plan must meet all requirements of the NPDES General Permit for Storm Water Discharge pertaining to this project.
   J. Provide hygroscopic materials for dust control that are acceptable to the Engineer.

Part 3—Execution

3.01 Erection, Installation, and Application:
   A. Provide suitable on-site field offices and sheds to accommodate Contractor needs.
      1. Facilities shall include adequate space for project meetings with tables, chairs, and utilities.
      2. Provide and maintain adequate sanitary facilities for use by all personnel.
   B. Maintain temporary facilities and controls to provide safe working conditions and prevent unauthorized persons from entering the site.

C. Remove temporary controls and facilities when use is no longer required and grade area. Provide erosion control to the graded surface and clean as required in Section 01710.

D. Follow provisions of Section 02140 to provide temporary drainage by dewatering.

E. Provide temporary protection to existing improvement including fences, poles, wires, pipes, property corners, and so on, that the Engineer deems necessary to preserve. If during the course of construction, the preserved items are injured, the Contractor shall restore them to original conditions immediately. When fences interfere with the Contractor's operation, permission must be requested to temporarily remove during construction and then immediately replace. No additional compensation shall be allowed to the Contractor for this work. Trees and shrubs so designated shall be saved and protected by the contractor from equipment, materials, and employee abuse.

F. Erect a temporary six foot high wire fence around unattended open trenches to seal off. Place bright plastic ribbon at random on the fence for warning and position hazard signs with blinking lights prominently around the perimeter for caution at conspicuous intervals.

G. Continuously maintain cleanliness and orderliness around storage areas during construction to allow safe passage, good drainage, and access.

   1. Pick up scrap, waste, debris, and spoil daily and place in temporary holding areas.

   2. Straighten and restack storage piles at least once each week.

   3. Remove debris, waste, spoil, and scrap material from the construction site and dispose of by approved means at least every other week, unless specifically directed in writing to do otherwise by the Engineer.

   4. Responsibility for the disposition of all transported waste material shall be that of the Contractor only.

   5. Inspect the construction site every 7 days or within 24 hours after a rainfall of 0.5 inch or more to determine if erosion has occurred and if sediment accumulations warrant removal.

H. Apply calcium chloride or other hygroscopic materials for dust control using application rates specified by the Engineer.

Information contained in this section is subject to change and is intended for educational purposes only. CSI, CSC, and CSRF did not prepare nor do they specifically endorse this material. The current format is contained in CSI/CSC's effective edition of MASTERFORMAT along with current CSI sponsored master guide specifications containing CSRF's SPECTEXT/SPECTEXT II.

End of Section 01500

SECTION 01730

OPERATION AND MAINTENANCE DATA

Part 1—General

1.01 Summary:

    A. Section includes provisions for providing information necessary to properly operate and maintain installed systems.

    B. Related Sections include but are not necessarily limited to:

        1. General Conditions.

        2. Supplementary Conditions.

        3. Sections contained in Division 1 of this Project Manual.

1.02 Submittals:

    A. Prepare and submit two copies of proposed Operation and Maintenance Manuals for each system requiring operation and maintenance for review by the Engineer.

    B. Prepare and submit seven copies of each Operation and Maintenance Manual, incorporating comments provided by the Engineer's review of the proposed manuals.

1.03 Quality Assurance:

    A. Utilize only knowledgeable personnel and manufacturers recommendations to prepare Operational and Maintenance Manuals.

Part 2—Products

2.01 Materials:

    A. Prepare Operation and Maintenance Manuals using only $8\frac{1}{2}$ inch $\times$ 11 inch white bond paper with text being electronically or mechanically printed and bound on the left margin with front and back cover sheets.

    B. Label each manual to indicate the applicable system for which the manual was prepared, the Contractor's name, and a place for the Engineer's approval.

2.02 Components:

    A. Include in each manual:

        1. Table of Contents.

        2. Complete operating instructions including assembly and disassembly procedures, lubrication procedures and lubricants, and required special tools and spare parts.

        3. Means to identify and describe all components of each piece of equipment.

        4. Parts listing of all components including supplier names and addresses.

        5. Manufacturer's data sheets and drawings.

        6. Troubleshooting guides for malfunctions.

Part 3—Execution

3.01 Preparation:

    A. Obtain manufacturer's recommended operation and maintenance procedures.

3.02 Application:
- A. Instruct designated operator(s) in the proper operation and maintenance of the system with appropriate manufacturer's representatives.

Information contained in this section is subject to change and is intended for educational purposes only. CSI, CSC, and CSRF did not prepare nor do they specifically endorse this material. The current format is contained in CSI/CSC's effective edition of MASTERFOR-MAT along with current CSI sponsored master guide specifications containing CSRF's SPECTEXT/SPECTEXT II.

End of Section 01730

Division 2—Sitework

SECTION 02110

SITE CLEARING

Part 1—General

1.01 Summary:
- A. Section includes procedures for clearing and grubbing areas shown on the Drawings and where required in these Specifications as part of the Work.
- B. Related Sections include but are not necessarily limited to:
  1. General Conditions.
  2. Supplementary Conditions.
  3. Sections contained in Division 1 of this Project Manual.

1.02 Quality Assurance:
- A. Detail sufficient numbers of qualified workmen that are experienced in executing the activities specified in this Section.
- B. Detail and maintain sufficient equipment capable of accomplishing the activities specified in this Section while producing sufficient progress on the Work.
- C. Operate all equipment with due caution.
- D. Instruct workmen in procedures specified in this Section.

1.03 Delivery, Storage, and Handling:
- A. Adhere to pertinent portions of Section 01500.

1.04 Maintenance:
- A. Restore material and property on which damage has been inflicted to predamaged conditions.

Part 2—Products

2.01 Materials:
- A. Provide incidental materials required for satisfactory completion of the activities specified in this Section.

Part 3—Execution

3.01 Examination:
   A. Examine areas where clearing and grubbing activities must be performed and verify that limits of clearing are field staked.
   B. Request underground utility location service to locate utilities within work areas.
   C. Determine prior to commencing clearing operations whether temporary drainage is required and incorporate in accordance with Section 01500 if needed.
   D. Review erosion control plan.

3.02 Preparation:
   A. Protect improvements from equipment, material, and employee abuse to include:
      1. Property corners, survey field markers, and monuments.
      2. Designated trees, shrubs, and grassed areas.
      3. Existing utilities, sedimentation and erosion control systems, property improvements, pavements, sidewalks, driveways, curbing, poles, wires, pipes, and fences unless otherwise noted on the Drawings or specifically designated for removal by the Engineer. When fences interfere with construction operations, request permission from the Owner to temporarily remove during the time work is being conducted in the area and immediately replace upon completion at no additional cost to the Owner.
      4. All areas not within the construction site boundary subject to falling limbs, trees, and dust.
      5. Persons and personal property in or adjoining the work area.
   B. Provide for dust control if measures are warranted in accordance with Section 01500.
   C. Insure sedimentation and erosion control procedures are utilized.

3.03 Erection, Installation, Application:
   A. Clear and grub the width of road rights of way, easements, drainage facilities, utility sites, and any other specified areas shown on the Drawings.
   B. Restrict clearing and grubbing activities to within five feet in back of median curbs where divided entrance curb and gutter street sections are to be constructed. Protect trees and shrubs in addition to areas specifically noted on the Drawings or designated by regulatory authorities.
   C. Burn trash, brush, roots, and unwanted trees in accordance with Contractor procured burn permit requirements, or remove cleared materials off-site in accordance with Section 01060 unless otherwise directed by the Engineer.
   D. Salable timber obtained from clearing operations shall become the property of the Contractor and shall be moved off-site while clearing operations are under way.
   E. Remove all roots and vegetation 1 inch and larger from cleared areas.
   F. Strip usable topsoil and stockpile in designated areas in accordance with Section 01500 for later landscaping use. Unusable or waste material will either be placed on-site in accordance with written instructions from the Engineer, or moved and be properly disposed of off-site by the Contractor at an approved location of the Contractor's choice in accordance with Section 01060.

Information contained in this section is subject to change and is intended for educational purposes only. CSI, CSC, and CSRF did not prepare nor do they specifically endorse this material. The current format is contained in CSI/CSC's effective edition of MASTERFOR-MAT along with current CSI sponsored master guide specifications containing CSRF's SPECTEXT/SPECTEXT II.

End of Section 02110

SECTION 02140

DEWATERING

Part 1—General

1.01 Summary:
 A. Section includes provisions for the installation and operation of dewatering facilities required to safely complete the Work in an acceptable manner and on schedule as part of the Contract Price.
 B. Related Sections include but are not necessarily limited to:
  1. General Conditions.
  2. Supplementary Conditions.
  3. Sections in Division 1 of this Project Manual.

1.02 Quality Assurance:
 A. Detail sufficient numbers of qualified workmen that are experienced in executing the activities specified in this Section.
 B. Detail and maintain sufficient equipment capable of undertaking the activities specified in this Section to produce sufficient progress on the Work.
 C. Operate all equipment with due caution.
 D. Instruct workmen in procedures specified in this Section.

Part 2—Products

2.01 Equipment:
 A. Provide equipment necessary to install, maintain, and remove dewatering facilities.

Part 3—Execution

3.01 Examination:
 A. Examine areas where dewatering activities must be performed and verify whether existing conditions are suitable to safely undertaking dewatering. Take any corrective action required prior to commencing dewatering.
 B. Request underground utility service to locate all utilities within work area.
 C. Determine whether temporary drainage is required and incorporate in accordance with Section 01500.

3.02 Preparation:
 A. Protect improvements from equipment, material, and employee abuse including:
  1. Property corners, survey field markers, and monuments.

  2. Designated trees, shrubs, and grassed areas.

  3. Existing utilities, sedimentation and erosion control systems, property improvements, pavements, sidewalks, driveways, curbing, poles, wires, pipes, and fences unless otherwise noted on the Drawings or specifically designated for removal by the Engineer. When fences interfere with construction operations, request permission from the Owner to temporarily remove during the time work is being conducted in the area and immediately replace upon completion at no additional cost to the Owner.

  4. All areas not within the construction site boundary subject to the discharge of pumped water.

  5. Persons and personal property in or adjoining the work area.

3.03 Installation:

  A. Provide well points, headers, piping, pumps, and other required equipment to safely remove unwanted water from the construction work area.

3.04 Protection:

  A. Discharge pumped water into an approved storm drain, natural drainage area away from the construction area, or into a specified area in accordance with Section 01060. Prevent dewatering discharges from being directed on construction surfaces such as road subgrades.

Information contained in this section is subject to change and is intended for educational purposes only. CSI, CSC, and CSRF did not prepare nor do they specifically endorse this material. The current format is contained in CSI/CSC's effective edition of MASTERFORMAT along with current CSI sponsored master guide specifications containing CSRF's SPECTEXT/SPECTEXT II.

End of Section 02140

SECTION 02150

SHORING AND UNDERPINNING

Part 1—General

1.01 Summary:

  A. Section includes procedures to provide shoring in all excavations not having sloped walls five feet or greater in depth to protect workers, material, real and personal property, and bystanders, unless poor soil conditions or overburden loads warrant shoring for excavations of less than five feet as part of the Contract Price.

  B. Related Sections include but are not necessarily limited to:

   1. General Conditions.

   2. Supplementary Conditions.

   3. Sections in Division 1 of this Project Manual.

1.02 System Description:

  A. Select an appropriate shoring and/or underpinning system to provide adequate support under all weather and work conditions.

  B. Provide securely held sheathing and bracing systems for trenches used for pipe laying that provide a clear bottom of the trench with a width that is at least 12 inches wider than the bell of the pipe unless otherwise permitted by the Engineer. In water bearing soil, no portion of the sheathing below the level of four feet above the pipe shall be removed.

1.03   Submittals:

    A.   Adhere to germane portions of Section 01340.

    B.   Obtain regulatory approved shoring design with a Contractor procured duly registered structural field engineer and submit the approved design to the Engineer for record purposes prior to installing the shoring system.

    C.   Provide the Engineer with a report of an inspection conducted by the responsible Contractor procured field engineer under whose direction the installed shoring system has been designed. This report should certify the shoring installation once installed.

1.04   Quality Assurance:

    A.   Provide workers who are experienced in the skills necessary to install and work around shoring.

    B.   Detail and maintain sufficient equipment capable of undertaking the activities specified in this Section to produce sufficient progress on the Work.

    C.   Operate all equipment with due caution.

    D.   Instruct workmen in procedures specified in this Section.

    E.   Employ an experienced licensed professional field engineer qualified to design and inspect a completed system.

    F.   Require shoring plan to be coordinated with other proposed improvements and work space requirements.

1.05   Warranty:

    A.   Warrant the safety, workability, adaptability, and utility of the installed shoring system in conjunction with the Contractor procured field engineer while holding other parties harmless.

1.06   Maintenance:

    A.   Monitor dewatering systems to insure proper operation and provide maintenance to the installed shoring system when required.

Part 2—Products

2.01   Components:

    A.   Provide design calculations with material and erection drawings for the proposed shoring system to include provisions necessary to provide lateral support to surrounding properties.

    B.   Submit and obtain required approvals from regulatory agencies for the proposed shoring system. Pay regulatory fees at no additional cost to the Owner.

2.02   Materials:

    A.   Provide materials required to install the shoring system as shown on Contractor procured field engineer's approved plans.

2.03   Equipment:

    A.   Provide equipment necessary to install, maintain, and remove the shoring system.

Part 3—Execution

3.01   Examination:

A.   Inspect the area where the shoring system is to be located and determine whether conditions are suitable for installation to begin.

B.   Inspect surrounding area to verify that conditions have not changed that would alter the shoring system design, such as newly placed overburden loads.

C.   Correct any deficiencies prior to commencing shoring installation.

3.02   Preparation:

A.   Request underground utility location service to locate underground utilities within the work area.

B.   Provide adequate access for soil handling and material.

C.   Provide adequate surface and subsurface drainage in accordance with Section 02140.

3.03   Installation:

A.   Install the proposed shoring system in accordance with the approved shoring plans.

Information contained in this section is subject to change and is intended for educational purposes only. CSI, CSC, and CSRF did not prepare not do they specifically endorse this material. The current format is contained in CSI/CSC's effective edition of MASTERFOR-MAT along with current CSI sponsored master guide specifications containing CSRF's SPECTEXT/SPECTEXT II.

End of Section 02150

SECTION 02210

GRADING

Part 1—General

1.01   Summary:

A.   Section includes provisions for the conduct of excavation, backfill, compaction, and grading of earthwork required as part of the Work.

B.   Related Sections include but are not necessarily limited to:

1.   General Conditions.

2.   Supplementary Conditions.

3.   Sections on Division 1 of this Project Manual.

4.   Section 00220 for Geotechnical Data.

1.02   Submittals:

A.   Submit satisfactory results of field density tests conducted in accordance with Section 01410 and reviewed by the Geotechnical Consultant in accordance with Section 00220 certifying densities attained prior to requesting approval from the Engineer.

B.   Submit shoring plan when required in accordance with Section 02150.

1.03   Quality Assurance:
   A.   Detail sufficient numbers of qualified workmen who are experienced in executing the activities specified in this Section.
   B.   Detail and maintain equipment capable of undertaking the activities specified in this Section to produce sufficient progress on the Work.
   C.   Check final graded surface for elevations and alignment. Use grades obtained from construction Drawings as finished grades unless directed otherwise. Account for pavement thickness requirements, foundation slab thicknesses, and other structural features to produce the final graded elevations.
   D.   Limit extent of earthwork to authorized areas designated on the Drawings. Fill no wetlands unless authorized by permit. Allow no excavation under existing structure foundations, unless authorized in writing by the Engineer.
   E.   Instruct workmen in procedures specified in this section and operate equipment with due caution.
   F.   Insure finished graded surfaces are free from roots, trash, and other foreign material.

1.04   Delivery, Storage, and Handling:
   A.   Adhere to germane portions of Section 01610.
   B.   Remove excess soil from the area being graded. Do not deposit or store soil closer than five feet from any drainage ditch, pond, or catchment.
   C.   Utilize temporary storage areas provided in Section 01500.

1.05   Sequencing and Scheduling:
   A.   Coordinate the requirements of this Section with Section 02110.

1.06   Maintenance:
   A.   Maintain completed areas by scarifying, reshaping, or recompacting to correct deficiencies attributed to construction related activities or weather.
   B.   Maintain dust free environment in accordance with Section 01500.
   C.   Maintain drainage.

Part 2—Products

2.01   Materials:
   A.   Classify soils into the following categories:
      1.   Embankment (fill) materials.
      2.   Unclassified (general) excavation materials.
      3.   Classified excavation materials, which are grouped into the following categories:
         a.   Muck
         b.   Rock
         c.   Topsoil
   B.   Embankment materials shall include:
      1.   Only soils that have a maximum four inch size and that are free from organic and deleterious material.
      2.   Only approved on-site soil or approved off-site borrow soil that is included as part of the Contract Price.
   C.   Suitable unclassified excavated materials shall be reused for embankments, with unsuitable material being handled as waste as part of the Contract Price.

D. Classified excavation materials include:
   1. Extensive unusable material encountered during grading operations, which is classified as muck.
   2. Soil extensively comprised of rocks larger than six inches in size; removal of this material is classified as rock excavation. No rock excavation is anticipated in the Work.
   3. Topsoil, which shall be:
      a. Fertile organic soil, preferably of a loamy nature to support the planting of cover vegetation, and free from noxious weeds, debris, roots, litter, and other undesirable material.
      b. Provided normally from approved stockpiled on-site strippings resulting from Section 02110, unless additional approved off-site borrow is required to complete the Work.

Part 3—Execution

3.01 Examination:
   A. Inspect areas to be graded and determine whether conditions are suitable to undertake the portions of work specified in this Section. Do not commence work under this Section until conditions are satisfactory.
   B. Request underground utility service to locate all utilities within work area.
   C. Verify line and grade stakes and report any obvious discrepancies to the Engineer for rectification.

3.02 Preparation:
   A. Identify, locate, and protect utilities located in areas to be graded.
   B. Protect traffic, pedestrians, and the public including pavements, sidewalks, utilities, and real and personal property by barricading open cuts with positive restraints and marking as hazardous the work area with signs, high visibility plastic ribbon, and blinking barricade lights for nighttime use.
   C. Provide adequate drainage to the work area as provided in Section 01500.
   D. Improve poor soil in areas shown on the Drawings or as directed by the Geotechnical Consultant using procedures contained in Section 02240.

3.03 Application:
   A. Excavate using these procedures:
      1. Conduct excavations required to provide the required grades and elevations shown on the Drawings and required by these specifications.
         a. Shape cross sections in accordance with the Drawings including swales, ditches, slopes, and other areas shown.
         b. Grade swale sections beginning with zero cut at the point farthest from the discharge outlet and gradually taper the invert to the discharge point where the cut does not exceed $2\frac{1}{2}$ feet unless otherwise directed or shown on the Drawings.
      2. Excavate, transport, deposit, spread, and compact suitable general excavation material in embankment areas. Stockpile or remove from the construction sites excess material as designated by the Engineer.
      3. Handle unusable general excavation material as waste in accordance with Section 01060.

4. Muck areas shown on plans. Mucked material will be excavated to limits established by the Geotechnical Consultant and handled as waste in accordance with Section 01060. For roadbeds, the roadway shall be removed for the entire width, ditch line to ditch line (or two feet in back of curb to back of curb) for a depth specified by the Geotechnical Consultant. Provide suitable backfill material and compact in accordance with these specifications to the desired grade.

5. Construct side slopes no greater than a 1:1 gradient unless directed otherwise by the Geotechnical Consultant or in the Construction Documents. Provide and maintain shoring until excavation is backfilled in accordance with Section 02150 if sloping of embankment is not possible.

B. Construct embankments using these procedures:
  1. Surface receiving fill shall be satisfactory to the Geotechnical Consultant and shall be free from frost, ice, mud, and extraneous material such as trash and debris.
  2. Install permanent bracing if required.
  3. Remove shoring as embankment progresses in accordance with the Contractor procured field engineer's recommendations.
  4. Scarify existing surface and compact to minimum acceptable field density.
  5. Place loose fill in lifts not exceeding eight inches thickness and aerate or moisten, if necessary, to attain optimum moisture conditions.
  6. Compact uniformly each layer to the required density and verify prior to proceeding with the next activity.
  7. Exercise care when placing and compacting fill next to structures.

C. Compact soil using these procedures:
  1. Obtain and evaluate field densities in accordance with provisions contained within AASHTO T 191-86.
  2. Maintain soil during compaction at optimum moisture content. Regulate by aeration and scarification or by adding water and mixing.
  3. Densify in accordance with the following requirements:
     a. Pavement subgrade areas, road shoulders, and areas under sidewalks and structures—95 percent Modified Proctor (AASHTO T 180-90) compaction to a depth of eight inches of subgrade and for each layer of placed fill.
     b. All other areas—90 percent Modified Proctor (AASHTO T 180-90) compaction to a depth of eight inches of subgrade and for each layer of placed fill.

D. Grade using these procedures:
  1. Grade areas external to building lines to conform with typical cross sections using field control markers established by Owner procured surveyor and elevations shown on the Drawings. Streets shall be fine graded to grade stakes set on the centerline and edge of right of way at intervals of not more than 100 feet on straight grades and not more than 50 feet on vertical curves. Create uniform surfaces as work progresses, insuring that no detrimental dips or bumps exist that might affect positive drainage resulting in ponding or pavement imperfections.
  2. Grade areas adjacent to building lines to provide positive surface drain-

age and prevent ponding. Shape surface areas to provide a uniform transition throughout.

3.04   Field Quality Control:

    A.   Coordinate and allow testing in accordance with Section 01410 for each compacted lift of fill material in embankments.

    B.   Undertake at least three representative field density tests using AASHTO T 191-86 or other methods approved by the Engineer for each 2500 square feet of area to be paved or constructed upon.

    C.   Recompact and regrade tested areas that indicate unsatisfactory densities in accordance with Section 01410.

    D.   Check completed surface to verify acceptable density, shape, and grade.

3.05   Protection:

    A.   Protect all graded surfaces from erosion and maintain condition of shaped surfaces free from rutting, potholes, and depressions.

    B.   Repair immediately any utilities, or other improvements damaged during grading operations at no cost to the Owner.

    C.   Report to the Engineer utilities found during grading operations that appear to require relocation. Engineer will investigate and coordinate in writing any required activities with the Contractor and utility owner.

    D.   Protect graded areas from pumped water discharges and storm water action. Maintain moisture content of soil at optimum.

    E.   Protect air from excessive dust by controlling dust production in accordance with provisions contained in Section 01500.

Information contained in this section is subject to change and is intended for educational purposes only. CSI, CSC, and CSRF did not prepare nor do they specifically endorse this material. The current format is contained in CSI/CSC's effective edition of MASTERFOR-MAT along with current CSI sponsored master guide specifications containing CSRF's SPECTEXT/SPECTEXT II.

End of Section 02210

SECTION 02220

EXCAVATING, BACKFILLING, AND COMPACTING

Part 1—General

1.01   Summary:

    A.   Section includes procedures for trench and structural excavations, including requirements for pipe bedding, backfilling, and pipe jacking and boring using the dry boring method with a smooth wall steel casing to accomplish the Work.

    B.   Related Sections include but are not necessarily limited to:

        1.   General Conditions.

        2.   Supplementary Conditions.

        3.   Sections in Division 1 of this Project Manual.

        4.   Section 00220 for Geotechnical Data.

1.02   Submittals:

A.   Prior to requesting approval from the Engineer submit satisfactory results of field density tests conducted in accordance with Section 01410 and reviewed by the Geotechnical Consultant in accordance with Section 00220 certifying densities attained.

B.   Provide shoring submittal requirements for all nonsloped wall excavations greater than five feet deep and excavations under five feet where conditions warrant and for jacking and boring installations as specified in Section 02150.

1.03   Quality Assurance:

A.   Detail sufficient numbers of qualified workmen who are experienced in executing the activities specified in this Section.

B.   Detail and maintain equipment capable of undertaking the activities specified in this Section to produce sufficient progress on the Work and operate equipment with due caution.

C.   Check final grades and alignment. Use grades obtained from the Drawings as finished grades unless directed otherwise.

D.   Limit extent of earthwork to authorized areas designated on the Drawings. Fill no wetlands unless authorized by permit. Allow no excavation under existing structure foundations, unless authorized in writing by the Engineer.

E.   Instruct workmen in procedures specified in this Section.

1.04   Delivery, Storage, and Handling:

A.   Adhere to germane portions of Section 01610.

B.   Remove excess soil from the area being graded. Do not deposit or store soil closer than five feet from any drainage ditch, pond, or catchment.

C.   Utilize temporary storage areas provided in Section 01500.

1.05   Sequencing and Scheduling:

A.   Coordinate the requirements of this Section with Sections 02110 and 02210.

B.   Excavations for trenching shall not extend more than 300 feet ahead of pipe laying, nor shall trenching be left open overnight unless permission is received from the Engineer and provisions are implemented in accordance with Section 01500.

C.   Coordinate excavations with the requirements of Section 02150.

1.06   Maintenance:

A.   Maintain completed areas by scarifying, reshaping, recompacting to correct deficiencies attributed to construction related activities or weather.

B.   Maintain dust free environment in accordance with Section 01500. Utilize power brooms, and water flushing as necessary to initially clean pavements where excavations require pavement cutting and maintain a relatively dust free environment until resurfaced.

C.   Maintain drainage.

D.   Correct immediately upon notification by the Engineer any settlement that occurs in trenches during the warranty period. Settlement occurring in asphalt pavement areas shall be repaired by resurfacing the entire width of the street with asphaltic concrete at the expense of the Contractor.

Part 2—Products

2.01   Materials:

A.   Use only new materials conforming to the requirements of these specifications and approved by the Engineer. All materials proposed to be used may be tested and inspected at any time during their preparation and use. If, after trial, it is found that the source of supply has not been approved, or does not furnish a uniform product, or if the product becomes unacceptable at any time, the Contractor shall furnish materials from another source. Approved materials that become unfit for use shall not be used in the Work and shall be removed from the job site. Materials indicated on the Drawings that are required in the Work but are not covered in detail in the Specifications shall be of the best quality available and shall conform to the current specification of the American Society for Testing and Materials (ASTM) or AASHTO, as approved by the Engineer.

B. Backfill shall conform to the requirements of embankment materials under Section 02210.

C. Pipe shall conform to the applicable requirements set forth in Sections 02600, 02720, and 02730.

D. Stone bedding for Class B, Class C, and special pipe bedding usage shall be standard 789 stone.

E. Steel casing for pipe jacking shall have a minimum yield strength of 35,000 pounds per square inch and a minimum wall thickness as shown on the Drawings.

Part 3—Execution

3.01 Examination:

A. Inspect areas to be excavated or jacked and bored to determine whether conditions are suitable for undertaking the portions of work specified in this Section. Do not commence work under this Section until conditions are satisfactory.

B. Request underground utility service to locate all utilities within work area.

C. Verify line and grade stakes and report obvious discrepancies immediately to the Engineer for rectification.

3.02 Preparation:

A. Identify, locate, and protect utilities located in areas to be graded.

B. Include adequate drainage to the work area as provided in Sections 01500 and 02140.

C. Protect traffic, pedestrians, and the public including pavements, sidewalks, utilities, real and personal property by barricading open cuts with positive restraints and adequately marking as hazardous the work area with signs, high visibility plastic ribbon, and blinking barricade lights for nighttime use.

3.03 Application:

A. Excavate using these procedures:

1. Conduct excavation of materials as required to provide the grades and elevations shown on the Drawings and as required by these Specifications. Shape all cross sections in accordance with the Drawings including swales, ditches, slopes, and other areas as provided in Section 02210.

2. Excavate the lower portion of the trench including the area two feet above the pipe to a width that is not greater than the diameter of the bell of the pipe plus 12 inches to prevent undercutting the trench walls. If the trenches are excavated wider than specified or if they collapse, the Contractor shall be required to use special methods of bedding the pipe included in this Section at no additional cost to the Owner.

Terminate mechanical excavations four inches above invert grade with the remaining depth being excavated manually, or after the mechanical excavation has been completed, return suitable fill material to the trench to achieve the proper grade. Mechanically tamp backfill material to insure proper compaction.

3. Provide excavations for manholes, junction boxes, catch basins, and appurtenances so that at least 12 inches to 24 inches outside clearance is obtained on all sides.

4. Transport, deposit, spread, and compact excess suitable general excavation material in embankment areas elsewhere on-site once excavation is backfilled. Stockpile or remove from the construction site excess material as designated by the Engineer.

5. Handle unusable general excavation material as waste in accordance with Section 01060.

6. Muck areas shown on Drawings. Mucked material will be excavated to limits established by the Geotechnical Consultant and handled as waste in accordance with Section 01060. Provide suitable backfill material and compact in accordance with these specifications to the desired grade.

7. Construct side slopes no greater than a 1:1 gradient unless directed otherwise by the Geotechnical Consultant or directed differently in the Contract Documents. Provide and maintain shoring until excavation is backfilled in accordance with Section 02150 if sloping of embankment is not possible.

B. Install piping using these procedures:

1. Comply with applicable requirements set forth in Sections 02600, 02720, and 02730.

2. Unless shown otherwise on the Drawings, provide a minimum of Class B bedding under all storm drainage (subject to traffic and other heavy loads), sanitary sewers, and potable water mains:

    a. Use compacted granular bedding with tamped backfill. Bed the pipe in compacted granular material placed on a flat trench bottom. Provide a minimum granular bed thickness of one-eighth the outside pipe diameter, but not more than four inches nor more than six inches between the barrel and the trench bottom, for the full width of the trench. The granular bedding must extend halfway up the pipe barrel on both sides. Fill the remainder of the sides to a minimum depth of 12 inches over the top of the pipe with carefully compacted material.

    b. The cost of providing Class B bedding shall be included in the cost of laying the pipe without extra payment being allowed.

3. Provide special bedding and tamping if the trench walls collapse or if the trench is excavated in excess of the specified depth. Shape the trench bottom to a depth of four inches below the bottom of the bell. Provide Class B bedding to place the pipe at the proper position. Remove or cut off the sheeting, as required, after the pipe is laid. Backfill the remainder of the trench in accordance with this Section. Be careful to avoid disturbing the pipe grade and alignment. No extra payment will be allowed for these bedding conditions.

Provide special concrete encasement where authorized by the Engineer in accordance with Section 03300. Pour the encasement on an undisturbed trench bottom and provide a minimum thickness of four inches at all points around the pipe's circumference. Payment for encasement shall be made using quantities determined by the Engineer.

4. Install storm drainage pipes (other than arch types) using Class C bedding where pipes are located where they will not be subjected to traffic or other heavy loads. Class C bedding can be provided:

a. Use bed of compacted granular material (789 stone) that provides a minimum of four inches beneath the pipe barrel (but not more than six inches) placed on a flat trench bottom. This material shall extend up the sides of the pipe one-sixth of the outside diameter of the pipe. Lightly compacted backfill for a depth of six inches over the top of the pipe is then added.

b. The cost of providing Class C bedding shall be included in the cost of laying the pipe without extra payment being allowed.

C. Install manholes and appurtenances once the soil is excavated where six inches of crushed stone bedding can be placed and compacted to allow the proper alignment and elevation to be established. Place a temporary eight inch pipe into the stone bed and extend it to the ground surface prior to backfilling to allow temporary dewatering once the backfill is installed in accordance with Section 02140. Remove the temporary pipe at job completion.

D. Soil that is unstable in the opinion of the Geotechnical Consultant shall be removed from under pipe and structures as directed and replaced as specified for fill.

E. Install backfill using these procedures:

Place suitable backfill that is approved by the Geotechnical Consultant over pipes immediately after the pipes have been laid and inspected by the Engineer. The initial backfill material shall be suitable to provide Class B bedding, except as modified above for certain storm drains, and shall be tamped on both sides to prevent pipe displacement. Exercise care to insure that compaction does not damage or deform the pipe. Backfill the remainder of the trench with general excavation material by placing it in six inch lifts and mechanically compacting it to 95 percent of its maximum density using the Modified Proctor (AASHTO T 180-90) method where pavements, sidewalks, structures, or heavy loads are expected. In all other areas place backfill in 12 inch lifts and mechanically compact to 90 percent of its maximum density using the Modified Proctor (AASHTO T 180-90) procedure. Provide in those areas not scheduled to be paved or covered by structures or sidewalks four inches of topsoil to sustain plant growth in accordance with Section 02900.

F. Mechanically compact soil using these procedures:

1. Obtain and evaluate field densities in accordance with provisions contained within AASHTO T 191-86.

2. Maintain soil at optimum moisture content during compaction. Regulate by aeration and scarification or by adding water and mixing.

G. Grade surface areas using these procedures:

1. Grade areas external to building lines using field control markers estab-

lished by Owner procured surveyor and elevations shown on the Drawings. Create uniform surfaces as work progresses, insuring no detrimental dips or bumps exist that might affect positive drainage, resulting in ponding, or pavement imperfections.

2.   Grade areas adjacent to building lines to provide positive surface drainage and prevent ponding. Shape all surface areas to provide a uniform transition throughout.

H.   Pipe jacking and boring:

1.   Install casing in accordance with the manufacturer's requirements and instructions unless otherwise specified. Provide the depth, length, and minimum wall thickness of the casing as shown on the Drawings.

2.   Jacking and boring shall be upgrade, unless prior approval is obtained from the Engineer. Make the bore diameter the same as the outside diameter of the casing plus the thickness of the protective coating. Butt weld each new section of encasement pipe to the section previously jacked into place as the boring progresses. Fill voids with a 1 : 3 portland cement pressure grout at 50 pounds per square inch to insure there will not be any settlement and to provide uniform bearing. Insure that the alignment and grade of the casing will allow the carrier pipe to be installed as shown on the Drawings. Fill all voids between the casing and carrier pipe with grout unless otherwise directed by the Engineer or shown on the Drawings.

3.04   Field Quality Control:

A.   Coordinate and allow field testing in accordance with Section 01410.

B.   Undertake at least three representative field density tests using AASHTO T 191-86 or other methods approved by the Engineer for each 2500 square feet of areas to be paved or constructed upon.

C.   Recompact and regrade tested areas indicating unsatisfactory densities in accordance with Section 01410.

D.   Check completed surface to verify acceptable density, shape, and grade.

E.   Check line and grade of all installed storm water mains, potable water mains, and sanitary sewer mains. Clean interior of all conduits and appurtenances. Conduct integrity tests for leaks and correct any found prior to notifying the Engineer of readiness to conduct record tests.

3.05   Protection:

A.   Protect graded surfaces from erosion and maintain condition of shaped surfaces free from rutting, potholes, and depressions.

B.   Repair immediately utilities or other improvements damaged during grading operations at no cost to the Owner.

C.   Report to the Engineer utilities found during excavation operations that appear to require relocation. Engineer will investigate and coordinate in writing any required activities with the Contractor and utility owner.

D.   Protect areas from pumped water discharges and storm water action. Maintain moisture content of soil at optimum.

E.   Protect air from excessive dust by controlling dust production in accordance with Section 01500.

F.   Use extra care when trenching along the edge of an existing pavement. Resurface the entire width of the street if the pavement edge is damaged for the entire length of the damaged area plus twenty feet past on each end in accordance with provisions of Section 02510.

Information contained in this section is subject to change and is intended for educational purposes only. CSI, CSC, and CSRF did not prepare nor do they specifically endorse this material. The current format is contained in CSI/CSC's effective edition of MASTERFOR-MAT along with current CSI sponsored master guide specifications containing CSRF's SPECTEXT/SPECTEXT II.

End of Section 02220

SECTION 02230

BASE COURSES

Part 1—General

1.01 Summary:
   A. Section includes provisions for macadam base courses composed of crushed stone filled and bound with screenings, placed on a prepared subgrade to the specified shape, dimensions, and elevations as part of the Work.
   Included are provisions for an alternate sand asphalt base course of hot mix asphalt (HMA) composed of fine aggregates and asphalt cement prepared in an approved plant and constructed on a prepared subgrade.
   B. Related Sections include but are not necessarily limited to:
   1. General Conditions.
   2. Supplementary Conditions.
   3. Sections in Division 1 of this Project Manual.
1.02 Submittals:
   A. Follow requirements set forth in Section 01340.
   B. Submit product data for materials and mix design to the Engineer for approval 30 days prior to commencing paving operations.
   C. Include manufacturer's recommended asphalt cement mixing temperature for a viscosity of 170 centistokes (± 20) and minimum compaction temperature for a viscosity of 280 centistokes (± 30).
1.03 Quality Assurance:
   A. Detail sufficient numbers of qualified workmen who are experienced in executing the activities specified in this Section.
   B. Detail and maintain sufficient equipment capable of undertaking the activities specified in this Section to produce sufficient progress on the Work.
   C. Check final graded surface for elevations and alignment. Use grades obtained from construction Drawings as finished grades unless directed otherwise. Account for pavement thickness requirements, foundation slab thicknesses, and other structural features to produce the final graded elevations.
   D. Limit extent of work to authorized areas designated on the Drawings. Fill no wetlands unless authorized by permit. Allow no excavation under existing structure foundations unless authorized in writing by the Engineer.

E.   Instruct workmen in procedures specified in this Section and operate equipment with due caution.

F.   Insure finished graded surfaces are free from foreign material.

1.04   Delivery, Storage, and Handling:

A.   Adhere to germane portions of Section 01610.

B.   Deliver sand asphalt base HMA at temperatures suitable for proper placement and compaction.

C.   Load delivery truck to prevent mix segregation during transport.

1.05   Sequencing:

A.   Coordinate activities of this Section with Sections 02210, 02220, and 02520.

Part 2—Products

2.01   Materials:

A.   Use only new materials conforming to the requirements of these Specifications and approved by the Engineer. All materials proposed to be used may be tested and inspected at any time during their preparation and use. If, after trial, it is found that the source of supply has not been approved, or does not furnish a uniform product, or if the product becomes unacceptable at any time, the Contractor shall furnish materials from another source. Approved materials that become unfit for use shall not be used in the Work. Materials indicated on the Drawings that are required in the Work but are not covered in detail in the Specifications shall be of the best quality available and shall conform to the current specification of the American Society for Testing and Materials (ASTM) or AASHTO, as approved by the Engineer.

B.   Macadam base materials:

1.   Use coarse aggregates retained on the No. 4 sieve that have hard durable particles of crushed stone and are reasonably free from thin or elongated pieces and organic or other deleterious materials. Include only aggregates that have an abrasion loss of not more than 45 percent when subjected to the Los Angeles Abrasion Test (AASHTO T 96-87).

2.   Provide choke stone consisting of fine aggregates passing the No. 4 sieve produced by crushing operations. Sand shall not be permitted. Use fine aggregates that do not have a liquid limit greater than 25 and a plasticity index greater than 6 when tested in accordance with AASHTO T 89-90 and AASHTO T 90-87 procedures.

3.   Use water that is potable and free from contaminants.

C.   Hot laid sand asphalt base materials shall include:

1.   Asphalt cement that is either AC-20 or AC-30, viscosity graded, and conforms to AASHTO M 226-80 (1986), Table 2.

2.   Rapid cure cutback asphalts that conform to AASHTO M 81-90 and medium cure cutback asphalts that conform to AASHTO M 82-75 (1990).

3.   Emulsified asphalts that conform to AASHTO M 140-88 or AASHTO M 208-87.

4.   Aggregates that are local sand or local sand containing crushed shell, blends of sand and stone, slag or limestone screenings, or other material approved by the Geotechnical Consultant. The approved material shall

be nonplastic and shall provide sharp, hard, and durable grains that are free from deleterious substances while being reasonably graded from coarse to fine.

2.02 Mixes:

A. Hot laid sand asphalt base mix designs shall conform to the procedures contained in AASHTO T 245-90 and shall meet the following:

AGGREGATE GRADATION

Sieve Size	Percent Passing
No. 4	100
No. 8	75–100
No. 16	60–90
No. 30	45–75
No. 50	15–45
No. 100	5–15
No. 200	2–6

Asphalt content: 4.5 to 10 percent

MARSHALL DESIGN CRITERIA USING 50 BLOWS

Minimum stability:	750
Flow, 0.01 inch:	8–18
Percent of voids in total mix:	3–5

Part 3—Execution

3.01 Examination:

A. Examine surface areas to insure that no noticeable moisture exists and that they are properly prepared for HMA. Commence paving operations when deficiencies are corrected.

B. Assess weather conditions for restrictions to paving operations. Sand asphalt base HMA courses can be placed anytime during the year except when the air temperature is below 40°F in the shade and falling or below 35°F in the shade and rising. Sand asphalt base HMA material shall not be placed on wet or frozen surfaces.

C. Inspect delivery of HMA to insure it is covered by tarpaulin and is at an acceptable temperature for proper placement and compaction.

3.02 Preparation:

A. Macadam base courses shall be constructed on a subgrade approved by the Engineer.

B. Sand asphalt base HMA courses shall be placed over a prepared subgrade that has been approved by the Engineer and primed with MC-30, MC-70, RC-30, or RC-70 cutback asphalt at a rate of 0.25 to 0.3 gallon per square yard. Once the prime has set, a tack coat of RS-1, RS-2, CRS-2, MS-1, MS-2, CRS-1, or CMS-2 emulsified asphalt shall be applied between 50°F and 160°F as directed by the Engineer depending on field conditions at a rate of 0.05 to 0.15 gallon per square yard. Allow emulsified asphalt to "break" prior to applying the sand asphalt base HMA.

3.03 Installation:
A. Macadam base courses shall be constructed by:
1. Spreading the coarse aggregate evenly to conform to the required shape, thickness, and elevation. Limit the maximum layer thickness to eight inches. For base depths greater than eight inches place in multiple layers of less than eight inches each.
2. Adding water and mixing the materials without disturbing the prepared subbase. Respread for compaction.
3. Compacting the base material by rolling from the edges and proceeding toward the center, except on superelevated curves where rolling shall proceed from the lower to the upper side and shall continue until the aggregates are firmly keyed to each other.
4. Placing and spreading choke stone on the compacted base surface in sufficient quantities to fill the voids. Wet and broom the surface prior to rolling, and roll until the base is thoroughly compacted to 100 percent maximum density by AASHTO T 180-90 (Method D) and bonded for the full width and depth of the layer.
B. Sand asphalt base HMA courses shall be constructed by:
Placing the sand asphalt base HMA on an approved primed surface using an approved paving machine to produce a compacted layer by rolling while the mixture is sufficiently hot. The minimum thickness after compaction shall be as shown on the Drawings.

3.04 Field Quality Control:
A. Check the finish of the macadam base course so that it does not vary more than $\frac{3}{8}$ inch from a straight edge 10 feet long placed parallel to the centerline, nor more than $\frac{1}{2}$ inch perpendicular to the centerline.
B. Test macadam densities every 1000 feet of roadbed.
C. Use HMA plants to produce sand asphalt base HMA that conforms to AASHTO M 156-89. Heat asphalt cement to the manufacturer's recommended mixing temperature for plant mixing. Compact hot mix in the field at temperatures above the minimum recommended by the manufacturer. Provide haul trucks that have smooth metal beds and are covered. Provide a hole in the bed of each delivery truck for checking mix delivery temperatures. Use a paving machine that includes mechanical devices such as equalizing runners, level arms, and screeds. Place sand asphalt base in layers that do not exceed three inches in compacted thickness. Compact HMA using maximum roller speeds of three miles per hour for static or vibratory rollers and of five miles per hour for pneumatic rollers. Utilize three wheeled steel wheel rollers of 10 to 12 tons in weight or 8 to 12 ton two axle tandem rollers. Provide pneumatic rollers capable of varying tire pressures from 40 to 90 pounds per square inch gauge by adjusting ballast and inflation pressure.
D. Final inspection by regulatory agencies will be scheduled by the Engineer after the work is approved by the Engineer.

3.05 Protection:
A. Maintain completed macadam surface to keep a smooth and true grade and to prevent raveling.
B. Keep traffic off of primed and tack coated surfaces.
C. Protect new HMA surfaces from traffic until sufficiently cooled, set, and cured to prevent rutting, scuffing, and pavement displacement.

Information contained in this section is subject to change and is intended for educational purposes only. CSI, CSC, and CSRF did not prepare nor do they specifically endorse this material. The current format is contained in CSI/CSC's effective edition of MASTERFOR-MAT along with current CSI sponsored master guide specifications containing CSRF's SPECTEXT/SPECTEXT II.

End of Section 02230

SECTION 02240

SOIL STABILIZATION

Part 1—General

1.01 Summary:
   A. Section includes provisions for stabilizing a prepared subgrade to increase its strength by adding crushed stone, gravel, or slag using granular stabilization techniques in areas delineated on the Drawings or as directed by the Engineer as part of the Work.
   B. Related Sections include but are not necessarily limited to:
      1. General Conditions.
      2. Supplementary Conditions.
      3. Sections in Division 1 of this Project Manual.

1.02 Submittals:
   A. Comply with provisions of Section 01340.
   B. Submit for review and approval to the Engineer within 30 working days of receipt of the Notice to Proceed a list of materials necessary for the work encompassed by this Section along with manufacturer's product data and specifications demonstrating compliance with these Specifications.

1.03 Quality Assurance:
   A. Detail sufficient numbers of qualified workmen who are experienced in executing the activities specified in this Section.
   B. Detail and maintain equipment capable of undertaking the activities specified in this Section to produce sufficient progress on the Work.
   C. Check final graded surface for elevations and alignment. Use grades obtained from the Drawings as finished grades unless directed otherwise. Account for pavement thickness requirements, foundation slab thicknesses, and other structural features to produce the final graded elevations.
   D. Limit extent of work to authorized areas designated on the Drawings. Fill no wetlands unless authorized by permit. Allow no excavation under existing structure foundations, unless authorized in writing by the Engineer.
   E. Instruct workmen in procedures specified in this Section and operate equipment with due caution.
   F. Insure all finished graded surfaces are free from foreign material.

1.04 Sequencing and Scheduling:
   A. Conduct stabilization procedures after the designated subgrade has been prepared to receive stabilized material and approved by the Engineer.
   B. After stabilized material has been incorporated into the subgrade material, complete the preparation of the subgrade in accordance with Section 02210.

Part 2—Products

2.01   Materials:

A.   Provide crushed stone composed of tough durable stone that is reasonably free from soft, thin, elongated, or laminated particles and that does not contain organic or deleterious material. Include only aggregates that have an abrasion loss of not more than 45 percent when subject to the Los Angeles Abrasion Test (AASHTO T 96-87) and have a gradation that conforms to:

Sieve Size	Percent Passing by Weight
2 inches	100
$1\frac{1}{2}$ inches	95–100
1 inch	70–100
$\frac{1}{2}$ inch	35–65
No. 4	10–40
No. 8	——

B.   Provide gravel that is composed of hard durable particles of clean stone and that does not contain excessive amounts of thin or elongated pieces and organic or other deleterious materials. Follow other requirements specified for crushed stone under Paragraph A above.

Part 3—Execution

3.01   Examination:

A.   Inspect prepared subgrade areas where stabilization is required including soil that will be under roadway pavements, sidewalks, curbing, curb and guttering, base courses, and shoulders to insure the requirements of this Section can be accomplished. Correct any deficiencies prior to commencing work under this Section.

3.02   Preparation:

A.   Protect all improvements of value from equipment, material, and employee abuse including:

1.   Property corners, survey field markers, and monuments.

2.   Designated trees, shrubs, and grassed areas.

3.   Existing utilities, sedimentation and erosion control systems, property improvements, pavements, sidewalks, driveways, curbing, poles, wires, pipes, and fences unless otherwise noted on the Drawings or specifically designated for removal by the Engineer. When fences interfere with construction operations, request permission from the Owner to temporarily remove during the time work is being conducted in the area, and immediately replace upon completion at no additional cost to the Owner.

4.   All areas not within the construction site boundary subject to the discharge of pumped water.

5.   Persons and personal property in or adjoining the work area.

3.03   Application:

A.   Place aggregate upon the approved subgrade and spread to conform to the required shape and elevations. Mix the aggregate subbase material to the depths shown on the construction drawings or directed by the Geotechnical Consultant.

B. Adjust and maintain the moisture content of the blended mixture to optimum conditions to achieve proper compaction.

C. Spread the mixture to a uniform thickness while at the optimum moisture content. Compact and consolidate the aggregates to provide bond with the subbase in accordance with subgrade requirements in Section 02210.

D. The completed surface shall be compacted to required density and shall be smooth and free from loose material.

3.04 Field Quality Control:

A. Follow provisions contained in Section 02210 for Field Quality Control.

3.05 Protection:

A. Protect the completed stabilized surface by rolling and recompacting to preserve its integrity until the base is applied.

Information contained in this section is subject to change and is intended for educational purposes only. CSI, CSC, and CSRF did not prepare nor do they specifically endorse this material. The current format is contained in CSI/CSC's effective edition of MASTERFORMAT along with current CSI sponsored master guide specifications containing CSRF's SPECTEXT/SPECTEXT II.

End of Section 02240

SECTION 02270

SLOPE PROTECTION AND EROSION CONTROL

Part 1—General

1.01 Summary:

A. Section includes provisions for temporary erosion controls required throughout the Work to control erosion and water pollution by using temporary construction entrances, controlled discharge outlets, berms, sediment basins, sediment dams, silt fences, brush barriers, baled hay and straw, mulching, and temporary seeding.

Provisions are also included for riprap and geotextile fabrics for permanent slope protection as part of the Work.

B. Related Sections include but are not necessarily limited to:
1. General Conditions.
2. Supplementary Conditions.
3. Sections in Division 1 of this Project Manual.

1.02 Submittals:

A. Comply with provisions of Section 01340.

B. At least 96 hours prior to breaking ground submit to the Engineer all information required by the NPDES General Permit for Storm Water pertaining to this project.

C. Submit for review and approval to the Engineer within 30 working days of receipt of the Notice to Proceed:

      1. A list of materials necessary for and the sequencing of activities to accomplish the work encompassed by this Section along with manufacturer's product data and specifications demonstrating compliance with these Specifications.

      2. Installation instructions from applicable manufacturers for products proposed to be included in the installation.

  D. Maintain on-site and keep current all required elements of the NPDES General Permit for Storm Water pertaining to this project.

1.03 Quality Assurance:

  A. Detail qualified workmen that are experienced in executing the activities specified in this Section.

  B. Detail and maintain equipment capable of accomplishing the activities specified in this Section while producing sufficient progress on the Work.

  C. Operate all equipment with due caution.

  D. Instruct workmen in procedures specified in this Section.

1.04 Sequencing and Scheduling:

  A. Install temporary construction entrances as part of the job mobilization activities.

  B. Install temporary perimeter dikes where shown on erosion control plans and where needed to prevent erosion.

  C. Install silt fencing prior to clearing and grubbing to prevent sediment runoff.

  D. Construct sediment traps as soon as practical to control sediments.

  E. Install straw bales and temporary sediment traps as work progresses.

  F. Utilize detention ponds during construction as temporary sediment basins. Clean and place into operation as storm drainage detention facilities when the Work is nearing completion.

1.05 Maintenance:

  A. Routinely inspect the site and clean out sediment traps at least every seven days or within 24 hours after a rainfall of 0.5 inch or more.

  B. Maintain sediment traps, silt fencing, and dikes in good working order.

  C. Correct washouts and eroded areas by refurbishing and temporarily seeding.

Part 2—Products

2.01 Materials:

  A. Use only new materials conforming to the requirements of these specifications and approved by the Engineer. All materials proposed to be used may be tested and inspected at any time during their preparation and use. If, after trial, it is found that the source of supply has not been approved, or does not furnish a uniform product, or if the product becomes unacceptable at any time, the Contractor shall furnish materials from another source. Approved materials that become unfit for use shall not be used in the Work and shall be removed from the job site.

      Materials indicated on the Drawings that are required in the Work but are not covered in detail in the Specifications shall be of the best quality available and shall conform to the current specification of the American Society for Testing and Materials (ASTM) or AASHTO, as approved by the Engineer.

B. For sedimentation and erosion control use:
1. Baled straw and hay that complies with the requirements contained in Section 02900.
2. Silt fences manufactured of fabric specifically made for such purpose shall conform to the requirements of AASHTO M 288-90, and approved by the Engineer.
3. Temporary CMP drainage pipes chosen by the Contractor that are adequate to perform under subject field conditions.
4. Temporary seeding that complies with requirements contained in Section 02900.
5. Three inch maximum size coarse aggregate that is free from deleterious materials and does not contain more than 10 percent materials finer than the No. 4 sieve.
C. Slope protection shall be provided by:
1. Using stones that weigh no more than 150 pounds with at least 20 percent weighing more than 60 pounds while no more than 20 percent weigh less than 20 pounds for hand placed riprap.
2. Using geotextile fabric that is manufactured engineering fabric conforming with AASHTO M 288-90 for reducing soil erosion and that is approved by the Engineer.

Part 3—Execution

3.01 Examination:
A. Examine all areas where slope protection and erosion control are required.
B. Request underground utility location service to locate all utilities within work areas.
3.02 Preparation:
A. Protect all improvements of value from equipment, material, and employee abuse to include:
1. Property corners, survey field markers, and monuments.
2. Designated trees, shrubs, and grassed areas.
3. Existing utilities, installed sedimentation and erosion control systems, property improvements, pavements, sidewalks, driveways, curbing, poles, wires, pipes, and fences unless otherwise noted on the Drawings or specifically designated for removal by the Engineer. When fences interfere with construction operations, request permission from the Owner to temporarily remove during the time work is being conducted in the area. Immediately replace upon completion at no additional cost to the Owner.
4. Persons and personal property in or adjoining the work area.
B. Provide for dust control if measures are warranted in accordance with Section 01500.
3.03 Installation:
A. Provide temporary construction entrances at all ingress/egress locations to reduce or eliminate transporting soil or mud by vehicles to roadways and public rights of way. Construct these entrances by installing a stone mat using

three inch maximum crushed stone. This mat shall be a minimum of 50 feet long by 12 feet wide and at least 6 inches thick. Drain wash water to spray the underside of vehicles and tires when parked on the stone mat. Provide spent wash water drainage into a sediment basin or trap.

B. Install temporary perimeter, diversionary, and interceptor dikes as shown on the Drawings and other places where required in the field. Their purpose is to divert sediment laden storm runoff to an on-site sedimentation trapping facility or to direct storm water away from exposed slopes.

C. Provide temporary discharge controls that limit concentrations of storm water flows by using level spreaders, gravel outlets, and temporary drop piping as shown on the Drawings.

D. Provide sediment traps using silt fences, straw bales, or sediment basins as shown on the Drawings. Utilize detention pond facilities initially as sediment basins by installing a temporary riser pipe as shown on the Drawings. Install a marker to indicate the maximum level of sediment accumulation to indicate when the facility requires cleaning. Remove and replace the riser piping when nearing the completion of the Work and approved by the Engineer. Remove accumulated residual sediment at that time and treat the residue as waste. Place the facility into a storm water detention mode of operation.

E. Hand place riprap a minimum of 12 inches thick on designated slopes to comply with the Drawings. Begin placement two feet below the toe of slope while firmly imbedding each piece.

F. Install geotextile fabric for slope stability in accordance with the manufacturer's recommendations. Unless otherwise directed, install the fabric with the long dimension parallel to the toe of slope, and lay where a smooth and wrinkle free surface is obtained that is not under tension. Overlap subsequent fabric layers by at least 18 inches. Secure fabric using approved fasteners and cover within 30 days using materials approved by the Engineer.

3.04 Protection:

A. Repair damaged sediment control devices.

B. Insure exposed soil surfaces are not eroding.

C. Inspect and maintain the temporary construction entrances' stone from accumulation of silt, mud, or other debris.

D. Prohibit vehicular traffic from crossing installed geotextile fabrics.

Information contained in this section is subject to change and is intended for educational purposes only. CSI, CSC, and CSRF did not prepare nor do they specifically endorse this material. The current format is contained in CSI/CSC's effective edition of MASTERFORMAT along with current CSI sponsored master guide specifications containing CSRF's SPECTEXT/SPECTEXT II.

End of Section 02270

SECTION 02510

ASPHALTIC CONCRETE PAVING

Part 1—General

1.01 Summary:
  A. Section provides for installation of asphaltic concrete base, binder, or surface courses of HMA or intermediate course composed of mineral aggregates and asphalt cement, mixed in an approved plant and constructed on an approved subgrade or base course.
       Each pavement course shall be produced and constructed in accordance with the specified gradation (i.e., base, binder, or surface requirements).
  B. Related Sections include but are not necessarily limited to:
    1. General Conditions.
    2. Supplementary Conditions.
    3. Sections contained in Division 1 of this Project Manual.
1.02 Submittals:
  A. Follow requirements set forth in Section 01340.
  B. Submit product data for materials and mix design to the Engineer for approval 30 days prior to commencing paving operations.
  C. Include manufacturer's recommended asphalt cement mixing temperature for a viscosity of 170 centistokes ($\pm$20) and compaction temperature for a viscosity of 280 centistokes ($\pm$30).
1.03 Quality Assurance:
  A. Detail sufficient numbers of qualified workmen who are experienced in executing the activities specified in this Section.
  B. Detail and maintain sufficient equipment capable of undertaking the activities specified in this Section to produce sufficient progress on the Work.
  C. Instruct workmen in procedures specified in this Section.
  D. Operate equipment with due caution.
  E. Check the final surface for elevation, smoothness, and compaction. Use grades obtained from the Drawings as finished grades unless directed otherwise. Account for pavement thickness requirements, foundation slab thickness requirements, and other structural features to produce the desired elevations.
  F. Limit extent of work to authorized areas designated on the Drawings. Fill no wetlands unless authorized by permit.
1.04 Delivery, Storage, and Handling:
  A. Adhere to germane portions of Section 01610.
  B. Require HMA delivery trucks to be properly loaded to prevent mix segregation during transport.
  C. Use only an approved release agent to prevent the HMA from sticking to the truck body.
  D. Receive HMA at temperatures above the required compacting temperature to allow proper placement and compaction.
1.05 Sequencing and Scheduling:
  A. Coordinate activities of this Section with Sections 02230 and 02520.

Part 2—Products

2.01 Materials:
    A. Use only materials conforming to the requirements of these Specifications and approved by the Engineer. All materials proposed to be used may be tested and inspected at any time during their preparation and use. If, after trial, it is found that the source of supply has not been approved, or does not furnish a uniform product, or if the product becomes unacceptable at any time, the Contractor shall furnish materials from another source. Approved materials that become unfit for use shall not be used in the Work.

        Materials indicated on the Drawings that are required in the Work but are not covered in detail in the Specifications shall be of the best quality available and shall conform to the current specification of the American Society for Testing and Materials (ASTM) or AASHTO, as approved by the Engineer.

    B. Use asphalt cements that are either AC-20 or AC-30 viscosity graded, and conform to AASHTO M 226-80 (1986), Table 2.

    C. Use rapid cure cutback asphalts that conform to AASHTO M 81-90, and medium cure cutback asphalts that conform to AASHTO M 82-75 (1990).

    D. Use emulsified asphalts that conform to AASHTO M 140-88 or AASHTO M 208-87.

    E. Use aggregates as follows:
       1. Mineral filler that conforms to AASHTO M 17-88.
       2. Fine aggregate that is uniformly graded from coarse to fine, and is hard, sharp, angular, and free from clay and other deleterious substances.
       3. Coarse aggregate that is clean, tough, angular, and free from excessive amounts of elongated particles or deleterious materials. Crushed stone and gravel must have an abrasion loss of not more than 45 percent when subjected to the Los Angeles Abrasion Test (AASHTO T 96-87).

2.02 Mixes:
    A. Blended aggregate mixtures shall conform to the following specification and be approved by the Engineer:

Sieve Size	Type I	Type II	Binder	Base
1.5 inches	100	100	100	100
1 inch	100	100	100	85–100
$\frac{3}{4}$ inch	100	100	40–100	——
$\frac{1}{2}$ inch	97–100	97–100	72–90	60–80
$\frac{3}{8}$ inch	80–100	80–100	——	——
No. 4	58–75	53–73	42–62	40–55
No. 8	42–60	38–53	30–48	30–45
No. 30	19–40	20–34	——	——
No. 100	8–20	8–16	——	——
No. 200	3–8	3–8	——	——
Percent of Asphalt Cement	4.8–6.8	4.8–6.8	3.5–6.0	3.5–5.5
Minimum Stability	1400	1400	1000	——
Percent Air	3.5–6.0	3.5–6.0	——	——

    B.   The mix design shall include the percent of asphalt cement determined by the Marshall Method for (50 or) 75 blow criteria using procedures in AASHTO T 245-90 and AASHTO R 12-85.

2.03   Source Quality Control:

    A.   Provide the Engineer with HMA test results on materials processed and produced hot mix. These results will include stability, flow, air voids, voids in the mineral aggregate, and voids filled with asphalt.

Part 3—Execution

3.01   Examination:

    A.   Examine surface areas to insure that no noticeable moisture exists and that they are properly prepared to receive HMA. Commence paving operations when deficiencies are corrected.

    B.   Assess weather conditions for restrictions to paving operations.

        1.   Bituminous plant mix surface courses generally shall not be placed between December 15 and March 1 (or other dates specified to conform with local regulations). Temperature and weather restrictions prohibit placement of surface courses when the air temperature is below 45°F in the shade and falling. Bituminous plant mix surface courses shall not be placed on a wet or frozen surface.

        2.   Bituminous plant mix base and binder courses can be placed anytime during the year except when the air temperature is below 40°F in the shade and falling. HMA base and binder courses shall not be placed on wet or frozen surfaces.

    C.   Inspect delivery of HMA to insure it is covered by tarpaulin and is at an acceptable temperature to permit proper placement and compaction.

3.02   Preparation:

    A.   Place prime coat on previously untreated macadam base surfaces using MC-30 or RC-30 cut back asphalt at a rate of 0.25 to 0.3 gallon per square yard and allow to cure. Under some circumstances, MC-70 or RC-70 may be used.

3.03   Application:

    A.   Apply a tack coat of RS-1, RS-2, CRS-2, MS-1, MS-2, CRS-1, or CMS-2 emulsified asphalt at a temperature between 50°F and 160°F as directed by the Engineer depending on field conditions and applied at a rate of 0.05 to 0.15 gallon per square yard. Permit emulsified asphalt to "break" prior to applying HMA.

    B.   Base, binder, and surface courses shall be laid hot using a standard asphalt paving machine to produce compacted layers shown on the plans. The compaction procedure shall consist of breakdown, intermediate, and finish rolling. Use a steel wheel roller and pneumatic tire roller to complete the compaction operation prior to reaching the allowable temperature limitations. Obtain required compaction along centerline joints and insure smooth surface transitions between the various layers.

    C.   Apply striping neatly in accordance with striping plans contained in the Drawings using approved paint.

3.04 Field Quality Control

    A.   Use HMA plants to produce HMA that conforms to AASHTO M 156-89. Heat asphalt cement to the manufacturer's recommended mixing temperature for plant mixing. Compact in the field above the manufacturer's minimum compacting temperature. Provide haul trucks that have smooth metal beds and are covered. Provide a hole in each side of the bed of each delivery truck for checking mix delivery temperature. Use a paving machine that includes mechanical devices such as skis, traveling string lines, equalizing runners, level arms, and automatic screeds. Place HMA in layers that do not exceed two inches (compacted thickness) for surface courses, three inches (compacted thickness) for binder courses, and four inches (compacted thickness) for base courses. Compact HMA using maximum roller speeds of three miles per hour for static or vibratory steel wheel rollers and five miles per hour for pneumatic rollers. Utilize three wheeled steel wheel rollers of 10 to 12 tons in weight or 8 to 12 ton two axle tandem rollers. Provide pneumatic rollers capable of varying tire pressures from 40 to 90 pounds per square inch gauge by adjusting ballast and inflation pressure.

    B.   Coordinate presence of materials testing firm to verify truck delivery temperatures adequate for proper placement and compaction as provided in Section 01410.

    C.   Coordinate three random samples per 10,000 square feet of paved surface area for Marshall samples to verify air voids, stability, and other required test properties.

    D.   In parking lots, recreational courts, or curb and gutter street sections identify and correct low areas of $\frac{1}{8}$ inch per 10 feet that result in ponding when subjected to surface water.

    E.   Tack coat surfaces of previously placed asphalt pavement layers prior to placing subsequent layers unless subsequent layers are placed on the same day, in which case the tack coat may be omitted.

    F.   Final inspection by regulatory agencies will be scheduled by the Engineer after the work is approved by the Engineer.

3.05 Protection:

    A.   Protect tacked surfaces from traffic.

    B.   Protect new HMA surfaces from traffic until sufficiently cooled, set, and cured to prevent rutting, scuffing, and pavement displacement.

End of Section 02510

SECTION 02600

WATER DISTRIBUTION

Part 1—General

1.01 Summary:
  A. Section includes provisions for the potable distribution system shown on the Drawings.
  B. Related Sections include but are not necessarily limited to:
    1. General Conditions.
    2. Supplementary Conditions.
    3. Sections in Division 1 of these Specifications.
1.02 References:
  A. The following references are applicable in part or in total to the conduct of the work associated with this Section:
    1. Manufacturer's recommended installation instructions approved by the Engineer.
    2. ANSI/AWWA C104/A21.4-90*.
    3. ANSI/AWWA C100/A21.10-87.
    4. ANSI/AWWA C111/A21.11-90.
    5. ANSI/AWWA C150/A21.50-91.
    6. ANSI/AWWA C151/A21.51-91.
    7. ANSI/AWWA C200-86.
    8. ANSI/AWWA C203-91.
    9. ANSI/AWWA C205-91.
    10. ANSI/AWWA C600-87.
    11. ANSI/AWWA C900-89.
1.03 Submittals:
  A. Comply with provisions of Section 01340.
  B. Submit for review and approval to the Engineer within 30 working days of receipt of the Notice to Proceed:
    1. A list of materials necessary for the work encompassed by this Section along with manufacturer's product data and specifications demonstrating compliance with these Specifications.
    2. Names and addresses of the closest repair facilities and spare parts supply for products proposed to be included in the installation.
    3. Installation instructions from applicable manufacturers for products proposed to be included in the installation.
1.04 Quality Assurance:
  A. Detail qualified workmen that are experienced in executing the activities specified in this Section.
  B. Detail and maintain equipment capable of undertaking the activities specified in this Section to produce sufficient progress on the Work.
  C. Instruct workmen in procedures specified in this Section.
  D. Insure water mains and appurtenances are installed below potential frost penetration depths.

*See references at end of chapter for complete titles.

1.05 Delivery, Storage, and Handling:
    A. Adhere to germane portions of Section 01610.
    B. Store and handle materials and products in accordance with manufacturer's recommended practices.

Part 2—Products

2.01 Materials:
    A. Use only new materials conforming to the requirements of these specifications and approved by the Engineer. All materials proposed to be used may be tested and inspected at any time during their preparation and use. If, after trial, it is found that the source of supply has not been approved, or does not furnish a uniform product, or if the product becomes unacceptable at any time, the Contractor shall furnish materials from another source. Approved materials that become unfit for use shall not be used in the Work.

        Materials indicated on the Drawings that are required in the Work but are not covered in detail in the Specifications shall be of the best quality available and shall conform to the current specification of the American Society for Testing and Materials (ASTM) or the American Water Works Association (AWWA), as approved by the Engineer.

    B. Utilize only materials that are National Sanitation Foundation (NSF) approved. Ductile-iron pressure pipe shall conform to ASTM A 377-89. The pipe shall be Class 150, cement lined, and have a bituminous coating. Mechanical or slip joints may be used. Ductile-iron pipe thickness shall be determined from plans in accordance with ANSI/AWWA C150/A21.50-91. It shall comply also with the following requirements:
        1. Use only pipe and fittings conforming to:
            a. Ductile-iron with cement mortar linings complying with ANSI/AWWA C104/A21.4-90.
            b. Polyvinyl-chloride (PVC) complying with ASTM D 1784-81, SDR 27 and ANSI/AWWA C900-89. PVC pipe used for water mains and fittings shall conform to ASTM Specification D-2241-89 for standard dimension ratios 160 pound per square inch pipe SDR 26. The pipe shall be extruded from clean virgin approved Class 1254-A PVC compound conforming to ASTM D 1784-81. Rubber rings shall conform to ASTM D 1869-78.
            c. Slip joints shall be used on all pipes where thrust blocks can be installed. Ductile-iron pipes requiring restraints shall be of the mechanical joint type with tie rod restraints. Ductile-iron flange connections shall be used where shown on the plans.
        2. Valves and boxes shall conform to the following:
            a. Gate valves shall be rated at a working pressure of 125 pounds per square inch and have standard connections that are compatible with the water main being connected. The valve shall provide a clear waterway area equal to the full nominal diameter of the valve when fully opened by counterclockwise stem rotation. This is permanently and prominently indicated on the top of the valve stem with a direction arrow. Valves shall conform to AWWA C500-86 requirements

for standard, bronze trimmed, nonrising stem, solid wedge disc valves.

   b. Valve boxes shall be of the adjustable type with an extension if required. Include the label "WATER" cast into the cover.

   c. Check valves shall be rated at a working pressure of 125 pounds per square inch and shall provide a clear waterway area equal to the full nominal diameter of the valve. Check valves shall be constructed to allow the flow of water in only one direction while closing tightly when pressures attempt a flow reversal. Check valves shall be of the nonslam type and include flange connections at each face using standard 125 flanges. Each check valve shall have the valve size, direction of flow, and working pressure cast on its casing.

  3. Tapping sleeves shall be of the sleeve coupling type with the outlet being a flange connection using a standard 125 flange configuration.

  4. Fire hydrants shall conform to ANSI/AWWA C502-85.

  5. Service lines, meters, and meter boxes shall conform to the Drawings.

  6. Joint restraints, tie rods, and thrust blocks shall conform to the Drawings.

  7. Back flow prevention devices shall conform to ANSI/AWWA C510-89 or C511-89 as indicated on the Drawings.

 C. Coordinate, procure, and pay for potable water required to complete the system including charging, flushing, and testing.

Part 3—Execution

3.01 Examination:

 A. Examine all areas where work under this Section is affected. Provide corrective actions to result in satisfactory working conditions prior to commencing work on the water distribution system.

 B. Insure new right of way areas have been rough graded so that construction of the system takes place in soil that is at the approximate finished elevations.

3.02 Preparation:

 A. Request underground utility location service to mark all utilities.

 B. Protect all improvements of value from equipment, material, and employee abuse to include:

  1. Property corners, survey field markers, and monuments.

  2. Designated trees, shrubs, and grassed areas.

  3. Existing utilities, sedimentation and erosion control systems, property improvements, pavements, sidewalks, driveways, curbing, poles, wires, pipes, and fences unless otherwise noted on the Drawings or specifically designated for removal by the Engineer. When fences interfere with construction operations, request permission from the Owner to temporarily remove during the time work is being conducted in the area. Immediately replace upon completion at no additional cost to the Owner.

  4. All areas not within the construction site boundary subject to falling limbs, trees, and dust.

  5. Persons and personal property in or adjoining the work area.

 C. Provide for dust control if measures are warranted in accordance with Section 01500.

D.   Insure sedimentation and erosion control procedures are utilized.

E.   Transport pipe and fittings to trenched areas to protect them from damage. Carry all pipe lengths without dragging. Protect ends to insure tight joint surfaces. Keep the interior clean and free from contamination.

F.   Squarely cut pipe to lengths where installation can be easily assembled without placing pipe and fittings under strain. Remove burrs created during cutting process and prepare the pipe end in accordance with the approved manufacturer's recommended method of installation. Repair any damaged interior linings prior to installation.

G.   Store rubber gaskets and pipe lubrication away from excessive heat and contamination such as dust and oil.

3.03   Installation:

A.   Pipes and fittings shall be installed in a prepared trench and Class B bedding in accordance with Section 02220 by:

1.   Ductile-iron pipe shall be installed in accordance with ANSI/AWWA C150/A21.5-91 and ANSI/AWWA C105/A21.5-88.

2.   PVC pipe shall be installed in accordance with ANSI/AWWA C900-89 and AWWA Manual M 23 by:

a.   Preparing a trench that provides sufficient space to permit tamping around the pipe.

b.   Backfilling the pipe immediately after bedding while leaving the joints exposed one foot on either side. Insure that the backfill is compacted. The installed line should be pressure tested as soon as ready. Each portion of the line to be tested should be complete with thrustblocks and temporary end blocking, if necessary. After the line is satisfactorily pressure tested, complete backfilling the joints. Backfill thrust blocks and valve fittings after inspection by the Engineer.

3.   Temporarily plug the open pipe end when securing incomplete work to prevent debris and small animals from entering the pipe.

B.   Valves and boxes shall be installed by:

1.   Positioning valve along centerline of water main piping and orienting valve stem in vertical position.

2.   Connecting pipes to valve faces, and backfilling five feet along the pipeline each way from the valve's position using compacted material. Position the valve box on top of the valve and tighten the stuffing boxes.

3.   Opening and closing the valve fully to insure that all parts are working properly.

C.   Fire hydrants shall be installed in accordance with the Drawings and AWWA Manual M 17.

D.   Thrust blocks or restraining joints shall be installed where water mains change directions (such as at ells and tees), change size, terminate, or where thrusts are expected to develop at valves and hydrants. Concrete used for thrust blocking shall conform to Section 03300.

1.   Place thrust blocking in accordance with locations and dimensions shown on the Drawings.

2.   Insure that block facing rests on undisturbed native soil for good bearing.

3.   When thrust blocking cannot be placed due to field limitations, relocate

thrust blocks with tie rods that anchor the joints or use tie rods in conjunction with restrained joints to link the pipe together.

E. Service lines, meters, and boxes shall be installed by:
1. Locating meter boxes to provide water service at each tap location shown on the Drawings.
2. Locating the box three feet from the property line within the road right of way. It shall be serviced by a $\frac{3}{4}$ inch line for a single service and a 1 inch line for a double service.
3. Installing a shutoff valve on supply side and linesetter in box. Meter will be set later by regulatory agency when dwelling service is requested.

F. Place air release valves at high points along the water mains as indicated on the Drawings.

G. Install blow offs in accordance with the Drawings.

H. Install back flow prevention devices in accordance with the drawings and AWWA Manual M 14.

3.04 Field Quality Control:
A. Final inspection by regulatory agencies will be scheduled by the Engineer after the work is approved by the Engineer.
B. Test the integrity of all watertight joints by conducting a hydrostatic pressure test.
1. Pressure test all potable water lines in accordance with ANSI/AWWA C600-87,* Section 4.1. Conduct the test for at least two hours using a test pressure of approximately 1.5 times the design pressure.
    a. Calculate the allowable leakage per hour for ductile-iron pipe using:

$$L = [SD\,(P)^{0.5}]/(133{,}200)$$

where:

L = The allowable leakage in gallons per hour.
S = The length of pipe tested in feet.
D = The nominal diameter of the pipe in inches.
P = The average test pressure in pounds per square inch.

    b. Calculate the allowable leakage per hour for PVC pipe using the *Unibell Handbook of PVC Pipe,** with:

$$L = (N \times D \times P^{0.5})/7400$$

where:

L = The allowable leakage in gallons per hour.
N = The number of joints in the length of line tested.
D = The nominal diameter of the pipe in inches.
P = The average test pressure in pounds per square inch.

2. Repair any leaks.

*Reprinted from AWWA Standard C600-87, by permission.
*Reprinted from Unibell Plastic Pipe Association (1991), by permission.

C. Test the delivery capability of the completed system by conducting hydrant flow test in conjunction with the Engineer and regulatory authorities.

D. Test the cleanliness of the installed system by conducting bacteria contamination tests. It is the responsibility of the Contractor to obtain two samples of potable water from the installed system and submit them to an approved sanitation lab for determination of contamination. Take the samples 24 hours apart. Obtaining satisfactory test results is the responsibility of the Contractor prior to scheduling the final water system inspection with the Engineer.

E. All work accomplished and materials installed shall be subject to inspection by the Engineer and all jurisdictional regulatory authorities. All improper work disclosed by such inspection shall be reconstructed at the Contractor's expense. All rejected materials shall be removed from the site immediately.

F. Provide the necessary separation between sanitary sewers and potable water mains. Lay sewers at least 10 feet horizontally from any existing or proposed water main. Should local conditions prevent a lateral separation of 10 feet, a sewer may be laid closer than 10 feet to the water line if:

1. It is laid in a separate trench.

2. It is laid in the same trench with the water mains located at one side on a bench of undisturbed earth.

3. In either case the elevation of the crown of the sewer is at least 18 inches below the invert of the water main.

   Whenever sewers must cross under water mains, the sewer shall be laid at such an elevation that the top of the sewer is at least 18 inches below the bottom of the water main. When the sewer cannot be buried deeply enough to meet the above requirements, the water main shall be relocated to provide this separation or reconstructed with slip on or mechanical joint ductile-iron pipe for a distance of 10 feet on either side of the sewer. One full pipe length of water main should be centered over the sewer so that both joints will be as far from the sewer as possible.

   When it is impossible to obtain proper horizontal and vertical separation as stipulated above, the water main shall be constructed of slip on or mechanical joint ductile-iron pipe and the sewer constructed of mechanical joint ductile-iron pipe. Pressure test both lines to insure watertightness.

3.05 Cleaning:

A. Do not charge any mains with water until a minimum of six working days after the last thrust block is installed.

B. Flush system slowly and discharge flushed water into an approved storm drainage system.

C. Disinfect system after satisfactory hydrostatic pressure and flow tests are conducted.

1. Disinfect all water mains by sterilizing with a solution containing not less than 50 parts per million available chlorine.

2. Sterilizing solution shall remain in the system for 24 hours. After the disinfection period, the residual chlorine shall not be less than 10 parts per million.

3. After sterilization, the solution shall be flushed with clean water until the residual chlorine is not greater than 0.2 part per million.

3.06   Protection:

A. Protect all installed potable water system components by placing protective flagging around fire hydrants and keeping heavy equipment away from the area.

Information contained in this section is subject to change and is intended for educational purposes only. CSI, CSC, and CSRF did not prepare nor do they specifically endorse this material. The current format is contained in CSI/CSC's effective edition of MASTERFORMAT along with current CSI sponsored master guide specifications containing CSRF's SPECTEXT/SPECTEXT II.

End of Section 02600

SECTION 02720

STORM SEWERAGE

Part 1—General

1.01   Summary:

A. Section includes provisions for the installation of the storm drainage system shown on the Drawings.

B. Related Sections include but are not necessarily limited to:
1. General Conditions.
2. Supplementary Conditions.
3. Sections in Division 1 of this Project Manual.
4. Section 02210 for ditch and swale construction.
5. State Highway Department's Standard Specifications for Highway Construction.

1.02   Submittals:

A. Comply with provisions of Section 01340.

B. Submit for review and approval to the Engineer within 30 working days of receipt of the Notice to Proceed the following:
1. A list of materials necessary for the work encompassed by this Section along with manufacturer's product data and specifications demonstrating compliance with these Specifications.
2. Installation instructions from applicable manufacturers for products proposed to be included in the installation.

1.03   Quality Assurance:

A. Detail qualified workmen who are experienced in executing the activities specified in this Section.

B. Detail and maintain equipment capable of accomplishing the activities specified in this Section while producing sufficient progress on the Work.

C. Instruct workmen in procedures specified in this Section.

D. Operate all equipment with due caution.

1.04   Delivery, Storage, and Handling:

A. Adhere to germane portions of Section 01610.

    B. Store and handle materials and products in accordance with manufacturer's recommended practices.

1.05 Sequencing and Scheduling:

    A. Coordinate activities under this Section with Sections 02110, 02210, 02220, 02270, and 02770.

    B. Temporarily utilize partially completed system and detention facilities for sediment control. Place detention facility into operation nearing completion of the Work as directed by the Engineer.

    C. Coordinate the inspection of pipe joints by the Engineer and regulatory inspectors prior to covering with backfill.

Part 2—Products

2.01 Materials:

    A. Use only new materials conforming to the requirements of these Specifications and approved by the Engineer. All materials proposed to be used may be tested and inspected at any time during their preparation and use. If, after trial, it is found that the source of supply has not been approved, or does not furnish a uniform product, or if the product becomes unacceptable at any time, the Contractor shall furnish materials from another source. Approved materials that become unfit for use shall not be used in the Work.

        Materials indicated on the Drawings that are required in the Work but are not covered in detail in the Specifications shall be of the best quality available and shall conform to the current specification of the American Society for Testing and Materials (ASTM) or AASHTO, as approved by the Engineer.

    B. Provide storm drainage piping as shown on the Drawings that conforms to the following:

        1. Reinforced concrete pipe shall conform to AASHTO M 170-89 Class II, Wall B and all aspects of the State Highway Department's Standard Specifications for Highway Construction.

        2. Corrugated metal steel pipe shall be made of zinc coated iron or steel sheets and conform to ASTM A 444-89. All rivets shall conform to ASTM A 31-89 Grade A, coated with zinc. Bolts shall be zinc coated, manufacturer's Grade C. All corrugations shall be one inch. All CMP culverts and connection bands shall be fully bituminous coated to a minimum thickness of 0.05 inch in accordance with AASHTO M 190-88.

        3. Corrugated aluminum pipe shall conform to AASHTO M 196-90 with thickness requirements shown on the Drawings and shall be bituminous coated.

    C. Provide 3000 pound per square inch concrete in accordance with Section 03300.

    D. Provide mortar made of portland cement, Type I that conforms to ASTM C 150-89, and hydrated lime, Type "S" that conforms to ASTM C 207-79(1988). Mortar sand shall conform to ASTM C 144-89, and the mix water shall be potable without contamination.

        Mortar, if not otherwise specified, shall be mixed in the proportions of one part portland cement, one part dry hydrated lime or lime paste, and six parts sand by volume.

   E.   Castings and fittings shall conform to requirements set forth in Section 05540.

   F.   Reinforcing steel shall conform to requirements set forth in Section 03300.

   G.   Riprap shall conform to Section 02270.

   H.   French drains shall include:
1. 789 crushed stone for drain filler.
2. PVC underdrain pipe meeting AASHTO M 278-87 for Class PS 50 or bituminous coated perforated CMP pipe meeting AASHTO M 36-90, Type III.
3. Geotechnical fabric suitable for underdrain applications approved by the Engineer must conform with AASHTO M 288-90.

   I.   Stone bedding for pipe shall conform to 789 aggregate.

Part 3—Execution

3.01   Examination:

   A.   Examine all areas where work under this Section is affected. Provide corrective actions to result in satisfactory working conditions prior commencing work.

   B.   Insure new right of way area has been rough graded so that construction of the system takes place in soil that is at the approximate finished elevations.

3.02   Preparation:

   A.   Request underground utility location service to locate utilities within the work area.

   B.   Protect all improvements of value from equipment, material, and employee abuse to include:
1. Property corners, survey field markers, and monuments.
2. Designated trees, shrubs, and grassed areas.
3. Existing utilities, sedimentation and erosion control systems, property improvements, pavements, sidewalks, driveways, curbing, poles, wires, pipes, and fences unless otherwise noted on the Drawings or specifically designated for removal by the Engineer. When fences interfere with construction operations, request permission from the Owner to temporarily remove during the time work is being conducted in the area. Immediately replace upon completion at no additional cost to the Owner.
4. Persons and personal property in or adjoining the work area.

   C.   Provide for dust control if measures are warranted in accordance with Section 01500.

   D.   Insure sedimentation and erosion control procedures are utilized.

   E.   Transport pipe and fittings to excavated area to protect them from damage. Carry all pipe lengths without dragging. Protect ends to insure tight joint surfaces. Keep the interior clean and free from contamination.

   F.   Squarely cut pipe to lengths where installation can be easily assembled without placing pipe and fittings under strain. Remove burrs created during cutting process and prepare the pipe end in accordance with the approved manufacturer's recommended method of installation. Repair any damaged interior linings prior to installation.

   G.   Store rubber gaskets and pipe lubrication away from excessive heat and contamination such as dust and oil.

3.03 Installation:
  A. Install pipes and fittings by:
    1. Excavating trenches in accordance with Section 02220.
    2. Lowering pipe sections into trenches to prevent the ends from becoming soiled and laying them in prepared trench bottoms to provide a minimum of Class B bedding with the bell end facing the direction of pipe laying. If the backfilled pipe will not be subject to wheel or other heavy loads, use a minimum of Class C bedding.
    3. Bedding the pipe and backfilling temporarily while leaving joints exposed until approved by the regulatory inspectors and the Engineer.
    4. Temporarily plugging the open pipe end when securing incomplete work to prevent debris from entering the pipe.
  B. Construct storm drainage structures in accordance with details shown on the Drawings. Slope and taper inverts to provide smooth hydraulic flow characteristics.
  C. Install underdrains (french drains) $2\frac{1}{2}$ feet in back of the curb and properly connect to a permanent type drainage outlet, such as a catch basin, junction box, or storm water manhole. Make the invert of the drain at least two feet below the bottom of the curb as shown on the street profile. Provide suitable outlets for the pipe underdrains at designed elevations and complete the installation prior to placing the base course. Install pipe underdrains on both sides of the street where mucking operations have taken place and where the road cut is of such depth that the water table will temporarily be within 24 inches of the centerline subgrade during extended periods of rainfall.

    Lay pipe underdrains on grade and in accordance with the State Highway Department's Standard Specification for highway construction. Do not cover until they have been inspected and approved by regulatory authorities and the Engineer.
  D. Place riprap specified in Section 02270 to form a protective covering on slopes of embankments, at ends of bridges, around culvert inlets and outlets, on slopes and bottoms of ditches, around foundations and other locations, when necessary for the purpose of preventing scour, erosion, or slipping of embankments.

    Install riprap at the open ends of all pipelines larger than 18 inches in diameter or span, at the open ends of all pipeline at cross ditches, at the outlet ends of pipelines, at outlet ditches, lakes, and so on, and other locations necessary due to soil conditions.

    Place riprap in open channels at points of intersection, at angle points where the angle of turn is twenty degrees or greater, and at other locations shown on the Drawings.
3.04 Field Quality Control:
  A. Final inspection by regulatory agencies will be scheduled by the Engineer after the system has been approved by the Engineer.
3.05 Cleaning:
  A. Inspect pipe interiors and structure inverts. Remove debris, trash, and unwanted material.
  B. Flush and scour the completed system to remove sediment.

3.06   Protection:

    A.   Prohibit unnecessary construction equipment from crossing installed pipe and structures to prevent displacement, cracking, and injury.

Information contained in this section is subject to change and is intended for educational purposes only. CSI, CSC, and CSRF did not prepare nor do they specifically endorse this material. The current format is contained in CSI/CSC's effective edition of MASTERFOR-MAT along with current CSI sponsored master guide specifications containing CSRF's SPECTEXT/SPECTEXT II.

End of Section 02720

SECTION 02730

SANITARY SEWERAGE

Part 1—General

1.01   Summary:

    A.   Section includes provisions for the installation of the sanitary sewerage system including services, together with foundations, manholes, and appurtenances as shown on the Drawings or specified.

    B.   Related Sections include but are not necessarily limited to:

        1.   General Conditions.

        2.   Supplementary Conditions.

        3.   Sections in Division 1 of this Project Manual.

1.02   References:

    A.   Construction of sanitary sewers shall be in accordance with American Society of Civil Engineers Manual of Engineering Practice No. 60 and the Water Pollution Control Federation Manual of Practice FD-5 (ASCE and WPCF, 1982).

1.03   Submittals:

    A.   Comply with provisions of Section 01340.

    B.   Submit for review and approval to the Engineer within 30 working days of receipt of the Notice to Proceed:

        1.   A list of materials necessary for the work encompassed by this Section along with manufacturer's product data and specifications demonstrating compliance with these Specifications.

        2.   Names and addresses of the closest repair facilities and spare parts supply for products proposed to be included in the installation.

        3.   Installation instructions from applicable manufacturers for products proposed to be included in the installation.

1.04   Quality Assurance:

    A.   Detail qualified workmen that are experienced in executing the activities specified in this Section.

    B.   Detail and maintain equipment capable of undertaking the activities specified in this Section to produce sufficient progress on the Work.

C. Instruct workmen in procedures specified in this Section.

D. Operate equipment with due caution.

1.05 Delivery, Storage, and Handling:

A. Adhere to germane portions of Section 01610.

B. Store and handle materials and products in accordance with manufacturer's recommended practices.

Part 2—Products

2.01 Materials:

A. Use only new materials conforming to the requirements of these Specifications and approved by the Engineer. All materials proposed to be used may be tested and inspected at any time during their preparation and use. If, after trial, it is found that the source of supply has not been approved, or does not furnish a uniform product, or if the product becomes unacceptable at any time, the Contractor shall furnish materials from another source. Approved materials that become unfit for use shall not be used in the Work.

Materials indicated on the Drawings that are required in the Work but are not covered in detail in the Specifications shall be of the best quality available and shall conform to the current specification of the American Society for Testing and Materials (ASTM), as approved by the Engineer.

Install only sanitary sewer pipe and fittings where indicated on the Drawings that conform to the following:

1. Use polyvinyl chloride (PVC) pressure sewer pipe that conforms to all requirements of ASTM D 1784-81 and ASTM D 2241-89. It shall be SDR 26. All PVC pipe shall be Class 1245-B (PVC 1120) suitable for 160 pound per square inch working pressure and designed to withstand without failure a pressure of 400 pounds per square inch for 1000 hours. Joints shall be push on type conforming to ASTM D 3139-89 using synthetic rubber ring gaskets and nontoxic water soluble vegetable type lubrication.

    Use PVC gravity sewer pipe and fittings made from clean virgin PVC compound conforming to ASTM resin specification D 1784-81. ASTM D 3034-89 shall be adhered to in all respects. The SDR shall be 35 in all diameters. Clean reworked material generated from the manufacturer's own pipe production may be used. Maximum allowable ordinate as measured from the concave side of the pipe shall not exceed $\frac{1}{16}$ inch per foot of length.

    All pipe shall be suitable for use as a gravity conduit. Provisions must be made for contraction and expansion at each joint with a rubber ring. The bell shall consist of an integral wall section stiffened with two PVC retaining rings that securely lock the solid cross section rubber ring into position.

    Use PVC plastic fittings and accessories that are manufactured and furnished by a pipe supplier and that have bell and spigot configurations identical to that of the pipe.

2. Use ductile-iron pipe that conforms to the latest requirements of the ANSI/AWWA C150/A21.50-91 Type III mechanical joint. The slip joint,

using rubber gaskets, also may be used. Flange piping used as shown on the Drawings must conform to ANSI/AWWA C151/A21.51-91 and ANSI/AWWA C110/A21.10-87. This pipe shall be Class 150 with a bituminous seal coat inside and out. Pressure pipe joints must conform to ANSI/AWWA C111/A21.11-90. Ductile-iron gravity sewer pipe must conform to ASTM A 746-86.

B. Use manholes that are precast concrete meeting ASTM C-478-88a and have a minimum interior diameter of 48 inches. Provide a maximum chimney height of 12 inches to adjust the rim and cover top elevation.

C. Provide 3000 pound per square inch concrete in accordance with Section 03300.

D. Provide mortar made of portland cement, Type I that conforms to ASTM C 150-89 and hydrated lime, Type "S" that conforms to ASTM C 207-79 (1988). Mortar sand shall conform to ASTM C 144-89, and the water shall be potable.

E. Use castings and fittings that conform to requirements set forth in Section 05540.

F. Use reinforcing steel that conforms to requirements set forth in Section 03300.

G. Use 789 stone for Class B pipe bedding.

Part 3—Execution

3.01 Examination:

A. Examine all areas where work under this Section is affected. Provide corrective actions to result in satisfactory working conditions prior to commencing work on the sanitary sewerage system.

B. Insure new right of way area has been rough graded so that construction of the system takes place in soil that is at the approximate finished elevations.

3.02 Preparation:

A. Request underground utility location service to mark all existing utilities.

B. Protect all improvements of value from equipment, material, and employee abuse to include:

1. Property corners, survey field markers, and monuments.

2. Designated trees, shrubs, and grassed areas.

3. Existing utilities, sedimentation and erosion control systems, property improvements, pavements, sidewalks, driveways, curbing, poles, wires, pipes, and fences unless otherwise noted on the Drawings or specifically designated for removal by the Engineer. When fences interfere with construction operations, request permission from the Owner to temporarily remove during the time work is being conducted in the area. Immediately replace upon completion at no additional cost to the Owner.

4. All areas not within the construction site boundary subject to falling limbs, trees, and dust.

5. Persons and personal property in or adjoining the work area.

C. Provide for dust control if measures are warranted in accordance with Section 01500.

D. Insure sedimentation and erosion control procedures are utilized.

E.   Transport pipe and fittings to excavated area to protect them from damage. Carry all pipe lengths without dragging. Protect ends to insure tight joint surfaces. Keep the interior clean and free from contamination.

F.   Squarely cut pipe to lengths where installation can be easily assembled without placing pipe and fittings under strain. Remove burrs created during cutting process and prepare the pipe end in accordance with the approved manufacturer's recommended method of installation. Repair any damaged interior linings prior to installation.

G.   Store rubber gaskets and pipe lubrication away from excessive heat and contamination such as dust and oil.

3.03   Installation:

A.   Lay pipes in trenches where the interior surface conforms reasonably with the grade and alignment as established by surveying field control markers. Pipe laying shall be done in a manner that will disturb as little as possible the pipe previously laid. Unless otherwise directed, the pipe shall be laid upgrade with the spigots pointing downgrade without any break in line or grade between manholes in accordance with the following procedures:

1.   Excavate trench in accordance with Section 02220.

2.   Before laying, the pipe shall be wiped clean of all soil and foreign matter, and the joints of all bells and spigots shall be clean and dry. The interior of the pipe shall be carefully freed from all soil and surplus material as the work proceeds.

3.   Lower pipe sections into trench to prevent ends from becoming soiled, and lay in prepared trench bottom to provide a minimum of Class B bedding with the bell end facing the direction of pipe laying.

4.   Bed pipe and backfill.

5.   Temporarily plug open pipe end when securing work to prevent debris from entering the pipe.

   Measure sewer lines on the basis of the horizontal distance between manhole centers with no deduction for the space occupied by the manholes.

B.   Install ductile-iron pipe in accordance with ANSI/AWWA C150/A21.50-90 and thermoplastic sewer pipe in accordance with ASTM D 2321-83a and ASTM D 2774-73 (1983). Also use the approved pipe manufacturer's recommendations except as modified in these Specifications.

C.   Install manholes in accordance with Section 02220.

D.   Install manhole drops in accordance with the Drawings.

E.   Install service lines in accordance with the Drawings, with precautions being taken to provide and secure end caps for air testing capability.

F.   Provide piers in accordance with the Drawings on undisturbed soil and in accordance with Section 03300.

G.   Provide thrust blocks in accordance with Section 03300.

H.   Install forcemain gate valves in accordance with Section 02600.

3.04   Field Quality Control:

A.   Lamp installed gravity mains to insure straight alignment from manhole to manhole.

B.   Verify grade of the installed lines to insure compliance with the Drawings.

C.   Install gravity pipe so that the maximum diametric deflection of the pipe will not exceed 5 percent with verification by pulling a mandrel through completed segments as part of the final inspection.

D. Conduct watertight integrity tests on gravity lines as determined by the Engineer using either:
1. Infiltration.
2. Exfiltration.
3. Air test.

E. Conduct watertight integrity tests by:
1. *Pressure Main Testing:* Make the pressure test after the forcemain has been laid and backfilled. Locate and repair any detected leaks at no expense to the Owner.

    The purpose of the leakage test is to establish that the section of line to be tested—including all joints, fittings, and other appurtenances—will not leak or that the leakage is within tolerance. Leakage, if any, usually will be found involving joints at saddles, valves, transition joints, adapters, and not usually in the pipe joints.

    Provide air release valves when setting up a section of line for testing. Air trapped in the line during testing will affect the results and can cause damage to the pipeline.

    Test forcemains in accordance with ANSI/AWWA Standard C600,* Section 4.1. Conduct the test for at least two hours with the test pressure approximately 1.5 times the design pressure.

    a. Calculate the allowable leakage per hour for ductile-iron pipe using:

$$L = [SD \ (P^{\{0.5\}}]/(133,200)$$

where:

L = The allowable leakage in gallons per hour.
S = The length of pipe tested in feet.
D = The nominal diameter of the pipe in inches.
P = The average test pressure in pounds per square inch.

    b. Calculate the allowable leakage per hour for PVC pipes using the *Unibell Handbook of PVC Pipe* * with:

$$L = (N \times D \times P^{0.5})/7400$$

where:

L = The allowable leakage in gallons per hour.
N = The number of joints in the length of line tested.
D = The nominal diameter of the pipe in inches.
P = The average test pressure in pounds per square inch.

*Reprinted from AWWA Standard C600-87, by permission.
*Reprinted from Unibell Plastic Pipe Association (1991), by permission.

Check the forcemain for blockage and obstructions by flushing with water at a rate approved by the Engineer with the spent flush water being diverted into an approved drainage facility.

Repair any leaks and retest.

2. *Gravity Main Testing:* Test every section of pipe between manholes after it has been laid and backfilled, using either an infiltration, exfiltration, or low pressure air test. When no groundwater exists at the time of the test, the pipe shall be subjected to an exfiltration test or an air test. The Engineer shall designate the type of test to be conducted and the manner in which it shall be performed. When infiltration testing is permitted, the Engineer shall give explicit instructions to be followed in carrying out the test. The maximum allowable amount of infiltration measured by testing shall be at a rate of not greater than 200 gallons per inch of pipe diameter per mile per 24 hours. Should any test on any section of pipe line disclose an infiltration rate greater than that permitted, the Contractor shall, at no expense to the Owner, locate and repair defective joints of pipe and retest until the infiltration is within specified allowance.

When an exfiltration test is required, the Engineer shall give explicit instructions to be followed in carrying out the test. The maximum allowable amount of exfiltration measured by the test shall be at a rate not greater than 200 gallons per inch of pipe diameter per mile per 24 hours. The average internal pressure of the system under test shall not be greater than five pounds per square inch (11.6 feet of head), and the maximum internal pressure in any part of the system under test shall not be greater than 10.8 pounds per square inch (25 feet of head).

Should any test on any section of pipe disclose an exfiltration rate greater than that permitted, the Contractor shall, at no expense to the Owner, locate and repair defective joints or pipes, and retest until the exfiltration is within specified allowances.

When an air test is required, in lieu of the water exfiltration test, the Engineer shall give explicit instructions to be followed in carrying out the test using Section 306-1.4.4 of *Standard Specifications for Public Works Construction* (Building News et al., 1988.)* as a guide.

Introduce air into the installed sewer main until 3 pounds per square inch gauge pressure is attained, then reduce the inflow of air to maintain an internal pressure between 2.5 and 3.5 pounds per square inch gauge for at least 2 minutes. Do not allow the internal pressure to exceed 5 pounds per square inch gauge at any time.

After the temperature of the air introduced into the pipeline has stabilized and the end plugs exhibit no leaks, the internal pressure can be allowed to drop to 2.5 pounds per square inch. Begin timing the number of seconds required for the pressure to drop from 2.5 pounds per square inch to 1.5 pounds per square inch gauge. If the observed lapsed time exceeds the allowable shown below, then the pipeline tested satisfactorily.

*Reprinted from Building News et al. (1978) by permission.

Main Line	Length (feet)	4 Inch House Connections			6 Inch House Connections		
		100 feet	200 feet	300 feet	100 feet	200 feet	300 feet
Diameter, 8 Inches	50	50	70	90	70	110	110
	100	90	100	100	110	120	110
	200	120	110	110	130	120	120
	300	130	120	110	130	120	120
	400	130	120	120	130	130	120
Diameter, 10 Inches	50	70	90	100	90	120	120
	100	130	120	110	140	130	130
	200	150	140	130	150	140	140
	300	160	150	140	160	150	140
	400	160	150	150	160	150	150
Diameter, 12 Inches	50	100	110	110	120	140	130
	100	170	150	140	170	150	140
	200	180	170	160	180	170	160
	300	190	180	170	190	180	170
	400	190	180	180	190	180	180

Should any test on any section of pipe disclose an air loss rate greater than that permitted, the Contractor shall, at his own expense, locate and repair the defective joints of pipe and retest until the air loss is within the specified allowances as shown above.

F. Before the sewer system is accepted it shall be tested and cleaned to the satisfaction of the Engineer. If any obstruction is found, the Contractor will be required to clean the sewer by means of rods or other equipment. The pipe line shall be straight with a uniform grade between manholes. The bottoms of all manholes will be smooth to ensure free flow of sewage and shall be sloped to insure drainage of any splashed material back into the line.

G. Final inspection by regulatory agencies will be scheduled by the Engineer after the work is approved by the Engineer.

3.05 Cleaning:

A. Inspect pipe interiors and structure inverts. Remove debris, trash, and unwanted material. Flush, scour, and drain the completed system to remove sediment.

3.06 Protection:

A. Provide all precautions necessary to protect and maintain the buildings, fences, water lines, storm drains, power and telephone lines, and cable, and other structures. These precautions include the furnishing of suitable bridges and footways across intersected streets, replacing of pavement, sidewalks, curb and gutters, and cleaning away of all debris and surplus materials and any other work necessary to put in complete working order the specified sewers and appurtenances. Any damage to any utility lines, including overhead or underground, services, or structures of any nature (including any trees or shrubs so designated by the Engineer to remain protected) shall be repaired or replaced by the Contractor at no expense to the Owner.

B.   Prohibit unnecessary construction equipment from crossing installed pipe and structures to prevent displacement, cracking, and injury.

Information contained in this section is subject to change and is intended for educational purposes only. CSI, CSC, and CSRF did not prepare nor do they specifically endorse this material. The current format is contained in CSI/CSC's effective edition of MASTERFOR-MAT along with current CSI sponsored master guide specifications containing CSRF's SPECTEXT/SPECTEXT II.

End of Section 02730

REFERENCES

American Association of State Highway and Transportation Officials (AASHTO). *AASHTO M 17-88, Standard Specification for Mineral Filler for Bituminous Paving Mixtures.* Washington: AASHTO, 1988.

American Association of State Highway and Transportation Officials (AASHTO). *AASHTO M 36-90, Standard Specification for Corrugated Steel Pipe, Metallic-Coated, for Sewers and Drains.* Washington: AASHTO, 1990.

American Association of State Highway and Transportation Officials (AASHTO). *AASHTO M 81-90, Standard Specification for Cut-Back Asphalt (Rapid Cure Type).* Washington: AASHTO, 1990.

American Association of State Highway and Transportation Officials (AASHTO). *AASHTO M 82-75 (1990), Standard Specification for Cut-Back Asphalt (Medium-Curing Type).* Washington, AASHTO, 1990.

American Association of State Highway and Transportation Officials (AASHTO). *AASHTO M 140-88, Standard Specification for Emulsified Asphalt.* Washington: AASHTO, 1988.

American Association of State Highway and Transportation Officials (AASHTO). *AASHTO M 156-89, Requirements for Mixing Plants for Hot-Mixed, Hot-Laid Bituminous Paving Mixtures.* Washington: AASHTO, 1989.

American Association of State Highway and Transportation Officials (AASHTO). *AASHTO M 170-89, Standard Specification for Reinforced Concrete Culvert, Storm Drain, and Sewer Pipe.* Washington: AASHTO, 1989.

American Association of State Highway and Transportation Officials (AASHTO). *AASHTO M 190-88, Standard Specification for Bituminous Coated Corrugated Metal Culvert Pipe and Pipe Arches.* Washington: AASHTO, 1988.

American Association of State Highway and Transportation Officials (AASHTO). *AASHTO M 196-90, Standard Specification for Corrugated Aluminum Pipe for Sewers and Drains.* Washington: AASHTO, 1990.

American Association of State Highway and Transportation Officials (AASHTO). *AASHTO M 208-87, Standard Specification for Cationic Emulsified Asphalt.* Washington: AASHTO, 1987.

American Association of State Highway and Transportation Officials (AASHTO). *AASHTO M 226-80 (1986), Standard Specification for Viscosity Grades Asphalt Cement.* Washington: AASHTO, 1986.

American Association of State Highway and Transportation Officials (AASHTO). *AASHTO M 278-87, Class PS 50 Polyvinyl Chloride (PVC) Pipe.* Washington: AASHTO, 1987.

American Association of State Highway and Transportation Officials (AASHTO). *AASHTO M 288-90, Standard Specification for Geotextiles.* Washington: AASHTO, 1990.

American Association of State Highway and Transportation Officials (AASHTO). *AASHTO R 12-85, Standard Recommended Practice for Bituminous Mixture Design Using the Marshall and Hveem Procedures.* Washington: AASHTO, 1985.

American Association of State Highway and Transportation Officials (AASHTO). *AASHTO T 89-90, Determining the Liquid Limit of Soils.* Washington: AASHTO, 1990.

American Association of State Highway and Transportation Officials (AASHTO). *AASHTO T 90-87, Determining the Plastic Limit and Plasticity Index of Soils.* Washington: AASHTO, 1987.

American Association of State Highway and Transportation Officials (AASHTO). *AASHTO T 96-87, Resistance to Abrasion of Small Size Coarse Aggregate by Use of the Los Angeles Machine.* Washington: AASHTO, 1987.

American Association of State Highway and Transportation Officials (AASHTO). *AASHTO T 180-90, Moisture Density Relations of Soils Using a 10-lb [4.54 kg] Rammer and an 18-in. [457 mm] Drop.* Washington: AASHTO, 1990.

American Association of State Highway and Transportation Officials (AASHTO). *AASHTO T 191-86 (1990), Density of Soil In-Place by the Sand-Cone Method.* Washington: AASHTO, 1990.

American Association of State Highway and Transportation Officials (AASHTO). *AASHTO T 245-90, Resistance to Plastic Flow of Bituminous Mixtures Using Marshall Apparatus.* Washington: AASHTO, 1990.

American National Standards Institute (ANSI) and American Water Works Association (AWWA). *ANSI/AWWA C104/A21.4-90, American National Standard for Cement-Mortar Lining for Ductile-Iron Pipe and Fittings for Water.* Denver: AWWA, 1990.

American National Standards Institute (ANSI) and American Water Works Association (AWWA). *ANSI/AWWA C105/A21.5-88, American National Standard for Polyethylene Encasement for Ductile-Iron Piping for Water and Other Liquids.* Denver: AWWA, 1988.

American National Standards Institute (ANSI) and American Water Works Association (AWWA). *ANSI/AWWA C110/A21.10-87, American National Standard for Ductile-Iron and Gray-Iron Fittings, 3 in. Through 48 in., for Water and Other Liquids.* Denver: AWWA, 1987.

American National Standards Institute (ANSI) and American Water Works Association (AWWA). *ANSI/AWWA C111/A21.11-90, American National Standard for Rubber-Gasket Joints for Ductile-Iron and Gray-Iron Pressure Pipe and Fittings.* Denver: AWWA, 1990.

American National Standards Institute (ANSI) and American Water Works Association (AWWA). *ANSI/AWWA C150/A21.50-91, American National Standard for the Thickness Design of Ductile-Iron Pipe.* Denver: AWWA, 1991.

American National Standards Institute (ANSI) and American Water Works Association (AWWA). *ANSI/AWWA C151/A21.51-91, American National Standard for Ductile-Iron Pipe, Centrifugally Cast in Metal Molds or Sand-Lined Molds, for Water or Other Liquids.* Denver: AWWA, 1991.

American National Standards Institute (ANSI) and American Water Works Association (AWWA). *ANSI/AWWA C200-86, Standard for Steel Water Pipe 6 in. and Larger.* Denver: AWWA, 1986.

American National Standards Institute (ANSI) and American Water Works Association (AWWA). *ANSI/AWWA C203-91, Standard for Coal-Tar Protective Coatings and Linings for Steel Water Pipes—Enamel and Tape–Hot Applied.* Denver: AWWA, 1991.

American National Standards Institute (ANSI) and American Water Works Association (AWWA). *ANSI/AWWA C205-91, Standard for Cement-Mortar Protective Lining and Coating for Steel Water Pipe—4 in. and Larger—Shop Applied.* Denver: AWWA, 1991.

American National Standards Institute (ANSI) and American Water Works Association (AWWA). *ANSI/AWWA C502-85, Standard for Dry-Barrel Fire Hydrants.* Denver: AWWA, 1985.

American National Standards Institute (ANSI) and American Water Works Association (AWWA). *ANSI/AWWA C510-89, Standard for Double Check Valve Backflow Prevention Assembly.* Denver: AWWA, 1989.

American National Standards Institute (ANSI) and American Water Works Association (AWWA). *ANSI/AWWA C511-89, Standard for Reduced Pressure Principal Backflow Prevention Assembly.* Denver: AWWA, 1989.

American National Standards Institute (ANSI) and American Water Works Association (AWWA). *ANSI/AWWA C600-87, Standard for Installation of Ductile-Iron Water Mains and Their Appurtenances.* Denver: AWWA, 1987, p.16.

American National Standards Institute (ANSI) and American Water Works Association (AWWA). *ANSI/AWWA C900-89, Standard for Polyvinyl Chloride (PVC) Pressure Pipe, 4 in. Through 12 in., for Water.* Denver: AWWA, 1989.

American Society for Testing and Materials (ASTM). *ASTM A 31-89, Specification for Steel Rivets and Bars for Rivets, Pressure Vessels.* Philadelphia: ASTM, 1989.

American Society for Testing and Materials (ASTM). *ASTM A 377-89, Standard Index of Specifications for Ductile Iron Pressure Pipe.* Philadelphia: ASTM, 1989.

American Society for Testing and Materials (ASTM). *ASTM A 444-89, Standard Specification for Steel Sheet, Zinc-Coated (Galvanized) by the Hot-Dip Process for Culverts and Underdrains.* Philadelphia: ASTM, 1989.

American Society for Testing and Materials (ASTM). *ASTM A 674-89, Standard Recommended Practice for Polyethylene Encasement for Gray and Ductile Cast Iron Pipe for Water or Other Liquids.* Philadelphia: ASTM, 1989.

American Society for Testing and Materials (ASTM). *ASTM A 746-86, Standard Specification for Ductile-Iron Gravity Sewer Pipe.* Philadelphia: ASTM, 1986.

American Society for Testing and Materials (ASTM). *ASTM C 144-89, Standard Specification for Aggregate for Masonry Mortar.* Philadelphia: ASTM, 1989.

American Society for Testing and Materials (ASTM). *ASTM C 150-89, Standard Specification for Portland Cement.* Philadelphia: ASTM, 1989.

American Society for Testing and Materials (ASTM). *ASTM C 207-79(1988), Standard Specification for Hydrated Lime for Masonry Purposes.* Philadelphia: ASTM, 1988.

American Society for Testing and Materials (ASTM). *ASTM C 478-88a, Standard Specification for Precast Reinforced Concrete Manhole Sections.* Philadelphia: ASTM, 1988.

American Society for Testing and Materials (ASTM). *ASTM D 1784-81, Standard Specification for Rigid Poly(vinyl Chloride) (PVC) Compounds and Chlorinated Poly(vinyl Chloride) (CPVC) Compounds.* Philadelphia: ASTM, 1981.

American Society for Testing and Materials (ASTM). *ASTM D 1869-78(1989), Standard Specification for Rubber Rings for Asbestos-Cement Pipe.* Philadelphia: ASTM, 1989.

American Society for Testing and Materials (ASTM). *ASTM D 2241-89, Standard Specification for Poly(vinyl Chloride) (PVC) Pressure-Rated Pipe (SDR-Series).* Philadelphia: ASTM, 1989.

American Society for Testing and Materials (ASTM). *ASTM D 2321-83a, Recommended Practice for Underground Installation of Flexible Thermoplastic Sewer Pipe.* Philadelphia: ASTM, 1983.

American Society for Testing and Materials (ASTM). *ASTM D 2774-73(1983), Recommended Practice for Underground Installation of Thermoplastic Pressure Piping.* Philadelphia: ASTM, 1983.

American Society for Testing and Materials (ASTM). *ASTM D 3034-89, Standard Specification for Type PSM Poly(vinyl Chloride) (PVC) Sewer Pipe and Fittings.* Philadelphia: ASTM, 1989.

American Society for Testing and Materials (ASTM). *ASTM D 3139-89, Standard Specification for Joints for Plastic Pressure Pipes Using Flexible Elastometric Seals.* Philadelphia: ASTM, 1989.

American Society of Civil Engineers (ASCE) and Water Pollution Control Federation (WPCF).

Gravity Sanitary Sewer Design and Construction—ASCE Manual No. 60, MOP FD-5. New York: ASCE, 1982, pp. 201–204. Referenced material is reproduced by permission of the American Society of Civil Engineers.

American Water Works Association (AWWA). *AWWA C500-86, Standard for Gate Valves for Water and Sewerage Systems.* Denver: AWWA, 1986.

American Water Works Association (AWWA). *AWWA C600-87, Installation of Ductile-Iron Water Mains and Their Appurtenances.* Denver: AWWA, 1987. Copyright © 1987 by American Water Works Association. Referenced material is reproduced by permission of the American Water Works Association.

American Water Works Association (AWWA). *Recommended Practice for Backflow Prevention and Cross-Connection Control, AWWA Manual M 14.* Denver: AWWA, 1990.

American Water Works Association (AWWA). *Installation, Field Testing, and Maintenance of Fire Hydrants, AWWA Manual M 17.* Denver: AWWA, 1989.

American Water Works Association (AWWA). *PVC Pipe—Design and Installation, AWWA Manual M 23.* Denver: AWWA, 1980.

Building News, Inc. and Joint Cooperative Committee of the Southern California Chapter of American Public Works Association and Southern California Districts—AGC of California. *Standard Specification for Public Works Construction.* Los Angeles: Building News, 1988, pp. 450–454. Referenced material is reprinted by permission of Building News Books, Los Angeles (1-800-873-6397).

Construction Sciences Research Foundation, Inc. (CSRF). *SPECTEXT.* ℅ Charles R. Carroll, Jr., 202 Wyndhurst Ave., Baltimore, Md.

Construction Sciences Research Foundation, Inc. (CSRF). *SPECTEXT II.* ℅ Charles R. Carroll, Jr., 202 Wyndhurst Ave., Baltimore, Md.

Construction Specifications Institute (CSI). *Manual of Practice.* Alexandria, Va.: CSI, 1985, Parts I and II.

Construction Specifications Institute (CSI) and Construction Specifications Canada (CSC). *MASTERFORMAT.* Alexandria, Va.: CSI, 1988, pp. 1–87.

National Society of Professional Engineers (NSPE), American Consulting Engineers Council, American Society of Civil Engineers (ASCE), and Construction Specifications Institute. *Standard General Conditions of the Construction Contract (EJCDC No. 1910-8).* New York: ASCE, 1990.

National Society of Professional Engineers (NSPE), American Consulting Engineers Council, American Society of Civil Engineers (ASCE), and Construction Specifications Institute. *Standard Form of Agreement Between Owner and Contractor on Basis of a Stipulated Price* (EJCDC No. 1910-8-A-1). New York: ASCE, 1990.

National Society of Professional Engineers (NSPE), American Consulting Engineers Council, American Society of Civil Engineers (ASCE), and Construction Specifications Institute. *Uniform Location of Subject Matter—Information in Construction Documents* (EJCDC No. 1910-16). New York: ASCE, 1981.

Smith, Robert J. "Construction Documents Revised," *Civil Engineering.* New York: ASCE, August 1991, pp. 68–69. Referenced material is reproduced by permission of the American Society of Civil Engineers.

Unibell Plastic Pipe Association. *Unibell Handbook of PVC Pipe,* 3rd ed. Dallas: Unibell Plastic Pipe Association, 2655 Villa Creek Drive, 1991, pp. 382–387. Copyright © 1991 by Unibell Plastic Pipe Association. Referenced material is reproduced by permission of Unibell Plastic Pipe Association.

United States Department of Agriculture—SCS. *Erosion and Sediment Control . . . in Developing Areas. Planning Guidelines/Design Aids.* Columbia, S.C., 1977.

14 Project Costs and Scheduling

Construction costs and schedules are factors vitally important to a successful development. They are necessary to effectively transform the existing site into a finished product. Plans and specifications reflect the amount of work the contractor must perform within an established period of time; this work is used as a basis for contractor compensation. Preliminary estimates formulated during the initial design phase do not contain details that are included in a final analysis because detailed drawings are not available when these estimates are made. The process of preparing preliminary and final estimates, however, is the same.

14.1 ESTABLISHING A DATA BASE

The estimator should establish a data base to reference project estimates. This base can draw from previous jobs (adjusted for economic trends), reference books (such as Mean's: *Building Construction Cost Data*), other organizations' current bid information, and/or manufacturer's quotations. The availability of local equipment and manpower, along with prevailing industrial activities, also must be considered.

14.2 COST ANALYSIS

The cost analysis of a job should begin with the geographic location of the project. Prevailing wage and material rates vary with location. Transportation linked to geographic location can be a factor relative to the supply of materials, labor, and equipment. The availability of local labor, local regulations including union requirements, and local assessments will affect the cost of construction. The project's size, composition, and available time for construction can vary costs. Projected economic trends, new products, and new technology also could have an effect.

14.2.1 Background Information

The estimator should take an inventory of existing conditions at the job site. The project synopsis (Figure 5.4) contains pertinent information that can assist the estimator immensely. It is recommended that the estimator conduct a site visit so that familiarity with the project can be obtained firsthand. Topographic and site conditions, existing utilities, drainage features, accessibility, and the impact on adjoining property should be observed. The specifications need to be read and understood as to how they relate to the plans. The development design engineer who also provides an estimate would already be familiar with these conditions.

574

14.2.2 The Cost Estimating Process

A sound procedure for job estimating is to use felt tipped pens of various colors for marking the plans and specifications. Their use can indicate each included item. Every trade usually is assigned a different color pen for clarity. The estimator should make sure that no information is omitted such as notes, addenda to specifications, and specialized reports. If pertinent factors are undefined, every effort should be made to quantify them using meaningful methods.

All estimating work should be neat and in logical order so that another person might easily review the estimate without guesswork. Forms are available for estimators to use; however, computer spreadsheets allow estimators more flexibility.

Normally the engineer's cost estimate is based on a unit price quantity that would include the contractor's cost of material, labor, overhead, profit, and other costs associated with construction. Any quotations solicited by the estimator should be examined for time limitations, escalation clauses, freight costs, taxes, and so on, prior to use. The divisions of cost entries vary, depending on whether a preliminary or final estimate is being made. Also entries vary based on the project's complexity, scope, and magnitude. Typical work units include square yards (S.Y.), cubic yards (C.Y.), linear feet (L.F.), and lump sum (L.S.). The estimate is prepared by determining the number of units in each category; these numbers are then multiplied by the corresponding unit prices and summed. An example of a cost estimate for an improved land development, prepared using a spreadsheet, is shown in Figure 14.1.

14.3 PROJECT SCHEDULING

To guide a project to completion in an orderly and timely manner, a work schedule must be formulated. Normally this schedule contains estimated costs and durations of various tasks associated with the job in their order of precedence. The contractor usually prepares the working project schedule based on developer provided start and completion dates; however, design professionals often prepare a theoretical schedule to evaluate possible cost and time savings that can be reflected in the final construction documents prior to bidding.

In the early 1900's Henry L. Gantt and Frederick W. Taylor produced charts that graphically depicted the scheduling of similar activities. The critical path method (CPM), another method widely used today for project management, evolved in 1956 within the E. I. du Pont de Nemours Company for improved company management. Another available method, performance evaluation and review technique (PERT), was developed to manage the Polaris Fleet Ballistic Missile program.

CPM is most commonly used for managing larger land development projects. Activities are usually segregated by trades. The method includes the sequencing of events and time, which encompasses planning for manpower and equipment. These data are required for any project scheduling, and using a method such as CPM provides a logical and systematic approach to planning, scheduling, and monitoring the project.

14.3.1 Scheduling Process

The estimator should go through this process at least twice. The first pass will in part be a learning process through which the estimator can establish a "feel" for the project. The

I. ROADS AND STORM DRAINAGE	UNITS	UNIT PRICES	NO.UNITS	COSTS
a. Clearing and Grubbing	Acres	$4,000.00	1.30	$5,200.00
b. Curb Inlets	Each	$750.00	2.00	$1,500.00
c. Junction Boxes	Each	$600.00		$0.00
d. Ditch Excavation	L.F.	$4.00	600.00	$2,400.00
e. French Drains	L.F.	$5.50		$0.00
f. Riprap Outfalls	Each	$400.00	2.00	$800.00
g. Pavement	S.Y.	$11.50	870.00	$10,005.00
h. Curb and Gutter	L.F.	$6.25	1200.00	$7,500.00
i. Road Signs	Each	$100.00		$0.00
j. 15" RCP	L.F.	$35.00		$0.00
k. 18" RCP	L.F.	$37.50		$0.00
l. 24" RCP	L.F.	$40.50	175.00	$7,087.50
m. 36" RCP	L.F.	$46.25	30.00	$1,387.50
n. 48" RCP	L.F.	$55.00		$0.00
o. 18" x 29" CPM Arch	L.F.	$20.00		$0.00
p. 22" x 36" CPM Arch	L.F.	$25.00		$0.00
q. 27" x 43" CPM Arch	L.F.	$30.00		$0.00
r. Erosion Control	L.S.			$2,000.00
s. Earthwork	C.Y.	$5.00	1400	$7,000.00
t. Clean Up (10%)	L.S.			$4,488.00
SUBTOTAL				$49,368.00
II. POTABLE WATER SYSTEM				
a. 2 Inch Main	L.S.	$3.00		$0.00
b. 4 Inch Main	L.S.	$6.00		$0.00
c. 6 Inch Main	L.S.	$9.50	600.00	$5,700.00
d. 8 Inch Main	L.S.	$11.00		$0.00
e. 6" Valve and Box	Each	$475.00	2.00	$950.00
f. 8" Valve and Box	Each	$550.00		$0.00
g. 3-Way Hydrant	Each	$1,150.00	1.00	$1,150.00
h. Thrust Blocking	Each	$150.00	2.00	$300.00
i. Service Lines	L.F.	$4.00	300.00	$1,200.00
j. Meters	Each	$175.00	12.00	$2,100.00
k. Testing/Cleanup(10%)	L.S.			$1,140.00
SUBTOTAL				$12,540.00
III. SANITARY SEWER SYSTEM				
a. 8" Main 0'-5' Cut	L.F.	$10.00	300.00	$3,000.00
b. 8" Main 5'-7' Cut	L.F.	$12.00	160.00	$1,920.00
c. 8" Main 7'-9' Cut	L.F.	$13.00		$0.00
d. 8" Main 9'-11' Cut	L.F.	$15.00		$0.00
e. 8" Ductile (add on)	L.F.	$12.00		$0.00
f. 4' Manhole 0'-5' Cut	Each	$975.00	2.00	$1,950.00
g. 4' Manhole 5'-7' Cut	Each	$1,050.00		$0.00
h. 4' Manhole 7'-9' Cut	Each	$1,100.00		$0.00
i. 4' Manhole 9'-11' Cut	Each	$1,200.00		$0.00
j. Drop to Manholes	Each	$500.00		$0.00
k. 6" Service Lines	L.F.	$6.50	300.00	$1,950.00
l. 8"x8"x6" Wyes	Each	$50.00	6.00	$300.00
m. Testing/Cleanup(10%)	L.S.			$912.00
SUBTOTAL				$10,032.00
SUMMARY OF COSTS				
I. ROADS AND STORM DRAINAGE		$49,368.00		
II. POTABLE WATER SYSTEM		$12,540.00		
III. SANITARY SEWER SYSTEM		$10,032.00		
IV. MISCELLANEOUS		$3,500.00		
TOTAL CONSTRUCTION COSTS		$75,440.00		
ENGINEERING FEES (6%)		$4,526.40		
SURVEYING FEES (2%)		$1,508.80		
INSPECTION FEES (2%)		$1,508.80		
GRAND TOTAL		$82,984.00		

Figure 14.1 An example of a cost estimate for a 5.13 acre development.

second pass will allow the estimator to scrutinize each project element and allow work to be checked. The key is to be relaxed and not rush the job. This estimate should reflect what a qualified contractor would deem reasonable, although in practice the engineer does not normally control the contractor's work force or material procurement.

14.3.2 Time of Project Completion

Normally a developer prescribes a project completion time based on outside influences and constraints or on the design and/or professional construction manager's recommendations. Suppliers' performance, weather, labor, economic factors, and site conditions can affect the progress of a job. In most land development projects using the customary approach for professional design services as shown in Figure 3.1, the design professional formulates a theoretical construction schedule based on the developer's timetable and prevailing local conditions. As an example the designer might select construction production rates that best represent what a typical contractor could produce if awarded the job. If the resulting construction schedule indicates the developer's completion date can be met adequately, then the completion time remains reasonable and can be used as a basis for bidding and construction contract terms.

However, if the design professional's theoretical construction time exceeds the time hoped for by the developer, the engineer may have to reexamine the project schedule to determine whether adjustments can be made to shorten the project. If the time allowed for completion cannot be lengthened, an increase of activity intensity (requiring possibly longer than normal working hours, more crews, etc.) may result in higher construction costs being required to complete the project quickly.

For projects that involve professional services based on the customary approach, that use either project management (Figure 3.2) or a professional construction manager (Figure 3.3), project milestone scheduling and job duration would be jointly established by the developer, the construction manager, and the design professional.

14.3.3 The Critical Path Method

Principally, CPM utilizes a logic diagram that shows the precedence of activities, with their durations and estimated costs. This logic diagram initially must be prepared manually; however, computer applications can be used for final drawings, calculations, and schedules. Once computerized, data may be utilized for project monitoring at the commencement of construction.

There are many considerations associated with CPM diagramming network logic. Most commonly in land developments, an activity on arrow network is utilized. Each project must be analyzed and dissected into incremental tasks or activities. An activity is shown in the network by an arrow that points in the direction that the work progresses relative to its duration. The scale of the arrow length is not necessarily related to the amount of time required to complete the activity. Normally in a network, the name or abbreviation of the activity is placed above the arrow shaft, and the estimated duration is placed in a box under the arrow, as shown in Figure 14.2.

An event is the instant in time when one or more activities start or end with no associated duration. Events are usually represented by circles at the beginning and end of each activity arrow. Various activities may share events, with several originating or terminating together. Each circle is called a node and each node contains either an "i" or

Clear and Grub Road

| 2 |

Figure 14.2 An example of an activity on an arrow.

"j" number. The "i" number represents the beginning of the activity, while the "j" is used to terminate an activity., An example of "i–j" numbering is shown in Figure 14.3.

Occasionally, activities require logical restraints to produce a workable and effective network. To produce this condition a dummy activity is utilized that has no duration. The dummy activity is depicted by a dotted or dashed arrow. The complete CPM network consists of interconnected activities, nodes, and dummies.

14.3.4 Network Logic

Following are some logic conditions that may be encountered in CPM networking, as illustrated in Figure 14.4.

Condition I: Activity *E* must be completed prior to commencing activity *G*.

Condition II: Activity *I* must be completed prior to commencing activity *J*. Activity *I* must also be completed prior to commencing activity *K*.

Condition III: Activity *G* must be completed prior to commencing activity *I*. Activity *H* must also be completed prior to commencing activity *I*.

Condition IV: Activities *A* and *B* precede activities *C* and *D*.

Condition V: Activity *C* must be completed prior to commencing activity *E*. Activity *D* must be completed prior to commencing activities *E* and *F*.

14.3.5 Diagramming Procedures

Network logic is assumed to flow in the direction of the arrow tip and can never flow against the arrow's direction. The logic also flows from left to right throughout the diagram, with each project beginning and ending at single nodes called the starting point and ending point, respectively. All activities that terminate at a node must be completed prior to the subsequent activities being started. To insure clarity and accountability, each activity must have unique "i" and "j" numbers, which, if multiples of 10 are utilized, will allow the network to be modified or edited later. To denote activity progression, each "i" number is made smaller than the corresponding "j" value.

Restraints or dummy activities are utilized to prevent duplicate "i" and "j" numbers from occurring, as shown between activities *I* and *L* in Figure 14.4.

Dummies also can be utilized to denote logical restraints between nodes in parallel activity lines or as logic splitters to prevent unwanted flows from occurring, similar to that illustrated in Figure 14.5. Without the use of logic splitters activity *A* must be completed prior to starting activity *B*. Activities *A* and *C* must be completed prior to starting activity *D*. Also, activities *A, C,* and *E* must be completed prior to commencing activity *F*. With

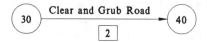

Figure 14.3 An example of activity nodes.

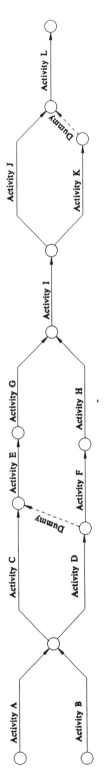

Figure 14.4 An example of critical path network logic.

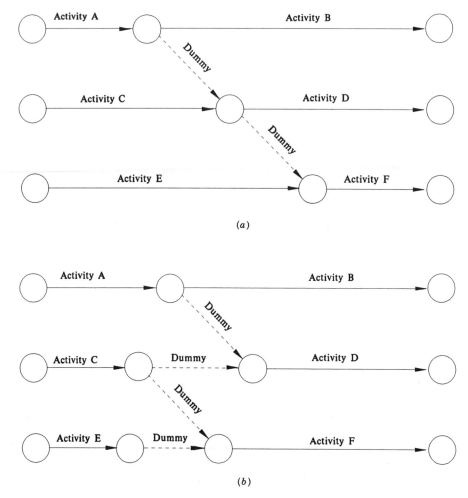

Figure 14.5   Examples of critical path network logic splitting. (*a*) Network without a logic splitter. (*b*) Network with a logic splitter.

the use of logic splitters, activity *A* still must be completed prior to starting activity *B*, and activities *A* and *C* still need to be completed prior to starting activity *D*. With logic splitters activities *C* and *E* must be completed prior to commencing activity *F* while activity *A* does not affect the start of activity *F*.

14.3.6   Critical Path

The critical path of a network is the path (by time duration) through the diagram that produces the longest lapsed time. All activities that comprise this path are termed critical, which means that they must be completed on schedule as estimated or the job will lag. Other activities not considered critical may run over their estimated time schedule without resulting job delays. The amount of time an activity can slip without delaying the scheduled finish of the job is called float. Sometimes a series of activities may share float with other activities. If the float for an activity can be expended on that activity without delaying the earliest possible start of any other activity, that float is called free float. All the float available to an activity is called total float. Only that part of the total float that can be used without delaying the early start of any other activity is free. Free float is always some part of total float. Therefore, the critical path contains no float.

14.3.7 CPM Calculations

Given the simple five activity network shown in Figure 14.6, various calculations will be performed to illustrate the critical path, early start, late start, early finish, late finish, and float. Information from the network is entered in tabular format to ease calculations. This is depicted in Figure 14.7a. Workdays are usually the time unit used; however, other units may be used as long as they are consistent throughout the diagram. When days are used, work will always start at the beginning of the first day and end at the end of the last day. Starts and finishes are instants in time (no duration) and therefore are associated with nodes rather than activity arrows.

From the network it can be seen that activity A is the first activity. Therefore it is assumed that it will commence on the first day of work at the beginning of the day and that the early start (ES) for that activity will be time 0. ES is the earliest time that the activity can be started.

Early finish (EF) is the earliest time that an activity can be completed. This is determined by the ES of the activity plus its duration. The EF of activity A is the end of day 2 ($ES + D = 0 + 2$).

The calculations are shown in Figure 14.7b. Note that the minimum length of time to complete the project is 10 days. Now the calculations must be continued in reverse order beginning with the EF of the last activity and using that as its late finish (LF). The LF is the latest time that an activity can be finished without delaying the completion of the project. If the LF of activity E is 10 days, then the late start (LS) of activity E is 9 ($LF - D$). LS is the latest time that an activity can be started without delaying the project. The calculations are continued in Figure 14.7c.

Now the float can be computed. Total float (TF) is the amount of time that the beginning or ending of an activity can be delayed without delaying the completion of the project ($TF = LF - EF$ or $LS - ES$). The TF of activity A is 0. Since there is no spare time activity A becomes critical and is part of the critical path. Free float (FF) is the amount of time that the finish of an activity can be delayed without delaying the earliest starting time for following activities ($FF = ES$ of next activity $- EF$ of this activity). The calculations are completed in Figure 14.7d.

Upon completion of the schedules and charts the network diagram should be redrawn, plotting the activities to a time scale with actual calendar dates used. Weekends and holidays should be left in for this plot as they can represent a form of float for all

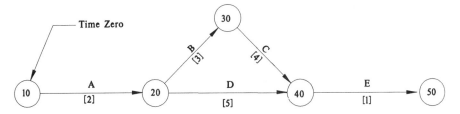

Figure 14.6   An example of a critical path network diagram using activity on arrow methods.

A. CPM Calculation Data									
Activity	i	j	Duration	ES	LS	EF	LF	TF	FF
A	10	20	2						
B	20	30	3						
C	30	40	4						
D	20	40	5						
E	40	50	1						
B. Early Start and Early Finish Calculations									
Activity	i	j	Duration	ES	LS	EF	LF	TF	FF
A	10	20	2	0		2			
B	20	30	3	2		5			
C	30	40	4	5		9			
D	20	40	5	2		7			
E	40	50	1	9		10			
C. Late Start and Late Finish Calculations									
Activity	i	j	Duration	ES	LS	EF	LF	TF	FF
A	10	20	2	0	0	2	2		
B	20	30	3	2	2	5	5		
C	30	40	4	5	5	9	9		
D	20	40	5	2	4	7	9		
E	40	50	1	9	9	10	10		
D. Float Calculations									
Activity	i	j	Duration	ES	LS	EF	LF	TF	FF
A	10	20	2	0	0	2	2	0	0
B	20	30	3	2	2	5	5	0	0
C	30	40	4	5	5	9	9	0	0
D	20	40	5	2	4	7	9	2	2
E	40	50	1	9	9	10	10	0	0

Figure 14.7   An example of critical path network calculations.

activities. On a plot of this sort, start dates for activities with float can be adjusted to best use the resources available. This type of plot can be used to record project progress and is more effective for this purpose than a bar chart.

Economical computer software, such as CRITIC (Smith, 1985) can be utilized to accomplish the above calculations efficiently and can incorporate a calendar to provide meaningful output. A sample network is shown in Figure 14.8 for a residential subdivision that was analyzed using this program.

14.3.8   CPM Computer Application

To illustrate the use of a computer to assist in the management of a small improved land development, the construction costs developed in Figure 14.1 are utilized.

The activity network shown in Figure 14.8 represents in a general manner the sequencing of construction events necessary to complete improvements on a 5.13 acre development. The activities shown are broadly described for illustration purposes; however, further subdivision of these activities could be provided to include more detail. Most computer application programs require data similar to that shown in Figure 14.9. The included code listing are for the various accounts for administrative costs (Code 1 in this example), road and storm drainage construction costs (Code 2), water and sewerage construction costs (Code 3), and surveying costs (Code 4).

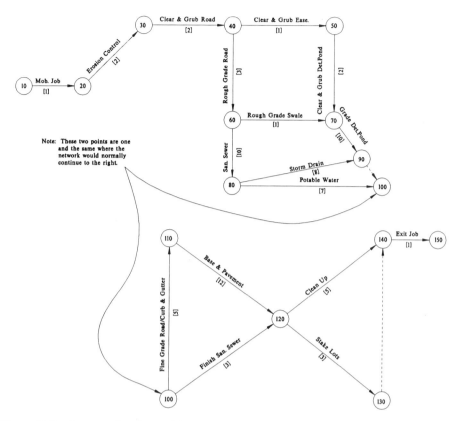

Figure 14.8 An example of an activity on arrow network for the sample 5.13 acre development site.

I	J	Activity	Duration	Cost	Code
10	20	Mobilize to Job	1	$1,000.00	1
20	30	Install Erosion Control	2	$2,000.00	2
30	40	Clear and Grub Road	2	$2,760.00	2
40	50	Clear and Grub Easement	1	$560.00	2
40	60	Rough Grade Road	3	$2,500.00	2
60	70	Rough Grade Swale	1	$2,400.00	2
50	70	Clear and Grub Det. Pond	2	$1,880.00	2
70	90	Grade Detention Pond	10	$4,500.00	2
60	80	Install Sanitary Sewer	10	$8,120.00	3
80	90	Install Storm Drainage	8	$10,775.00	2
80	100	Install Potable Water	7	$11,400.00	3
90	100	Dummy	0	$0.00	1
100	110	Fine Grade; Curb & Gutter	5	$7,500.00	2
110	120	Road Base and Pavement	12	$10,005.00	2
100	120	San. Sewer Invert & Covers	3	$1,000.00	3
120	130	Stake Lots	3	$1,500.00	4
120	140	Clean-Up and Inspections	5	$6,540.00	1
130	140	Dummy	0	$0.00	1
140	150	Exit Job Site	1	$1,000.00	1

Figure 14.9 An example of critical path network data for a 5.13 acre development site. Source: Harrison S. Smith. *CRITIC*, 1985.

Nodes I	J	Activities	Duration	Early Start	Late Start	Early Finish	Late Finish	Total Float	Free Float	
		5.13 Acre Development Project								
10	20	Mob Job	1	1 06 MAY 91	1	1	1 06 MAY 91	0	0	Critical
20	30	Erosion Control	2	2 07 MAY 91	2	3	3 08 MAY 91	0	0	Critical
30	40	C & G Road	2	4 09 MAY 91	4	5	5 10 MAY 91	0	0	Critical
40	50	C & G Easement	1	6 13 MAY 91	14	6	14 23 MAY 91	8	0	
40	60	Rough Grade Rd.	3	6 13 MAY 91	6	8	8 15 MAY 91	0	0	Critical
50	70	C & G Det. Pond	2	7 14 MAY 91	15	8	16 28 MAY 91	8	1	
60	70	Rough Grade Swale	1	9 16 MAY 91	16	9	16 28 MAY 91	7	0	
60	80	San Sewer Mains	10	9 16 MAY 91	9	18	18 30 MAY 91	0	0	Critical
70	90	Grade Det. Pond	10	10 17 MAY 91	17	19	26 11 JUN 91	7	7	
80	90	Storm Drain	8	19 31 MAY 91	19	26	26 11 JUN 91	0	0	Critical
80	100	Potable Water	7	19 31 MAY 91	20	25	26 11 JUN 91	1	1	
100	110	Fine Grade & Curb and Gutter	5	27 12 JUN 91	27	31	31 18 JUN 91	0	0	Critical
100	120	Fin San Sewer	3	27 12 JUN 91	41	29	43 05 JUL 91	14	14	
110	120	Base & Pavement	12	32 19 JUN 91	32	43	43 05 JUL 91	0	0	Critical
120	130	Stake Lots	3	44 08 JUL 91	46	46	48 12 JUL 91	2	0	
120	140	Clean Up & Inspection	5	44 08 JUL 91	44	48	48 12 JUL 91	0	0	Critical
140	150	Exit Job	1	49 15 JUL 91	49	49	49 15 JUL 91	0	0	Critical

Figure 14.10 An example of a critical path network working schedule. Source; Harrison S. Smith. *CRITIC*, 1985.

Figure 14.10 shows the results of the computer analysis in tabular form. This job consists of a total of 49 work days with work beginning the morning of 06 May 1991 and ending the afternoon of 15 July 1991. The total cost of the project is $75,440.00, and the work consists of 17 activities and 2 dummy activities. This particular program does not include in its construction calendar weekends and normal holidays.

Most programs provide bar charts similar to that shown in Figure 14.11. This enables one to see the progression of activities relative to the passing of time.

Work schedules for the various accounts also can be obtained, with Figure 14.12 depicting the activities of two accounts.

The job's progress relative to the accumulated percentage of cost can be shown in an S-curve similar to the one in Figure 4.13. This curve allows projections of expenditures and provides a means of monitoring progress based on cash flow.

14.3.9 CPM Submission

Construction contracts may require the successful bidder to submit a work schedule upon award. This schedule, for example, could be required as part of the NPDES General Permit for Storm Water that pertains to the project. The engineer (construction manager)

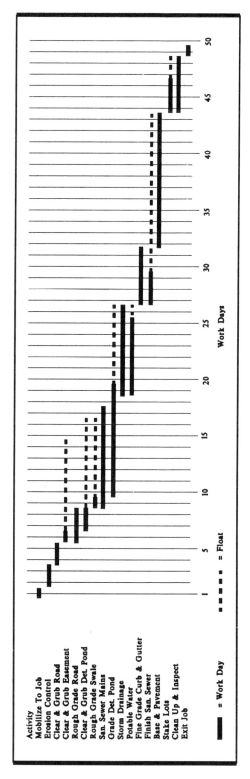

Figure 14.11 Schedule bar chart for the sample project schedule. Source: Harrison S. Smith. *CRITIC*, 1985.

5.13 Acre Development For Account No. 1						
Activity	Scheduled Start	Actual Start	Scheduled Finish	Actual Finish	Extimated Cost	Costs to Date
Mob Job	Day 1 06 May 91		Day 1 06 May 91		$1,000.00	
Clean Up & Inspect	Day 44 08 July 91		Day 48 12 July 91		$6,540.00	
Exit Job	Day 49 15 July 91		Day 49 15 Jul 91		$1,000.00	

5.13 Acre Development For Account No. 2						
Activity	Scheduled Start	Actual Start	Scheduled Finish	Actual Finish	Extimated Cost	Costs to Date
Erosion Control	Day 2 07 May 91		Day 3 08 May 91		$2,000.00	
C & G Roads	Day 4 09 may 91		Day 5 10 May 91		$2,760.00	
C & G Ease.	Day 6 13 May 91		Day 6 13 May 91		$560.00	
R. Grade Roads	Day 6 13 May 91		Day 8 15 May 91		$2,500.00	
C & G Det. Pond	Day 7 14 May 91		Day 8 15 May 91		$1,880.00	
R. Grade Swale	Day 9 16 May 91		Day 9 16 May 91		$2,400.00	
Grade Det. Pond	Day 10 17 May 91		Day 19 31 May 91		$4,500.00	
Storm Drain.	Day 19 31 May 91		Day 26 11 Jun 91		$10,775.00	
F. Grade C & Gutter	Day 27 12 Jun 91		Day 31 18 Jun 91		$7,500.00	
Base & Pavement	Day 32 19 Jun 91		Day 43 05 July 91		$10,005.00	

Figure 14.12   An example of a critical path network work schedule by account numbers. Source: Harrison S. Smith. *CRITIC*, 1985.

should insure that this schedule reflects a reasonable approach and that the project's activities will be proportioned without inordinate time and expenses being allocated for certain activities. This schedule will probably vary from the theoretical schedule initially provided by the design professional as the contractor has actual control over specific labor, equipment, and materials. As an example, the contractor may plan on utilizing two construction crews to accomplish an activity that the design professional initially estimated would be accomplished with one. The contractor's construction schedule can be utilized to assist the developer and engineer in monitoring project progress. This is also helpful when construction management services are being provided and the contractor is obligated to attain various "milestone" events at predetermined times to maintain productivity. Should actual construction progress lag in the field, the engineer can confer and make recommendations to the contractor to insure timely job completion. These recommendations are not to be construed as binding on the contractor because the engineer (construction manager) does not control the contractor's work force or equipment.

14.3.10 Progress Monitoring

Throughout the construction period the engineer (construction manager) is able to monitor the job's progress based on site observations and reports from the contractor. Should

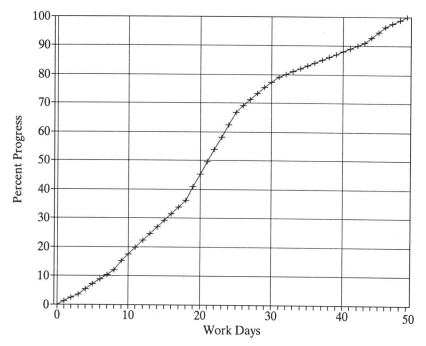

Figure 14.13   S-Curve for the sample critical path network. Source: Harrison S. Smith. *CRITIC*, 1985.

progress stall or lag behind the submitted contractor's schedule, alterations to the CPM network can be made by adding new activities or altering previously used ones to update the schedule. This is accomplished by changing all durations associated with completed activities to zero and modifying the durations of current activities to reflect actual remaining times until their estimated completion. These alterations originate from the contractor or are suggestions made by the engineer (construction manager) and adopted voluntarily by the contractor to reflect the contractor's desire to improve productivity. Reanalyzing the network will yield an updated schedule that might cause the critical path to change. This process can be repeated if the results are unacceptable and do not indicate punctual project completion.

REFERENCES

American Society of Civil Engineers (ASCE). *Construction Cost Control—ASCE Manual No. 65.* New York: ASCE, 1985.

Fisk, Edward R. *Construction Project Administration,* 3rd ed. New York: Wiley, 1988.

Robert Snow Means. *Building Construction Cost Data.* Duxbury, MA. (Annually).

Smith, Harrison S. *CRITIC.* Kehilan Farm, Route 2, Box 1500, Columbus, N.C.: H. S. Smith, 1985.

15 Development Guidelines

Development guidelines aid in project theme reinforcement by forcing landowners to coordinate individual preferences. These guidelines, which augment other local regulatory codes such as those required by zoning, are usually incorporated into project restrictions or protective covenants. They possibly include provisions for building setback distances, tree protection, and landscape regulations (Rubenstein, 1987).

15.1 PURPOSE OF GUIDELINES

Project guidelines, essential for successful improved land developments and beneficial to staged projects, are intended to protect a project's marketability and integrity during construction and post-construction. Site improvements undertaken after the sale of improved property to others require control of quality, aesthetics, and conformity of use. In addition, acceptable improvements must be compatible with the development's installed infrastructure.

15.2 CONTROLLING ELEMENTS

Development guidelines are usually tied to the project's site through the development's master plan. For small projects, the final plan may suffice. Guidelines contained in covenants and restrictions usually include architectural, landscape, signage, lighting, and maintenance controls. Enforcement of these elements is usually accomplished through the actions of the homeowners' association. The developer may require the services of an architect, landscape architect, and attorney to assist in the preparation of guideline documents.

15.2.1 Homeowners' Association

The homeowners' association is an incorporated, nonprofit organization comprised of project property owners. Association rights and regulations are recorded in the public records office along with deeds of conveyance for common property from the developer to the association. This conveyance is completed prior to dwelling unit or parcel sales. The association is charged with the operation, maintenance, control, and ownership of all development common areas, recreational areas, streets and easements not dedicated to the public. Property owners within the development are voice and vote members of this association with undivided interest in all association controlled property. The developer usually maintains voice and vote based on the number of dwelling units remaining in the development. Each member is annually assessed a fee for operating revenue. Initial operational fees, however, as completely borne by the developer, while subsequent financial participation is reduced according to the developer's proportional dwelling unit holdings. Special assessments also are levied in case of emergency expenses or the need for capital improvement projects such as bottom and side replastering or pump and filter replacement associated with the association's swimming pool. Within the homeowners' association is usually a board that reviews all improvements that involve the development.

15.2.2 Board of Review

This board, initially appointed by the developer, insures that new construction within the development meets established guidelines and criteria. As the project nears completion, the developer's involvement dwindles while the homeowners' involvement increases. This allows the homeowners an opportunity to appoint board members of their own choosing to carry out the organization's charge; the board should include an architect, landscape architect, or engineer to provide other board members with technical insight. Proposed plans and specifications associated with various development phases are reviewed for compliance with the project's approved master development plan and restrictions. Each site's submission is evaluated for architectural compatibility. Building plans that show the shape, size, color, and exterior treatments in addition to proposed fencing and walls are reviewed along with the proposed structure positioning on the parcel in relation to setback requirements. The proposed improvements also must be compatible with overall development transportation plans for vehicular, pedestrian, and bicycle travel. Parking areas must comply with development regulations.

The board of review insures other types of site improvements are acceptable including lighting for pedestrian, vehicular, and accent applications. Proposed grading and drainage plans must conform to the development's overall plan, as well as the utilization and protection of easements and facilities.

Site landscape plans pertaining to buildings, parking lots, and open space are reviewed for compatibility with drainage and maintenance master plans.

15.2.3 Project Restrictions or Protective Covenants

Restrictions should be initially prepared for the entire development and recorded prior to parcel or dwelling unit sales. Large phased developments that contain restriction variations that are not definable at the project's beginning should have each subsequent phase buffeted from previous phases to minimize conflicts. Each phase's restrictions are recorded prior to ownership transfers within that phase. Since protective covenants are recorded, they run with the land and are binding to all owners. Covenants, therefore, are an express agreement between the developer and purchaser with enforcement provisions. In addition to regulating site improvements, covenants are used to control noise and nuisances including erection of temporary shelters, tents, trailers, sheds, outbuildings, oil drilling, mining operations, kennels, and any harboring of farm animals.

· Covenants can provide safety by regulating any alteration and excavation of slope control areas requiring stabilization and protection. In addition site control easements where corner lots must allow clear vision at street intersections for safe vehicular travel can be enforced through covenants. Finally, covenants can specify types of allowable business operations in commercial areas that are included in the development.

15.3 ADDITIONAL CONTROLS

The use of guidelines provides a means for activity coordination at the conclusion of improved land and total development activities; however additional controls are sometimes helpful. For example, development guidelines provide guidance for the dwelling unit's builder as well as for the end user; however, Figure 15.1 illustrates a form of additional control that minimizes problems during building construction.

DWELLING UNIT BUILDER'S RESPONSIBILITY
1. Preservation of Improvements. The builder shall be responsible to the owner for preserving any and all land improvements made in the development. Employees of the builder shall be directed to use good housekeeping procedures to insure positive lot drainage and to minimize hazards.
2. Lot Improvements. After project contractors (e.g. road/sewer,etc) have completed their work, all remaining lot improvements shall be constructed in a self-sustained manner by the builder. No borrow material shall be obtained from any place on site unless such area is held in interest by the builder and removal does not alter the project plans. Waste material resulting from lot clearing shall be properly disposed of by the builder and not placed on adjoining lots or pushed into storm drainage facilities and subsequently covered with soil.
3. Service Taps. The tie-in area shall be maintained to allow the proper flow of stormwater when water and wastewater taps are exposed to make connections and visual inspections by the local regulatory authorities.
4. Delivery of Materials. It shall be the responsibility of the builder to insure that building material deliveries do not deface development improvements. For example, bricks shall be off-loaded from the delivery truck far enough back on the lot so as to not interfere with storm drainage facilities and ready-mix concrete trucks shall not deface curbing.
5. Responsibility of Maintenance. Each builder shall be responsible for maintaining all improvements which are on or contiguous to the building lot. All storm drains shall be kept open to run free, and all debris will be disposed of properly.
6. Conformance. Each builder shall conform to the requirements of all applicable project construction requirements which may include: a) Zoning b) Building, electrical, and plumbing codes. c) Sedimentation and erosion control plans. d) Site grading plans. e) Project restrictions and covenants.
Date: Builder's Signature   Owner's Signature

Figure 15.1 Form of understanding between builder and developer.

REFERENCES

Rubenstein, Harvey M. *A Guide to Site and Environmental Planning,* 3rd ed. New York: Wiley, 1987, pp. 377–397.

U.S. Department of Housing and Urban Development (HUD). *Manual of Acceptable Practices,* 1973 ed., 4930.1 Vol. IV. Washington: HUD, 1973, Data sheets 40, 41.

16 Completion of Design Activities

The final stage of the design process includes the assembly of all documents related to project construction, and their submission to regulatory authorities for approval. Some of these documents are plans, specifications, and applications for permits to construct.

16.1 FINAL PLANS FORMAT

Engineering drawings are normally placed on standard size sheets having the following dimensions:

A Size: 9 inches × 12 inches
B Size: 12 inches × 18 inches
C Size: 18 inches × 24 inches
D Size: 24 inches × 36 inches
E Size: 36 inches × 48 inches

Land development plans customarily utilize the standard D size format because available drawing space is usually sufficient to adequately display information but not so large as to be cumbersome in the field. Record drawings are usually inked on vellum or mylar drafting materials.

The presentation of lines normally follows acceptable characteristics portrayed in the alphabet of lines (French et al., 1970). In addition, emphasizing certain lines such as borders and rights of way edges by using various width pens produces a more pleasing display.

Engineering lettering is accomplished using freehand lettering guides, such as Keuffel & Esser Co.'s Leroy®, or CAD applications. Letter size conventions vary according to the importance of information being displayed, with the most general information being shown largest and the most specific information being shown smallest.

Depiction of information usually follows established graphic standards such as those contained in *Architectural Graphic Standards* (Hoke, 1988). Graphic presentations should embody a consistent style throughout the set of plans including the use of:

(a) A standard border format.
(b) A standard title block, placed toward the right side of each drawing, that includes the project's name and location, date, scale, sheet number, revision table, engineer's signature and seal, engineer's address, and any other design professional's seal and signature.

16.1.1 Title Sheet

The cover sheet identifies the drawing set by the title, which is affixed using the largest size letters. The project's location is usually included along with a vicinity map that would aid a visitor in finding the site. The developer's name and address are also shown along with a listing of the internal drawings by sheet and topic. This cover sheet includes the engineer's name, address, seal, and signature.

Title sheets can have regulatory authority statements and project approval certifications affixed for endorsement and dating. Other information can include revision data, field survey book references, and general notations such as special conditions for emphasis to the contractor.

Other title page entries may be a summary sheet of the project's land use areas, including the number of various categories of building sites. Figure 16.1 shows the general layout of a typical title sheet.

16.1.2 Project Configuration Sheets

The first sheet in a set of construction plans is usually the approved preliminary plan upon which the set of drawings are based. This drawing should contain the data cited in Section 5.3.1.

Following the preliminary plan are drawings that contain lot and parcel dimensions similar to the final recorded development plat. This sheet is useful for orientation and field work.

16.1.3 Road and Storm Drainage Sheets

Design information required to construct the road and street system normally follows the lot layout drawings. These road drawings are normally prepared on standard plan and profile sheets. The plan view represents the street's horizontal control nomenclature including: centerline beginning and ending points; points of intersecting street centerlines (PI's) with tie equality stationing; centerline bearings; horizontal curve data; superelevation limits; stationing; right of way width data; lot line and property line intersections with the rights of way; easements; drainage facilities including catch basins, culverts, junction boxes, french drains, curb and guttering or ditching; bench mark locations and identification; horizontal scale; north arrow and meridian reference; and adjoining landowners if appropriate. A storm drainage culvert schedule may be included that summarizes in tabular form the pipe's identification, diameter, length, material, slope, and end invert elevations. If the job site is subject to flooding, the 100 year flood limits are often delineated. Where jobs are located close to existing facilities, or in areas where the work must be confined, a line indicating the limits of construction (or clearing) can be provided in the plan view.

Each profile section includes road centerline stationing, existing ground profile, proposed road profile (being careful to specify either the subgrade or the finished grade), vertical curves, design elevations at each station (or half station) and at all key horizontal control stations (curve tangent points and tie equalities), culverts, catch basins, french drains, and the horizontal and vertical scales.

Following the road plan and profile sheets are the road detail sheets. These sheets can

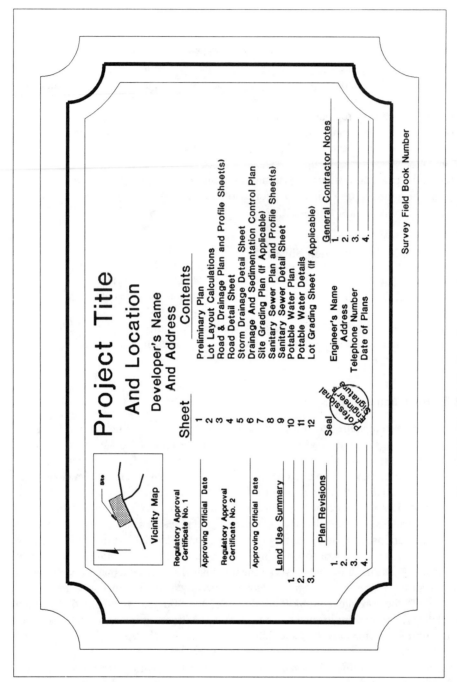

Figure 16.1 Typical title sheet for a set of construction plans.

include typical geometric cross sections, typical structural cross sections showing the pavement materials and thicknesses, cul de sac details, curb (and false curb) details, driveway and intersection details, signage details, and other standard details necessary for the complete construction of the road system.

Because a number of items included on road plan and profile sheets are associated with drainage, a storm drainage detail sheet is usually included. This sheet can include typical details for the construction of curb inlets, drop inlets, junction boxes, french drains, culvert headwalls, installation procedures for culverts (including bedding requirements), riprap placement, and typical channel cross sections.

16.1.4 Drainage and Sedimentation Control Plan

The drainage and sedimentation control drawings can be included in the set of plans after the road drawings since they are closely related. This plan typically shows the location of drainage easements, channels (including typical cross sections, slope, and material lining), culverts (including length, diameter, material, strength classification, invert elevations in and out, and slope), french drains, riprap placement, retention/detention facility locations and configurations, and outfall structure locations that can include stilling basin and splash pool locations.

Sedimentation and erosion control provisions can also be shown on this plan. These provisions can include locations where various control practices are to be employed during construction. For example, the location of a temporary construction entrance can be delineated along with how spray water will be provided (with a back flow preventer) for washing a vehicle's underside. Silt fencing locations can also be shown so that the contractor can install the fencing before major clearing and grubbing operations begin. Temporary perimeter, interceptor, and diversionary dikes are included along with the location of temporary level spreaders for dispersing runoff. Other sediment control provisions can be shown such as straw (or hay) bale barrier locations, as well as sediment trap and basin locations. A sediment and erosion control detail sheet is helpful to show the contractor how to construct and install the various erosion control devices shown on the plans. If an NPDES General Permit for Storm Water is in effect, this drainage and sediment control plan must be kept current and on file at the construction site.

16.1.5 Site Grading Plan

If the project warrants a site grading plan, it is conveniently placed after the drainage and erosion control drawings because site grading closely relates to road and storm drainage facility construction. The site grading plan shows existing and proposed contours along with existing spot elevations and any finished floor elevations. Areas determined during the geotechnical investigation to contain poor soils, such as expansive clays, can be delineated on the site grading plan for removal or modification. Other on-site areas requiring protection from alteration, such as designated wetlands, can also be delineated on this plan to caution the contractor's field personnel.

16.1.6 Sanitary Sewer Sheets

The plan and profile sheets for the wastewater collection system number as many as needed to adequately display the project's system. The plan view shows necessary horizontal control data including bearings and distances between manhole centers, line diameters and material used for manufacture. In addition, existing system structures are shown, along with information on new manholes, including stationing, rim and cover elevations (set above anticipated flood elevations), invert elevations, and drop heights. If the site is subject to flooding, the 100 year flood limits are often shown. Service lines, diameters, material, and tap locations also are shown in the plan view.

The profile includes the system's stationing originating from centerline of manhole and progressing to subsequent manhole centers *while being completely independent of road stationing*. The profile shows the finished ground profile, the invert and crown line of the proposed main with the slope included along with any pipe crossings. Invert elevations for all manholes are included in addition to those associated with drops. A sanitary sewer detail sheet follows that shows approved construction techniques. Manhole details (including drops), pipe installation procedures (including pipe bedding requirements), piers, cleanouts, and service connections can be included.

Should a sanitary sewer pumping station and force main (or wastewater treatment plant) be part of the project's scope, these drawings are included after the collection system and could entail several sheets just for the pumping station construction.

16.1.7   Potable Water System Sheets

The potable water distribution system drawings usually follow those for the sanitary sewer because the water system is normally constructed after the deeper sanitary sewer system is installed. Water lines allow the contractor more flexibility than gravity sewer lines during installation. The water system is pressurized and does not have to be installed on a highly controlled grade where constant fall is required to provide efficient gravity flow conditions, such as are required in a gravity flow sanitary sewer collection system. The water system can be shown in a plan view where water main length, diameter, strength classification, and make are labeled along with valving, fire hydrants, thrust block locations, service lines, and tap locations. Blow offs and air release valves locations are also shown. If a profile of the water main is required, the finished surface profile must be shown in addition to pipe crossings, elevations, and water main stationing. The plans should also include any existing water works structures and facilities, and any underground improvements that could affect the water system such as storm drains and sanitary sewers that are located close by. Where water mains cross streams, the normal high and low water elevations should be shown along with the 100 year flood elevation located on site. A potable water detail sheet provides information on installation and construction techniques required to properly install the water system. These details can include pipe installation procedures, valve and box details, fire hydrant details and their installation, thrust block positioning and sizing for various line sizes and working pressures, service and meter connections, and procedures for disinfecting the installed system including the required pounds of chlorine per unit length of watermain diameter. If potable water connections are to be made where possible contamination can occur, such as at a sanitary sewer pumping station, then details pertaining to back flow preventers must be included. Drawings associated with any water source development or treatment would also be included in the set of project plans.

16.1.8   Lot Grading Sheet

Often, in an improved land development, individual parcel owners are responsible for lot clearing and landscaping. To provide guidance that coordinates with the development's

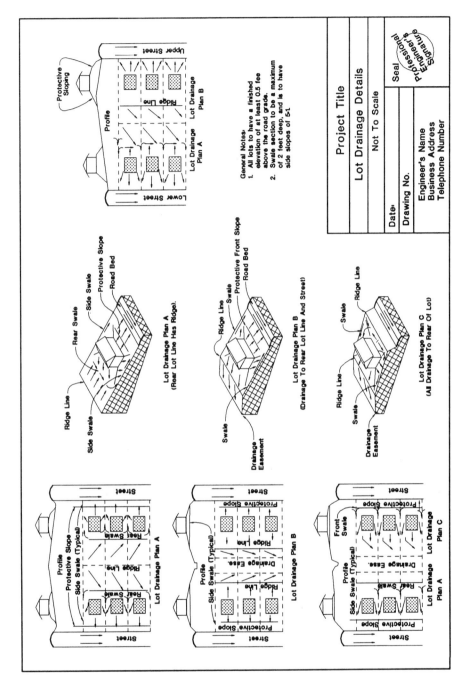

Figure 16.2 Typical lot drainage detail sheet. Source: U.S. Department of Housing and Urban Development.

storm drainage plans, a lot grading sheet can be prepared to show building site grading. This is accomplished by assigning each lot a particular generalized grading plan, either "A," "B," or "C," that is keyed to a lot drainage detail sheet similar to the one shown in Figure 16.2.

16.1.9   Document Reproduction and Assembly

Once the set of plans has been prepared and assembled, an initial set of blue or black line prints should be made for information verification by the engineer. This process includes the use of a felt-tipped marker or colored pencil to strike through each verified entry on every drawing sheet. If errors are detected, the engineer can note the corrections in red, and they can later be made on the original drawing. This check process is considered ordinary care to minimize errors and omissions. Once the plans have been checked, and all necessary corrections made, corrected copies must be prepared for a final scan by the engineer.

16.2   COMPLETED PROJECT MANUAL

The project manual is normally bound into pamphlet or book form, having a cover sheet, table of contents, invitation to bid, bid information and proposal forms, conditions of the contract, insurance requirements, and construction specifications. Each segment requires checking for clarity, accuracy, and completeness. Any noted errors and omissions must be corrected prior to a final scan by the engineer.

16.3   PERMIT APPLICATIONS TO CONSTRUCT

Most regulatory agencies require the engineer to submit various forms, drawings, design calculations, engineering reports, and several sets of construction plans and specifications for review when seeking approval for project construction. Submittal requirements for these regulatory reviews must be obtained by the engineer so that when the applications are prepared by the engineer, they are complete. Often the developer must endorse the completed applications and simultaneously submit fees with these review documents when they are delivered to reviewing authorities.

16.4   FINAL DOCUMENT SCAN BY ENGINEER

After correcting any detected errors and adding any omitted information, the engineer should make a final scan of all project documents for a final check to insure that nothing is being overlooked. This last verification should include double-checking the accuracy of all information shown on the plans. In addition, the engineer should verify that ample room is available for the contractor to maneuver equipment and store materials on the job site while conducting the required work. If additional temporary construction easements are required, the engineer should communicate this need to the developer. The engineer also should verify that the plans and specifications stipulate who is to provide any temporary utility services necessary during the construction phase, and, if overlooked, to discuss the matter with the Owner where appropriate action can be taken. Jobs that require temporary storage areas and stockpile areas that are provided for in the plans and specifications should be delineated on the drawings. Also, all specified material and equipment should be cross-checked for availability to minimize supply problems, unless specialty items are mandatory.

The engineer should consider during this final review the types of construction required to complete the job, and probable methods that might be employed by the contractor to accomplish the work. Minor modifications to the plans and specifications might allow the contractor additional flexibility that could result in reduced overall construction costs. An example of this might be the inclusion of sanitary sewer manholes on each end of an oversized jacked and bored casing placed under a roadbed. These manholes would allow the contractor added flexibility to compensate for boring inaccuracies in line and grade, which are critical for gravity sewer mains, resulting in reduced boring costs.

The engineer should reread all specifications to insure clarity and continuity between what is shown on the plans and what is included in the specifications. This will help alleviate future confusion and grounds for possible construction claims.

16.5 DEVELOPER'S REVIEW

Prior to any final plans, project manuals (specifications), cost estimates, or applications for permits to construct being promulgated by the engineer, the developer should review and approve in writing all documents. This review will allow the engineer to make any minor last minute alterations for the developer and allow the developer to consider the estimated construction costs based on completed drawings.

16.6 FINANCIAL INSTITUTION'S REVIEW

The developer usually forwards completed plans, project manuals (specifications), and the estimate of construction costs to the financial institution to complete application requirements. Pending regulatory notification of the documents' approval for construction, the institution can complete loan processing.

16.7 REGULATORY REVIEWS

After the developer has approved all documents, the engineer is free to submit the required number of copies of the various documents to the jurisdictional reviewing authorities. The review process normally requires the engineer to periodically answer questions, provide additional information, make alterations, and confer with the reviewing authorities, along with the developer, to gain plan approval. Once approval is received, the developer can obtain a contractor for construction.

REFERENCES

French, Thomas E., Charles J. Vierck, and Robert J. Foster. *Graphic Science and Design,* 4th ed. New York: McGraw-Hill, 1970, p. 17.

Hoke, John, Ed. *Architectural Graphic Standards,* 8th ed. New York: Wiley, 1988.

U.S. Department of Housing and Urban Development (HUD). *Land Planning Principles for Home Mortgage Insurance, A HUD Handbook, 4140.1.* Washington: HUD, 1973, Section 6.

PART IV
Construction Related Activities

17 Contracts and Award

Once all construction documents have been approved, the developer has gained title to the development site, and all construction loan financing has been completed, the project enters the construction phase. During this phase a contractor is engaged to convert the plans and specifications into reality. On occasion the developer may not have received approval from all regulatory agencies or the financial lender, but may wish to proceed with preparations to complete the job. If public funds are involved, the developer will be required to solicit competitive bids from several contractors. The competitive bid process is beneficial even if private funding is utilized, although a negotiated contract can be made.

For improved land development work, where several different types of construction are involved requiring different trades, *experience has shown that it is essential that only one prime contractor be awarded* (unless prohibited by law), who would then have the responsibility to coordinate and supervise the various subcontractors for the entire job. Cases where several individual prime contracts were let for work in a common area resulted in confusion, loss of efficiency, and loss of project control. Should several different development sections (that are separated from each other) be simultaneously developed, different prime contractors can be procured for each section; however, it is recommended that prime contracts not be awarded to any contractor who is simultaneously providing subcontract work in another section to minimize construction conflicts.

17.1 TYPES OF CONSTRUCTION CONTRACTS

Essential elements of an enforceable contract for engineering services presented in Chapter 3 also apply to contracts related to construction. Construction contracts can be awarded either as competitive bid or negotiated contracts. If public funds are involved, the impending work must be advertised to solicit competitive bids. Specified components within the project must include "or equal" provisions. The contractor is, in addition, usually obligated to provide minimum insurance amounts and include bid, performance, and payment bonds.

If the time in which the work is to be completed is not specified in the contract, it is understood that it will be performed within a "reasonable time." If "time is of the essence," it is clear that the completion date is a part of the consideration of the agreement, and the failure of a party to complete obligations within the time specified may result in contractual liability. The amount of damages may be determined by agreement; however, such damages often are difficult to determine, and usually an amount is stipulated to be a specific amount per day of delay as liquidated damages in lieu of determining actual damages. The injured party must be prepared to justify the amount stipulated. The contractor may be excused from liquidated damages if the contractor is entitled to a time extension or if the owner adds extra work to the contract. Bad weather conditions are assumed to be part of the anticipated working conditions and are not considered sufficient reason for waiving liquidated damages unless it can be demonstrated using climatological data that such weather was unseasonable, and therefore an extraordinary problem.

Bonus payments of a specified amount per day for early completion of the contract and penalty assessments of an equal amount sometimes are included in the contract provisions as an incentive for the contractor to complete work in a timely manner.

17.1.1 Competitive Bid Contracts

Competitive bid contracts are usually based on a fixed price and can be classified as lump sum or unit price contracts. The lump sum contract provides that the contractor shall be paid a lump sum of money for all work covered by the plans, specifications, and issued permits; this lump sum amount shall include all material and labor costs, overhead, profit, and so forth. The unit price contract provides for a breakdown of the number of units of each type of construction and a price for each unit.

Lump sum contracts are used on projects requiring a variety of operations where it is impractical to break down the work into units. It is critical that finalized plans and specifications be comprehensive for this type of contract; they should show in complete detail the requirements of the work. If the plans are incomplete, change orders and extra work orders after the contract is signed are expensive and could lead to controversies. The contractor bidding on the job must allow ample funds to cover contingencies, which usually drives the bid price upward. Any performance risks, therefore, fall on the contractor.

When projects consist of large quantities of relatively few types of construction and the volume or limits of the work cannot be accurately determined in advance, the unit price contract may be advantageous. The unit price contract allows for minimum alterations in the scope of the work without the need of formal change orders.

On occasion, a combination of these two forms of contracts may be used, the lump sum basis to account for a known base amount of work and the unit price basis for unknown amounts of work. For example, a lump sum contract could be solicited for the construction of a development's clubhouse structure, while the foundation work may be based on unit prices as a result of the uncertainty of subsurface conditions. The EJCDC has prepared a set of guidelines, designated *Recommended Competitive Bidding Procedures for Construction Projects (EJCDC No. 1910-9-D)* (Smith, 1987), that more fully addresses competitive bid activities.

17.1.2 Negotiated Contracts

A negotiated contract is awarded to a selected contractor who is chosen from a field of contractors. This method of award is not normally used when public funds are involved except under unusual conditions.

Negotiated contracts can include variations from cost plus a percentage of the cost to costs plus a fixed fee. One advantage of using a cost plus contract is that construction can begin before plans are completed. Cost plus contracts also allow the developer to make changes in the project at actual audited costs. The final project cost, however, is not guaranteed and a good deal of effort is required to audit the contractor.

Other cost plus type contract variations can include incentives with the contract, such as cost plus a fixed fee with profit-sharing, a bonus clause, or a sliding fee scale. Cost plus a guaranteed maximum price (upset price contract) also allows certain contractual incentives. In this form of contract the contractor agrees to perform the required work with reimbursements being made on a basis of cost plus either a percent of the cost or a fixed fee up to an agreed to threshold or guaranteed maximum. The contractor is required to complete all work even if costs exceed the agreed to maximum.

A final type of negotiated contract would be a design-build contract. This type of arrangement provides for the prime contractor to design and construct the project under one contract. The construction portion of the agreement can be accomplished using either lump sum, cost plus, or a guaranteed maximum price contract. A turn-key contract is a variation of the design-build contract because it usually includes provisions present in a design-build contract while possibly incorporating land acquisition, financing, and/or lease arrangement activities.

17.2 ENGINEER'S DUTIES

The engineer's contract for professional services requires the performance of various duties during the construction phase, which could include assisting the developer in securing, tabulating, and analyzing the bids, culminating in a recommendation for contractor engagement. In addition, the engineer can assist in the preparation of the formal contract documents.

17.3 JOB BIDDING AND/OR NEGOTIATION

The developer can solicit competitive bids for contract award by placing notice of the impending work in local newspapers, trade journals, and local trade offices. Also, the developer can notify contractors of the impending work and verbally invite their participation in the bid process. Should the project involve private funds, the developer may wish to negotiate or informally solicit proposals from one or several contractors who desire to perform the required work, from which one can be chosen.

17.3.1 Advertisement for Bids

The typical notice includes a description of the work along with its scope and location. Information concerning the type of bid (with formalities or the reserved right to reject any and all bids) and the bid opening date, place, and time are included along with the method that will be used to base the contract award. Requisite information pertaining to bonds, deposits, insurance, and payment schedules may be included. For further information, the notice lists, along with the owner, the engineer's firm, where copies of the applicable documents can be reviewed or obtained, and the amount of security deposit required.

All interested contractors who desire copies of the bid documents should be required to sign for a numbered document set when placing their document security deposit. This contractor roster, which includes the date when documents were issued, shows an accounting of what documents (and addenda) were issued, and includes the contractor's address and telephone number in case subsequent addenda are issued to clarify questions or to modify the contract document's scope.

Advertising for bids is not in itself a contractual offer on the part of the developer; however, a bid response to such an advertisement is an offer that creates no legal rights for the bidder until it is accepted by the developer.

17.3.2 Pre-Bid Conference

On occasion, a pre-bid conference is helpful to provide contractors with general information and an overview of the job requirements. Also, specific elements of the job can be emphasized, and other pertinent information can be issued. Care must be exercised to refrain from making comments contrary to material contained in the issued plans and specifications. An audio recording of the meeting from which a transcription can be prepared is invaluable. Copies should be made available to all potential bidders. The purpose of this conference is to promulgate information.

Should contractors encounter problems in interpreting the plans or specifications, the engineer can render an interpretation in the form of numbered addenda that can be provided to the party from which the issue arose. Copies are sent to all other potential bidders, and acknowledgment is documented, possibly by certified mail.

17.3.3 The Bid Package

The bid package issued to each bidder usually includes a copy of the invitation to bid, instructions for the bidder, a bid form to be used when preparing for submission, and all previously issued addenda. Included forms must be used by the bidder; otherwise the bid might be considered nonresponsive.

The instruction to bidders is a document that informs all bidders of the various aspects of the job. This provides uniform distribution of material and knowledge to all bidders from which they may prepare unbiased bids. The EJCDC has prepared a *Guide to the Preparation of Instructions to Bidders (EJCDC No. 1910-12)* (NSPE et al.) that can be of value to the engineer when preparing these instructions.

Usually the instruction for bidders requests the bidder to provide evidence of construction experience along with a financial statement and equipment inventory or to have submitted prequalification data. The Association of General Contractors (AGC) has available a standard form, *Contractors' Experience Questionnaire and Financial Statement—Standard Form No. 28,* that can be used to obtain pertinent information about potential bidders.

The instructions to bidders can include procedures for the bidder to use when preparing and submitting the bid that reference the construction plans, specifications, and issued permits. If the contract is on a unit price basis, the estimated quantities of each unit of work are given; however, if a lump sum contract is solicited, an exact definition of the scope of work must be included. Although the lump sum contract relies on the total cost of the job for bid purposes, it is usually good practice for the engineer to provide a form that has the contractor account for the lump sum bid. This information can be useful in determining whether the bid reflects consideration of all aspects of the job. Normally included would be a statement as to the accuracy of the information being provided relative to subsurface data, test borings, and so on. Information on bid formalities and provisions for the rejection of any bid or bids should be included. The information for bidders also includes the allowable length of time to complete the work.

To insure that all bids will be prepared essentially in the same format, an appropriate bid form should be included. This will help in the bid analysis to detect any informalities. The bid form should include the price for which the contractor will perform the work. Bid surety requirements can be included. Two standard form EJCDC documents for bid bonds are available. One is the *Bid Bond—Penal Sum Form (EJCDC No. 1910-28-C),* and the other is the *Bid Bond—Damages Form (EJCDC No. 1910-28-D)* (NSPE et al.). Bid bonds are not an expense to the developer. Bonding companies, hoping to later provide performance and payment bonds (if an award is made to their bid bonded contractor), usually provide the bid bond as a service to the bidding contractor.

Performance bonds, if required, insure that the contractor will complete all terms of the contract in a satisfactory manner. Payment bonds are different in that they insure that the contractor will pay all bills and will hold the developer harmless for any liens or financial claims relating to the job after the developer has satisfied all financial obligations with the contractor. An available EJCDC form, designated as *Construction Performance Bonds (EJCDC No. 1910-28-A)* (NSPE et al.), can be used as part of the contract documents when performance bonds arc being stipulated, along with the form *Construction Payment Bonds (EJCDC No. 1910-28-B)* (NSPE et al.) when required. Bid forms should list all addenda included with the plans and specifications. The contractor also should include a list of proposed subcontractors to be used on the job, if required. The form usually includes a certification that the site was examined and that the plans and specifications are understood by the bidder, in addition to certifying that no fraud or collusion was committed in the bid preparation. This form must be signed and witnessed prior to bid submission.

17.3.4 Bid Acceptance and Opening

On competitive bid jobs sealed bids are accepted by a designated receiver up to a stipulated deadline. For each bid that is received a written acknowledgment is provided the originator. At the designated time and place of bid opening, the developer, usually with the assistance of the engineer, opens all submitted sealed bids and reads aloud the bid contents. A record of all bid submittals is made that includes each participating contractor and the amount bid. Each bid will initially be examined for conformance with job specifications and checked to insure that all addenda have been received and acknowledged on the required bid form. All bid bond information must be included. Also, any specification requirements for the bidder to provide potential subcontractors must be noted for subsequent review. If the bid complies with all criteria, it is accepted and tabulated. Once the tabulations are complete, the apparent low bidder is determined and announced, subject to later bid review and verification.

17.4 BID REVIEW AND ANALYSIS

The developer usually awards the contract to the lowest responsible bidder, while considering the contractor's experience in performing work similar to the solicited work. In addition, financial and organizational stability may be considered, along with the firm's reputation and past performance.

17.4.1   Verification of Bids

At a convenient time following the bid opening, the engineer reviews the submitted bid material and verifies all cost breakdowns and totals. If the contract is based on unit prices, care must be exercised by the engineer to insure that the bid is balanced. Occasionally contractors will raise the price on certain types of work and make corresponding reductions on other types, which produces an unbalanced bid.

All remaining information is double-checked, as a summary is prepared for the developer's use. Once the summary is completed, the engineer can make a considered determination of which contractor to recommend for the job to the developer, who will make the final determination.

17.5   CONTRACT AWARD

The developer usually awards the contract to the lowest responsible bidder on a competitive bid job. Circumstances may require the developer to abandon the project if the bid prices exceed a project's estimated construction cost, which is one reason for the developer to include in the bidding documents the right to reject any and all bids.

17.5.1   Post-Bid Activity

The developer notifies the successful bidder with either a letter of intent to award the contract and proceed with the project or a notice of award similar to the EJCDC prepared document *Notice of Award (EJCDC No. 1910-22)* (NSPE et al.). The engineer is now free to return all unsuccessful bidders any bid bond that was specified except the bonds pertaining to the remaining two next lowest bidders and to notify all bidders of the successful bidder's name and bid price.

17.5.2   Pre-Construction Conference

The engineer normally schedules a pre-construction conference with the successful bidder to discuss various housekeeping details associated with the job, such as payment requests, change orders, submittals, sedimentation and erosion control plan requirements, and shop drawings reviews. Topics could include a review of the project's scope and permit requirements along with the anticipated contract administration procedure that will be utilized. Temporary staging areas, storage areas, and construction easements are discussed. The engineer usually discusses the procedure to be used for field control establishment and the importance of marker protection. The contractor is usually requested to provide the engineer with an anticipated work and progress schedule that can be used to monitor the job and to satisfy regulatory requirements such as those associated with certain sedimentation and erosion control permits. Arrangements for a field office are discussed if specified as a requirement. During this conference the contractor is reminded of any contract administration submittals to be made prior to commencing work.

17.5.3   Review of Contractor's Submittals

The contractor must submit to the engineer for review all information specified in the bid documents. This process takes place between the notice of award being made and the time that actual work begins. Documents normally include performance and payment bonds, insurance certificates, the preliminary construction schedule, the name of the job superintendent, the field office telephone number, and the name of the designated job safety supervisor.

17.5.4 Contract Execution

As soon as reasonably possible, pending last minute project approvals, a formal contract is delivered to the successful contractor, which must be executed along with the developer. Two typical forms of contracts prepared by EJCDC that can be utilized are:

(a) *Standard Form of Agreement Between Owner and the Contractor on the Basis of a Stipulated Price (EJCDC No. 1910-8-A-1).*

(b) *Standard Form of Agreement Between Owner and Contractor on the Basis of Cost Plus (EJCDC No. 1910-8-A-2).*

The agreement, which must be signed and witnessed by the developer and contractor, establishes the rules of engagement between the two parties to finish the project. The contractor's bid bond is now returned upon execution of the contract. A "notice to proceed" can now be issued by the developer to the contractor similar to the one prepared by EJCDC and designated as *Notice To Proceed (EJCDC No. 1910-23)* (NSPE et al.). This notice establishes the beginning time from which the allowable contract duration is reckoned (to achieve both substantial and final completion) and alerts the contractor to commence work. Once the agreement is signed, the remaining two next lowest bidders' bid bonds can be returned. Should the agreements not be signed by the successful bidder, then the next lowest responsible bidder can be awarded the job, while the owner utilizes the bid bond protection provided by the initial successful bidder.

REFERENCES

Abbett, Robert W. *Engineering Contracts and Specifications,* 4th ed. New York: Wiley, 1965.

Associated General Contractors of America, Inc. (AGC). *Contractors' Experience Questionnaire and Financial Statement—Standard Form No. 28.* Washington: AGC.

National Society of Professional Engineers (NSPE), American Consulting Engineers Council, American Society of Civil Engineers (ASCE), and Construction Specifications Institute (CSI). *Standard Form of Agreement between Owner and Contractor on the basis of Cost-Plus* (EJCDC No. 1910-8-A-2). New York: ASCE, 1990.

National Society of Professional Engineers (NSPE), American Consulting Engineers Council, American Society of Civil Engineers (ASCE), and Construction Specifications Institute (CSI). *Standard Form of Agreement between Owner and Contractor on the basis of a Stipulated Price, EJCDC No. 1910-8-A-1.* New York: ASCE, 1990.

National Society of Professional Engineers (NSPE), American Consulting Engineers Council, American Society of Civil Engineers (ASCE), and Construction Specifications Institute (CSI). *Guide to the Preparation of Instructions to Bidders (EJCDC No. 1910-12).* New York: ASCE, 1990.

National Society of Professional Engineers (NSPE), American Consulting Engineers Council, American Society of Civil Engineers (ASCE), and Construction Specifications Institute (CSI). *Notice of Award (EJCDC No. 1910-22).* New York: ASCE, 1990.

National Society of Professional Engineers (NSPE), American Consulting Engineers Council, American Society of Civil Engineers (ASCE), and Construction Specifications Institute (CSI). *Notice to Proceed (EJCDC No. 1910-23)*. New York: ASCE, 1990.

National Society of Professional Engineers (NSPE), American Consulting Engineers Council, American Society of Civil Engineers (ASCE), Construction Specifications Institute (CSI), Surety Association of America, Associated General Contractors of America (AGC), and American Institute of Architects (AIA). *Construction Performance Bond (EJCDC No. 1910-28-A)*. New York: ASCE, 1984.

National Society of Professional Engineers (NSPE), American Consulting Engineers Council, American Society of Civil Engineers (ASCE), Construction Specifications Institute (CSI), Surety Association of America, Associated General Contractors of America (AGC), American Institute of Architects (AIA), American Subcontractors Association, and Associated Specialty Contractors. *Construction Payment Bond (EJCDC No. 1910-28-B)*. New York: ASCE, 1984.

National Society of Professional Engineers (NSPE), American Consulting Engineers Council, American Society of Civil Engineers (ASCE), and Construction Specifications Institute (CSI). *Bid Bond—Penal Sum Form (EJCDC No. 1910-28-C)*. New York: ASCE, 1990.

National Society of Professional Engineers (NSPE), American Consulting Engineers Council, American Society of Civil Engineers (ASCE), and Construction Specifications Institute (CSI). *Bid Bond—Damages Form (EJCDC No. 1910-28-D)*. New York: ASCE, 1990.

Smith, Robert J. (prepared for National Society of Professional Engineers, American Consulting Engineers Council, American Society of Civil Engineers (ASCE), and Construction Specifications Institute (CSI). *Recommended Competitive Bidding Procedures for Construction Projects (EJCDC No. 1910-9-D)*. New York: ASCE, 1987.

18 Layout, Observation, and Monitoring

The engineer's role during the construction phase may appear to conflict with interests of the developer or contractor. However, the engineer possesses professional knowledge that requires unbiased performance if the project is to be successful and safe for use. In addition, the developer should engage the engineer during construction to assist in the implementation of the plans that are inherently close to the engineer. This can help to protect the developer from inferior materials and workmanship. Having the engineer involved during the construction phase helps to maintain the construction schedule by resolving problems rapidly. Contract documents should therefore reserve the right for the engineer to make periodic job observations.

18.1 SCOPE OF ENGINEERING ACTIVITIES

The engineer's activities can fall under the categories of contract administration and construction administration during the construction phase. These activities are regulated by the terms of the engineering services contract contained in Chapter 3, depending on whether the customary approach or the customary approach with project management is used.

18.1.1 Customary Approach

After the construction contract award has been made, duties associated with its administration include the engineer's review of all job related mill and equipment performance reports, documents, and submittals including construction, shop, and erection drawings. The EJCDC *Standard General Conditions of the Construction Contract (EJCDC No. 1910-8)* (NSPE et al.) require the contractor to submit a schedule of values along with three preliminary schedules for progress, shop drawings, and sample submittals. If questions arise during construction, the engineer makes interpretations of the plans and specifications in a manner that is equitable for both the developer and the contractor to attempt to resolve potential claims. The engineer can also provide plan modifications to resolve conflicts caused by actual field conditions. These modifications could be reflected in change orders formulated by the engineer. The engineer makes periodic field visits to observe the progress of work for the purpose of processing payment requests and to view the work for general conformance with the approved plans, specifications, and issued permits. These intermittent visits do not place any responsibility on the engineer for the contractor's performance. It is the contractor's responsibility to maintain quality of materials, workmanship, and performance within the limits of the contract. According to Smith

(1991), the EJCDC *Standard General Conditions of the Construction Contract (EJCDC No. 1910-8)* (NSPE et al.) stipulate that the contractor is responsible for the inspection of work. The engineer, however, warrants that the design will work. The engineer also observes any specified field tests to insure compliance with the construction documents. Upon completion of the final inspection, the engineer prepares record drawings for reference purposes, and certifies that the constructed project conforms to the approved plans and specifications based on the engineer's knowledge of the project including periodic observations and final inspection results.

18.1.2   Customary Approach with Project Management

Should the engineer be required to provide project management services in addition to design services, the engineer would in essence become the construction administrator. Construction administration includes all duties associated with contract administration and additional duties entailing the coordination, planning, scheduling, and closing out of the project. These duties are not undertaken to allow the engineer to dictate job activities, construction methods employed, or to relieve the contractor of responsibility but are undertaken to insure that certain milestone events are met. A resident project representative could be used to observe the contractor's prosecution of the work and to report any deviations from the approved construction plans, specifications, and construction permits.

The frequency and need for a resident project representative's services is governed by the project's size. For larger jobs a full time representative may be justified; however, for smaller jobs, a part time one might be sufficient.

18.1.3   Other Approaches

If a professional construction manager is engaged by the developer, the engineer would provide auxiliary services to support the project's management, as indicated in Chapter 3. If, however, the developer has engaged the engineer for a design-build or turn-key project, all factors would require the efforts of the engaged firm.

18.2   CONSTRUCTION LAYOUT

If the design professional has the capability and responsibility of providing field surveying services for the contractor to verify and use, then these services must be coordinated. These auxiliary surveying services must otherwise be acquired, either by the developer or the contractor, depending on the contract document provisions.

18.2.1   Clearing and Grubbing Control

The initial construction stakes set in the field for the contractor are flat stakes, usually 24 inches tall, to delineate the limits of clearing and grubbing (removal of roots and stumps) for all road rights of way, easements, and other areas requiring clearing. These stakes can also serve as limits for various sedimentation and erosion control practices. These stakes are marked with high visibility material to aid the equipment operator's perception of areas to be cleared. Often these stakes are destroyed during the clearing and grubbing operation.

18.2.2 Rough Road Grading Stakes

Once the clearing and grubbing operation has been completed and all necessary topsoil has been stripped from the roadbed, the contractor requires control stakes to be set in the field for both line and grade. These stakes usually include centerline hubs (stakes set flush with the ground surface to locate a tack point of known position) and flat stakes (stakes similar to the ones used for clearing to guard and identify with inscribed information the adjacent hub). Both types are shown in Figure 18.1. Accurate staking of the centerline is essential because all horizontal and vertical control for other improvements relates to the road's centerline. Figure 18.2 illustrates a technique that can minimize field errors resulting from the use of short instrument backsights when establishing the proposed road's horizontal centerline control. Offset hubs are positioned normal to the centerline at each station or half station in addition to any intermediate control points such as PI's, PC's, and PT's. Cut or fill data are placed on the flat stake so that the contractor can shape the roadbed. If poor soil is encountered, the limits of mucking and backfill must be established to document the amount of classified excavation work required to provide an acceptable subgrade in that area. Slope stakes, if required, are set at this time.

18.2.3 Sanitary Sewer Collection System

The gravity sanitary sewer collection system installation usually begins after the contractor has rough graded the roads. This is necessary because the sanitary system usually requires the deepest excavations. If the contractor is utilizing laser control equipment, hubs and tack points are usually set for each manhole's center with a cut distance marked on the accompanying flat stake to the manhole's invert. An offset stake or stakes may be required to preserve the manhole's centerline information, as the centerline stakes are removed when excavation begins. Tap locations and the terminus of service lines must also be marked in the field.

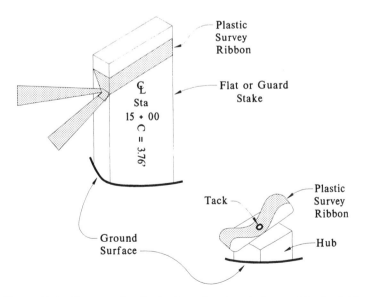

Figure 18.1 Survey stakes for horizontal and vertical control in the field.

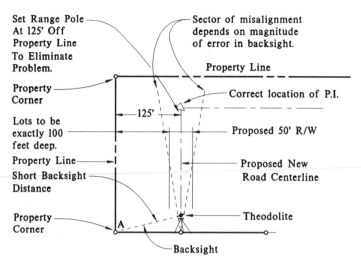

Figure 18.2 Field procedure to correct for sighting error caused by using a short instrument backsight.

If the sanitary sewer collection system is a pressure type, it can be installed using the same general procedures that apply to potable water lines.

18.2.4 Storm Drain Collection System

The storm drainage collection system is usually the second deepest system installed in a development. It normally is installed after the gravity sanitary sewer collection system is in place. The centerlines of all proposed catch basins, drop inlets, and junction boxes are staked in the field, with cut data being supplied on flat stakes to the inverts. If offset hubs are required, they too must be established.

Also, the engineer must establish control stakes along all culvert and open ditch locations for the contractor to use.

18.2.5 Potable Water System

One of the last systems to be installed is the potable water system, mainly because it is fairly shallow and has pressure flow that is not dependent on critical elevations other than at pipe crossings. Valve, hydrant, thrust block, blow off, and service tap locations must be established in the field, along with the locations where pipe diameters change and terminate.

18.2.6 Fine Grading Road Stakes

Once the potable water system is installed, the engineer can reestablish all centerline control stakes for the roads and place offset blue topped stakes along the proposed edge of the subgrade. These stakes are set at a known horizontal and vertical position and are called blue tops. Other necessary control stakes, including those for curb and gutter placement, can be established in the field to guide the contractor's activities.

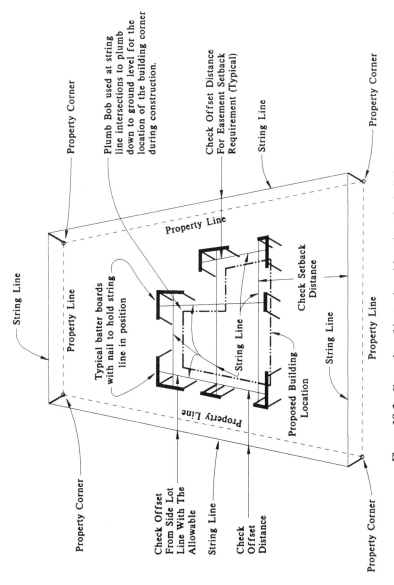

Figure 18.3 Examples of batter board locations for building layout.

Property Corner

Property Corner

String Line

Property Line

Plumb Bob used at string line intersections to plumb down to ground level for the location of the building corner during construction.

Check Offset Distance For Easement Setback Requirement (Typical)

String Line

Property Corner

Property Line

Typical batter boards with nail to hold string line in position

Property Line

Check Setback Distance

String Line

Proposed Building Location

String Line

Property Line

Property Corner

Check Offset From Side Lot Line With The Allowable

String Line

Check Offset Distance

615

18.2.7   Miscellaneous Field Control

Other field control markers can be set by the engineer to establish batter boards, and finished floor elevations, if required. Batter boards are construction stakes that are commonly used in the field to locate building corners and walls. Figure 18.3 illustrates the use of these control devices.

18.3   ADMINISTRATIVE DUTIES

Certain activities are required of the engineer to minimize confusion, provide accurate accounting of events, and to insure project success. These activities include administrative duties that are as important as other engineering duties associated with a development project.

18.3.1   Records Keeping

The engineer should maintain various types of records that are used for documentation and reference purposes including a daily activity log. Telephone communication logs are important to record progressive activities. In addition, field observation records, delivery ticket stubs, and material test results normally performed by outside testing firms should be catalogued by the engineer. Each record should include various modifying information to render a complete and accurate portrayal of information. Other records kept for accurate administration of the contract include change order and extra work records. Additional records include payment requests, progress reports, and material deliveries.

18.3.2   Submittal Review

The engineer must review all working, shop, and erection drawings submitted by the contractor to insure that they are in compliance with the approved plans and specifications. In addition, all "cut sheets" (manufacturers' specification sheets) and samples that are submitted must be reviewed for approval. Should the job require the contractor to submit concrete or asphalt mix designs, the engineer must carefully check these to insure compliance with appropriate design criteria.

The specifications may require the contractor to submit for administrative purposes copies of all contractor procured engineering calculations and drawings pertaining to sheeting, bracing, and shoring systems that will be used on site.

18.4   CONSTRUCTION CONFLICTS AND EXTRA WORK

Conflicts usually result in alterations and modifications being made to the construction drawings, which may translate into change orders. These change orders are formulated by the engineer through authority provided in the contract documents. The engineer must be vigilant to avoid any apparent or actual conflict of interest (or fraud) when exercising these duties. Figure 18.4 illustrates a typical change order document.

CHANGE ORDER

No._____

PROJECT ....

DATE OF ISSUANCE .... EFFECTIVE DATE ....

OWNER ....

OWNER's Contract No. ....

CONTRACTOR .... ENGINEER ....

You are directed to make the following changes in the Contract Documents.

Description:

Reason for Change Order:

Attachments: (List documents supporting change)

CHANGE IN CONTRACT PRICE:	CHANGE IN CONTRACT TIMES:
Original Contract Price	Original Contract Times
	Substantial Completion: _____
$ _____	Ready for final payment: _____ *days or dates*
Net changes from previous Change Orders No. ____ to No. ____	Net change from previous Change Orders No. ____ to No. ____
$ _____	_____ *days*
Contract Price prior to this Change Order	Contract Times prior to this Change Order
	Substantial Completion: _____
$ _____	Ready for final payment: _____ *days or dates*
Net Increase (decrease) of this Change Order	Net Increase (decrease) of this Change Order
$ _____	_____ *days*
Contract Price with all approved Change Orders	Contract Times with all approved Change Orders
	Substantial Completion: _____
$ _____	Ready for final payment: _____ *days or dates*

RECOMMENDED: APPROVED: ACCEPTED:

By: _____   By: _____   By: _____
    Engineer (Authorized Signature)   *Owner (Authorized Signature)*   *Contractor (Authorized Signature)*

Date: _____   Date: _____   Date: _____

Figure 18.4 Change order form. Prepared by the Engineers Joint Contract Documents Committee and endorsed by The Associated General Contractors of America. Source: National Society of Professional Engineers, American Consulting Engineers Council, American Society of Civil Engineers, The Construction Specifications Institute, Engineers' Joint Contract Documents Committee, (EJCDC No. 1910-8-B), 1990. Reproduced by permission of the American Society of Civil Engineers.

18.4.1   Differing Site Conditions

Contract plans and specifications usually include information on subsurface site conditions. Should this information prove not to be significantly representative of the actual field conditions, the contractor may be entitled to additional remuneration or time if unanticipated additional work is involved. On occasion, specifications may state that supplied subsurface information is for the developer's benefit and that the contractor is responsible for making whatever investigations are necessary to determine the extent of subsurface working conditions prior to bid.

According to Smith (1991), the revised EJCDC general conditions provide "that the owner (developer) should indemnify the contractor, its subcontractors, the engineer and its consultants for consequences of unanticipated hazardous conditions on the site." These conditions can be attributed to the discovery of previously unknown asbestos, hazardous waste, petroleum, PCB's, or radioactive substances on site. If these types of materials are encountered, the contractor can stop work until the situation is rectified by the developer.

18.4.2   Plans and Specification Changes

The engineer may be required to alter the approved plans and specifications as a result of several reasons. If errors are detected in the plans or specifications during construction, or if necessary information was omitted or is found to deviate substantially from that portrayed in the documents, changes may be required for amplification and correction. Other changes can result from jurisdictional regulatory review authority requirements or from the developer revising a portion of the project to address new needs. These changes may increase or decrease the scope of the project resulting in additional compensation and/or time being given to the contractor.

18.4.3   Other Conditions

Other conditions affecting the contracted work might include a change in substantial or final completion times, resulting in the contractor being asked to accelerate work production. In addition, the scope can be altered because of damage being inflicted to completed work outside the contractor's control. Should the developer interfere with the contractor's prosecution of work, or suspend work, the outcome could result in contract changes.

18.5   WORK DIRECTIVES

A work directive is notice given to the contractor to concentrate efforts in a certain area due to emergency conditions. An example of a work directive is shown in Figure 18.5. This directive normally is used as a basis upon which a future change order is issued to account for the included work.

18.6   DELAYS

Time delays result in the passing of time when insufficient job progress is being made. Delays can result in the contractor either being granted an extension to complete the job or being suspended from work, depending on the cause of delay.

WORK CHANGE DIRECTIVE

No._____

PROJECT ....

DATE OF ISSUANCE .... EFFECTIVE DATE ....

OWNER ....

OWNER's Contract No. ....

CONTRACTOR .... ENGINEER ....

You are directed to proceed promptly with the following change(s):

Description:

Purpose of Work Change Directive:

Attachments: (List documents supporting change)

If a claim is made that the above change(s) have affected Contract Price or Contract Times any claim for a Change Order based thereon will involve one or more of the following methods of determining the effect of the change(s).

Method of determining change in Contract Price:	Method of determining change in Contract Times:
☐ Unit Prices	☐ Contractor's records
☐ Lump Sum	☐ Engineer's records
☐ Other _____	☐ Other _____

Estimated increase (decrease) in Contract Price:
$ _____
If the change involves an increase, the estimated amount is not to be exceeded without further authorization.

Estimated increase (decrease) in Contract Times:
Substantial Completion:_____ days;
Ready for final payment: _____ days.
If the change involves an increase, the estimated times are not to be exceeded without further authorization.

RECOMMENDED: AUTHORIZED:

_____   _____
ENGINEER   OWNER

By: _____   By: _____
(Authorized Signature)   (Authorized Signature)

Figure 18.5 Work directive form. Source: Prepared by the Engineers Joint Contract Documents Committee and endorsed by the Associated General Contractors of America. NSPE et al., Engineer's Joint Contract Documents Committee (EJCDC No. 1910-8-F), 1990. Reproduced by permission of the American Society of Civil Engineers.

18.6.1 Owner Caused Delays

Owner caused delays are compensable to the contractor. Such delays result from problems associated with construction plans and specifications that require revisions. In addition, the lack of available funds may result in the contractor not being paid on time, causing various delays. Other owner caused delays can be attributed to inordinate time being required to respond to pertinent issues associated with the prosecution of the work, for example, not gaining access to the site, not issuing a timely notice to proceed, not certifying tests and evaluations in a timely manner, or not providing "owner furnished" materials when required.

18.6.2   Contractor Caused Delays

Delays caused by the contractor can result from poor procurement of equipment and materials, manifested by inadequate lead times, or from financial problems. Delays resulting from poor scheduling and supervision or from the lack of subcontractor coordination can also cause the contractor problems. In addition, delays associated with contractor procured permits for the job may result in problems.

18.6.3   Other Delays

Other delays that can keep the contractor from prosecuting project work in a timely manner range from unusual weather conditions (such as that associated with a hurricane) to labor strikes at the job site or supplier's location. Riots, wars, and natural disasters also cause unavoidable delays.

18.6.4   Disposition of Delays

Most delays caused by some event or someone other than the contractor result in the contractor seeking more compensation or an extension in time to complete the project. The engineer must use equity in resolving situations such as these by undertaking negotiations to issue appropriate change orders that reflect a mutual agreement between the developer and contractor. It is usually to the best interest of all parties involved to find agreement. Should negotiations fail, the contractor has the right to file a timely protest while performing the work in accordance with the contract documents. The unsettled issue can be arbitrated or settled in court as a result of litigation.

18.7   CONSTRUCTION PROGRESS

Progress made when prosecuting the contractual work requires the engineer periodically to certify payment requests for completed work and material delivered. Requests for work payments are submitted only after work has been performed, using a form similar to that illustrated in Figure 18.6. The contractor's project schedule also is used by the engineer to monitor progress. The engineer reviews contractor prepared revisions to this schedule to help insure that the project will progress to completion in a timely manner.

18.7.1   Engineer's Observation

The contractor normally has responsibility for scheduling, prosecuting, and completing all work in accordance with the contractual documents unless a contract administrator is designated the coordinator. These documents usually call for the contractor to be responsible for job safety and to appoint a safety supervisor. The engineer who observes a job in progress has the obligation to cite anything that would contribute to loss of life, property, or limb regardless of responsibilities delineated within the contractual documents.

APPLICATION FOR PAYMENT NO. _____

To _____ (OWNER)

Contract for _____ .

OWNER's Contract No. _____ . ENGINEER's Project No. _____ .

For Work accomplished through the date of _____ .

ITEM	CONTRACTOR's Schedule of Values			Work Completed	
	Unit Price	Quantity	Amount	Quantity	Amount
	$		$		$
Total (Orig. Contract) C.O. No. 1 C.O. No. 2			$		$

Accompanying Documentation:

_____

_____

_____

_____

GROSS AMOUNT DUE .... $ _____
LESS _____ % RETAINAGE .... $ _____
AMOUNT DUE TO DATE .... $ _____
LESS PREVIOUS PAYMENTS .... , $ _____
AMOUNT DUE THIS APPLICATION .... $ _____

CONTRACTOR'S Certification:

The undersigned CONTRACTOR certifies that: (1) all previous progress payments received from OWNER on account of Work done under the Contract referred to above have been applied to discharge in full all obligations of CONTRACTOR incurred in connection with Work covered by prior Applications for Payment numbered 1 through _____ inclusive; (2) title to all Work, materials and equipment incorporated in said Work or otherwise listed in or covered by this Application for Payment will pass to OWNER at time of payment free and clear of all liens, claims, security interest and encumbrances (except such as are covered by Bond acceptable to OWNER indemnifying OWNER against any such lien, claim, security interest or encumbrance); and (3) all Work covered by this Application for Payment is in accordance with the Contract Documents and not *defective* as that term is defined in the Contract Documents.

Dated _____ , 19 _____   _____
CONTRACTOR

By _____
(Authorized Signature)

Payment of the above AMOUNT DUE THIS APPLICATION is recommended.

Dated _____ , 19 _____   _____
ENGINEER

By _____
(Authorized Signature)

Figure 18.6 Application for payment form. Source: Prepared by the Engineers Joint Contract Documents Committee and endorsed by the Associated General Contractors of America. NSPE et al., Engineer's Joint Contract Documents Committee (EJCDC No. 1910-8-E), 1990. Reproduced by permission of the American Society of Civil Engineers.

The observing engineer acts as the eyes and ears of the developer to monitor job progress and activities. While the contractor has the obligation to guarantee the quality of workmanship and installed materials, the developer has the responsibility of offering to the end user a product that is suitable for its intended use. This is part of the reason that the developer should secure a resident project representative and at the same time engage a reputable contractor for the job. The engineer has the obligation to report observed findings of defective or unacceptable work to the contractor so that corrective action can be undertaken as quickly as possible. Harboring critical information until the final inspection on the part of the engineer is unreasonable if the job is expected to be completed in a timely manner.

18.7.2 Regulatory Inspections

In most cases the installed improvements within a development are dedicated to the public for operation, maintenance, control, and ownership. Thus, the developer is normally not the end user or long term "owner." Many municipalities undertake some type of inspection and quality control program (possibly to include the developer paying for these inspection services) to insure the finished project is acceptable. Should deficiencies be noted by the potential end user, the engineer and developer must rectify the unacceptable condition through the contractor such that the end product can be successfully dedicated (Baltz and Richardson, 1985).

REFERENCES

Baltz, Louis J., III, and James W. Richardson, Jr. "Inspection Costs Versus the Benefits Derived," *Avoiding Contract Disputes—Proceedings of a Symposium Sponsored by the Construction Division of the American Society of Civil Engineers (ASCE) in Conjunction with the ASCE Convention in Detroit, Michigan.* New York: ASCE, 1985, pp. 92–99.

National Society of Professional Engineers (NSPE), American Consulting Engineers Council, American Society of Civil Engineers (ASCE), and Construction Specifications Institute (CSI). *Standard General Conditions of the Construction Contract (EJCDC No. 1910-8).* New York: ASCE, 1990.

National Society of Professional Engineers (NSPE), American Consulting Engineers Council, American Society of Civil Engineers (ASCE), and Construction Specifications Institute (CSI). *Change Order (EJCDC No. 1910-8-B).* New York: ASCE, 1990. Referenced material is reproduced by permission of the American Society of Civil Engineers.

National Society of Professional Engineers (NSPE), American Consulting Engineers Council, American Society of Civil Engineers (ASCE), and Construction Specifications Institute. *Application for Payment (EJCDC No. 1910-8-E).* New York: ASCE, 1990. Referenced material is reproduced by permission of the American Society of Civil Engineers.

National Society of Professional Engineers (NSPE), American Consulting Engineers Council, American Society of Civil Engineers (ASCE), and Construction Specifications Institute. *Work Change Directive (EJCDC No. 1910-8-F).* New York: ASCE, 1990. Referenced material is reproduced by permission of the American Society of Civil Engineers.

Smith, Robert J. "Construction Documents Revised," *Civil Engineering.* New York: ASCE, August 1991, pp. 68–69. Referenced material is reproduced by permission of the American Society of Civil Engineers.

19 Completion and Start-Up

As the contractor nears completion of an improved land project, the engineer must check the various installed systems to see that they are operational. This requires the observation of tests and possibly making various field measurements to certify the job is installed in accordance with the approved construction documents. Usually these tests are coordinated with the overseeing regulatory agencies and ultimate system owner.

Total development activities include closeout procedures that are similar to those used for improved land projects and others associated with the additional improvements and structures.

19.1 SYSTEM TESTING AND START-UP

Utility systems controlled by health officials normally require testing for certification purposes. When system testing and evaluation is undertaken, most regulatory agencies send observers to insure that the systems are in compliance prior to placing them into service.

19.1.1 Potable Water System

The water distribution system must meet various test criteria to be serviceable. First, the system must be capable of withstanding a predetermined internal pressure for a prescribed duration when a hydrostatic pressure test is conducted. Maintaining the system test pressure indicates that no leaks exist. Once the system has been determined to be tight, a flow test is conducted to insure that proper fire fighting flows are available for protection. This test is similar to the hydrant flow test described in Chapter 11. Should this test's results prove unsatisfactory, the engineer should require the contractor to check to insure that all new valves are in the open position. If this fails to produce satisfactory results, a recheck of the flow and pressure at the point of supply is necessary to determine if the current performance of the water system is approximately what it was when initially flowed while obtaining design field data. If the contractor tapped the existing system's water main, it is possible the contractor could have inadvertently used a drill bit appreciably smaller than the diameter of the new water line extension, resulting in a constriction. It is good practice to have the contractor wire the removed plug of pipe to the tapping sleeve when the connection is made to the existing line so that visual proof is available to show that the proper size hole is available to supply water to the new extension. If the hole is of proper size but the flows are still inadequate, the new system may have an obstruction such as a rag, root, or hand tool. The contractor must reexcavate sequential segments of the line to systematically flush the system to attempt to find the problem and correct it.

A final test requirement prior to placing the water system into operation is to check the newly charged water main for bacteria-free water. Contamination of water mains resulting from open air exposure during construction can be eliminated by chlorination. Grab samples must be obtained and tested to insure that a minimum chlorine residual exists and that there are no pathogenic organisms present. Should bacteria tests indicate contamination, the contractor must resterilize the system. If results are still unsatisfactory, the contractor must dismantle segments of the installed system to flush the mains, because small animals occasionally become trapped in the line during construction.

19.1.2 Sanitary Sewer Collection System

The watertight integrity of the completed sanitary sewer collection system must be checked by the contractor. This will insure that no leaks exist in the system that would allow wastewater to leach into the surrounding area, especially at ditch crossing. It also prevents groundwater from entering the collection system. Any such groundwater would then have to be treated. Tests for the system's integrity include infiltration, exfiltration, or air pressure testing between manholes. These tests should indicate satisfactory joint and fitting tightness. If not, the contractor must reexcavate the line to locate and repair the leak. If a portion of the installed wastewater system consists of pressure forcemains, then a hydrostatic time test must be conducted similar to the one used on the potable water system.

Regulatory authorities often require the engineer to certify that the installed gravity system is in accordance with the approved plans and specifications. To certify alignment the engineer must lamp (look through using a light beam) each line to insure that a uniform and straight bore exists from manhole to manhole. Grade certification can be accomplished only if the engineer observes each invert elevation using a level and measures the horizontal distance between installed manholes. Should any section of line indicate less than minimum grade or any misalignment, the contractor must take corrective action. Other agency requirements may include a mandrel test, in which a steel sleeve is pulled through the installed sewer main to prove that the pipes have not deflected out of tolerance when certain pipe materials are used.

19.1.3 Sanitary Sewer Pumping Station

Should the project include a sanitary sewer pumping station, various component tests must be conducted to insure proper operation. Initially, the electrical system must be checked for circuit continuity and resistance testing. Pump motors, controls, and appurtenances, such as sump pumps, ventilators, dehumidifiers, and so forth, must also be tested. Usually the equipment supplier is charged in the construction specifications with providing start-up supervision and inspection. Observational tests to insure that the equipment is rotating properly without excessive vibration or overheating, and without drawing excessive power, are essential, as is a determination that the equipment is meeting the specified pumping rate and head.

19.2 FINAL INSPECTION

All installed improvements must be inspected by regulatory authorities that have indicated a willingness to accept the completed systems. Approval allows the developer to petition for dedication of the system, and, at the same time, usually establishes the point in time from which the warranty duration is reckoned. As with system testing and start-up, regulatory agency representatives normally participate in the final inspection. At this inspection most authorities request copies of the prepared plats delineating all rights of way, easements, and other dedicated areas that will provide space for later operational and maintenance activities.

19.2.1 Road and Storm Drainage Systems

Most road and storm drainage systems are inspected after final cleanup and after rights of way have been dressed with topsoil and seeded. The roadbed is inspected for uniformity and to insure that low spots do not exist where water might puddle, especially if curb and gutter sections are used. Pavement core samples strategically obtained may be required to ascertain pavement thickness and base material depth. Storm drainage culverts, junction boxes, and catch basins are inspected for alignment and grade; these inspections are similar to those conducted for the gravity wastewater collection system. Inspectors also insure that all features included on the plans are present, such as french drains, riprap, energy dissipators, and detention/retention facility components.

19.2.2 Potable Water System

Final inspection of the water system includes receiving copies of certified system test results and conducting a visual inspection to insure that all valves are properly installed with their boxes, that fire hydrants are situated properly and anchored with adequate drain pit areas, and that thrust blocking is in place. This last item is accomplished by inspecting random locations or by requiring the contractor to keep the thrust blocks exposed until verified.

19.2.3 Sanitary Sewer Collection System

The sanitary sewer collection system must be flushed clean of sediment and debris and must be fully completed when the final inspection is conducted. All integrity test results and certificates by the engineer that the installed mains are on line and grade are given to the local jurisdictional authorities. The entire system is visually inspected by lamping, and all manhole drops and inverts are inspected for smoothness and uniformity. In addition, position of each manhole's rim and cover, relative to the surrounding area, is inspected along with the manhole joints to insure watertightness that will prevent infiltration. Should the rim be located in an area where appreciable storm water might enter, a watertight gasket may be required on that manhole.

19.3 APPROVAL OF IMPROVEMENTS

If all improvements are acceptable to the developer, engineer, and regulatory agencies, then certificates of approval are issued by the engineer and the regulatory agencies having jurisdictional control. This is usually followed by a permit from the regulatory agency to place into operation the new system. The developer can file a certificate of completion with the local public records office for liens notice. The engineer is free to accept and process the final progress payment request from the contractor, and to notify the developer that the facility is ready for use. If the deadline date is overrun during a "time is of the essence" construction contract, liquidated damages can be assessed by the developer and deducted from the money due the contractor.

19.3.1 Discrepancy Corrections

If discrepancies are detected during testing or inspection, the contractor must take corrective action to eliminate them. Usually these discrepancies are entered on a list, called a punch list, that is used as a basis for a follow-up inspection. Should the engineer issue a "Certificate of Substantial Completion," as illustrated in Figure 19.1, it would represent

---

CERTIFICATE OF SUBSTANTIAL COMPLETION

PROJECT ....

DATE OF ISSUANCE ....

OWNER ....

OWNER's Contract No. ....

CONTRACTOR .... ENGINEER ....

This Certificate of Substantial Completion applies to all Work under the Contract Documents or to the following specified parts thereof:

TO ....
OWNER

And To ....
CONTRACTOR

The Work to which this Certificate applies has been inspected by authorized representatives of OWNER, CONTRACTOR and ENGINEER, and that Work is hereby declared to be substantially complete in accordance with the Contract Documents on

....
DATE OF SUBSTANTIAL COMPLETION

A tentative list of items to be completed or corrected is attached hereto. This list may not be all-inclusive, and the failure to include an item in it does not alter the responsibility of CONTRACTOR to complete all the Work in accordance with the Contract Documents. The items in the tentative list shall be completed or corrected by CONTRACTOR within _____ days of the above date of Substantial Completion.

Figure 19.1 Certificate of substantial completion. Source: Prepared by the Engineers Joint Contract Documents Committee and endorsed by The Associated General Contractors of America. NSPE et al., Engineer's Joint Contract Documents Committee (EJCDC No. 1910-8-D), 1990. Reproduced by permission of the American Society of Civil Engineers.

From the date of Substantial Completion the responsibilities between OWNER and CONTRACTOR for security, operation, safety, maintenance, heat, utilities, insurance and warranties and guarantees shall be as follows:

RESPONSIBILITIES:

OWNER: _____

_____

_____

CONTRACTOR: _____

_____

_____

The following documents are attached to and made a part of this Certificate:

[For items to be attached see definition of Substantial Completion as supplemented and other specifically noted conditions precedent to achieving Substantial Completion as required by Contract Documents.]

This certificate does not constitute an acceptance of Work not in accordance with the Contract Documents nor is it a release of CONTRACTOR's obligation to complete the Work in accordance with the Contract Documents.

Executed by ENGINEER on .... , 19 .....

....
ENGINEER

By: ....
(Authorized Signature)

CONTRACTOR accepts this Certificate of Substantial Completion on .... , 19 .....

....
CONTRACTOR

By: ....

OWNER accepts this Certificate of Substantial Completion on .... , 19 .....

....
OWNER

By: ....
(Authorized Signature)

Figure 19.1 (continued)

that the inspection revealed minor problems that are shown on a punch list; however, all other work is approved as is. With the issuance of this certificate the contractor continues to maintain responsibility and control over the job until it is fully acceptable.

If the engineer issues a "Beneficial Occupancy" certificate instead, it allows the developer to occupy and use the incomplete facility. The contractor, however, maintains the obligation to complete the contracted work while the developer assumes the responsibility of operation, and maintenance of the facility including the procurement of insurance, heat, and utilities. Product warranties for improvements installed in the facility usually begin at this time, although, the contractor's warranty normally does not begin until a certificate of completion is issued.

19.4   OPERATIONAL NEEDS

If the installed improvements include systems that require operational guidance, the contractor must insure that such guidance has been provided prior to completing the job.

19.4.1   Instruction Manuals

All instruction and operation/maintenance manuals must be supplied and reviewed with the end user to insure that the facility will function properly. If the specifications provide for operator training, which is necessary when mechanical equipment is involved, then provisions for on- or off-site instruction must be undertaken (usually by the equipment supplier).

19.4.2   Delivery of Specified Materials

Often with the installation of mechanical equipment various special tools, spare parts, and lubricants are provided by the supplier as part of the construction specifications. These items should be inventoried and turned over to the end user for safe storage. Facility keys should be logged and turned over to the owner for security purposes.

19.4.3   Peripheral Document Review

The contractor must supply the developer with all product warranties, such as the roof bond, and certificates of any other tests or inspections that pertain to the facility. In addition, the contractor must supply the engineer with all recorded measurement and field notations from which the engineer can prepare copies of record drawings if specified. On a small project this information may be recorded on blueprints by the contractor in color with copies submitted to the developer, engineer, and regulatory authorities. Most jobs, however, require several sets of record drawings that are prepared either by the engineer or by the contractor depending on the stipulations contained in the contract specifications. These drawings are usually prepared on reproducibles from which multiple copies can be made for distribution to all involved parties.

19.5   FINAL STAKE OUT OF PARCELS

After all improvements have been installed and the project has been cleaned up and dressed, the area is prepared for the final property monuments to be installed by a registered land surveyor. These monuments must be installed to established standards of precision and must reflect accurately on the ground the property corners that are represented on the final development plat to be recorded.

19.6   RELEASE OF RETAINAGE

Prior to releasing all retained funds due the contractor, a check of the public records office must be made to insure that no liens have been filed against the project. In addition, the contractor may be requested to submit a waiver of liens and a certificate of surety before a recommendation for final payment can be made by the engineer. Upon receiving all monies due from the developer, the contractor can issue a certificate of completed payment for all work performed. Completion of these details allows the developer to obtain an unencumbered project that is free for dedication or conveyance.

19.7   ENGINEER'S ADMINISTRATIVE CLOSEOUT

The engineer must make a final accounting to the developer once the job is completed. This usually is accomplished in a final report. This report includes a summary of notices, records, and inspections. It constitutes a record file for the project that can be updated throughout the warranty period and closed out when the warranty period has passed. For development projects that fall under the jurisdiction of NPDES General Permits for Storm Water, a "Notice of Termination" (NOT) can be filed once the site has been finally stabilized.

19.8   WARRANTY PERIOD

The warranty period for all improvements is normally included within the job specifications, during which time performance bond coverage is usually reduced to cover contingencies. The engineer should periodically inspect the completed job during this period to insure that all is in order. Should corrective action be required, resulting from trench settlement or soil erosion for example, the contractor must be contacted immediately to take corrective action. At the end of the warranty period, the engineer should receive final acceptance notices from all involved agencies, and copies should be sent to the developer to close out the project.

REFERENCES

National Society of Professional Engineers (NSPE), American Consulting Engineers Council, American Society of Civil Engineers (ASCE), and Construction Specifications Institute. *Certificate of Substantial Completion (EJCDC No. 1910-8-D).* New York: ASCE, 1990. Referenced material is reproduced by permission of the American Society of Civil Engineers.

APPENDIX
Conversion Factors

The following tables are from the source: Elwyn E. Seelye. *Data Book For Engineers—Design Volume 1*. Adapted from Tables of Conversion Factors for Engineers by Dorr Oliver, Inc., Milford, Conn. Copyright © 1960 by John Wiley & Sons, Inc. Reprinted by permission of John Wiley & Sons, Inc. and Dorr Oliver Inc.

GENERAL CONVERSION FACTORS – I

TABLES OF CONVERSION FACTORS FOR ENGINEERS

Data are arranged alphabetically.

Unless designated otherwise, the English measures of capacity are those used in the United States, and the units of weight and mass are avoirdupois units.

The word gallon, used in any conversion factor, designates the U. S. gallon. To convert into the Imperial gallon, multiply the U. S. gallon by 0.83267. Likewise, the word ton designates a short ton, 2,000 pounds.

The figures 10^{-1}, 10^{-2}, 10^{-3}, etc. denote 0.1, 0.01, 0.001, etc. respectively.

The figures 10^{1}, 10^{2}, 10^{3}, etc. denote 10, 100, 1000, respectively.

With respect to the properties of water, it freezes at 32°F., and is at its maximum density at 39.2°F. In the conversion factors given below using the properties of water, calculations are based on water at 39.2°F. in vacuo, weighing 62.427 pounds per cubic foot, or 8.345 pounds per U. S. gallon.

"Parts Per Million," designated as P.P.M., is always by weight and is simply a more convenient method of expressing concentration, either dissolved or undissolved material. Usually P.P.M. is used where percentage would be so small as to necessitate several ciphers after the decimal point, as one part per million is equal to 0.0001 per cent.

As used in the Sanitary field, P.P.M. represents the number of pounds of dry solids contained in one million pounds of water, including solids. In this field, one part per million may be expressed as 8.345 pounds of dry solids to one million U. S. gallons of water. In the Metric system, one part per million may be expressed as one gram of dry solids to one million grams of water, or one milligram per liter.

In arriving at parts per million by means of pounds per million gallons or milligrams per liter, it may be mentioned that the density of the solution or suspension has been neglected and if this is appreciably different from unity, the results are slightly in error.

Multiply	By	To Obtain
Acres	43,560	Square feet
"	4047	Square meters
"	1.562x10⁻³	Square miles
"	4840	Square yards
Acre-feet	43,560	Cubic feet
"	325,851	Gallons
"	1233.49	Cubic meters
Atmospheres	76.0	Cms. of mercury
"	29.92	Inches of mercury
"	33.90	Feet of water
"	10,333	Kgs./sq. meter
"	14.70	Lbs./sq. inch
"	1.058	Tons/sq. ft.
Barrels–oil	42	Gallons–oil
–cement	376	Pounds–cement
Bags or sacks–cement	94	Pounds–cement
Board–feet	144 sq. in. x 1 in.	Cubic inches
British Thermal Units	0.2520	Kilogram-calories
" " "	777.5	Foot-lbs.
" " "	3.927x10⁻⁴	Horse-power-hrs.
" " "	107.5	Kilogram-meters
" " "	2.928x10⁻⁴	Kilowatt-hrs.
B.T.U./min	12.96	Foot-lbs./sec.
" / "	0.02356	Horse-power
" / "	0.01757	Kilowatts
" / "	17.57	Watts
Centares (Centiares)	1	Square meters
Centigrams	0.01	Grams
Centiliters	0.01	Liters
Centimeters	0.3937	Inches
"	0.01	Meters
"	10	Millimeters
Centimtrs. of mercury	0.01316	Atmospheres
" " "	0.4461	Feet of water
" " "	136.0	Kgs./sq. meter
" " "	27.85	Lbs./sq. ft.
" " "	0.1934	Lbs./sq. inch

Multiply	By	To Obtain
Centimeters/second	1.969	Feet/min.
" / "	0.03281	Feet/sec.
Centimeters/second	0.036	Kilometers/hr.
" / "	0.6	Meters/min.
" / "	0.02237	Miles/hr.
" / "	3.728x10⁻⁴	Miles/min.
Chain (Gunters)	66	Feet
Cms./sec./sec.	0.03281	Feet/sec./sec.
Cubic centimeters	3.531x10⁻⁵	Cubic feet
" "	6.102x10⁻²	Cubic inches
" "	10⁻⁶	Cubic meters
" "	1.308x10⁻⁶	Cubic yards
" "	2.642x10⁻⁴	Gallons
" "	10⁻³	Liters
" "	2.113x10⁻³	Pints (liq.)
" "	1.057x10⁻³	Quarts (liq.)
Cubic feet	2.832x10⁴	Cubic cms.
"	1728	Cubic inches
"	0.02832	Cubic meters
"	0.03704	Cubic yards
"	7.48052	Gallons
"	28.32	Liters
"	59.84	Pints (liq.)
"	29.92	Quarts (liq.)
Cubic feet/minute	472.0	Cubic cms./sec.
" " / "	0.1247	Gallons/sec.
" " / "	0.4720	Liters/sec.
" " / "	62.43	Pounds of water/min.
Cubic feet/second	0.646317	Million guls./day
" " / "	448.831	Gallons/min.
" " / "	13.8 / sq. mi.	drainage area Inches/year
Cubic inches	16.39	Cubic centimeters
" "	5.787x10⁻⁴	Cubic feet
" "	1.639x10⁻⁵	Cubic meters
" "	2.143x10⁻⁵	Cubic yards
" "	4.329x10⁻³	Gallons
" "	1.639x10⁻²	Liters
" "	0.03463	Pints (liq.)
" "	0.01732	Quarts (liq.)
Cubic meters	10⁶	Cubic centimeters
" "	35.31	Cubic feet
" "	61.023	Cubic inches
" "	1.308	Cubic yards
" "	264.2	Gallons
Cubic meters	10³	Liters
" "	2113	Pints (liq.)
" "	1057	Quarts (liq.)
Cubic yards	7.646x10⁵	Cubic centimeters
" "	27	Cubic feet
" "	46,656	Cubic inches
" "	0.7646	Cubic meters
" "	202.0	Gallons
" "	764.6	Liters
" "	1616	Pints (liq.)
" "	807.9	Quarts (liq.)
Cubic yards/min	0.45	Cubic feet/sec.
" " / "	3.367	Gallons/sec.
" " / "	12.74	Liters/sec.
Decigrams	0.1	Grams
Deciliters	0.1	Liters
Decimeters	0.1	Meters
Degrees (angle)	60	Minutes
" "	0.01745	Radians
" "	3600	Seconds
Degrees/sec.	0.01745	Radians/sec.
" / "	0.1667	Revolutions/min.
" / "	0.002778	Revolutions/sec.
Dekagrams	10	Grams
Dekaliters	10	Liters
Dekameters	10	Meters
Drams	27.34375	Grains
"	0.0625	Ounces
"	1.771845	Grams
Fathoms	6	Feet
Feet	30.48	Centimeters
"	12	Inches
"	0.3048	Meters
"	1/3	Yards

GENERAL CONVERSION FACTORS — 2

Multiply	By	To Obtain
Feet of water	0.02950	Atmospheres
" " "	0.8826	Inches of mercury
" " "	304.8	Kgs./sq.meter
" " "	62.43	Lbs./sq.ft.
" " "	0.4335	Lbs./sq.inch
Feet/min......	0.5080	Centimeters/sec.
" "	0.01667	Feet/sec.
" "	0.01829	Kilometers/hr.
" "	0.3048	Meters/min.
" "	0.01136	Miles/hr.
Feet/sec......	30.48	Centimeters/sec.
" "	1.097	Kilometers/hr.
" "	0.5921	Knots
" "	18.29	Meters/min.
" "	0.6818	Miles/hr.
" "	0.01136	Miles/min.
Feet/sec./sec	30.48	Cms./sec./sec.
" " / "	0.3048	Meters/sec./sec.
Foot-pounds....	1.286×10^{-3}	British Thermal Units
"	5.050×10^{-7}	Horse-power-hrs.
"	3.241×10^{-4}	Kilogram-calories
"	0.1383	Kilogram-meters
"	3.766×10^{-7}	Kilowatt-hrs.
Foot-pounds/min.	1.286×10^{-3}	B. T. Units/min.
" " / "	0.01667	Foot-pounds/sec.
" " / "	3.030×10^{-5}	Horse-power
" " / "	3.241×10^{-4}	Kg.-calories/min.
" " / "	2.260×10^{-5}	Kilowatts
Foot-pounds/sec.	7.717×10^{-2}	B. T. Units/min.
" " / "	1.818×10^{-3}	Horse-power
" " / "	1.945×10^{-2}	Kg.-calories/min
" " / "	1.356×10^{-3}	Kilowatts
Gallons........	3785	Cubic centimeters
"	0.1337	Cubic feet
"	231	Cubic inches
"	3.785×10^{-3}	Cubic meters
"	4.951×10^{-3}	Cubic yards
"	3.785	Liters
"	8	Pints (liq.)
"	4	Quarts (liq.)
Gallons, Imperial.	1.20095	U.S. gallons
U.S.....	0.83267	Imperial gallons
Gallons water....	8.3453	Pounds of water
Gallons/min..	2.228×10^{-3}	Cubic feet/sec.
" / "	0.06308	Liters/sec.
" / "	8.0208	Cu. ft./hr.
" / "	8.0208	Overflow rate (ft./hr.) Area (sq. ft.)
Gallons water/min	6.0086	Tons water/24 hrs.
Grains (troy)....	1	Grains (avoir.)
"	0.06480	Grams
"	0.04167	Pennyweights (troy)
"	2.0833×10^{-3}	Ounces (troy)
Grains/U.S. gal.	17.118	Parts/million
/U.S. gal.	142.86	Lbs./million gal.
/Imp. gal.	14.286	Parts/million
Grams........	980.7	Dynes
"	15.43	Grains
"	10^{-3}	Kilograms
"	10^3	Milligrams
"	0.03527	Ounces
"	0.03215	Ounces (troy)
"	2.205×10^{-3}	Pounds
Grams/cm.....	5.600×10^{-3}	Pounds/inch
Grams/cu. cm..	62.43	Pounds/cubic foot
" / " / "	0.03613	Pounds/cubic inch
Grams/liter....	58.417	Grains/gal.
" / "	8.345	Pounds/1000 gals.
" / "	0.062427	Pounds/cubic foot
" / "	1000	Parts/million
Hectares........	2.471	Acres
"	1.076×10^5	Square feet
Hectograms......	100	Grams
Hectoliters......	100	Liters
Hectometers......	100	Meters
Hectowatts......	100	Watts

Multiply	By	To Obtain
Horse-power........	42 44	B.T. Units/min.
"	33,000	Foot-lbs./min.
"	550	Foot-lbs./sec.
"	1.014	Horse-power (metric)
"	10.70	Kg.-calories/min.
"	0.7457	Kilowatts
"	745.7	Watts
Horse-power (boiler)	33,479	B.T.U./hr.
"	9.803	Kilowatts
Horse-power-hours..	2547	British Thermal Units
"	1.98×10^6	Foot-lbs.
"	641.7	Kilogram-calories
"	2.737×10^5	Kilogram-meters
"	0.7457	Kilowatt-hours
Inches..............	2.540	Centimeters
Inches of mercury..	0.03342	Atmospheres
"	1.133	Feet of water
"	345.3	Kgs./sq. meter
"	70.73	Lbs./sq.ft.
"	0.4912	Lbs./sq. inch
Inches of water......	0.002458	Atmospheres
"	0.07355	Inches of mercury
"	25.40	Kgs./sq. meter
"	0.5781	Ounces/sq. inch
"	5.202	Lbs./sq. foot
"	0.03613	Lbs./sq. inch
" " " /year	$\dfrac{\text{sq. mi.}}{13.8}$	Cu. ft. /sec.
Kilograms..........	980.665	Dynes
"	2.205	Lbs.
"	1.102×10^{-3}	Tons (short)
"	10^3	Grams
Kilograms-calories..	3.968	British Thermal Units
"	3086	Foot-pounds
"	1.558×10^{-3}	Horse-power-hours
"	1.162×10^{-3}	Kilowatt-hours
Kilogram-calories/min.	51.43	Foot-pounds/sec.
" / "	0.09351	Horse-power
" / "	0.06972	Kilowatts
Kgs./meter........	0.6720	Lbs./foot
Kgs./sq. meter.....	9.678×10^{-5}	Atmospheres
" / "	3.281×10^{-3}	Feet of water
" / "	2.896×10^{-3}	Inches of mercury
" / "	0.2048	Lbs./sq. foot
" / "	1.422×10^{-3}	Lbs./sq. inch
Kgs./sq. millimeter	10^6	Kgs./sq. meter
Kiloliters.........	10^3	Liters
Kilometers........	10^5	Centimeters
"	3281	Feet
"	10^3	Meters
"	0.6214	Miles
"	1094	Yards
Kilometers/hr.....	27.78	Centimeters/sec.
" / "	54.68	Feet/min.
" / "	0.9113	Feet/sec.
" / "	0.5396	Knots
" / "	16.67	Meters/min.
" / "	0.6214	Miles/hr.
Kms./hr./sec......	27.78	Cms./sec./sec.
" / " / "	0.9113	Ft./sec./sec.
" / " / "	0.2778	Meters/sec./sec.
Kilowatts........	56.92	B.T. Units/min.
"	4.425×10^4	Foot-lbs./min.
"	737.6	Foot-lbs./sec.
"	1.341	Horse-power
"	14.34	Kg.-calories/min.
"	10^3	Watts
Kilowatt-hours....	3415	British Thermal Units
"	2.655×10^6	Foot-lbs.
"	1.341	Horse-power-hrs.
"	860.5	Kilogram-calories
"	3.671×10^5	Kilogram-meters
Link (Gunters) 66		Feet
Liters............	10^3	Cubic centimeters
"	0.03531	Cubic feet
"	61.02	Cubic inches
"	10^{-3}	Cubic meters
"	1.308×10^{-3}	Cubic yards
"	0.2642	Gallons
"	2.113	Pints (liq.)
"	1.057	Quarts (liq.)

Multiply	By	To Obtain
Liters/min..........	5.886×10^{-4}	Cubic ft.-sec.
" / "	4.403×10^{-3}	Gals./sec.
Lumber Width (in.) x Thickness (in.) $\dfrac{}{12}$	Length (ft)	Board Feet
Meters..............	100	Centimeters
"	3.281	Feet
"	39.37	Inches
"	10^{-3}	Kilometers
"	10^3	Millimeters
"	1.094	Yards
Meters/min........	1.667	Centimeters/sec.
" / "	3.281	Feet/min.
" / "	0.05468	Feet/sec.
" / "	0.06	Kilometers/hr.
" / "	0.03728	Miles/hr.
Meters/sec........	196.8	Feet/min.
" / "	3.281	Feet/sec.
" / "	3.6	Kilometers/hr.
" / "	0.06	Kilometers/min.
" / "	2.237	Miles/hr.
" / "	0.03728	Miles/min.
Microns............	10^{-6}	Meters
Miles..............	1.609×10^5	Centimeters
"	5280	Feet
"	1.609	Kilometers
"	1760	Yards
Miles/hr...........	44.70	Centimeters/sec.
" / "	88	Feet/min.
" / "	1.467	Feet/sec.
" / "	1.609	Kilometers/hr.
" / "	0.8684	Knots
" / "	26.82	Meters/min.
Miles/min.........	2682	Centimeters/sec.
" / "	88	Feet/sec.
" / "	1.609	Kilometers/min.
" / "	60	Miles/hr.
Milliers...........	10^3	Kilograms
Milligrams........	10^{-3}	Grams
Milliliters........	10^{-3}	Liters
Millimeters........	0.1	Centimeters
"	0.03937	Inches
Milligrams/liter....	1	Parts/million
Million gals./day...	1.54723	Cubic ft./sec.
Miner's inches.....	1.5	Cubic ft./min.
Minutes (angle)....	2.909×10^{-4}	Radians
Ounces............	16	Drams
"	437.5	Grains
"	0.0625	Pounds
"	28.349527	Grams
"	0.9115	Ounces (troy)
"	2.790×10^{-5}	Tons (long)
"	2.835×10^{-5}	Tons (metric)
Ounces, troy......	480	Grains
"	20	Pennyweights (troy)
"	0.08333	Pounds (troy)
"	31.103481	Grams
"	1.09714	Ounces, avoir.
Ounces (fluid).....	1.805	Cubic inches
"	0.02957	Liters
Ounces/sq. inch....	0.0625	Lbs./sq. inch
Overflow rate (ft./hr.)	0.12468 x	
1 area (sq. ft.)		Gals./min.
"	8.0208	Sq. ft./gal./min.
Overflow rate (ft./hr.)		
Parts/million.......	0.0584	Grains/U.S. gal.
"	0.07016	Grains/Imp. gal.
"	8.345	Lbs./million gal.
Pennyweights (troy).	24	Grains
"	1.55517	Grams
"	0.05	Ounces (troy)
"	4.1667×10^{-3}	Pounds (troy)
Pounds............	16	Ounces
"	256	Drams
"	7000	Grains
"	0.0005	Tons (short)
"	453.5924	Grams
"	1.21528	Pounds (troy)
"	14.5833	Ounces (troy)

GENERAL CONVERSION FACTORS—3

Multiply	By	To Obtain
Pounds (troy)	.5760	Grains
" "	.240	Pennyweights (troy)
" "	12	Ounces (troy)
" "	.373.24177	Grams
" "	0.822857	Pounds (avoir.)
" "	13.1657	Ounces (avoir.)
" "	3.6735x10⁻⁴	Tons (long)
" "	4.1143x10⁻⁴	Tons (short)
" "	3.7324x10⁻⁴	Tons (metric)
Pounds of water	0.01602	Cubic feet
" " "	27.68	Cubic inches
" " "	0.1198	Gallons
Pounds of water/min.	2.670x10⁻¹	Cubic ft./sec.
Pounds/cubic foot	0.01602	Grams/cubic cm.
" / "	16.02	Kgs./cubic meter
" / "	5.787x10⁻⁴	Lbs./cubic inch
Pounds/cubic inch	27.68	Grams/cubic cm.
" / "	2.768x10⁴	Kgs./cubic meter
" / "	1728	Lbs./cubic foot
Pounds/foot	1.488	Kgs./meter
Pounds/inch	178.6	Grams/cm.
Pounds/sq. foot	0.01602	Feet of water
" / " "	4.883	Kgs./sq. meter
" / " "	6.945x10⁻³	Pounds/sq. inch
Pounds/sq. inch	0.06804	Atmospheres
" / " "	2.307	Feet of water
" / " "	2.036	Inches of mercury
" / " "	703.1	Kgs./sq. meter
Quadrants (angle)	90	Degrees
" "	5400	Minutes
" "	1.571	Radians
Quarts (dry)	67.20	Cubic inches
Quarts (liq.)	57.75	Cubic inches
Quintal, Argentine	101.28	Pounds
" Brazil	129.54	Pounds
" Castile, Peru	101.43	Pounds
" Chile	101.41	Pounds
" Mexico	101.47	Pounds
" Metric	220.46	Pounds
Quires	25	Sheets
Radians	57.30	Degrees
"	3438	Minutes
"	0.637	Quadrants
Radians/sec.	57.30	Degrees/sec.
" / "	0.1592	Revolutions/sec.
" / "	9.549	Revolutions/min.
Radians/sec./sec.	573.0	Revolutions/min/min.
" / " "	0.1592	Revolutions/sec./sec.
Reams	500	Sheets
Revolutions	360	Degrees
"	4.	Quadrants
"	6.283	Radians
Revolutions/min.	6	Degrees/sec.
" / "	0.1047	Radians/sec.
" / "	0.01667	Revolutions/sec.
Revolutions/min./min.	1.745x10⁻³	Rads./sec./sec.
" / " "	2.778x10⁻⁴	Revs./sec./sec.
Revolutions/sec.	360	Degrees/sec.
" / "	6.283	Radians/sec.
" / "	60	Revolutions/min.
Revolutions/sec./sec.	6.283	Radians/sec./sec.
" / " "	3600	Revolutions/min/min.
Rods	16.5	Feet
Seconds (angle)	4.848x10⁻⁶	Radians
Square centimeters	1.076x10⁻³	Square feet
" "	0.1550	Square inches
" "	10⁻⁴	Square meters
" "	100	Square millimeters
Square Chains (Gunters)	16	Square rods

Multiply	By	To Obtain
Square feet	2.296x10⁻⁵	Acres
" "	929.0	Square centimeters
" "	144	Square inches
" "	0.09290	Square meters
" "	3.587x10⁻⁸	Square miles
" "	1/9	Square yards
1	8.0208	Overflow rate (ft./hr.)
Sq. ft./gal./min.		
Square inches	6.452	Square centimeters
" "	6.944x10⁻³	Square feet
" "	645.2	Square millimeters
Square kilometers	247.1	Acres
" "	10.76x10⁶	Square feet
" "	10⁶	Square meters
" "	0.3861	Square miles
" "	1.196x10⁶	Square yards
Square meters	2.471x10⁻⁴	Acres
" "	10.76	Square feet
" "	3.861x10⁻⁷	Square miles
" "	1.196	Square yards
Square miles	640	Acres
" "	27.88x10⁶	Square feet
" "	2.590	Square kilometers
" "	3.098x10⁶	Square yards
Square millimeters	0.01	Square centimeters
" "	1.550x10⁻³	Square inches
Square yards	2.066x10⁻⁴	Acres
" "	9	Square feet
" "	0.8361	Square meters
" "	3.228x10⁻⁷	Square miles
Temp. (°C.) + 273	1	Abs. temp. (°C.)
"	17.78 1.8	Temp. (°F.)
" (°F.) + 460	1	Abs. temp. (°F.)
" — 32	5/9	Temp. (°C.)
Tons (long)	1016	Kilograms
" "	2240	Pounds
" "	1.12000	Tons (short)
Tons (metric)	10³	Kilograms
" "	2205	Pounds
Tons (short)	2000	Pounds
" "	32000	Ounces
" "	907.18486	Kilograms
Tons (short)	2430.56	Pounds (troy)
" "	0.89287	Tons (long)
" "	29166.66	Ounces (troy)
" "	0.90718	Tons (metric)
1		Area
	(sq. ft.)	Sq. ft./ton/24 hrs.
Tons dry solids/24 hrs.		
Tons of water/24 hrs.	83.333	Pounds water/hour
" " " /	0.16643	Gallons/min.
" " " /	1.3349	Cu. ft./hr.
Watts	0.05692	B. T. Units/min.
"	44.26	Foot-pounds/min.
"	0.7376	Foot-pounds/sec.
"	1.341x10⁻³	Horse-power
"	0.01434	Kg.-calories/min.
"	10⁻³	Kilowatts
Watt-hours	3.415	British Thermal Units
"	2655	Foot-pounds
"	1.341x10⁻³	Horse-power-hours
"	0.8605	Kilogram-calories
"	367.1	Kilogram-meters
"	10⁻³	Kilowatt-hours
Yards	91.44	Centimeters
"	3	Feet
"	36	Inches
"	0.9144	Meters

About the Author

Thomas R. Dion, a registered professional engineer and land surveyor, is an Associate Professor of Civil Engineering at The Citadel in Charleston, South Carolina, where he has taught since 1976. Mr. Dion has obtained considerable experience in activities relating to land development, both as a graduate of The Citadel and at Clemson University, where he was awarded a Master of Science degree in Civil Engineering with a minor in City and Regional Planning.

(Continued)

A former engineer for the South Carolina State Highway Department and later for a consulting engineering firm that focused on land development, Mr. Dion gained practical experience in civil engineering and land surveying. More recently he has acted as a consultant on these subjects to various governmental agencies and private organizations.

Mr. Dion was a charter member of the board of Architectural Review for the Town of Summerville, S.C., and served as a member of the Commissioners of Public Works where he gained experience in the operation, maintenance, and ownership of municipal water and wastewater systems. As a result, he recently obtained a U.S. Patent pertaining to water flow systems and is currently involved with other patent applications.

Mr. Dion is a member of the American Society of Civil Engineers and has held various positions at both state and local levels. As faculty advisor to The Citadel's student chapter, he received outstanding national service awards each year from 1976 to 1986. He has served as an educational consultant to the South Carolina State Board of Engineering and Land Surveying Examiners since 1988, and is active in various other professional organizations including Tau Beta Pi. He is listed in *Who's Who in the South and Southwest 1991–1992*, *Who's Who in American Education 1992–1993*, *Who's Who in Science and Engineering 1992–1993*, and *Who's who in Finance and Industry*, 1992–1993.

INDEX